Springer Polar Sciences

Series editor
James Ford, Priestley International Centre for Climate, University of Leeds, Leeds, West Yorkshire, UK

Springer Polar Sciences

Springer Polar Sciences is an interdisciplinary book series that is dedicated to research in the Arctic, sub-Arctic regions, and the Antarctic. In recent years, the polar regions have received increased scientific and public interest. Both the Arctic and Antarctic have been recognized as key regions in the regulation of the global climate, and polar ecosystems have been identified to be particularly susceptible to the ongoing environmental changes. Consequently, the international efforts in polar research have been enhanced considerably, and a wealth of new findings is being produced at a growing rate by the international community of polar researchers.

Springer Polar Sciences aims to present a broad platform that will include state-of-the-art research, bringing together both science and humanities to facilitate an exchange of knowledge between the various polar science communities. The Series offers an outlet to publish contributions, monographs, edited works, conference proceedings, etc. Topics and perspectives will be broad and will include, but not be limited to: climate change impacts, environmental change, polar ecology, governance, health, economics, indigenous populations, tourism and resource extraction activities. Books published in the series will appeal to scientists, students, polar researchers and policy makers.

More information about this series at http://www.springer.com/series/15180

Eva Pongrácz • Victor Pavlov • Niko Hänninen
Editors

Arctic Marine Sustainability

Arctic Maritime Businesses and the Resilience of the Marine Environment

Springer

Editors
Eva Pongrácz
Water, Energy and Environmental Engineering Research Unit
University of Oulu
Oulu, Finland

Victor Pavlov
Water, Energy and Environmental Engineering
University of Oulu
Oulu, Finland

Niko Hänninen
Water, Energy and Environmental Engineering
University of Oulu
Oulu, Finland

ISSN 2510-0475 ISSN 2510-0483 (electronic)
Springer Polar Sciences
ISBN 978-3-030-28403-9 ISBN 978-3-030-28404-6 (eBook)
https://doi.org/10.1007/978-3-030-28404-6

This Springer imprint is published by the registered company Springer Nature Switzerland AG.
The registered company address is: Gewerbestrasse 11, 6330 Cham, Switzerland

Foreword

Dear reader,

The last piece of summer sea ice in the Arctic is expected to melt away in only a few decades. We know the consequences are manifold. New sea routes will open up for shipping, and economic activities will increase. Less ice along the shorelines will increase erosion and force communities to move to safer locations. Less permafrost will release methane into the atmosphere. Ecosystems in the Arctic Ocean will change as new species enter and old ones vanish.

We know the Arctic marine environment will change fundamentally, but we don't know how well it will maintain its functionality. The resilience of the Arctic marine environment is a key question when we look at the future of the Arctic as a whole.

When I went to school in the 1970s, the Arctic was a white spot on the top of the map hanging in the classroom. The Arctic was something hostile, harsh and unfriendly. The fascinating stories of the polar explorers Amundsen, Nansen and Nordenskiöld nurtured our minds. We were not taught indigenous people had been living in the Arctic for millennia. Today, that white spot is turning black and shrinking.

As black carbon falls on snow and ice, it accelerates the melting of these reflective surfaces. Black carbon is the most significant contributor to Arctic warming after carbon dioxide. It is also an air pollutant with serious public health effects. An effective way to reduce the climate impacts of black carbon in the Arctic would be to strictly limit the flaring of surplus gas in the oil industry.

The Arctic is a sink of environmental pollutants and harmful substances. The Arctic Ocean has been contaminated with persistent organic pollutants and mercury that make their way up the food chain. Plastic pollution in the Arctic Ocean underlines the unpleasant fact that the Arctic has become a trash barrel for our planet.

The Arctic States and Permanent Participants of the Arctic Council have sought decisive steps to stop the environmental degradation in the Arctic and to promote sustainable development in the region. Since its inception the Council has produced groundbreaking scientific assessments; worked to improve the living conditions of the people of the Arctic; negotiated legally binding agreements, e.g. on marine oil spill response and search and rescue; and facilitated international conventions like the Polar Code.

The Council has adopted a wealth of recommendations on mitigating climate change, protecting biodiversity and safeguarding the well-being of the Arctic Ocean. The recommendations are among the best outputs of the Arctic Council but at the same time its weakest link. The recommendations are adopted with consensus but not always implemented in unison by the states.

The question remains: How do we build the resilience of the Arctic marine environment and adapt to the changes ahead of us?

The first step in this direction is cutting the emissions of greenhouse gases and short-lived climate forcers. We have to reduce the use of fossil fuels and replenish our energy mix with clean solar, wind and wave technology. The second step is to develop the governance of the Arctic Ocean with conventions and agreements like the Polar Code and the Central Arctic Ocean Fisheries Agreement.

The future of the Arctic region will be decided by the actions of the international community. The scientific community, local communities and non-governmental organisations play a crucial role to inform policy- and decision-makers of the necessary actions. Scientists need to fill in the gaps of knowledge and uncover the unknowns. Decision-makers need well-reasoned recommendations for actions. Local communities and NGOs are in the position to put pressure on governments and regional authorities to implement the recommendations.

Arctic Marine Sustainability is a welcome step in informing policy- and decision-makers but also the broad public on Arctic maritime businesses and resilience of the marine environment. I hope you will find the stories interesting and inspirational. I wish to thank the publishers for the comprehensive compilation of outstanding articles.

We have the common task to work for a sustainable Arctic not only because of the pristine environment, the stunning landscapes or the fascinating fauna. We have this task because the survival of our planet depends on it. Without the Arctic, we lose.

Senior Arctic Official for Finland, Arctic Council, 2017-2019 René Söderman
Ministry for Foreign Affairs of Finland
Rene.Soderman@formin.fi
Oulu, Finland
May 2019

Preface

'Scramble for the Arctic'. 'Exploring the uncharted regions and claiming the remaining white spots of the globe by placing a flag on it'. 'Harvesting and extracting previously unattainable resources'. 'Go north, young man!'

Judging by the headlines, slogans and other conceptions that have been associated with the Arctic recently, it seems that history is repeating itself. Are we entering (or have we actually already entered) an era which resembles a lot of what happened in the later parts of the nineteenth century and the early twentieth century, the age of colonialism? During that stage of history, European and Western powers divided the less developed world between each other and exploited their natural riches. One might ask, are we there again? Many object to this idea and are not willing to draw parallels between these two cases, but what if that's exactly what is happening now in the Arctic? As we reach towards the 'final frontier', are we doomed to repeat the same mistakes again, or have we evolved as a species? Are we capable of addressing the vast Arctic marine resources, which are becoming more accessible to us due to global warming, in a sustainable way? This book will not provide a definite answer for that question for sure, but, hopefully, it will offer some ideas, food for thought and perhaps even suggestions on how it could be achieved.

The seed for this book was sown in 2017 during an 'Interdisciplinary PhD Course in Marine Sustainability', which was organised in Northern Norway. The organisers included various actors, including Future Earth Norway, Nord University, Nordland Research Institute and UiT The Arctic University of Norway, and students were not from the same mould either. They represented different disciplines and research approaches to marine sustainability, but that had been the goal of the organisers from the beginning: to look at things from different angles and integrate different perspectives and students with totally different research topics and tasks. The problems of the twenty-first century are complex, and the problem-solvers have to be complex too; simple solutions won't get us there. They must be able to put together solutions that are multidisciplinary in nature, take into account various approaches and deal thoroughly with the problems and challenges at hand.

This idea had taken a root among the students during the course and evolved since then into a book about Arctic marine sustainability at the University of Oulu, edited at the Water, Energy and Environmental Research Group at the Faculty of Technology. The research group was leading a Northern Periphery and Arctic Programme-funded project called APP4SEA – Arctic Preparedness Platform for Oil Spill and Other Environmental Accidents – and this book became a side product of that project. Some of the students who had taken part in the PhD course in Norway contributed to the book as did partners of the APP4SEA project. Other researchers working with Arctic marine sustainability issues, both close and further away, had joined the cause as well. You are holding in your hand the product of their love and labour.

This compilation addresses Arctic marine sustainability from many perspectives. The first part of the book sets the scene and addresses the Arctic Ecosystems in general and especially the sustainability issues related to it. The second and third parts of the book draw attention to the economic drivers of Arctic development: transport and oil and gas. These have received a lot of attention and have raised great expectations, but as always, there is more than meets the eye.

The fourth part of the book pulls the local communities under the spotlight, which are quite often overlooked and perhaps even totally ignored as changes and above all decisions about the Arctic are being discussed or made. However, these are the people, who will feel the impacts of these changes, who must adapt to them in order to survive. New opportunities will emerge, but they are not necessarily always positive; they might even challenge the fundamentals of the livelihoods, customs and culture. And in the end, no matter how hard you try to adapt to the changes, that might not be enough. They are in the receiving end.

This brings us to the final question, to that of sustainable governance. The discourse today is more polyphonic than in the days of the bygone old colonial masters, politicians and industrialists, and we have access to enormous amounts of information, which should influence us and lead us into making better decisions. Still, the same fundamentals of law, power and influence are present in the Arctic waters and have an impact on what happens in high north.

This multidisciplinary companion into the sustainability of Arctic marine has barely scratched the tip of the Arctic ice; there is much more below the surface, which we cannot see. Still, this book should serve as a good introduction to those multiple layers and issues of Arctic marine resources, which need to be addressed if we don't want to repeat the mistakes of the past generations. Most of us do not live by the Arctic waters; in fact, just a tiny fraction of the world's population calls Arctic as their home. But what happens in the Arctic affects all of us. Arctic is the thermometer of the globe, and as recent evidence suggests, the Arctic is warming up much faster than the rest of the globe. The melting Arctic ice will increase sea levels and can also affect ocean circulation patterns, both of which would have grave consequences around the world.

The forecast is not very promising, but there is still time to act. Perhaps, the youth will be delivering the change. Greta Thunberg has given a face to the anxiety that many millennials and younger ones feel about the climate change. They are dead serious about it, and they demand that we do something. Climate strikes won't stop the climate change, only actions will, but it can only take one to start a snowball effect. Perhaps, we need our very own Arctic 'Greta Thunberg' to summon the troops to act on behalf of the Arctic.

Oulu, Finland

Niko Hänninen

Acknowledgements

The editors would like to express their appreciation to those making this book possible. Firstly, our gratitude to the Northern Periphery and Arctic (NPA) Programme, for the funding provided for the Arctic Preparedness Platform for Oil Spill and Other Environmental Accidents (APP4SEA) project. APP4SEA provided us the motivation, inspiration and knowledge base for this book. We are also grateful to Mr. Adeleye Adetunji, who has acted as publication secretary of this book and coordinated the work of the 41 authors. Last but not least, we thank the valuable contribution of all the authors.

Oulu, Finland
11.12.2019

Eva Pongrácz
Victor Pavlov
Niko Hänninen

Contents

Part I
Arctic Ecosystems and Sustainability

Chapter 1
Sustainability in an Arctic Context: Resilience of the Arctic Marine Environment

Eva Pongrácz

Abstract This chapter provides the theoretical basis of the book by outlining the framework of sustainability in an Arctic marine context. The chapter presents the Sustainable Development Goals, especially goal 14 on conserving and sustainably use the oceans, seas, and marine resources, as well as Arctic States' commitments to goal 14. The chapter explains the sustainability framework conditions and the concept of resilience, in the context of the Arctic marine environment. The chapter further discusses the concerns of cumulative impacts to the Arctic marine environment from multiple and concurrent natural and human perturbations, and the consequent weakening of the resilience of Arctic marine environment. Finally, the chapter summarizes the status of processes that influence the resilience of Arctic marine ecosystems.

Keywords Sustainability · Sustainable development goals · Arctic · Marine environment · Resilience

1.1 Introduction

Since the introduction of the sustainability concept in 1987 (United Nations General Assembly), humans have struggled to comprehend the notion of sustainability. Even though the concept has moved to the forefront of consciousness to the degree of almost being passé, the understanding of what sustainability actually means still eludes many. The purpose of this chapter is to provide the theoretical basis of the book by providing the sustainability framework for main themes of the book.

Sustainable development defined as "development that meets the needs of the present without compromising the ability of future generations to meet their needs" (United Nations General Assembly 1987) speaks of the process. The report, '*Our Common Future*', highlighted that satisfying human needs is a major aspiration and

E. Pongrácz (✉)
Water, Energy and Environmental Engineering Research Unit, University of Oulu, Oulu, Finland
e-mail: eva.pongracz@oulu.fi

E. Pongrácz et al. (eds.), *Arctic Marine Sustainability*,
Springer Polar Sciences, https://doi.org/10.1007/978-3-030-28404-6_1

does not set limits to economic growth, but indicates that natural resources are to be exploited within the limits of their regenerative capability. It concludes that sustainable development is a process of change in which resource use, technological development and institutional change are in balance. The Millennium Development Goals, later replaced by the Sustainable Development Goals (United Nations General Assembly 2015), give goals to meet. The end-result, sustainability, could then be argued to be human-ecosystem equilibrium state (Shaker 2015). This ideal is viewed unattainable by many, pointing out the inherent unsustainability of unlimited growth in a closed system.

The United Nations 2030 Agenda for Sustainable Development emphasized the need to strengthen resilience across human and natural systems, with special emphasis on the sustainable management of marine ecosystems to avoid significant adverse impact and strengthening their resilience. Understanding and moderating human impacts on marine ecosystems is in the forefront of ecology research and new paradigms are surfacing. A cumulative assessment framework needs to be adopted, in order to respond to the diversity of cumulative impacts in the Arctic environment.

1.2 The Arctic

The Arctic may be considered a single region, but it can be defined and bordered in many different ways. First, I would like to introduce the different ways the Arctic can be and has been defined by different scholars and organizations.

1.2.1 Defining the Arctic

The Arctic is most typically defined by the Arctic Circle, 66° 33′ 44" North, which is the Northernmost latitude at which the sun can remain continuously above or below the horizon for 24 h. Another view is the Arctic tree line boundary: the northernmost latitude in the Northern hemisphere where trees can grow. Further north, it is too cold all year round to sustain trees. The low average temperature is an important indicator of the Arctic as well. The third definition is the 10 °C July isotherm, which is the area where the average temperature for the warmest month is below 10 °C and is often used by biologists as a definitional boundary of the Arctic. There are also other definitions, or more like areas of interest defined by different working groups of the Arctic Council. Figure 1.1 by the Arctic Portal illustrates some of these boundaries.

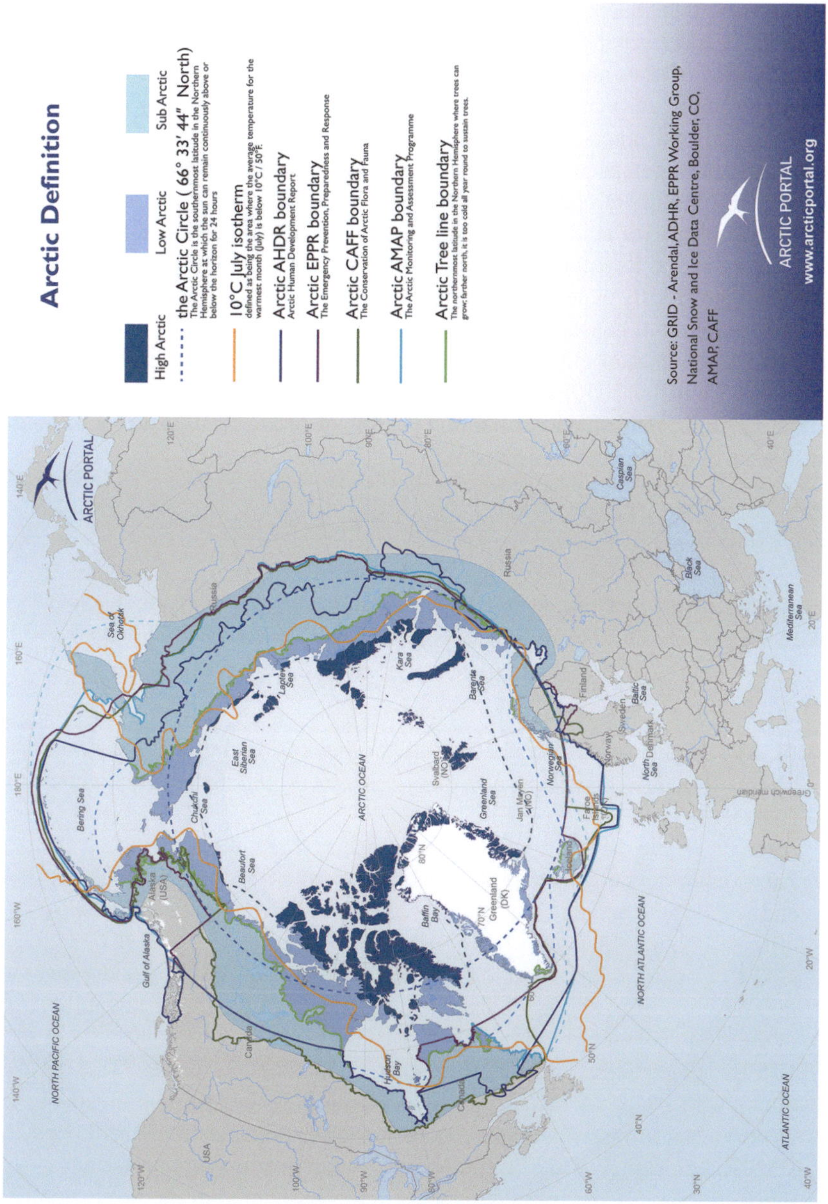

Fig. 1.1 The boundary of the Arctic by different definitions

1.2.2 The Changing Arctic

The ever-accelerated pace of change in the Arctic is a concern most scientists share. The issue is both the speed of change and the variety of changes, both ecological and societal. Climate change is the strongest driver of change, worsened by widespread pollution of seas, and the unprecedented interest in the Arctic, both in terms of natural resources and ecosystem services. These changes threaten the integrity of both ecological and human systems in the Arctic. Understanding Arctic change requires a systemic perspective that integrates human and natural dynamics. (Arctic Council 2016). Studies at open ocean and coastal sites around the world show that current levels of marine acidity have increased by about 26% on average since the start of the Industrial Revolution. Moreover, marine life is being exposed to conditions outside previously experienced natural variability. Global trends point to continued deterioration of coastal waters due to pollution and eutrophication. Without concerted efforts, coastal eutrophication is expected to increase in 20% of large marine ecosystems by 2050. These concerns call for the investigation of sustainability in an Arctic context, and examining the role of Arctic business in weakening the resilience of the marine environment.

The Arctic Council

The Ottawa Declaration (1996) lists the following countries as Members of the Arctic Council: Canada, the Kingdom of Denmark, Finland, Iceland, Norway, the Russian Federation, Sweden and the United States. In order to ensure sustainable development in the Arctic region, work is carried out in six Working Groups:

- **ACAP:** Arctic Contaminants Action Program is encouraging national actions to reduce emissions and other releases of pollutants (acap.arctic-council.org)
- **AMAP:** Arctic Monitoring and Assessment Programme monitors the Arctic environment, ecosystems and human populations, and provides scientific advice to support governments as they tackle pollution and adverse effects of climate change (www.amap.no)
- **CAFF:** Conservation of Arctic Flora and Fauna Working Group addresses the conservation of Arctic biodiversity, working to ensure the sustainability of the Arctic's living resources (www.caff.is)
- **EPPR:** Emergency Prevention, Preparedness and Response Working Group works to protect the Arctic environment from the threat or impact of an accidental release of pollutants or radionuclides (eppr.arctic-council.org)
- **PAME:** Protection of the Arctic Marine Environment Working Group is the focal point of the Arctic Council's activities related to the protection and sustainable use of the Arctic marine environment (www.pame.is)
- **SDWG:** Sustainable Development Working Group works to advance sustainable development in the Arctic and to improve the conditions of Arctic communities as a whole (www.sdwg.org)

1.3 The United Nation's Sustainable Development Goals

As a starting point, the Sustainable Development Goals (SDGs) of the United Nations General Assembly are introduced, which were intended to change societal values ((United Nations General Assembly) 2015). The United Nations adopted the 17 new Goals (Fig. 1.2) and their 169 targets in September 2015. The primary objectives are to end poverty, protect the planet, reduce inequality and, generally, improve the well-being of everyone in the world. The 2030 Agenda outlines an ambitious action plan for people, planet and prosperity, to strengthen peace and freedom.

SDG14 aims toward the conservation and sustainable use of oceans, seas and marine resource for sustainable development. Advancing the sustainable use and conservation of the oceans continues to require effective strategies and management to combat the adverse effects of overfishing, growing ocean acidification and worsening coastal eutrophication. The expansion of protected areas for marine biodiversity, intensification of research capacity and increases in ocean science funding remain critically important to preserve marine resources. Table 1.1 summarized the targets and indicators of SDG14.

The high-level United Nations Conference to Support the Implementation SDG14 (The Ocean Conference) was convened at United Nations Headquarters in New York in June 2017. The Conference adopted the "Our Ocean, Our Future: Call for Action" to support the implementation of SDG14. At the Conference, close to 1400 voluntary commitments for concrete action to advance implementation of SDG 14 were made by governments, the United Nations, civil society organiza-

Fig. 1.2 The Sustainable Development Goals of Agenda 2030 United Nations (2016)

Table 1.1 Sustainable development goal 14: Conserve and sustainably use the oceans, seas, and marine resources for sustainable development. Targets and indicators (UN 2016)

Targets		Indicators
14.1	By 2025, prevent and significantly reduce marine pollution of all kinds, in particular from land-based activities, including marine debris and nutrient pollution	Index of coastal eutrophication and floating plastic debris density
14.2	By 2020, sustainably manage and protect marine and coastal ecosystems to avoid significant adverse impacts, including by strengthening their resilience, and take action for their restoration in order to achieve healthy and productive oceans	Proportion of national exclusive economic zones managed using ecosystem-based approaches
14.3	Minimize and address the impacts of ocean acidification, including through enhanced scientific cooperation at all levels	Average marine acidity (pH) measured at agreed suite of representative sampling stations
14.4	By 2020, effectively regulate harvesting and end overfishing, illegal, unreported and unregulated fishing and destructive fishing practices and implement science-based management plans, in order to restore fish stocks in the shortest time feasible, at least to levels that can produce maximum sustainable yield as determined by their biological characteristics	Proportion of fish stocks within biologically sustainable levels
14.5	By 2020, conserve at least 10% of coastal and marine areas, consistent with national and international law and based on the best available scientific information	Coverage of protected areas in relation to marine areas
14.6	By 2020, prohibit certain forms of fisheries subsidies which contribute to overcapacity and overfishing, eliminate subsidies that contribute to illegal, unreported and unregulated fishing and refrain from introducing new such subsidies, recognizing that appropriate and effective special and differential treatment for developing and least developed countries should be an integral part of the World Trade Organization fisheries subsidies negotiation	Progress by countries in the degree of implementation of international instruments aiming to combat illegal, unreported and unregulated fishing
14.7	By 2030, increase the economic benefits to Small Island developing States and least developed countries from the sustainable use of marine resources, including through sustainable management of fisheries, aquaculture and tourism	Sustainable fisheries as a percentage of GDP in small island developing states, least developed countries and all countries

(continued)

Table 1.1 (continued)

Targets		Indicators
14.A	Increase scientific knowledge, develop research capacity and transfer marine technology, taking into account the intergovernmental oceanographic commission criteria and guidelines on the transfer of marine technology, in order to improve ocean health and to enhance the contribution of marine biodiversity to the development of developing countries, in particular small island developing states and least developed countries	Proportion of total research budget allocated to research in the field of marine technology
14.B	Provide access for small-scale artisanal fishers to marine resources and markets	Progress by countries in the degree of application of a legal, regulatory, policy, institutional framework which recognizes and protects access rights for small-scale fisheries
14.C	Enhance the conservation and sustainable use of oceans and their resources by implementing international law as reflected in UNCLOS, which provides the legal framework for the conservation and sustainable use of oceans and their resources, as recalled in paragraph 158 of the future we want	Number of countries making progress in ratifying, accepting and implementing through legal, policy and institutional frameworks, ocean-related instruments that implement international law, as reflected in the united nation convention on the law of the sea, for the conservation and sustainable use of the oceans and their resources

tions, the scientific community, and the private sector. The nine Communities of Ocean Action are:

1. Coral reefs
2. Implementation of the United Nations Convention on the Law of the Sea
3. Mangroves
4. Marine and coastal ecosystems management
5. Marine pollution
6. Ocean acidification
7. Scientific knowledge, research capacity development and transfer of marine technology
8. Sustainable blue economy
9. Sustainable fisheries

Currently, there are over 1500 voluntary commitments. Some of those made by Arctic nations and organizations are (https://oceanconference.un.org/commitments/):

- OceanAction16721: Iceland commits to reduce marine litter
- OceanAction16733: Addressing acidification by Iceland
- OceanAction18373: Study on Marine Litter including Microplastics in the Arctic by Sweden, Norway, Iceland

- OceanAction18818: Ban plastic microbeads in cosmetics by Sweden
- OceanAction18424: Adoption of Fisheries Management Plans with long term precautionary Harvest Control Rules for commercially harvested fish stocks in Icelandic waters
- OceanAction19375: Measures to reduce marine pollution and microplastics in Norway by targeting their sources
- OceanAction19509: Industry and research driven development and introduction of selective and low impact fishing gears by Sweden
- OceanAction18382: Identification of Ecologically or Biologically Significant Marine Areas in the Baltic Sea by HELCOM
- OceanAction17174: Strengthening the implementation of the HELCOM Baltic Sea Action Plan to support ocean-related SDGs (HELCOM)
- OceanAction 20,500: Reducing marine litter

There is significant attention on pollution prevention, with marine plastics on top of the list. Actions also concern sustainable fishing, as well as ecosystem-based management approaches.

1.4 Arctic nations' commitment to the Agenda 2030 and SDG14

Nordic environment and climate ministers are urging firmer action to combat plastic and microplastic pollution in seas and oceans. At their meeting on 10 April 2019, they signed a declaration of 11 key commitments. In it, the ministers ask the Nordic Council of Ministers to prepare a study to consider which specific elements should be included in a global agreement to combat microplastics and plastic waste in the marine environment. (Nordic Council of Ministers for the Environment and Climate 2019).

1.4.1 Norway

As a nation reliant on resources from the sea, Norway has taken a leading to combat marine litter. The latest status report on eutrophication (2016) classifies Norwegian offshore and outer coastal areas as non-problem areas. Norway monitors and records marine litter including plastic and micro plastic in our three oceans. More susceptible to ocean acidification than temperate waters, the pH surface layer of the Norwegian Sea has decreased by 0.13 pH units the past 30 years, compared to the global average of 0.1 pH units. Norway keep on monitoring ocean acidification and increase knowledge on its effects. Plastic waste in the oceans is a pressing global concern. Put forward by Norway, in December 2017 the UN Environment Assembly agreed on a vision for zero release of plastics into the ocean. Norway has also allocated NOK 150 million to combatting marine litter and microplastics in the oceans of developing countries. A High-Level Panel on Building a Sustainable Ocean Economy was also established, to increase global awareness of the relationship between clean and healthy oceans, sustainable use of ocean resources and economic growth and development. (Norwegian Ministry of Finance and Norwegian Ministry of Foreign Affairs, 2018).

1.4.2 Sweden

Sweden's 2030 agenda (2018) put forward an action for the conservation and sustainable use of the seas and marine resources. Sweden has introduced a ban on microplastics in certain cosmetic products and is working on reducing the amount plastic waste ending up in the sea and lakes. Action has also been taken to reduce pollution and eutrophication. The Marine and Water Authority draws up proposals based on the ecosystem approach. There is great emphasis on international cooperation on the implementation of SDG14. The Government's global strategy for the environment, climate and the sea and sustainable use of natural resources 2018–2022 is central to this work. Sweden is also working to establish an ambitious implementation agreement for the UN Convention on the Law of the Sea for the Protection of Biodiversity in Areas beyond National Jurisdictions. In its resolution 69/292 of 19 June 2015, the General Assembly decided to develop an international legally binding instrument under the United Nations Convention on the Law of the Sea (UNCLOS) on the conservation and sustainable use of marine biological diversity of areas beyond national jurisdiction. (Regeringskansliet 2018).

1.4.3 Finland

Finland was one of the first countries reporting their 2030 agenda. It mentions Food security, access to water and energy, and the sustainable use of natural resource as one of the priority areas, but SDG 14 is not mentioned as one of the focus areas. The report, however, has assessed that, as a starting level, the progress of Finland in terms of SDG14 goals has been moderate. The Ocean Health Index was labelled green; however, percentage of marine sites important to biodiversity that are completely protected was marked red. At this time Finland put focus on SDGs 8 and 13. (Prime Minister's Office, Finland 2016).

1.4.4 Iceland

The Iceland 2020 – governmental policy statement for the economy and community (Iceland Prime Minister's Office 2011) contains visions and measurable objectives targeted at improving welfare, knowledge and sustainability. In the preparation phase for Agenda 2030, Iceland actively promoted key areas such as sustainable management of oceans at home and abroad and stated that utilizing marine resources in a responsible manner based on scientific advice plays a vital role in ensuring food security and prosperity.

1.4.5 Denmark

The Danish government has formulated an Action Plan (The Danish Government 2017) to adapt the 2030 targets to national circumstances. The Action Plan is centered on the 5 P's: Prosperity, People, Planet, Peace and Partnerships. The government has formulated 37 targets. They reflect the government's wish to prioritize building on existing positions of strength as well as on areas where improvement is needed. Each target has one or two national indicators, which are in large part measurable and quantifiable.

1.4.6 Canada

Many of the Government of Canada's priorities and programs, both domestically and internationally, are already well aligned with the 2030 Agenda. Canada's 2016–2019 Federal Sustainable Development Strategy, which sets out Canada's sustainable development priorities, is linked to many SDGs, including SDG 14. Furthermore, in Budget 2018, the Government of Canada announced that it would provide $49.4 million over 13 years to establish an SDG unit and fund monitoring and reporting activities by Statistics Canada (Government of Canada 2018)

1.4.7 USA

Bertelsmann Stiftung and the Sustainable Development Solutions Network have been following the progress of all 193 countries in achieving the SDGs since 2016. Their report (2018) focusing on the G20 countries survey to gauge how strongly the SDGs were integrated into institutions and policy. In their evaluation, the United States ranked right at the bottom.

1.4.8 Russia

Bobylev and Solovyeva (2017) analyzed the compliance of the SDGs with the development goals of Russia and found that SDGs 12–15 are not reflected in the areas of activity outlined in Strategy 2020. Attention currently focuses on social and economic SDGs.

1.4.9 The Sustainable Development Working Group (SDWG) of the Arctic Council

The SDWG plays a lead role in addressing the human dimension of the Arctic within the Arctic Council. The goal is building self-sufficient, resilient and healthy Arctic communities for present and future generations while protecting the environment. SDWG work falls under six broad thematic areas, including sustainable economic activities and management of natural resources. During the Finnish chairmanship between 2017–2019, the SDWG was engaged in 20 projects, and reported on their outcome in May 2019. Among these were the progress report on the Arctic Resilience Action Framework; Good Practices for Environmental impact Assessment and Meaningful Engagement in the Arctic and The Arctic as a Food-Producing Region Final Report. (Arctic Council 2019).

1.5 Sustainability Framework Conditions

The vast and growing array of concepts, methods and tools in the sustainability field imply a need for a structuring and coordinating framework, including a unifying and operational definition of sustainability. One attempt at such framework began over 25 years ago and is now widely known as the Framework for Strategic Sustainable Development (Missimer et al. 2017). The Framework is the result of a long-term effort of several scientist led by Karl-Henrik Robért. One of the main feature of FSSD is the four system conditions for sustainability. To become a sustainable society we must (Robért et al. 2002):

1. Eliminate our contribution to systematic increases in concentrations of substances from the Earth's crust
2. Eliminate our contribution to systematic increases in concentrations of substances produced by society
3. Eliminate our contribution to the systematic physical degradation of nature
4. Contribute as much as we can to the meeting of human needs in our society and worldwide

In terms pertinent to Arctic marine sustainability, the first could be applicable to oil and gas exploration in the Arctic. We should try to manage with more renewable resources or such that do not contribute to climate change. The second condition points to the prevention of pollution of the seas, due to human incidents such as oil and chemical spills, runoff of wastewater or dumping other wastes to the sea that do not readily degrade in the marine environment. The third would warn both about the dangers of overfishing as well as the significant alteration of the physical environment that takes away the space from biological species and weakens marine biodiversity. On the other hand, we need to maintain the wellbeing of arctic communities and ensure that they can practice their traditional livelihood.

The framework follows from principles for how a system is constituted (ecological and social principles) and contains principles for a favourable outcome for the system (sustainability), as well as principles for the process to reach this outcome (sustainable development). Broman and Robèrt (2017) have concluded that essential aspects that need to be sustained include (a) assimilation capacity, (b) purification capacity, (c) food production capacity, (d) climate regulation capacity, and (e) diversity. In terms of the Arctic marine environment, many authors of this book argue that human activities have weakened all of these capacities of the Arctic marine ecosystem. While climate change is having direct effects on Arctic ecosystems, the dynamics of pollutants within Arctic ecosystems are also being affected, enhancing pollutant mobility and effects in some cases (Gamberg 2019).

With regards to sustainability framework conditions, an essential element in the Arctic is the multiple and concurrent perturbations from anthropogenic activities, and the limited capability of Arctic ecosystems to absorb them and regenerate. This prompts the discussion about resilience of ecosystems and its limitations in the Arctic.

1.6 The Concept of Resilience

Resilience is a popular narrative for conservation and it implicates the possibility that ecosystems can recover and rebound from disturbances. The term resilience captures two dynamic processes: the ability of ecosystems to resist and absorb disturbance, and their ability to recover. (Darling and Côte 2018) Resilience was originally introduced by Holling (1973) as a concept to help understand the capacity of ecosystems with alternative attractors to persist in the original state subject to perturbations. Walker et al. (2004) define resilience as the capacity of a system to absorb disturbance and reorganize in ways that retain essentially the same functions, structures, identities, and feedbacks. Interest in the concept of resilience has grown dramatically in recent years, and it is featured prominently in the Paris Agreement on climate change, the United Nations Sustainable Development Goals, and the Sendai Framework for Disaster Risk Reduction, among others (Carson and Peterson 2016).

1.6.1 Resilience of Marine Ecosystems

The intensity and frequency of climate-driven disturbances are increasing in coastal marine ecosystems, driven by disturbances associated with ocean warming, acidification, sea-level rise and extreme weather events (O'Leary et al. 2017). Whether marine ecosystems resist, recover, restructure, or vanish, hinges on how extreme future climate change is (Darling & Côte 2018). For example, in the case of coral reefs, Darling and Côte concluded that they will most likely transform beyond recognition in the coming decades. Such ecological shifts will, in turn, force people depending on marine ecosystems to change how they use and depend on ecosystem services. Similar conclusions were made in this book also by Koeningstein (2020), for the case of the Barents Sea. The implication is that we also need to improve the resilience of people and communities to help dampen coming climate shocks. The extensive review of O'Leary et al. (2017) also concluded that whilst "bright spots" of ecological resilience exist, indicating that ecosystems can be resilient even facing long-term chronic climatic stress. There was, however, also a high frequency of reporting "local stressors", both anthropogenic and biotic, that was preventing resilience. In general, genetic diversity seems to be the most important positive, and human interaction the most serious negative factor. The authors concluded that the escalating impacts of climatic change on marine ecosystems and ecosystem services require that the conditions and processes enabling resilience are understood and supported. The results indicate that the reduction of additional local stressors and the use of marine spatial planning, may be the most effective approaches to promoting resilience. Reducing the cumulative impacts to biogenic ecosystems during climatic disturbance is essential for maintaining at least some biogenic structure and source populations that can provide for post-disturbance recruitment and regrowth. The results indicate that although marine ecosystems face growing cumulative stress from coupled human perturbations and climatic instabilities, they still harbour enormous capability for resilience. Maintaining and rebuilding this capacity should be a major focus of marine science and management. (O'Leary et al. 2017).

1.6.2 Resilience in an Arctic Context

The Arctic Resilience Interim Report (2013) defines resilience as a "systems' capacity to cope with disturbances and recover in such a way that they maintain their core function and identity". It also relates to the capacity to learn from and adapt to changing conditions and, when necessary, to transform. The Arctic Resilience Action Framework (ARAF), was approved in 2017. The final report of the ARAF implementation project (Arctic Council SDWG 2019) highlights that the Arctic is warming twice as fast at the rest of the planet (Overland et al. 2018), and many researchers call the Arctic as the indicator of climate change. Substantial social, environmental, and economic changes have also taken place. There are 4 million people who call the Arctic their home (Larsen and Fondahl 2014), many of them Indigenous peoples who have lived in the Arctic for centuries and have a long history of navigating environmental changes. However, the current rate of change and the potential for surprises and shocks creates unprecedented challenges for Arctic residents.

The Arctic Climate Impact Assessment of 2014 was an eyeopener in highlighting the rapid changes occurring in the Arctic. Since then, the Arctic Council has continued to study the physical, ecological, and social changes that are impacting the people and the natural systems of the Arctic. During the Swedish Chairmanship of the Arctic Council (2011–2013), the Arctic Council, deeply concerned about climate impacts and other transformations happening in the region, initiated the Arctic Resilience Report (ARR) project. Its final report (Carson and Peterson 2016), concludes that rapid change is the norm in the Arctic and that the main drivers of this change are largely outside the Arctic. Climate change caused by greenhouse gas emissions plays a particularly large role, but migration, resource extraction, tourism, and shifting political relationships are also reshaping the Arctic in significant way. The report also identified "regime shifts", or large, abrupt changes in social-ecological systems, and evaluated characteristics of resilient Arctic communities. Resilience, as the capacity to buffer and adapt to stress and shocks, and thus navigate the large and rapid changes occurring in the Arctic, is immensely relevant to the people of the Arctic, its ecosystems, and the management and governance of the region's natural resources.

The Arctic Council plays an important role in building a collective understanding of Arctic change and resilience, promoting dialogue, and providing information, especially in connection with climate-related risks in the Arctic region (SDWG 2019). The Adaptation Actions for a Changing Arctic (AACA) project projected potential adaptation responses in three regions and complemented the work of the ARR project.

The first Arctic Resilience Forum in Rovaniemi in 2018 (Halonen et al. 2018) noted the multiple risks to Arctic livelihood, due to climate risks on ecosystem service. The report suggests that the changes in some cases are so dramatic and unavoidable that transformation of livelihoods remains the sole option.

1.7 Cumulative Impacts Assessment

Reports on Arctic resilience point to the multiple impacts from human activities, which calls for cumulative impact assessment. Cumulative impacts have been defined as impacts that result from incremental changes caused by other past, present or reasonably foreseeable actions together with the project (Walker and Johnston 1999). Examples listed were: (a) incremental impact from a number of separate developments; (b) combined effect of different impacts on the same receptor; (c) several individually insignificant impacts which together have a cumulative effect. It is required by legislation that, when conducting Environmental Impact Assessment, cumulative impacts and impact interactions should be included. It is because the environmental impacts resulting from cumulative impacts, and impact interactions can be significant. Cumulative assessment methods have been used to assess the accumulation of ecological impacts (Franks et al. 2010) as well as social sustainability benefits (Fedorova and Pongrácz, 2019) within a region. The indication from the literature in resilience is that cumulative impacts from several stressors are inhibiting the ability of ecosystems to resist and recover. As an illustration, Table 1.2 presents some of the most critical disturbances to Arctic marine ecosystems (marine pollution, ocean acidification, perturbance to genetic diversity and disturbance to

Table 1.2 Cumulative impacts to Arctic marine ecosystems and impact categories

	Coastal activities	Fishing	Transportation	Oil and gas exploration	Tourism
Marine pollution	Runoff spills, wastewater nutrients, plastics	Lost fishing lines	Waste from ships, oil spill	Seepage and spills	Waste from ships, potential spills
Ocean acidification	CO_2 emissions from human activities	Emissions from fishing fleet	Emissions from cargo ships	Burning fossil fuels	Emissions from cruise ships
Perturbance in genetic diversity	Fishfarming	Preference of commercial fish species	Invasive species from ballast water		
Disturbance to habitats	Stress to coastal ecosystems	Potential overfishing	Disruption due to ship traffic	Pollution and physical disruption	Diving and recreational activities

Table 1.3 The severity of disturbances to Arctic marine ecosystems

	Coastal activities	Fishing	Transportation	Oil and gas exploration	Tourism
Marine pollution	↑↑↑↑	↑	↑↑	↑↑↑	↑
Ocean acidification	↑↑↑	↑	↑↑↑	↑↑↑↑	↑↑
Perturbance in genetic diversity	↑	↑↑↑	↑		
Disturbance to habitats	↑↑↑↑	↑↑↑	↑	↑↑↑	↑

Legend:	↑↑↑↑	↑↑↑	↑↑	↑	
	Highest impact	High impact	Moderate impact	Low impact	No or minor impact

habitats) and some of the chief causes (coastal activities, fishing, transportation, oil and gas exploration, tourism). Further, Table 1.3 attempts to illustrate the severity of these impacts.

This method of visualization in Table 1.3 is to be considered only indicative and the scaling intuitive, the purpose being merely illustrative. The idea presented here is that although some of the activities may have currently moderate or no impact, the cumulative impact of multiple disturbances is significant. In most cases there are also activities that generate high or even very high impact. It is also to be noted that, due to increased global interest in the Arctic, in terms of tourist destination, shipping route, and source of fossil fuels, their impact is expected to increase and potentially aggravate the already stressed Arctic marine environment – thus weakening their resilience.

1.8 Planetary Boundaries and the Arctic

As a final notion to be introduced, the concept of planetary boundaries is presented. This was proposed by scientists lead by Johan Rockström from the Stockholm Resilience Centre and Will Steffer from the Australian national University. The idea (illustrated in Fig. 1.3) has been presented in a special feature of Nature (Rockström et al. 2009).

The planetary boundaries concept presents a set of nine planetary boundaries within which humanity can continue to develop and thrive for generations to come. These boundaries define the safe operating space for humanity with respect to the Earth system and are associated with the planet's biophysical subsystems or pro-

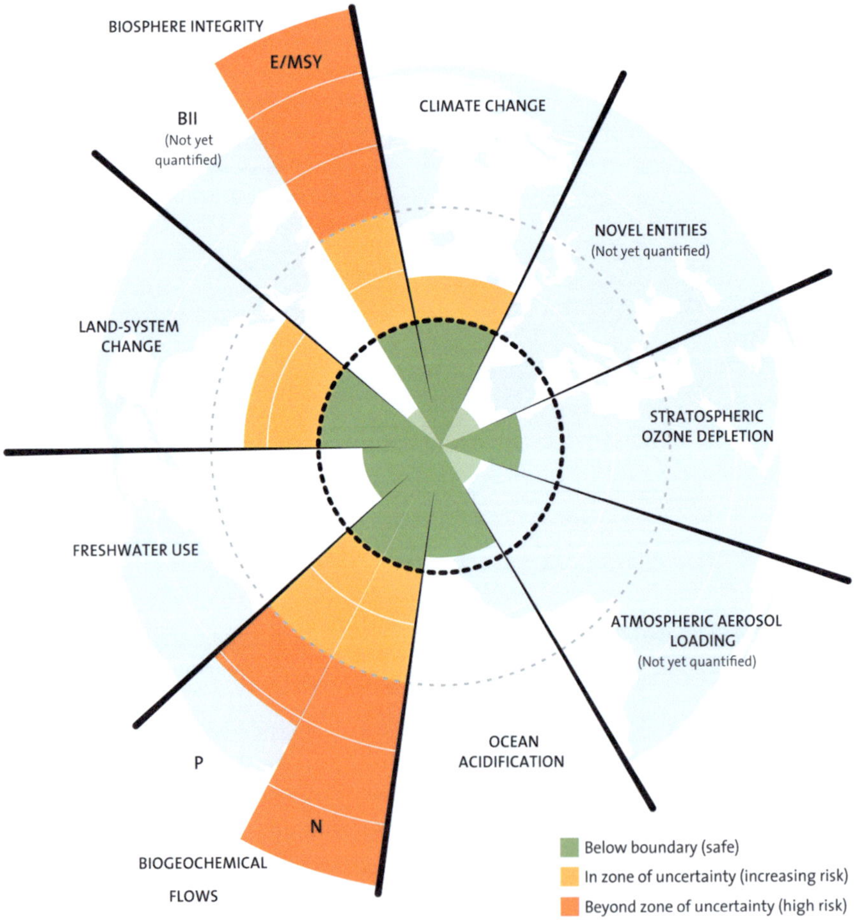

Fig. 1.3 The planetary boundaries concept Pharand-Deschênes (2015), based on Rockström et al. (2009)

cesses (Rockström et al. 2009). The scientists attempted to quantify the biophysical boundaries outside which, they believe, the Earth risks moving into a different system state. The purpose of identifying critical planetary boundaries was to inform society's decisions about sustainability and be potentially used in the societal decisionmaking process.

The wedges represent an estimate of the position for each variable, with green shading indicating safe operating space. The boundaries in three systems: rate of biodiversity loss, climate change and human interference with the nitrogen cycle, the planetary boundaries have already been exceeded (Rockström et al. 2009).

Based on this concept, and inspired by Nash et al. (2017), Fig. 1.4 provides a suggestion on the status of these aspect in the Arctic marine context.

Globally, genetic diversity and biogeochemical flows, especially that of nitrogen are disturbed the most by human activities. Phosphorous pollution follows closely, reaching critical levels. Functional diversity is not yet quantified. In the

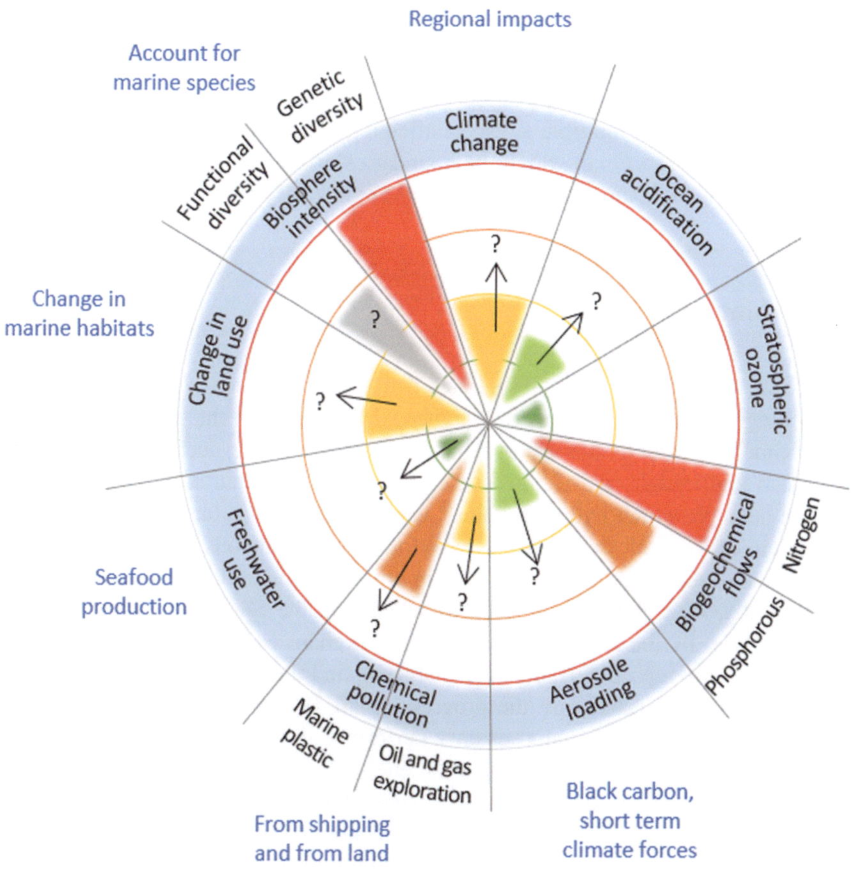

Fig. 1.4 Status of processes that influence the resilience of Arctic marine environment based on Nash et al. (2017)

marine context, one of the greatest concerns of today is marine plastics (PAME 2019), closely followed by the pollution risks of oil and gas exploration, which is could be a critical pollution in Arctic seas (Pavlov 2020). Multiple human stressors have added to the cumulation of impacts. Ocean acidification is also closing critical levels, and is closely monitored in the Arctic, the observation being that acidification levels are higher in Arctic seas than globally. While globally, aerosole loading is not yet quantified, in the Arctic seas, short term climate forcers, such as black carbon are a cause for concern (Shindell and Faluvegi 2009). Additionally, climate change is a major stressor and it is expected that its impact may accelerate. Land-use change in the arctic marine context is interpreted here as change in marine habitats. The indication is that human activities have altered marine habitats to a significant degree and several sectors provide cumulative impacts, as seen in Table 1.2. While water use in the global context refers to freshwater use, in the Arctic marine context this could be used to identify the impact of seafood production. The impacts to date are moderate, although the expectation is that the Arctic would become also a food producing region (Natcher et al. 2019). The arrows indicate potentially increasing impact, with the question marks indicating uncertainty.

The objective of this visualization in Fig. 1.4 was to indicate that, due to the cumulative impact of multiple stressors affecting the same receptor; in the Arctic, there are several aspects in which the boundaries of safe operation have been surpassed, and the indication is that the impacts are intensifying.

1.9 Conclusions

The Arctic cannot support the accelerated demand for resources and also assimilate the ecological impacts associated with these demands. Both the state of the Arctic environment as well as the use of Arctic resources indicate deepening unsustainability. We are caught in a vicious cycle of resource exploitation that leads to ecosystem damage, which will undermine livelihoods, which in turn will force Arctic communities to adapt and perhaps look for other livelihoods, leading to further exhaustion of resources, and a spiral of continuing degradation continues.

Currently, no country is on track to achieve all the Sustainable Development Goals (SDGs) by the target date of 2030, but perhaps SDG14 has been the most overlooked. The report of Bertelsmann Stiftung and the Sustainable Development Solutions Network (2018) on the progress of all 193 countries in achieving the SDGs found that the G20 countries have performed the worst in terms of failing to achieve SDG14 on protecting the marine environment. This will need to change. Most Arctic states have made efforts, focusing mainly on marine plastics, pollution prevention and sustainable fishing and blue economy. On May 9th, 2019, the general assembly of United Nations have adopted the resolution that the 2020 United Nation Conference will support the implementation of SDG14, to conserve and sustainably use the oceans, seas and marine resources for sustainable development (United Nations General Assembly 2019).

The message of this chapter is also that Arctic nations should better integrate Cumulative Impact Assessment as both a tool for planning and for evaluation. It should be considered as part of a precautionary sustainability strategy, not only as a standalone solution for dealing with development-related environmental impacts (Jones 2016).

Arctic marine sustainability is a "wicked problem". This term has been introduced to describe "*ill-formulated social systems, where the information is confusing, where there are many clients and decisionmakers with conflicting values, and where the ramifications of the whole system are confusing, and solutions often turn out to be worse than the symptoms*" (Churchman 1967). The moral of wicked problems is that it is wrong to attempt to tackle only one part of a wicked problem. Thus is the purpose of this book, to outline the many competing sectors, businesses and interests, aiming to use the Arctic, as a place to live, explore, exploit and bypass; as well as recognizing their current and potential future impacts. We need a dialogue across these sectors to avoid the cumulation of impacts, scientists to assess impact interaction and sustainable governance in Arctic States and of Arctic resources. The Arctic Council will no doubt lead the way to inform and guide, providing best practices to progress sustainable development and environmental protection in the Arctic.

References

Arctic Council. (2016). In M. Carson & G. Peterson (Eds.), *Arctic Resilience Report*. Stockholm: Stockholm Environment Institute and Stockholm Resilience Centre. http://www.arctic-council.org/arr.

Arctic Council. (2019). *Sustainable Development Working Group (SDWG)*. https://arctic-council.org/index.php/en/about-us/working-groups/sdwg

Arctic Council SDWG. (2019). *Arctic Resilience Action Framework (ARAF) 2017-2019 implementation project*. Final project report, May 2019.

Bertelsmann Stiftung and the Sustainable Development Solutions Network. (2018). *SDG index and dashboard report 2018*. Global responsibilities. Implementing the goals. http://www.sdg-index.org/assets/files/2018/01 SDGS G20 Summary 2018 WEB V6 110918.pdf

Bobylev, S. N., & Solovyeva, S. V. (2017). Sustainable development goals for the future of Russia. *Studies on Russian Economic Development, 28*(3), 259–265.

Broman, G. I., & Robèrt, K.-H. (2017). A framework for strategic sustainable development. *Journal of Cleaner Production, 140*, 17–31.

Carson, M., & Peterson, G. (Eds.). (2016). *Arctic resilience report*. Stockholm: Arctic Council, Stockholm Environment Institute and Stockholm Resilience Centre. http://www.arctic-council.org/arr.

Churchman, C. W. (1967). Wicked problems. *Management Science, 14*(4), B141–B146.

Darling, E. S., & Côté, I. M. (2018). Seeking resilience in marine ecosystems. *Science, 359*(6379), 986–987.

Fedorova, E., & Pongrácz, E. (2019). Cumulative social sustainability effects of local energy production value chains. *Renewable Energy, 131*, 1073–1088.

Franks, D. M., Brereton, D., & Moran, C. J. (2010). Managing the cumulative impacts of coal mining on regional communities and environments in Australia. *Impact Assess Project Appraisal, 28*(4), 299–312.

Gamberg, M. (2019). *Threats to Arctic ecosystems*. Reference module in earth systems and environmental sciences.

Government of Canada. (2018). *The 2030 agenda for sustainable development*. https://international.gc.ca/world-monde/issues_development-enjeux_developpement/priorities-priorites/agenda-programme.aspx?lang=eng
Halonen, M., Sepponen, S., Mikkola, J. & Descombes, L. (2018) *Report of the 1st Arctic resilience forum 10–11 September 2018 in Rovaniemi*. Finland, Gaia Consulting Ltd.
Holling, C. S. (1973). Resilience and stability of ecological systems. *Annual Review of Ecology and Systematics, 4*, 1–23.
Jones, F. C. (2016). Cumulative effects assessment: Theoretical underpinnings and big problems. *Environmental Reviews, 24*(2), 187–204.
Koeningstein, S. (2020). Chapter 3: Arctic marine ecosystems, climate change impacts and governance responses: An integrated perspective from the Barents sea. In P. Pongracz & Hänninen (Eds.), *Arctic marine sustainability. Arctic maritime business and the resilience of the marine environment* (pp. 45–71). Springer Polar Sciences, etc.
Larsen, J. N., & Fondahl, G. (2014). Major findings and emerging trends in Arctic human development. In J. N. Larsen & G. Fondahl (Eds.), *Arctic human development report II*. Copenhagen: Nordic Council of Ministers.
Missimer, M., Robért, K.-H., & Broman, G. (2017). A strategic approach to social sustainability – Part 1: Exploring the social systems. *Journal of Cleaner Production, 140*(1), 32–41.
O'Leary, J. K., Micheli, F., Airoldi, L., Boch, C., De Leo, G., Elahi, R., Ferretti, F., Graham, N. A. J., Litvin, S. Y., Low, N. H., Lummis, S., Nickols, K. J., & Wong, J. (2017). The Resilience of Marine Ecosystems to Climatic Disturbances. *Bioscience, 67*(3), 208–220.
Nash, K. L., Cvitanovic, C., Fulton, E. A., Halpern, B. S., Milner-Gulland, E. J., Watson, R. A., & Blanchard, J. L. (2017). Planetary boundaries for a blue planet. *Nature Ecology & Evolution, 1*(2017), 1625–1634.
Natcher, D., Yang, Y., … Valsdóttir Th. (2019). *The Arctic as a food producing region*. Final project report prepared for the Arctic Council's SDWG. February 2019.
Nordic Council of Ministers for the Environment and Climate. (2019). *Nordic ministerial declaration on the call for a global agreement to combat Marine plastic litter and microplastics*. URL: https://www.norden.org/en/declaration/nordic-ministerial-declaration-call-global-agreement-combat-marine-plastic-litter-and
Norwegian Ministry of Finance and Norwegian Ministry of Foreign Affairs. (2018). *One year closer 2018. Norway's progress towards the implementation of the 2030 agenda for sustainable development*. URL: https://www.regjeringen.no/globalassets/departementene/ud/vedlegg/utvikling/oneyearcloser_2018.pdf
Ottawa Declaration. (1996). *Declaration on the establishment of the Arctic council joint communique of the governments of the Arctic countries on the establishment of the Arctic council*. Ottawa: Arctic Council. https://oaarchive.arctic-council.org/bitstream/handle/11374/85/EDOCS-1752-v2-ACMMCA00_Ottawa_1996_Founding_Declaration.PDF?sequence=5&isAllowed=y.
Overland, J. E., Hanna, E., Hanssen-Bauer, I., Kim, S. -J., Walsh, J. E., Wang, M., Bhatt, U. S., & Thoman, R. L. (2018). *Surface air temperatures [in Arctic report card 2018]*. U.S. National Oceanic and Atmospheric Administration, NOAA. https://www.arctic.noaa.gov/Report-Card/Report-Card-2018.
PAME. (2019). *Desktop study on Marine litter including microplastics in the Arctic*. 11th Arctic council ministerial meeting, Rovaniemi, Finland, 7 May, 2019.
Pavlov, V. (2020). Chapter 11: Arctic marine oil spill response methods: Environmental challenges and technological limitations. In P. Pavlov & Hänninen (Eds.) *Arctic marine sustainability. Arctic maritime business and the resilience of the marine environment* (pp. 213–248). Springer Polar Sciences, etc.
Pharand-Deschênes, F. (2015). *Planetary boundaries*. Available at https://www.stockholmresilience.org/research/planetary-boundaries.html. Stockholm Resilience Insitute/Globaïa.
Prime Minister's Office Iceland. (2011). *Iceland 2020 – governmental policy statement for the economy and community knowledge*, sustainability, welfare. https://www.government.is/media/forsaetisraduneyti-media/media/2020/iceland2020.pdf

Prime Minister's Office Finland. (2016). *National report on the implementation of the 2030 agenda for sustainable development FINLAND*. Prime Minister's Office Publications 10/2016.

Regeringskansliet. (2018). *Handlingsplan agenda 2030 (agenda 2030 action plan*. In Swedish.) 2018–2020. URL: https://www.regeringen.se/49e20a/contentassets/60a67ba0ec8a4f27b04cc4098fa6f9fa/handlingsplan-agenda-2030.pdf.

Robèrt, K.-H., Schmidt-Bleek, B., de Larderel, J. A., Basilede, G., Jansen, J. L., Kuehr, R., Price Thomas, P., Suzuki, M., Hawken, P., & Wackernagel, M. (2002). Strategic sustainable development — Selection, design and synergies of applied tools. *Journal of Cleaner Production, 10*(3), 197–214.

Rockström, J., Steffen, W., et al. (2009). A safe operating space for humanity. *Nature, 461*, 472–475. https://www.nature.com/articles/461472a.

SDWG. (2019). *Arctic Resilience Action Framework (ARAF) 2017 – 2019 implementation project, final project report*. Arctic Council, Sustainable Development Working Group, May 2019.

Shaker, R. R. (2015). The spatial distribution of development in Europe and its underling sustainability correlations. *Applied Geography, 63*, 304–314.

Shindell, D., & Faluvegi, G. (2009). Climate response to regional radiative forcing during the twentieth century. *Nature Geoscience, 2*, 294–300.

The Danish Government. (2017). *Report for the voluntary national review*. Denmark's implementation of the 2030 Agenda for Sustainable Development. Ministry of Finance. June 2017.

UNGE (United Nations General Assembly). (2015). *Transforming our world: The 2030 agenda for sustainable development*. UN Document A/70/L, 1. https://sustainabledevelopment.un.org/content/documents/21252030 Agenda for Sustainable Development web.pdf

UNGE (United Nations General Assembly). (2019). *2020 United Nations conference to support the implementation of sustainable development goal 14: Conserve and sustainably use the oceans, Seas and Marine resources for sustainable development*, UN Document A/RES/73/292. https://undocs.org/en/A/RES/73/292

United Nations General Assembly. (1987). *Report of the world commission on environment and development: Our common future; transmitted to the general assembly as an annex to document a/42/427 – Development and international co-operation: Environment; our common future*, Chapter 2: Towards sustainable development; Paragraph 1.

Walker, L. J., Johnston, J. (1999). *Guidelines for the assessment of indirect and cumulative impacts and well as impact interactions*. European Communities, EC DG XI, May 1999.

Walker, B., Holling, C. S., Carpenter, S. R., & Kinzig, A. (2004). Resilience, adaptability and transformability in social—ecological systems. *Ecology and Society, 9*(2), 5.

Chapter 2
Thinking Like an Ocean: A Climate Ethic for the Arctic Marine Environment

Øyvind Stokke

> *"This is the new climate democracy: Of the people, by the people, for the planet"*
>
> (Senator Edward J. Markey launching Green New Deal resolution with representative Alexandria Ocasio-Cortez in the U.S. Congress 7.02.2019.)

Abstract An appropriate climate ethic has to take into account the collective action problems inherent in most environmental problems affecting the Arctic, like global warming and increasing acidification of the Barents Sea. But as all the environmental degradation is taking place in and through human practices and institutions – especially by isolated individuals and corporations acting strategically in the market – it is the task of citizens of the different Arctic communities to *think like an ocean*. Thinking like an ocean means overcoming the alienation from our environment by taking responsibility for it, ethically and politically. I start by describing the dire consequences of climate change for the Arctic environment, focusing on ocean acidification, only to confront the supply side of fossil fuel production. After discussing some ethical issues lying at the core of resource governance beyond the state, I outline a critical environmental theory. The recent awakening of green populism, in the forms of school strikes and climate lawsuits around the world indicates environmental democracy having become radical democracy: citizens engaging in communal and democratic practices through which they can build their own environmental futures and energy futures. But it is also an epistemic democracy formed within the context of an epistemic Arctic consisting of different but related epistemic communities.

Ø. Stokke (✉)
The Arctic University of Norway, Tromso, Norway
e-mail: oyvind.stokke@uit.no

E. Pongrácz et al. (eds.), *Arctic Marine Sustainability*,
Springer Polar Sciences, https://doi.org/10.1007/978-3-030-28404-6_2

Keywords Climate ethics · Global warming · Tragedy of the commons · Acidification · Non-exploitation · Environmental democracy · Vogel · Communicative action

2.1 The Problem: Democracy and the Oceanic Commons

In this chapter I argue that a proper climate ethic for the Arctic marine environment should explain to us – the *citizens* of the states encircling the Arctic Ocean and the Barents Sea – how the different kinds of threats toward this environment is in fact produced by *us*. The environment is not produced by something outside our practices and institutions, or by greedy politicians or corporate interests around the world beyond our control. It is built by us, for good and for bad (Vogel 2015). This statement is trivial until we address *the way* we produce our environment. Take the phenomenon that is identified as the biggest challenge in human history: global warming. In 2007 the Russian flag was planted on the seabed at the North Pole. That event reinforced the imaginary of the Arctic as a resource frontier being up for grabs due to ownership of the resources being somewhat undetermined (Steinberg et al. 2015, pp. 91–92). The target of this competition, oil and gas, are the crown jewels of the Arctic resources, but they are also the main source of carbon dioxide emissions which is the central cause of global warming. Climate change as the result of this development is one of the great threats to the Arctic marine environment, causing both acidification and melting of the Arctic sea-ice. A climate ethic should explain to us what is the right thing to do about global warming.

As argued by Simon Caney, dangerous climate change could be said to give rise to three broadly defined ethical responsibilities (Caney 2009, p. 126): Firstly, we have a responsibility to *mitigate* climate change with all the means available to us. We need to cut our emissions, maintain our natural carbon stocks, increase carbon capture through technological solutions, increase the use of alternative energy sources, and develop and transfer green technology to the poor countries. Secondly, we have a responsibility to make those hit hardest by climate change (read: the poor countries in the south) able to adapt to them. Thirdly, there is a high chance that limitation and adaptation will not extend, giving rise to a third commitment to compensate for adverse effects threatening the rights of those affected. However, even though an agreement was achieved regarding how the burdens of these responsibilities are to be distributed among states, corporations and individuals, global warming leave us with a collective action problem named "The tragedy of the commons" (Hardin 1968; Ostrom 1990). The tragedy of the commons denotes a situation where a number of agents (individuals, corporations, states) act in isolation from each other, planting their flags and claiming sovereign resource rights – and where these actions appear as both legitimate and rational. But their private actions and claims more often than not come with a price: resource depletion and environmental degradation.

Few have framed this kind of collective action problem more clearly than the environmental philosopher Steven Vogel (Vogel 2012, 2015). Vogel takes as his point of departure the work of Karl Marx who famously analysed the condition of the modern individual as one characterised by *alienation*. Alienation occurs when a human-made object, or a social phenomenon is conceived as something beyond human knowledge and control, thus becoming an alien power standing over and against us. In capitalism, humans become alienated from their economic and social relations because these relations, according to Marx, become external to us, appearing as they do as relations between *things*, not between persons. Consequently, the crisis produced by capitalism, like the financial crisis and the climate crisis, appear as *facts of nature*, and not as the results of our own doings (Vogel 2012, p. 303).

The strange thing is that precisely in this age bearing our name – Anthropocene – humans don't recognise their own shaping and re-shaping of the environment they inhabit. Thus, they blame global warming on the politicians, on greedy individuals and companies, or on natural variations in temperatures. But Anthropocene is the name of the age when the social and the natural world *are* made by humans. While alienation in environmental philosophy has normally been described in terms of our alienation towards nature, it is now time to realise that what we really are alienated from is the *environment* – not nature – and that this specific kind of alienation has fatal consequences in terms of what we are doing to that environment. Marx analysed this situation as one of isolated and strategically acting individuals in a market, unable to coordinate their actions in a rational way. Take an example from Steven Vogel: An environmentally concerned industrialist or fisherman or commuter all face the problem of the commons: Their praiseworthy acts of investing in green technology, protecting the fish-stocks, and of starting to use collective transport to work will in fact have no significant impact on achieving the goals of preventing pollution, saving the fish-stocks and preventing global warming. (op.cit., p. 309). This is the problem of inconsequentialism: My decision to stop flying for reasons of mitigating climate change has no consequences for the climate as long as the government makes no effort to limit the number of flights or refuse to introduce carbon taxes on aviation companies. But if my individual decision is inconsequential with respect to the climate, "I might still be under the obligation *to build the sort of community* capable of averting further climate change". This is the ethical obligation linking climate change (and climate ethics) to *democracy*: To move from the realm of the market to the realm of politics (op.cit., p. 214).

I start by describing the dire consequences of climate change for the Arctic environment, focusing on ocean acidification, only to confront the supply side of fossil fuel production. After discussing some ethical issues lying at the core of resource governance beyond the state, I outline a critical environmental theory that explains how our environmental problems are produced through the aggregation of unintended consequences of more or less rational decisions by agents in the market. Further, I argue that our main task as citizens in the Arctic communities is to engage in communal and democratic practices through which it is possible to build our own environmental futures and energy futures, thereby resisting the irresponsible policies of leaving these futures to the free market forces. I also argue that a main task for democracy is to resist irresponsible political-ideological pathways where individuals and corporations in the *market* are replacing speakers achieving a common understanding in the *forum*.

My account of a critical environmental theory takes Jürgen Habermas' concept of *communicative action* as a point of departure. Habermas has accomplished the unique task of combining critique of the ways monetarisation and commodification undermine the very conditions of our social and natural life-world with a positive theory explaining the linguistically mediated construction of our everyday social order (see Habermas 1981). And as nature does not speak, we – as speakers and listeners coordinating our actions through language – must answer the questions raised by the environmental problems (Vogel 1996, p. 200). If the environment is built by us, through our social practices, and those practices are unsustainable in terms of destructive effects on the environment, then we have to criticize and eventually change those practices.

The current rise of a global movement of school-strikes for the climate is used to illustrate this kind of linguistically mediated contruction, and eventually change, of a social order as a form of resistance against the market-driven depletion of natural resources. Thus, my critical account provides the ground for explaining the current awakening of green populism being formed in autonomous public spheres all over the world. One example is the Arctic Oil Lawsuit filed against the Norwegian Ministry of petroleum by three environmental organisations, a case that serves to highlight the tension between market-driven resource governance and democratic environmental discourse in the Arctic.

2.2 Some Consequences of Climate Change for the Arctic Environment

The current scenarios for the Arctic marine environment are dramatic, the latest report on the Svalbard Archipelago projecting a rise in the mean temperature on up to 10 ° C (Hanssen-Bauer et al. 2019, p. 9) which is likely to cause up to 65% increase in precipitation followed by increased rain-floods, as well as increased combined snowmelt-, glacier melt- and rain-floods – just to mention a few of the main findings in the report. The findings are in line with a recent paper in *Nature Climate Change* documenting that the Arctic has become a hotspot of global warming, with greatest temperature increases observed in the northern Barents Sea (Lind et al. 2018). That paper concludes that the northern Barents Sea may soon complete the transition from a cold and stratified Arctic to a warm and well-mixed Atlantic-dominated climate regime, the main finding being that the warming sea is related to decline in sea-ice import from the Arctic Ocean. Add to this the findings in a new, third report from The Arctic Monitoring and Assessment Programme (AMAP) anticipating that "ocean acidification, particularly coupled with ocean warming and deoxygenation, will drive changes in marine ecosystems and impact Arctic biota… it is likely that ocean acidification will drive changes at a magnitude that will affect people living in the Arctic and surrounding regions. These changes pose risks to commercial, subsistence, and recreational fisheries, as well as to the provision of other ecosystem services in the region." (AMAP 2018, p. 49). Anthropogenic ocean acidification refers to the component of pH reduction that is caused by human activity (IPCC 2018, p. 223; AMAP 2013, p. 1). In the Arctic, ocean acidification is intensified due to low temperatures, increased freshwater supply (river runoff and ice melt) and low pH Pacific water inflow. Generally, acidification damages sea life because acid dissolves shells and corals.

These are dramatic examples of how anthropogenic emissions of greenhouse gas emissions (GHGs) are warming the Arctic twice as fast compared to the global mean temperature increase of about 1 °C since pre-industrial time. Nevertheless, in 2016 the Ministry of Petroleum and Energy in Norway enacted an administrative legal decision allocating ten production licences in the 23rd licensing round on the Norwegian continental shelf. All ten production licences were located in the Barents Sea, including new exploration acreage made available to the industry in the south-eastern part (Norwegian Petroleum Directorate 2016). In sharp contrast to this decision, Christophe McGlade and Paul Ekins in their article on geographical distribution of fossil fuels unused when limiting global warming to 2 °C conclude – based on cost-benefit calculations and considerations of equity – that fossil fuel resources in the Arctic should be left in the ground. Furthermore, developing countries – if any – should be prioritized when distributing the portions left for extraction compatible with the 2 °C target towards 2050. Research on the global 'carbon budget' associated with the probability of successfully keeping the global temperature rise below a certain level, estimates that to have at least a 50% chance of keeping warming below 2 °C throughout the twenty-first century, the cumulative carbon emissions between 2011 and 2050 need to be limited to around 1100 gigatonnes of carbon dioxide (McGlade and Ekins 2015, p. 187).

However, the greenhouse gas emissions contained in present estimates of global fossil fuel reserves are around three times higher than this, opening the door to a business as usual scenario which could be expected to trigger temperature rises between 2.6 and 4.8 °C by the end of the century. Even temperature rises of 2 °C would likely cause increased drought, desertification, food insecurity and heightened vulnerability to water-borne diseases. More specifically McGlade and Ekins argue that development of oil and gas resources in the Arctic and any increase in unconventional oil production are incommensurate with efforts to limit average global warming to 2 °C, and "(…) that policy makers' instincts to exploit rapidly and completely their territorial fossil fuels are, in aggregate, inconsistent with their commitments to this temperature limit." (Mc Glade and Ekins 2015, p. 187).

That warning is reflected – perhaps somewhat surprisingly – in a recent report from the World Economic Forum finding that increasing divisions between the world's major powers is the most urgent global risk we face because it stymies vital collective action on climate change (Global Risks Report 2019). If the financial crisis in 2008–2009 and the migration crisis in 2015 have altered European and US politics along right-wing populist and protectionist lines, then this is bad news for the kind of international effort needed in order to confront the challenges affecting the Arctic marine environment.

2.3 Resource Governance Beyond the State

Another important collective action problem is revealed when we ask the important question of how we should govern natural resources beyond the state. The minerals or fossil fuels located under the Arctic seabed is one of a number of extraterritorial resources, like the water column itself functioning as a huge carbon sink, the ecosystems and icecap of Antarctica, and the atmosphere itself. In recent literature on territorial rights, ownership of resources are grounded in ideas of improvement of resources that happen to be located within the boundaries of the state, or in attachment-based claims related to projects and plans that depend on continued access to, and control over, natural resources. Finally, on some accounts, jurisdictional rights to and control over natural resources are seen as an integral part of the political self-determination of a 'people' (for a recent account, see Moore 2015). But neither political self-determination, nor attachment or improvement, seem to do the work when what we aim for is sustainable governance of global common resources beyond the state. The carbon sink of the water-column or the atmosphere are collective goods sharing two traits: (1) their size make it difficult (but not impossible) to exclude other users from extracting resource units from them (like the capacity to assimilate CO_2-molecules), and (2) any use of the resource subtracts a portion of the resource units to the detriment of other users, i.e. they are rivalrous.

Collective goods are extremely vulnerable to overuse because individuals and countries overconsume natural resources, especially GHGs, by leaving the question of the environment to the market: Typically the Sunday SUV-driver knows that the emissions generated by his Sunday SUV-trip are inconsequential with regard to global warming: the Sunday SUV-driver actually does nothing wrong (Sinnott-Armstrong 2005; for a critical discussion, see also Vogel 2015). But, what looks like a rational way of acting from the point of view of the (self-interested) individual or multinational, leads to a situation nobody wants: Environmental degradation, depletion of stocks, ocean acidification and dangerous climate change. Global warming now having become *the* environmental problem in the Arctic has to be analysed as a problem of the commons, and a proper climate ethic for the Arctic therefore has to be less a matter of developing new virtues than of finding ways to transform the institutional context within which human practices take place. This places upon us as citizens a responsibility to take political action with regard to how we want the natural, social and built environment to be.

As pointed out by Chris Armstrong, the case of "unburnable" fossil fuels "gains interest from the point of view of justice because those who refrain from exploiting a resource might also be considered to accrue a kind of cost. (…):

> Indeed on the surface of it, there seems to be no profound moral difference between the sacrifices an agent makes when justice requires her to actively protect a resource from a threat, and the sacrifice she makes when justice obliges her not to exploit a resource which she otherwise could. In both cases her interests are set back relative to a counterfactual world in which conservation was not required." (Armstrong 2017, p. 234)

However, it is not the case that whenever an agent is unable to exploit a resource, he has a claim to compensation. The normative issue here is fair distribution of opportunity costs when a community ends up with stranded assets in the ground. And in a world characterized by unequal access to the benefits flowing from natural resources, we have reason to mitigate the opportunity costs out of a general concern with some agents' disadvantage. For example, if a poor community loses out on opportunities for development because of stranded assets, justice requires that (some) countries with the ability to pay share (some of) these opportunity costs. First, citizens in countries like Congo or Cameroon have considerably less access to wellbeing compared to for example Norway. Second, it is reasonable to expect that the income generated from fossil fuel exports could make a substantial difference to their wellbeing. Third, opportunities to successfully diversify economies away from their present dependence upon natural resource exports appear to be rather poor for many of these countries. (See Armstrong 2017, pp. 237–238 on the *burdens of conservation* in this context). In contrast, it doesn't look morally troubling if a country like Norway lost out on opportunities when keeping the oil deposits in the ground. Norway has earned the world's largest sovereign wealth fund through exploitation of fossil fuels. At the same time, the country has not only continued, but in fact intensified its exploitation of these resources – like most developed oil-producing countries (Caney 2012) – since 1990 when the harmful effects of greenhouse gases in the atmosphere are considered publicly known according to IPCC.

Thus, from the point of view of historical (in)justice, because of a precarious access to wellbeing, and because of the opportunity costs and environmental risks related to exploitation of oil in the Arctic, there are good reasons to share the opportunity costs of poor, but resource rich countries in Sub-Saharan Africa. However, if the wellbeing of citizens is our goal, the loss of oil revenues could and should be reduced by altering what Armstrong calls the "opportunity structure" of these countries (Armstrong 2017, p. 239). This could be done through assisting these countries in achieving "economic diversification", thereby opening up a viable path for wellbeing in a post-carbon future. Economic diversification would also require the abolishment of Special Economic Zones (SEZ), erected to attract business through generally offering clear, predictable and lucrative terms such as less demanding environmental protection, less strict labour laws, and, most importantly: lower tax payment than the state in general. That is why developed countries like Norway should initiate an aggressively up-scaling of research on, and implementation of, renewable energy sources, thereby helping the developing, resource-dependent countries through programs of green technology-transfer.

2.3.1 Seabed Minerals in Areas Beyond National Jurisdiction

Based on the rich diversity of concepts and theories of resource rights in philosophy, there's a need to develop a normative framework for regulating access to seabed minerals or marine genetic resources, accompanied by an associated benefit-sharing scheme. Most theories of resource rights are based on a concept of property which becomes questionable with regard to the ocean floor for two reasons: First, it is based on human settlement, but at some distance adjacency to settlement seems to become morally arbitrary. Minerals or biodiversity on the surface of, or under, the oceanfloor are not the product of any state, nor are their effective use well served by resource grab by individual states. Second, it expresses the Lockean idea of full, individual ownership rights, disregarding the plurality, and combination of, differentiated resource rights which can be shared and managed in a variety of ways in order to promote equal access to, and sharing of, the ocean's benefits. As argued by Margaret Moore, jurisdictional rights over the seabed can't be based on a claim to collective self-determination, but have to take on an instrumental form. Therefore, control over occupation, transfer and sale of natural resources under the seabed should be delegated to an international level (Moore 2015, p. 169).

2.3.2 When Non-exploited Resources Will Continue to Deliver Important Collective Goods to Outsiders

The United Nations' Convention on the Law of the Seas expansively provides a framework by way of which states can lodge claims over non-living resources contained within 'extended continental shelves reaching out up to 350 nautical miles from the countries' shores. Nevertheless, exploitation of fossil fuels in the Arctic and Barents Sea is a contested case due to economic and environmental risks, but also because of the precarious agreements dividing the area into exclusive zones, and because of the struggle for access and control over the seabed under the high seas, i.e. to extended shelves beyond the zones. The Norwegian government has offered access to 87 pre-defined areas for oil exploitation in its last licencing round, including 53 areas in the Barents Sea. However, the implementation of a policy committed to efforts to limit average global warming to 2 °C "(…) would also render unnecessary continued substantial expenditure on fossil fuel exploration, because any new discoveries could not lead to increased aggregate production". (McGlade and Ekins 2015: 187). The government has "moved" the ice-edge opportunistically for political reasons several times – and against the scientific advice from the Norwegian Polar Institute (see Kristoffersen 2015 for a critical analysis). The Government's petro-strategy runs three risks at the same time: it is environmentally risky, as the areas defined as the "ice-edge" are highly vulnerable in terms of biodiversity and ecosystem services of fundamental significance to humanity; it is financially risky due an estimate of oil prices per barrel up to $ 80; and it is politically risky because the strategy might initiate a resource race with negative geo-political and environmental consequences (Heier 2019). In addition, exploitation will contribute substantially to global warming and ice-melt due to outlets of black carbon. On the other hand, non-exploitation will deliver non-excludable and subtractable collective goods like protecting vulnerable but high-value fish stocks in the Barents Sea, high-value biodiversity and the atmosphere's sequestering capacity – to the benefit of humanity.

The exploitation of important natural resources beyond the state involves collective goods (like the atmosphere's assimilative capacity) vulnerable to overuse. From the point of view of justice it actualizes a plurality, and combination of, differentiated resource rights which can be shared and managed in a variety of ways in order to promote equal access to, and sharing of, the ocean's benefits. In order to promote a both just and sustainable management of these reources in the Arctic, it is necessary to bring that management out of the market and into the political forum where scientific, pragmatic and moral discourses at different political levels could sustain informed democratic decision-making.

2.4 Between the Market and the Forum: The Climate Crisis Is a Democratic Crisis

I have been arguing that the environment is built by us. This simple fact explains the urgent need for ethical guidelines that can orient our individual and collective choices with regard to the Arctic environment. In this section I want to outline a critical environmental theory that explains how our environmental problems are produced through the aggregation of unintended consequences of more or less rational decisions by agents in the market. If our goal is to reduce the impact of these problems we need to shift the coordination of human action from the market forces to the force of the better argument in a democratic environmentalism speaking to citizens in the forum, not consumers in the market. The political theorist Theresa Scavenius has illustrated how this dilemma is played out in the field of climate politics, or the lack thereof: Today's dominant climate policy employs market-based solutions instead of measures directly regulating societal functions like climate policy or environmental policy. Few policies aim at introducing or maintain regulating policy-measures. Instead, the preferred political strategy is to delegate the responsibility and the control of the climate policy to agents in the market, the most well-known examples being international trade agreements like the Kyoto Protocol and the EU Emissions Trading System. Although climate policy-measures are often framed as a question of the individual behaviour and moral conscience of citizens, simple calculations reveal that "the customers' climate-motivated choice in the super-market cannot halt climate change… It remains a fact… that the global oil-consume is *rising*" (Scavenius 2016).

How are the prospects for green democracy? Today the forum, embodied in autonomous publics of a whole, new generation, has become the locus of a world-wide school-strike among children and teenagers protesting against a political class apparently unable to take action to prevent anthropogenic climate change. Having been invited to the Norwegian National Broadcasting in March 2019, children ranging from the age of 13–16 challenged politicians face-to face by pointing out the contradictions between supply side of fossil fuels production and Norway's national determinate contributions to cut emissions by 45% before 2030. What these children are doing, and are planning to do in the future, is to take action to build their own environmental future – or better, as they lack voting rights in democratic elections – to persuade their parents' generation to provide the tools necessary to build such a future. As speakers in the forum (though without the right to vote) these children perform what the philosopher Jürgen Habermas has termed *communicative action* underlying our everyday construction of a social order. Their protests are based on their common definition of a situation by relating to knowledge, social norms and their individual life projects. From the performative perspective of the speaker these children draw on reports from the Intergovernmental Panel on Climate Change (IPCC), egalitarian principles of intergenerational justice, as well as on their lived experience of fear of an endangered future. By exercising their communicative freedom to say "no" to the governments politics of extractivism and environmental degradation, these children are about to build a green political public sphere from below, an impressive task that culminated in 1.4 million children globally striking under the heading "Friday's for Future" on the 15th of March 2019. From the third person perspective of the social scientist, they are integrating scientific knowledge, moral justice and their personal fear of a future climate disaster, thereby making modernity's cultural systems of abstract knowledge – science, morality and and the inner sphere of personal experience – relevant for our everyday practices in civil society and the public sphere. One of the challenges of modern society is the formation of "expert cultures" encapsulating and closing, both from one another and from lay people's everyday lives. The encapsulation has two negative consequences. First, the potential of knowledge within the various knowledge speres, i.e. like science and post-conventional morality, does not benefit society – neither in the public nor in the private sphere, nor in the political institutions.[1] Secondly, there is a fragmentation of people's consciousness. When the expert cultures lose connection with people's daily life-world, it is very difficult for the individual to develop an understanding of society in general and the specific relationships that it includes. Therefore, communicative spaces need to open up between the various expert cultures, and between the knowledge spheres and people's everyday lives. To create and maintain this kind of space is presicely what the global schoolstrikes are about to achieve. However, it is the responsibility of their adults – the citizens – and the politicians of the Arctic communities to sluice public opinions formed within these communicative spaces into the formal channels of the parliamentary system.

Moreover, in the communicative construction of their green public sphere, the children have paid special attention to the moral principle of justice between generations. Being a principle at the core of the philosophy of climate justice, it has been justified at a more general level by the political philosopher John Rawls. In his theory of justice, Rawls emphasizes that future generations must be given due consideration in the justification of the principles we choose when allocating important goods such as rights, income, and natural resources.[2] But it is not enough that the

[1] Habermas denotes this "cultural impoverishment".

[2] In order to prevent us from benefiting our own generation, generational affiliation should be hidden in the negotiations about principles of distributive justice (Rawls 1999, p. 256).

parties agree on the normative validity of the principles guiding our distribution of goods. They must achieve this agreement in good faith in the sense of being convinced that they themselves will be able to live by the principles in practice. For example, the current generation's consumption of natural resources must not inflict such heavy burdens on future generations that *they* will not be able to meet the requirements of a reasonable distribution of the goods mentioned above. Or vice versa: Taken together, over-taxing fish stocks, destroying bio-diversity, or overconsuming the atmosphere's capacity to assimilate CO_2 could threaten the very stability of peoples because unbearable constraints lead to a society characterized by withdrawal, rage, contempt and violent conflicts. Our current environmental crisis can be said to have given Rawls' framing of justice under unbearable constraints a new and frightening relevance. To support and continue this critical environmental conversation in the new green communicative space where the childrens arguments can be brought into dialogue with moral philosophy and scientific research, is a way to fulfil our ethical obligation *to build the sort of community* capable of averting further climate change.

However, as isolated and strategically acting individuals of the *market* are replacing speakers achieving a common understanding in the *forum*, the communicative spaces where speakers can bring together abstract knowledge and everyday experiences are shrinking. What from a market perspective may look like the successful liberation of egoistic motives within a capitalist economy, looks to the actors themselves as complex risks (Ulrich Beck), fragmentation and "natural" forces. To the critical environmental theorist, finally, the development of an objectifying attitude towards the natural world through the market is conceptualised as the Tragedy of the commons (Vogel 2015).

At the same time there is a problem with the long tradition of a critique of the market in critical theory from Marx and onwards, and that problem is that there never was, and never will be, such a thing as The Free Market where the actions of isolated individuals following their own interests are coordinated by an invisible hand. Such a market has to be institutionalised through law, and that process is a political one, mediated through some kind of political will. This means that blaming the environmental problems on the market is too quick a move. More precisely, the side-effects of the individuals' self-interested orientations in the market might very well explain how we end up with consequences nobody wanted, as in the tragedy of the commons, but the main problem seems to be politicians who step down from their mandate to govern our commons, and leave that governance – our use of nature – to the market forces. So the problem here is not The Free Market (which never existed), but the strange non-death of neoliberalism, the ideological faith in the economic policy of "laisser-faire", privileging the market in exchange for democracy and the shrinking state in exchange for ecological and social protection of life-conditions. This is a willed and planned development in Western democracies in the last decades. It is an expression of ideology because the market, famously argued by Karl Polanyi in *The great Transformation*, is always embedded: "Such an institution could not exist for any length of time without annihilating the human and natural substance of society; it would have physically destroyed man and transformed his surrounding into a wilderness." (Polanyi 2001 [1944], p. 3). Again, if there is no such thing as a free market, we have a problem in the very moment we start to blame the environmental crisis on that market. More precisely, we have to blame those who are responsible for the willed dismantling of democratic politics due to their faith in the ideology of the laisser-faire market.

2.4.1 Politicising the Arctic Marine Environment

The climate crisis in the Arctic is a democratic crisis to the extent that politicians step down from their democratic duty to regulate important policy areas like extractive industries, aquaculture and fishery policy. Take the privatisation of fishing-quotas in Norway. One of the problems for coastal and fjord fishermen after this ideological shift in fishery policies is that the quotas follow the vessel, so that large shipowners can accumulate large quotas by buying up smaller vessels with a quota – a process that is impossible to reverse in practice. Thus, shipowners can make big money, but for the cost of low employment for fishermen and a reduction in local processing and value creation in the coastal villages. The horror example of how too much concentration of economic power and control of quotas might end up in the fisheries is Iceland. Here privatization of fishing quotas resulted in a number of coastal communities where the local fishermen were deprived of their rights to fish outside their own ports.

The fisheries and aquaculture industry is Norway's largest export industry after oil and gas, products from captured and farmed fish being exported to more than 150 countries (Norwegian Seafood Federation 2019). In Senja, the second largest island along the Norwegian coast, farmed salmon is exported for billions of Euros – but for the price of a huge carbon footprint thanks to transport by trucks and airlines. However the Norwegian government takes no responsibility to build a railway to the High North which would be a significant investment in cutting CO_2 –emissions on Norwegian territory. Instead the politicians leave the seafood industry with no other choice than fossil fuels driven transport, like trucks, airlines or seatranport.

On the other hand, what we might be witnessing at the moment in several climate lawsuits, in the streets of Paris and among school children in countries all over the world is *green* populism – a social mobilisation against a state privileging oil revenues and market based politics for the prize of rapidly accelerating climate change – a prize to be paid for by future generations. Populism arises among those who lose out whenever the state lacks the ability to govern major policy areas like climate change, welfare and economic globalisation. All political areas are being depicted as where they solely to comprise private activities in the market. However, there's an alternative pathway to the market-based, inequality-driving pathway leading to populism and apathy, namely democratising energy system transitions, partly through fossil fuel divestment and "keep it in the ground" strategies, and partly by the innovation and politics of smart infrastructure. A key concept in this context is "energy justice", highlighting the challenge that both consumers and politicians tend to de-ethicize and depoliticize the socio-energy system, as well as ignore that "a just transformation of the socio-energy system is also a decision to live in a different type of society, not simply a low-carbon version of the current one." Healy and Barry argue that

> change in energy systems –or regimes must address inequalities in power and especially recognise the power of multiple sources of the power of incumbent energy-producing actors. In the US, for example, fossil fuel corporations help shape US energy policies and influence energy transition options, effectively ensuring carbon lock-in from which they of course benefit. Overcoming this carbon lock-in requires confronting corporate energy power… alongside "disruptive technological innovation" we also need disruptive and confrontational political action. The divestment movement stands as a prime example of such confrontational and disruptive political innovations (Healy and Barry 2017, p. 453).

Given the dire consequences of climate change for the Arctic marine environment outlined above, the politics of divestment (and non-exploitation) is imperative on two accounts:

First, it seems to be implied of the normative content embodied in the Environmental Paragraph § 112 in the Norwegian Constitution (see argument below). Second, the social construction of nature (or Nature) in the Anthropocene places upon humanity the burden of responsibility: Our material environment is no longer "external" to us because we produce and reproduce our material surroundings through social practices. That is why this material environment isn't neutral to us: The material infrastructure creates path-dependencies that often come with high costs. Building highways instead of railways binds the whole population to driving or going by airways instead of travelling by train. Likewise, investments in and building of infrastructure for fossil fuel extraction on Norway's or Alaska's outer continental shelves bind not only governments but the whole socio-energy system – including supplier industry and labour forces – to drilling for almost two decades, on average (Helland-Hansen 2019). That is also why we need research on energy justice "across entire energy lifecycles, from extraction to final use, to offer an analytically richer and more accurate picture of the (in)justice impacts of energy policy decisions" (Healy and Barry 2017, p. 451). Politicising the Arctic marine environment means acknowledging the *normativity of a natural environment* that is changing due to human impacts, and to take responsibility for a low- or post-carbon energy future. While we are still waiting for the members of the Arctic Council to take that responsibility at the state level, we are witnessing a green momentum of social movements of democratically organized citizens confronting the supply side of fossil fuel production.

2.4.2 *Thinking Like an Ocean: The Law and Politics of Non-exploitation*

When we as agents in a market develop an objectifying attitude towards the natural environment, including the marine environment, we become alienated from it. We do not recognize our own actions – our energy consumption, our overuse of fish stocks, or our plastic-polluting behaviour – as contributing to degradation of the marine environment. But we have seen that climate change and environmental degradation *is* the product of our own individual or corporate actions. The necessary transition out of this collection action problem could be likened to "thinking like an ocean", that is: overcoming the alienation occurring at the individual level of market transactions and starting to act *in concert* at the collective level – by transforming our practices and institutions through politics. The recent awakening of green populism, in the forms of climate lawsuits[3] and school strikes in several countries around the world against the current generation's inaction with regard to anthropogenic climate change, exemplifies strategies for overcoming alienation from the environment: These efforts all represent different efforts at creating "a social order where decisions about the social practices we engage in are made communally and consciously, via democratic (and legal) discourse, and not left to the nature-like workings of a free market." (Vogel 2012, p. 312). Specifically, I contend that the Arctic Oil Lawsuit – also referred to as The Trial of the Century – could provide a constitutional moment in the history of climate democracy.

In May 2016 the Norwegian Ministry of Petroleum and Energy issued an offer of ten new production licences in the 23rd licensing round on the Norwegian continental shelf under the Barents Sea. Five months later three Norwegian environmental organisations filed a suit against the government based on § 112. The Italian lawyer Ugo Mattei has asked whether our liberal constitutions protect private property far more strongly compared to common goods – including our natural environment upon which we depend, but which in comparison has a weak protection – apparently confirmed through environmental harms like ocean acidification, global warming and the rapid loss of biodiversity. However, amended 2 years ago, the Constitution's Environmental paragraph 112 reads:

> "Every person has the right to an environment that is conducive to health and to a natural environment whose productivity and diversity are maintained. Natural resources shall be managed on the basis of comprehensive long-term considerations which will safeguard this right for future generations as well. In order to safeguard their right in accordance with the foregoing paragraph, citizens are entitled to information on the state of the natural environment and on the effects of any encroachment on nature that is planned or carried out. The authorities of the state shall take measures for the implementation of these principles."

Following Habermas I take it that the fundamental task of judicial review is to defend and protect the fundamental rights enshrined in the Constitution by way of rejecting statutory law that will foreseeably block future citizens' opportunity to participate and have a voice in democratic will-formation, or to claim their basic right to justification for any statutory law or ruling (Habermas 1992, ch. VI).

[3] According to a report by the United Nations, 654 cases have been filed in the United States as of March 2017, with more than 230 cases being filed in all other countries combined. Outside of the United States, the majority come from Australia, the United Kingdom and the European Union (Climate Liability News: https://www.climateliabilitynews.org/2018/12/26/legal-strategy-climate-lawsuits/).

This kind of deontological approach to adjudication has two implications for the legitimacy of law which are of relevance for the court: First, its rulings should reflect the voices of the silenced majority, i.e. to the members of future generations, as far as possible. Second, it should pay attention to the influence from social and economic power-elites on legislation and policy-making: This kind of influence blurs the line between legislation and social power and "is no less in need of the disciplining formation by the constitutional state than the administrative power of the government" (Habermas 1992, p. 320, transl. by the author). The interests of the citizens could be harmed through the power exercised by persons in certain social and economic positions (and not only by encroachments by the government). This places upon the court a specific duty to adjudicate contested norms related to the communicative and procedural conditions of the legislative process. To what extent did the oil companies, the supplier industry and other subcontractors influence the decision about new production licenses? This question is also underlying Healy and Barry's concluding remarks in their article: "While fostering technological innovation is important, we argue more critical research is needed on specific political economies of socio-technical energy transitions – in particular *how existing fossil fuel actors obstruct decarbonisation efforts*, further embedding carbon lock in." (Healy and Barry 2017, p. 456, italics added by the author). This is perhaps the right moment to ask whether the threat to global common goods – the atmosphere, the ocean and the cryosphere – points towards a historical constitutional moment where we as citizens ask our constitutional courts to bring out the normative surplus of our democratic constitutions, by adjudicating an ecology of law that protects these goods for current and future generations. In this way constitutional politics, including the higher courts, can become a tool protecting global common goods against commodification and privatisation – but without making new law itself, thereby respecting the separation of powers and the sovereign legislation of the parliament.

According to Habermas the theory of democracy must relate to those social and political developments that undermine the normative idea of the constitutional state by "confronting the institutions of the well-ordered society with a rather scornful mirror image." (Op.sit. Ch. 2, II). As we have seen, that scornful image is currently mirroring an atmospheric and oceanic injustice produced by a carbon socio-energy system constantly out of balance with climate democracy.

2.5 The Territorial Turn in Climate Ethics: What About the Ocean?

If what humans are doing to the global climate has implications for the ocean – if climate justice and ocean justice are linked – then that mirrors a territorial turn currently going on in the discussions of climate justice. This discussion can in turn be said to have two aspects, mitigation and adaptation. The discussion about mitigation has mostly revolved around the just distribution of emission shares. But what should be the focus of sound arguments about emission shares? Or, what counts as the same: what do we actually distribute in a fair scheme for distribution of emission shares?[4] Mitigation is usually thought to be about the reduction of GHG's through setting a cap on the total amount of emissions which is in turn to be distributed on an equal basis among states and their inhabitants. The leading concept behind this perspective is *atmospheric justice*. In this equal shares discourse the atmosphere is thought of as a global commons with a limited capacity to absorb GHG emissions produced by human activities. The usual approach to this is to distribute equal shares of emission quotas per capita, i.e. quotas of atmospheric absorptive capacity.

However, by closer inspection, GHG's are not the only good to be distributed in a proper scheme for mitigation measures. As part of the atmosphere they are only one aspect of a natural resource system that is more earthbound than we often think. With reference to Elinor Ostrom's analysis of common pool resources (CPR's) like fish stocks and ground water, the philosopher Megan Blomfield identifies the resource system that we should be viewing as a global CPR in discussions of climate change mitigation. The resource unit individuals appropriate when they engage in polluting activities, Blomfield tells us, is assimilative capacity, and the resource system is the global GHG assimilation system (Blomfield 2013). It is not the atmosphere alone that assimilates emissions of, say, CO_2. Carbon is part of a cycle that removes carbon atoms from the air through chemical processes like the photosynthesis in vegetation – be it pastureland, soil, plants and forests, which function like carbon sinks. But carbon is first and foremost absorbed into the sea. Because CO_2 simply dissolves from the air into the seawater, the oceans constitute a huge carbon sink (Broome 2012, p. 20). Other important sinks are rocks containing fossil fuels, and limestone formed of shells accumulated at the bottom of shallow seas. During what we call civilisation, there has been a balance in the carbon cycle – a balance that is now disrupted by the effects of the industrialisation process that started 200 years ago. According to Blomfield, it is when the capacity of oceanic and terrestrial sinks is overused that CO_2 exceeds its climatological limits in the atmosphere.

But who are responsible for the terrestrial and oceanic sinks? Terrestrial sinks, significantly, but to a great extent also their oceanic counterparts, are located within the borders of states which might be thought to have territorial rights over them, and it might be thought that citizens of those states have special claims over their sequestering capacities. (Armstrong 2015, pp. 2–3) Furthermore, some countries ruin their ecosystems through deforesting, some produce smart and green technology and apply it in environmental projects in local communities.[5] And so on. On principle, all these facts could count when we work out a scheme for allocation of emission

[4] My discussion in this section draws heavily on Megan Blomfields excellent discussion of the just distribution of GHG emissions with regard to terrestrial sinks, see Blomfield 2013.

[5] Since 2018 the Arctic center for Sustainable Energy (ARC) at the Arctic University of Norway coordinates two projects implementing Carbon Capture and Utilisation (CCU) and Smart Renewable Energy Infrastucture respectively in two coastal communities.

rights. More important, agents may sometimes be responsible for the assimilative capacity of the terrestrial sinks that at the moment happen to be under their control. Interestingly, this also goes for the oceanic sink since both turning the Barents Sea into a hot-spot and increasing acidification can harm ecosystems and the production of both algae and plankton (Assmy et al. 2017). As for the high seas, i.e. areas beyond national jurisdiction, responsibility should be shared by an international and global agent, and the Arctic Council is doing an important job in monitoring and researching the Arctic marine environment through AMAP.

2.6 Concluding Remarks: A Smart Arctic Future? Building Community Energy Futures

An appropriate climate ethic has to take into account the collective action problems inherent in most environmental problems. I have argued that the warming Arctic is the ultimate example of the Tragedy of the commons, where the overexploitation by each individual state leads to a situation nobody wants: dangerous climate change and acidification. A specific challenge relating to the Arctic is a rapid technological development making it possible to control and extract resources from extremely remote, uninhabitable areas, leading states to claim oceanic territorial rights along their respective extended continental shelves toward the North Pole. The result is political destabilisation and sovereignty conflicts. This is bad news for the Arctic as there is only a political solution to the Tragedy of the commons. I therefore contend that a sustainable climate ethic must work bottom-up and address those inhabiting the Arctic as *citizens* of different but connected *communities*, not as *consumers* using the Arctic as a storehouse for 'corporate greediness' in the *market*. The task of these citizens is to bring local, regional and state authorities to (1) reduce the emission of greenhouse gases, (2) to create and maintain (natural) carbon sinks, and (3) increase the use of alternative energy sources, as well as create and transfer clean technology, across the Arctic. The fate of the Arctic marine sustainability is to a great extent a warming Arctic and the increasing acidification of the Barents Sea. But as all the environmental degradation is taking place in and through human practices and institutions – especially by isolated individuals and corporations acting strategically in the market – it is the task of citizens of the different Arctic communities to *think like an ocean*. Recall the great momentum of taking action to stop marine plastic pollution that swept over many Arctic communities: Thinking like an ocean means making that momentum into a permanent way of thinking with respect to what we do to our environment. Environmental democracy must become radical democracy in terms of democratic citizens engaging in communal and democratic practices through which they can build their own environmental futures and energy futures. But it is also an epistemic democracy formed within the context of an epistemic Arctic consisting of different but related epistemic communities. Universities in the Arctic should form alignments with stakeholders, local communities and corporations to promote the formation of multiple publics in civil society. Publics can foster deliberation in the form of multiperspective inquiry through which environmental knowledge, practices, virtues, institutions and projects of self-realisation become increasingly woven into our common lifeworld. Strong communities in this sense can develop into environmental democracy and resource democracy managing the fine balance between resource management and environmental protection. The Arctic centre for Sustainable Energy (ARC) at the Arctic University of Norway coordinates two projects implementing Carbon Capture and Utilisation (CCU) and The Invention of the Renewable and Smart Rural Power System community (RENEW) respectively in two coastal communities in Northern Norway. In Canada, the partnership Community Appropriate Sustainable Energy Security in Indigenous and Northern Communities (CASES) is an initiative at the University of Saskatchewan collaborating with 14 Indigenous communities in Alaska, Sweden, Norway and Canada, and with academic institutions, utility, and industry. Common to these initiatives are a focus on decentralised energy-solutions, and private and/or shared off-grid solutions. The Arctic Council could possibly function as a *forum* for this kind of Arctic community-based initiatives.

References

AMAP. (2013). *AMAP Assessment 2013: Arctic Ocean Acidification*. Arctic Monitoring and Assessment Programme (AMAP), Oslo, Norway. viii + 99 pp.

AMAP. (2018). *AMAP assessment 2018: Arctic Ocean acidification*. Tromsø: Arctic Monitoring and Assessment Programme (AMAP).

Armstrong, C. (2015). Climate justice and territorial rights. In J. Moss (Ed.), *Climate change and justice* (pp. 59–72). Cambridge: Cambridge University Press.

Armstrong, C. (2017). *Justice and natural resources*. Oxford: Oxford University Press.

Assmy, P., et al. (2017). Leads in Arctic pack ice enable early phytoplankton blooms below snow-covered sea ice. *Scientific Reports, 7*. https://doi.org/10.1038/srep40850.

Blomfield, M. (2013). Global common resources and the just distribution of emission shares. *Journal of Political Philosophy, 21*(3), 283–304.

Broome, J. (2012). *Climate matters. Ethics in a warming world*. New York: W.W. Northon & Company.

Caney, S. (2009). Justice and the distribution of greenhouse gas emissions. *Journal of global ethics, 5*(2), 125–146.

Caney, S. (2012). Just emissions. *Philosophy and Public Affairs, 40*(4), 255–300.

Habermas, J. (1981). *Theorie des kommunikativen Handelns*. Band I-II. Frankfurt am Main: Suhrkamp Verlag.

Habermas, J. (1992). *Faktizität und Geltung. Beiträge zur Diskurstheorie des Rechts und des demokratischen Rechtsstaats*. Frankfurt am Main: Suhrkamp Verlag.

Hanssen-Bauer, I., Førland, E.J., Hisdal, H., Mayer, S., A.B. Sandø, Sorteberg, A. (eds.). (2019). Climate in Svalbard 2100. *A knowledge base for climate adaptation*. The Norwegian Centre for Climate Services (NCCS). Report 1/2019.

Hardin, G. (1968). The tragedy of the commons. *Science, 162*(3859), 1243–1248.

Healy, N., & Barry, J. (2017). Politicizing energy justice and energy system transitions: Fossil fuel divestment and a "just transition". *Energy Policy, 108*, 451–459.

Heier, T. (2019). *Svalbard i en ny kald krig*. High North News. Uploaded 12.4.2019.

Helland-Hansen, W. (2019). *«Jeg har nådd grensen for hva som er etisk forsvarlig forskning*. Nok er nok.» Harvest Magazine Online 5.2.2019.
IPCC. (2018). *IPCC special report: Global warming of 1.5 °C*.
Kristoffersen, B. (2015). Opportunistic adaptation. New discourses on oil, equity, and environmental security. In K. O'Brian & E. Selboe (Eds.), *The adaptive challenge of climate change*. Cambridge: Cambridge University Press.
Lind, S., Ingvaldsen, R., & Furevik, T. (2018). Arctic warming hotspot in the northern Barents Sea linked to declining sea-ice import. *Nature Climate Change, 8*, 634–639.
McGlade, C., & Ekins, P. (2015). The geographical distribution of fossil fuels unused when limiting global warming to 2 °C. *Nature, 517*, 187–190.
Moore, M. (2015). *A political theory of territory*. Oxford: Oxford University Press.
Norwegian Seafood Federation. (2019). https://sjomatnorge.no/norwegian-seafood-federation/. Accessed 24 Mar 2019.
Ostrom, E. (1990). *Governing the commons. The evolution of institutions for collective action*. Cambridge: Cambridge University Press.
Polanyi, K. (2001 [1944]). *The great transformation. The political and economic origins of our time*. Boston: Beacon Press.
Rawls, J. (1999 [1971]). *A theory of justice*. Oxford: Oxford University Press.
Scavenius, T. (2016). Klimaet kræver politik. Ræson. 26. 26.05.2016.
Sinnott-Armstrong, W. (2005). It's not my fault: Global warming and individual moral obligations. In W. Sinnott-Armstrong & R. B. Howarth (Eds.), *Perspectives on climate change: Science, economics, politics, ethics* (Advances in the economics of environmental research) (Vol. 5, pp. 293–315). Bingley: Emerald Group Publishing Limited.
Steinberg, P. E., Tasch, J., & Gerhardt, H. (2015). *Contesting the Arctic: Politics and imaginaries in the circumpolar North*. London/New York: I.B. Tauris.
Vogel, S. (1996). *Against nature. The concept of nature in critical theory*. New York: SUNY Press.
Vogel, S. (2012). Alienation and the commons. In A. Thompson & J. Bendik-Keymer (Eds.), *Ethical adaptation to climate change* (*Human virtues of the future*). Cambridge, MA: The MIT Press.
Vogel, S. (2015). *Thinking like a mall: Environmental philosophy after the end of nature*. Cambridge, MA: The MIT Press.

Chapter 3
Arctic Marine Ecosystems, Climate Change Impacts, and Governance Responses: An Integrated Perspective from the Barents Sea

Stefan Koenigstein

Abstract Arctic and sub-Arctic marine ecosystems and their living resources are especially sensitive to climate drivers. Under progressing climate change, ocean warming, sea ice melting, changing oceanic currents, and ocean acidification will lead to shifts in seasonal timing, spatial distribution and productivity of fish species, and affect plankton composition, marine mammals and seabirds. Shifts of boreal and sub-Arctic species into the Arctic and ensuing changes in species composition and biodiversity are already impacting a range issues from ecosystem services to human societies, e.g. fisheries, coastal tourism, cultural services, and biological carbon uptake and cycling. Small-scale fishers may be unable to adapt to the occurring shifts. Decreases in seabirds, marine mammals, and iconic Arctic species could have negative consequences for marine ecotourism and cultural values in the high north. Increasing anthropogenic impacts, such as fisheries and pollution, will interact with climate impacts and exacerbate the pressure on Arctic marine ecosystems.

The Arcto-boreal Barents Sea can serve as a model system for understanding future shifts in marine ecosystems, impacts on human users, and marine governance responses in a changing Arctic. Adaptive, ecosystem-based, internationally collaborative and participatory governance mechanisms will help to address the upcoming challenges in climate change adaptation for Arctic marine social-ecological systems.

Keywords Climate change · Marine ecosystems · Ecosystem-based management · Marine living resources · Barents Sea · Arctic Ocean

---

S. Koenigstein (✉)
Institute of Marine Sciences, University of California Santa Cruz, Santa Cruz, CA, USA
e-mail: stkoenig@ucsc.edu

E. Pongrácz et al. (eds.), *Arctic Marine Sustainability*,
Springer Polar Sciences, https://doi.org/10.1007/978-3-030-28404-6_3

3.1 Introduction: Climate Change Impacts on Marine Ecosystems: The Arctic as an Early Warning System

The oceans provide a wealth of biological resources and services to human societies, such as food provision from fisheries and aquaculture, carbon uptake and climate regulation, bioremediation, nutrient cycling, or recreation and cultural services (Beaumont et al. 2007; Allison and Bassett 2015). Marine ecosystems and the productivity of these resources are known to be susceptible to climate-related drivers. As increasing levels of CO_2 and other greenhouse gases in the atmosphere are changing the earth's climate system, the oceans and marine ecosystems are also impacted by climate change, but our understanding of their response to global climate change is incomplete (Pörtner et al. 2014; Hoegh-Guldberg et al. 2014). Under progressing global climate change, changes in ecological dynamics of marine systems have been observed in recent decades, and are expected to further progress in the course of the twenty-first century. Climate-related drivers such as ocean warming, ocean acidification (decreasing pH values), and deoxygenation (insufficient oxygen levels) are anticipated to affect marine organisms, drive changes in marine ecosystem structure and dynamics, and have profound impacts on the productivity of marine living resources and the provision of marine ecosystem services to human societies (Gattuso et al. 2015; Brander 2012; Allison and Bassett 2015). The most relevant climate drivers as well as the responses and resilience will vary among ocean regions, and the Arctic is a hotspot of marine climate change (Hoegh-Guldberg et al. 2014).

Ocean warming, due to increasing atmospheric temperature and heat input, is already observed to lead to poleward shifts in the spatial distribution of marine organisms, causing local changes in the composition of marine ecological communities, with regional extinctions of organisms by exceeding the thermal tolerance limits (Poloczanska et al. 2013). As species are generally moving their distribution range towards the poles, temperate-water species are entering the Arctic areas, and some polar species might disappear altogether. As a result, under ocean warming and diminishing sea ice, Arctic food webs are progressively changing towards boreal communities, and pronounced changes in sub-Arctic fish stocks and impacts on fisheries are expected (Kortsch et al. 2015; Hollowed and Sundby 2014).

Increased atmospheric CO_2 levels also cause a direct chemical interaction with the surface ocean, which is termed ocean acidification (The Royal Society 2005; CBD Secretariat 2014). Ocean acidification is anticipated to affect the productivity of lower trophic levels, particularly calcifying organisms, and some fish stocks, and thus potentially alter food web structure and energy transfer in Arctic regions (AMAP 2013).

Thus, a wide range of human uses and activities linked to the oceans will be affected by climate change, and the associated ecological changes are thus a big challenge for governance of marine systems (Charles 2012; Perry et al. 2010). However, economic and nutritional dependence on marine resources, and thus vulnerability towards climate change effects, differs strongly among countries (Allison et al. 2009). Societies have a range of options to adapt to changes in marine systems, and these depend on economic, social and cultural conditions (Haynie and Pfeiffer 2012; Perry et al. 2011). At the same time, anthropogenic impacts, such as overexploitation by fisheries, can exacerbate climate change impacts on marine ecosystems (Brander 2012). As a result of the growing recognition of the multi-faceted interactions of marine ecosystems with societal uses, increased efforts are undertaken worldwide to establish ecosystem-based management approaches of the ocean and its resources (Browman and Stergiou 2005; Katsanevakis et al. 2011; Long et al. 2015).

Continuing ocean warming, ice melting, and changes in biological productivity are leading to movement of boreal species into Arctic oceans. Ensuing changes in biodiversity, habitat conditions and competition for polar fish species pose special challenges for future Arctic fisheries management, biodiversity conservation, and interdisciplinary science (Bluhm et al. 2011; Christiansen et al. 2014).

The Barents Sea is a well-researched Arctic shelf sea, with high productivity on higher trophic levels and socio-economically important fish stocks (Wassmann et al. 2006; Loeng and Drinkwater 2007; Olsen et al. 2010). It is also a 'warming hot-spot' of the Arctic ocean. As a large amount of ecological survey data is available for the Barents Sea, it represents an ideal study region for climate impacts on marine ecosystems and living resources (Michalsen et al. 2013). In this chapter, a regional focus on the Barents Sea will provide a systems perspective on the interacting impacts of different climate change and anthropogenic drivers on Arctic marine ecosystems. The aim is to provide an integrated regional perspective, and explain the potential of ecosystem-based management processes under progressing climate change. Marine ecosystems, resources and expected climate-related shifts are similar in the other sub-Arctic marginal seas and areas along the sea ice border of the Arctic ocean.

3.2 Marine Ecosystems in the Barents Sea: Observed and Expected Shifts Under Climate Change

3.2.1 Overview of Marine Habitats, Ecosystems and Living Resources

The Barents Sea is the largest of the Arctic marginal seas. The Northern half of this shallow shelf sea is partially ice-covered during winter, while the Southern half remains ice-free throughout the year and is influenced by the warm-water inflow of the North Atlantic current (Loeng and Drinkwater 2007; Wassmann et al. 2006). While primary production is low, the Barents Sea has a high and socio-economically important fish production. The ecosystems, resources and ongoing shifts in the

Barents Sea are comparable to the seasonally ice-covered areas of the other marginal Arctic seas and adjacent subpolar areas, such as the Bering Sea.

The water column in the Barents Sea is a habitat for several commercially important demersal (near-bottom) and pelagic (inhabiting the water column) marine fish stocks. The main capture species for fisheries in the Arcto-boreal area are the demersal fish species Atlantic cod (*Gadus morhua*), haddock (*Melanogrammus aeglefinus*) and saithe (*Pollachius virens*), which have all been certified as sustainable fisheries by the MSC (Marine Stewardship Council), the pelagic fish Atlantic capelin (*Mallotus villosus*) and herring (*Clupea harengus*), and deep-sea fish species such as beaked redfish (*Sebastes mentella*), golden redfish (*Sebastes norvegicus*) and Greenland halibut (*Reinhardtius hippoglossoides*; Nakken 1998; Olsen et al. 2010; Haug et al. 2017).

Benthic (sea-bottom) habitats in the Barents Sea are typically inhabited by a large variety (> 3000 species) of invertebrates, such as bivalves (mussels and clams), polychaetes (bristle worms), crustacean crabs, echinoderms (sea stars and sea urchins), holothurians (sea cucumbers) and sponges (Wassmann et al. 2006). Coastal benthic habitats, especially the shallow Southeastern Barents Sea, are influenced by inflow of freshwater and nutrients from rivers. Northern prawns (*Pandalus borealis*) and crabs such as the Red king crab (*Paralithodes camtchaticus*) and the Snow crab (*Chionoecetes opilio*) represent near-bottom and benthic fisheries resources.

In the sea ice covered areas of the Barents Sea, under-ice biological communities form another significant Arctic ecosystem. Here, ice algae attached to the underside of sea ice provide the basis of the food chain, contributing a relevant portion to primary production in the Arctic. Zooplankton groups feed on these algae and contribute importantly to energy transfer to benthic systems (bentho-pelagic coupling). Polar cod (*Boreogadus saida*) and other smaller fish species feed and reproduce under the Arctic sea ice, and are an important prey for marine mammals such as ringed seals (*Pusa hispida*) and beluga whales (*Delphinapterus leucas*).

Other marine mammals that have adapted to life on and under the Arctic sea ice include bearded seals (*Erignathus barbatus*), walrus (*Odobenus rosmarus*), bowhead whales (*Balaena mysticetus*), and narwhales (*Monodon monoceros*). The Arctic sea ice margins are a habitat for harp seals (*Pagophilus groenlandicus*), harbour seals (*Phoca vitulina*), grey seals (*Halichoerus grypus*) and polar bears (*Ursus maritimus*). Also, a high variety of seabirds feeds on small fish and zooplankton in the Barents Sea, such as Black-legged Kittiwakes (*Rissa tridactyla*), Glaucous Gulls (*Larus hyperboreus*), Northern Fulmar (*Fulmarus glacialis*), Atlantic Puffins (*Fratercula arctica*), Little Auks (*Alle alle*) and different species of Guillemots (Wassmann et al. 2006; Loeng and Drinkwater 2007).

3.2.2 Climate Fluctuations, Climate Drivers and Trends in the Barents Sea

Climate variability on multi-annual and decadal scales is known to influence biological productivity and ecosystem dynamics in the Barents Sea (Yaragina and Dolgov 2009; Drinkwater et al. 2010). The climatic fluctuations are linked to the North Atlantic Oscillation (NAO; Ottersen and Stenseth 2001; Orlova et al. 2005; Dalpadado et al. 2012) and the Atlantic Meridional Oscillation (AMO; Sutton and Hodson 2005; Drinkwater et al. 2014). These climate oscillations cause changing oceanic regimes in the Barents Sea with regard to Atlantic and Arctic water masses, resulting in fluctuations in water temperature, plankton biomasses, and thus food availability for fish stocks (Sakshaug et al. 1994; Drinkwater 2006; Loeng and Drinkwater 2007). As a result, the most important pelagic and demersal fish populations in the Barents Sea, e.g. Atlantic cod, Norwegian spring-spawning herring and Atlantic capelin, show climate-dependent and interdependent fluctuations in productivity and abundance (Hamre 1994; Cury et al. 2008; Bogstad et al. 2015). Also, sea-bottom communities show climate-related fluctuations in species composition and distribution (Wassmann et al. 2006).

As a result of this climate-dependence, the substantial warming which observed in recent decades in the Barents Sea, which is projected to continue in the course of this century, is expected to change productivity of living marine resources, and composition and functioning of ecosystems. The earth system model projections used by the IPCC (Intergovernmental Panel on Climate Change) show increases of average surface temperatures for the Barents Sea of around 5 degrees until the year 2100 under continuously high greenhouse gas emissions (Representative Concentration Pathway 8.5; Bopp et al. 2013; Collins et al. 2013). Increasing ocean temperatures lead to changes in evaporation and sea level rise, melting of sea ice, deoxygenation, and changes in salinity, oceanic currents and stratification (stability of vertical water layers).

Atlantic water masses are spreading further northeastwards, leading to decreased sea ice coverage, and the Barents Sea is projected to become completely ice-free during summer from mid-century on (Onarheim et al. 2014; Lind et al. 2018). Furthermore, salinity could increase in the areas influenced by Atlantic waters due to increasing salinity in the Atlantic water inflow, but could decrease in coastal areas in the East due to increasing river water influx (Filin et al. 2016).

A further climate change driver expected to impact Arctic marine ecosystems is ocean acidification. Ocean acidification is a direct chemical interaction of the atmosphere with the marine environment, caused by the chemical dissolution of the increasing amount of atmospheric CO_2 into surface ocean water, which lowers its pH value (The Royal Society 2005). Because cold waters take up more CO_2 and the increasing ice-free water surface area, high latitude oceans are expected to be among the earliest impacted areas. With CO_2 levels expected to rise from 400 ppm (parts per million) to 700–1000 ppm at the end of the century, acidification is a serious concern for Arctic marine ecosystems (Denman et al. 2011; AMAP 2013; Skogen et al. 2014).

In the following subsections, an overview of the state of scientific knowledge about the impacts of climate change drivers on the productivity and survival of marine organism groups in the Barents Sea will be given.

3.2.3 Changes in Biological Productivity: Primary and Secondary Production

In marine ecosystems, the base of the food-web is formed by microscopic, planktonic algae which produce biomass by fixing CO_2 (primary production). Annual primary production in the Arctic is generally expected to increase, due to an increasing seasonally ice-free water area (Arrigo et al. 2008; Manizza et al. 2013). In the Arctic areas of the Barents Sea, primary production is thus projected to increase slightly, with increasing contribution of ice algae production due to decreasing ice thickness (Ellingsen et al. 2007; Wassmann and Reigstad 2011). Yet, an increase in total primary production could be counter-acted by increasing stratification (layering) of waters, which limits nutrient availability for phytoplankton. Furthermore, ice algae production contributes only a small part of primary production in the Barents Sea, of which a high percentage is exported vertically (Sakshaug et al. 1994; Slagstad et al. 2011; Wassmann and Reigstad 2011).

Changes in phytoplankton productivity and composition will have repercussions at the level of zooplankton. Potential changes in zooplankton advection and heat transport (temperature oscillations) from Atlantic waters and from Arctic water zooplankton species (*Calanus glacialis* and ice-associated amphipods) entering from the North and East determine energy influx into the higher Barents Sea food-web (Hunt et al. 2013). A concomitant increase in zooplankton production will be hindered by the seasonal light cycle which limits the phytoplankton spring bloom, unless arctoboreal zooplankton such as *Calanus finnmarchicus* can adapt to the temperatures, or temperate zooplankton species such as *C. helgolandicus* immigrating from the South can adapt their life cycles to resemble those of Arctic species, which overwinter at depth with energy stored in lipid reserves (Sundby et al. 2016).

These changes in planktonic systems may affect the biological carbon pump, which converts some of the CO_2 taken up by the ocean into organic matter. By this process, the dissolved inorganic carbon in the water is taken up through photosynthesis by marine microalgae (phytoplankton), transformed into phytoplankton biomass and then further trans- ported into the food web. A part of the plankton biomass sinks down into the deep layers of the ocean, where it is recycled by bacteria or, to a small fraction, buried forever in the marine sediments. Climate change effects on lower trophic level productivity can lead to repercussions in the higher food web of the Barents Sea. Changes in primary production by phytoplankton and ice algae, and changes in species composition of zooplankton groups, e.g. from Arctic copepods and amphipods to their boreal congeners, may further affect energy content of prey for fish.

3.2.4 Ocean Warming and Changing Productivity of Marine Fish Stocks

Recent warming of the Barents Sea has been linked to observed increases in the biomass of fish and Northern krill (Eriksen et al. 2017). This has for instance, together with good fisheries management, led to high biomass of the Atlantic cod stock in the Barents Sea (Kjesbu et al. 2014). Continuing warming projected for the coming decades is expected to lead to further shifts in the spatial distribution, productivity and biomass of fish stocks (Stenevik and Sundby 2007; Hollowed and Sundby 2014). Fish communities in the Barents Sea are changing towards typical boreal species, while Arctic species are retreating, leading to changes in food-web structure with a higher portion of larger generalist feeders (Fossheim et al. 2015; Frainer et al. 2017; Kortsch et al. 2015). Similarly, changes in zooplankton species composition from those associated with Arctic waters to predominantly Atlantic species are expected under continuing climate change (Dalpadado et al. 2012). These changes affect the provision of food energy to the fish stocks and may alter food-web functioning and dynamics in the Barents Sea (Johannesen et al. 2012).

The response of fish populations to ocean warming is determined by a combination of the species' evolved physiological preferences, other interacting environmental drivers, and limited or facilitated by interactions in the food-web and habitat availability (Pörtner and Peck 2010; Koenigstein et al. 2016a). An especially sensitive process for fish populations is the recruitment of early life stages. Correlations of recruitment with environmental parameters, mainly water temperature, have been identified for several fish stocks in the Barents Sea (Stige et al. 2006; Ottersen and Stenseth 2001; Cury et al. 2008). Future recruitment success can thus be expected to be driven by changes in environmental parameters, but also life-stage specific habitat availability and connectivity, e.g. suitable bottom topography for spawning. The spawning habitat of Barents Sea cod, off the Northern Norwegian coast and the Lofoten islands, is known to shift northwards during warm climate periods (Sundby and Nakken 2008; Langangen et al. 2019).

Highly specialized, ice-associated Arctic fish species, such as Polar cod, will be severely affected by retreating ice, when the distance to spawning habitats on the shelf increases, and competition for food with boreal species intensifies (Hop and Gjøsæter 2013). Pelagic fish species in the Barents Sea such as capelin may be able to shift their habitat into the Arctic ocean, while demersal species are expected to already have reached their Northward limit and probably not be able to enter the deeper Arctic ocean basins (Darnis et al. 2012; Haug et al. 2017). With a similar planktonic diet and a preference for slightly higher temperatures, capelin is a direct competitor to Polar cod (McNicholl et al. 2015). Although Polar cod could show a short-term increase in recruitment under warming temperatures and earlier seasonal sea ice break-up (Bouchard et al. 2017), it is expected to be replaced by sub-Arctic and boreal species in the mid- and long-term (Hop and Gjøsæter 2013; Renaud et al. 2011). Atlantic mackerel (*Scomber scombrus)* has been recently recorded in the

Southern Barents Sea area (Berge et al. 2015). In benthic systems, the recent expansion of the snow crab has been linked to increasing water temperatures, causing issues for management (Kaiser et al. 2018).

3.2.5 *Marine Mammals and Seabirds: Impacts of Sea Ice Retreat and Food Web Changes*

Marine mammals that depend on the sea ice habitat are well-known examples for species threatened by climate change in the Arctic. Seal species that breed and rest on sea ice, such as harp seals and ringed seals, are forced to migrate further north and increase their foraging effort (Kovacs et al. 2010; Hamilton et al. 2015). Decreases in feeding habitat and fat-rich food species like krill lead to reduction in body fat storage, and can cause breeding failures which impact the seal populations (Øigård et al. 2013). Polar bears hunting for seals along the ice border are forced to conduct longer Northward migrations in search for their prey under ongoing summer sea ice reductions (Derocher et al. 2004; Durner et al. 2009). More mobile marine mammals, such as Minke whales and other cetaceans, may be able to expand their habitat further into the Arctic ocean without significant energetic costs (Haug et al. 2017).

As warm-blooded animals, marine mammals and seabirds are not directly physiologically affected by ocean warming. Yet, through changes in their food availability, many seabird and marine mammal populations are observed to be impacted by changes in marine fish or planktonic food abundance, and can indeed be sensitive ecological indicators for marine food-web shifts under climate change (Simmonds and Isaac 2007; Durant et al. 2009). Warming-related changes in fish abundance, together with the reduction in sea ice, are affecting bird and mammal species in the Barents Sea region (Descamps et al. 2017).

In the Barents sea region, a negative correlation of adult survival rates with higher sea surface temperatures observed for Atlantic puffins, Common guillemots (*Uria aalge*) and razorbills (*Alca torda*), and recently observed negative effects of shifts in food composition for black-legged kittiwake and Brünnich's guillemot (*Uria lomvia*) suggest that these bird species could be negatively impacted by ecological shifts under continuing climate change in the Barents Sea (Sandvik et al. 2005; Barrett 2007; Fluhr et al. 2017).

Marine mammals in the Barents Sea, such as harp seals and minke whales, suffer negative effects on body condition in years when the abundance of both capelin and herring is low (Bogstad et al. 2015). This situation may become more common under continuing warming of the Barents Sea, as a continuation of the high production and biomass of Atlantic cod and haddock is expected due to faster growth and maturation. It is acknowledged that other processes, e.g. increased cannibalism and changes in prey availability for these fish, may act against this effect (ICES AFWG 2016). Orcas and other tooth whales have been increasing in recent years near the Northern Norwegian coast, and the proportion of pelagic fish (mackerel and herring) in their diet has increased, while amphipods and krill have decreased (Nøttestad et al. 2013).

3.2.6 Ocean Acidification: Potential Impacts on Marine Food-Webs

Ocean acidification, the decrease in water pH via increasing solution of atmospheric CO_2, is anticipated to impact marine ecosystems in the course of this century (The Royal Society 2005; CBD Secretariat 2014). Arctic waters are projected to undergo among the highest pH changes of world oceans until the end of the century, with waters to become corrosive to some shell-producing organisms by the middle of the twenty-first century (AMAP 2013; Skogen et al. 2014). This could affect survival, growth and metabolic performance of marine organisms under future ocean pH values (Cooley and Doney 2009; Denman et al. 2011). Various potential impacts of ocean acidification on marine organisms and ecosystems have been found in laboratory studies, including problems for shell-building organisms, potentially leading to negative impacts on mollusks (shellfish, marine snails), starfish and sea urchins, corals, and calcifying microalgae (Wittmann and Pörtner 2013; Kroeker et al. 2013).

The impact of ocean acidification on total primary production by phytoplankton in the ocean is still unclear. While photosynthesis should be generally positively impacted by a higher amount of CO_2 available, increased stability of depth layers might decrease nutrient input into the light zone at the surface and limit primary production. It is unclear to what extent the different groups of phytoplankton will be negatively affected by increasingly stressful conditions caused by combined warming and acidification. As two important phytoplankton groups (coccolithophores and foraminiferans) and some of the zooplankton (e.g. pteropods, or sea butterflies) have calcareous shells or structures, it is suspected that they will be negatively impacted by strong acidification (Le Quéré and Metzl 2004; Kroeker et al. 2013). While experimental work has shown Arctic phytoplankton communities to compensate for ocean acidification effects, the interactions with increasing temperatures, UV radiation and other drivers remain to be investigated further (Gao and Häder 2017; Hoppe et al. 2018).

Another issue of considerable scientific uncertainty is whether ocean acidification will impact the future recruitment of fish populations, as early life stages are bottlenecks of population sensitivity to multiple environmental drivers (Pörtner and Peck 2010; Koenigstein et al. 2016a). Experimental work has shown negative impacts of end-of-century CO_2 levels on eggs and larvae of Barents Sea cod (Stiasny et al. 2016; Dahlke et al. 2017). The combined effects of continuing warming and acidification could thus lead to severe reductions in average recruitment success of Barents Sea cod, but the potential for evolutionary adaptation to these physiological stressors is presently unclear (Koenigstein et al. 2018). Conversely, mesocosm community studies have shown for herring larvae that some fish larvae could profit indirectly from increased primary production under increased CO_2 availability (Sswat et al. 2018).

Cold-water coral reefs in deep waters of sub-Arctic areas are among the first habitats threatened by ocean acidification, as it can cause the dissolution of the dead calcium carbonate skeletons which form the base of reefs, and lead to reduced growth of corals (Roberts 2006; Büscher et al. 2017). Combined ocean acidification, deep-water warming and possibly deoxygenation thus threaten the existence of these sensitive deep-sea ecosystems at the entrance to the Barents Sea. The loss of deep-water coral reefs is expected to have consequences for marine food webs, as they are home to rich associations of benthic organisms and also provide habitat, spawning and feeding grounds for some demersal fish species (Turley et al. 2007).

Thus, many scientific uncertainties remain on the ecosystem-level effects of ocean acidification, especially when assessed in interaction with other climate change drivers (Riebesell and Gattuso 2015). The comparatively simple Arctic marine food webs are expected to be more vulnerable to impacts on certain keystone or bottleneck species than ecosystems with higher species diversity (Wassmann et al. 2006; Duarte et al. 2012), and functional redundancy in Arctic fish communities in the Barents Sea has recently decreased (Aune et al. 2018). Caution is thus warranted with regard to the potential impacts of ocean acidification on Arctic marine ecosystems, and further research is necessary to understand the responses of Arctic marine organisms and ecosystems to combined ocean warming and acidification.

3.3 Climate Change Impacts on Human Users of Biological Resources, and Resulting Challenges for Sustainable Governance

3.3.1 Impacts on Marine Fisheries and Adaptation Issues

The Barents Sea is one of the Arctic marginal seas where fisheries presently play a larger socio-economic role than in the inner Arctic ocean. Yet, with distribution shifts under continuing warming and sea ice reduction, fisheries at the margins of the Arctic ocean are expected to increase (Christiansen et al. 2014; Haug et al. 2017). In recent decades, distribution shifts to the Northeast, i.e. further into the Arctic, have been observed in the Barents Sea, and it is generally expected that this trend will continue under progressing warming of waters and melting of the Arctic sea ice. This will increase the potential for marine fisheries in the Arctic and its marginal seas (Hollowed and Sundby 2014; Haug et al. 2017). The Barents Sea plays a major role for the Norwegian and the Russian fisheries, and their capture fishery can be divided between industrial offshore vessels and small-scale coastal fisheries.

The fisheries sector will thus be forced to adapt to changes in location and distribution range, size and seasonal migration of the sub-Arctic fish stocks. Fishers may have higher fuel costs and may have to adjust fishing gear and method to size of new target fish and spatial occurrence. Societally relevant differences in adaptive capacity exist among different fishing fleets. The industrial offshore fleet based in West and South Norway harvests the largest share of the fish catch, targeting pelagic and demersal stocks and following fish migrations over long distances (FAO 2013). In contrast, the more coastal and small-scale fisheries in Northern Norway represent over 4000 employed fishermen, almost half of fishermen in Norway (Fiskeridirektoratet 2018). This fleet could be challenged by stock distribution range shifts in the Barents Sea to a higher extent, as their smaller vessels may not be able to follow stocks further away from the coasts. Additionally, local stocks in the fjords, such as the coastal cod, represent highly fragmented sub-populations with a low degree of interchange, slowing down recovery of depleted local stocks (Myksvoll et al. 2013). Yet, these coastal fisheries are of high regional socio-economic importance in smaller communities in Northern Norway. External pull for educated workers from the oil industry, livelihoods of fishers, employment alternatives and social structures are relevant factors for the stability of Northern communities (West and Hovelsrud 2010; Dannevig and Hovelsrud 2015; Koenigstein and Goessling-Reisemann 2014). Thus, changes in fish stocks could have locally and regionally dramatic impacts on communities, even when economic impact on a national level is limited.

From a management perspective, range shifts of fished stocks represent an adaptation challenge and may necessitate technical as well as regulative adjustments. The national and international distribution of catch quota shares, fishing gear regulations (e.g. type and mesh size of nets), as well as regulations of vessel sizes or closed areas for certain types of fishing, e.g. bottom trawling, may have to be adjusted. Where commercially exploited marine fish stocks shift their distribution across national borders, this results in special governance challenges, e.g. re-negotiation of quota agreements (Pinsky et al. 2018). The co-management of the Barents Sea cod stock by Norway and Russia is a good example for international cooperation that has succeeded in accommodating the life history and environmental variability of the target catch species across political borders (Eide et al. 2013). In contrast, in the recent cases of Atlantic mackerel and Norwegian herring stocks in the Northeast Atlantic, shifts of the stocks have led to political disputes about the allocation of catch quota between the EU and Norway on one side and Iceland and the Faroe Islands on the other.

Rising ocean temperatures could also promote the development of marine aquaculture installations along the coastal margins of the Arctic seas. Stakeholders from the fisheries sector in the Barents Sea region generally expect fish market prices to increase in the future; and with climate change possibly causing problems for food supply by agriculture in some world regions, the marine sector would have to deliver a growing share of the world food production, maximize long-term fisheries yield and increase the aquaculture share (Koenigstein and Goessling-Reisemann 2014).

3.3.2 Impacts on Coastal Tourism

The oceans play a substantial role in tourism and recreation. Recreational saltwater fishing has approximately 8–10 million practitioners and is a considerable industry with socio-economic relevance (FAO 2012). Recreational fisheries and associated tourism can provide alternative livelihoods for small-scale fishers, but tourism activities are also competing for space with professional fishery in some coastal areas. Tourism in Northern Norway is strongly linked to the Arctic and sub-Arctic fjords and coastal areas of the Barents Sea. The tourism industry, including transport, accommodation and gastronomy services, travel and tour companies, is an important employer especially in Northern Norway, where it provides 18,000 jobs and 6% of total added value. Recreational sea fishing alone creates a value of about € 11 million in the North (Klima- og Miljødepartementet 2011).

Among the most popular game fish are halibut, spawning Atlantic cod *(skrei)*, catfish, plaice and saithe. Sports fishing tourism is especially relevant on the Lofoten and Vesterålen islands in Northern Norway, where occurrence of some game fish species is reportedly linked to the annual cod spawning migrations, recently profiting from high spawner numbers in this stock. Apart from sea fishing, coastal tourism is centered around nature-related activities such as whale-watching, seal and seabird tours, kayaking, hiking and camping, as well as to small-scale fisheries and maritime activities as a cultural heritage of the Norwegian people (NMTI 2012). Tourism in Northern Norway is thus strongly linked to small-scale fishing, both because boats, harbors and the connected activities (e.g. production of stockfish) are a strong pull-factor for tourists, and because many small-scale fishing boat owners use their vessel seasonally for professional fishing as well as for sport fishing or other recreational activities linked to tourism.

Climate-related changes in fish, mammals and seabirds are thus likely to have consequences for tourism business in the high north. The occurrence of whales is linked to their prey, e.g. small fish and krill. The commercially successful whale-watching enterprises in the Lofoten, Vesterålen and Tromsø regions depend mostly on sperm and humpback whales. While the occurrence of killer whales (*Orcinus orca*) in some recent years in the Troms region has spawned commercial whale-watching activities in the region, it is currently too unreliable and variable to serve as a mainstay for tourism businesses. With declines in seabirds observed in many locations and potentially becoming more frequent under climate change, alternatives for providing tours are further decreasing for tour operators. Thus, tourism operators along the coasts of the Barents Sea are to a high degree dependent on certain locally abundant species (sperm, humpback and killer whales, cod, halibut, seals). Important parameters for tour providers are sighting probability and distance to be travelled by tour boats from the coasts, so decreasing abundance or shifts in migrations or distribution of target species too far from the coast make it impossible to provide trips. Changes in the abundance of herring, krill and other prey species, and possibly increasing marine noise by shipping and exploration activities, could thus have long-term impacts on ecotourism businesses in the Arctic.

3.3.3 Further Impacts: Cultural Services and Values, Biodiversity and Carbon Sequestration

While the tourism economy can serve as an indirect indicator for recreation value, value for local recreation and other cultural values linked to marine ecosystems are notoriously difficult to quantify. Whales, seabirds, seals and other top predators are the most visible parts of the marine ecosystem for the biggest part of the population and visitors, and thus also possess high aesthetic and educational significance. Apart from recreation, the coastal marine ecosystems provide aesthetic services, religious and spiritual services, cultural identity, as well as options for education and research. Most of these services are difficult to quantify on a monetary basis, but nevertheless have societal value.

It is important to consider that in many Arctic communities, impacts on fisheries not only have local economic repercussions, but also distinctly affect cultural and educational ecosystem services in the area. Atlantic cod has an exceptional cultural significance on the Lofoten islands and the Northern provinces Nordland, Troms and Finnmark. For many coastal communities, economic activities are closely connected to marine environments, and cod fishing in the fjords and coastal areas has a cultural and historical significance far beyond that. Furthermore, climate change adaptation will have to happen under the challenges by demographic aging and the remoteness of these communities (West and Hovelsrud 2010; Dannevig and Hovelsrud 2015).

Furthermore, the general functioning and biodiversity of marine ecosystems, and rare ecosystems such as cold-water coral reefs may be threatened by climate change (see Sects. 3.2.3, 3.2.6). Highly specialized, sensitive ecosystems like this can have a high significance for local biodiversity and local nutrient and carbon cycles. Furthermore, they represent a cultural (existence) value, providing unique ecological examples relevant for education and research. As a result of their uniqueness, valuation studies point to a very high willingness-to-pay of the Norwegian public to conserve cold-water coral reefs (Aanesen et al. 2015).

Another marine ecosystem service of high societal relevance is the climate regulation by sequestration and export of carbon from the atmosphere (Le Quéré and Metzl 2004; Beaumont et al. 2007). The biological carbon pump takes up CO_2 via primary production and exports it to the depth mainly via sinking particles. First economic estimates for Norway have indicated that the costs associated with an acidification–mediated reduction in biological carbon uptake may be several orders of magnitude higher than effects on fisheries and aquaculture (Armstrong et al. 2012). Yet, high scientific uncertainty remains with regard to the biogeochemical parameters and biological processes in lower trophic levels that will determine future changes in biological carbon uptake. Reduced sea-ice cover in the Arctic ocean may lead to increased biological CO_2 uptake in the previously ice-covered areas. Yet, increasing stratification due to increasing freshwater influx from rivers could act against this (Wassmann et al. 2006; Wassmann and Reigstad 2011).

3.4 Marine Ecosystem Shifts under Climate Change as a Challenge for Marine Governance in the Arctic

3.4.1 Ecosystem Responses and Interactions with Anthropogenic Drivers

While the description in this chapter has focused on the Barents Sea, similar shifts with regard to plankton composition, productivity of fish stocks and indirect impacts on top predators are observed and projected for other Arctic seas (Darnis et al. 2012; Hunt et al. 2002; Hollowed and Sundby 2014). It has become clear and has to be emphasized that the impacts of climate change on Arctic marine organisms are not adequately described by linear cause-effect pathways. Rather, drivers such as ocean warming, ocean acidification and sea ice reduction will interact in combination with oceanographic changes, and other drivers such as oxygen loss and increased UV radiation, to shape the future of Arctic marine ecosystems.

Direct environmental effects on marine organisms are then modulated by species interactions in the food web, leading to indirect effects on other species and changing dynamics of ecological communities (Kordas et al. 2011). Thus, investigation of climate change effects on marine fish stocks and ecosystems has to integrate the biological disciplines and different ecological levels of organization (organisms, populations and community levels). In addition to this, it has to build upon an understanding of potentially affected biological processes, including interactions and feedbacks (Doney et al. 2012; Sydeman et al. 2015; Pörtner and Peck 2010), in order to improve projections of future shifts and to describe management-relevant ecological trade-offs (Metcalfe et al. 2012; Koenigstein et al. 2016a).

Within marine ecosystems, ecological feedbacks may buffer impacts on marine species to some extent, but can also cause sudden and unexpected regime shifts. Cascading effects triggered by changes in one element of the marine food web, leading to food web restructuring under climate change, are especially likely in the simple food webs of the Arctic, and in systems with some degree of top-down (predator) control, like the Barents Sea (Mangel and Levin 2005; deYoung et al. 2008). Ecological interactions can override the physiological temperature preference of a fish stock in determining its distribution. For instance, the experienced ambient temperatures of two fish species in the Barents Sea, mackerel and haddock, have actually decreased on average under recent warming conditions (Landa et al. 2014; Sundby et al. 2016). Thus, the future distribution, abundance and productivity of Arctic species should not be simply extrapolated from the current sub-Arctic species expected to move into the Arctic, but an ecological perspective is necessary.

Another important co-determinant of the dynamic behavior of marine ecosystems is the degree of exploitation by fisheries (Brander 2012; Rijnsdorp et al. 2009; Fogarty et al. 2016). Fishing pressure will affect the response of Arctic marine foodwebs to environmental drivers, e.g. through impacts on productivity, age structure and evolution of fish stocks (Jorgensen et al. 2007; Perry et al. 2010). Furthermore, organisms will be impacted by indirect impacts through changes in anthropogenic impacts under continuing climate change. As fisheries are expected to move further into the Arctic, following the north- and eastward distribution of target species in the Barents Sea, bottom trawl fisheries will enter previously ice-covered areas and potentially damage highly sensitive benthic habitats, impacting sessile organisms such as sponges, sea stars, sea fans, and soft corals, and bottom-dwelling organisms such as deep-sea octopi, sea spiders and sea cucumbers (Jørgensen et al. 2015). Shifts in benthic organisms may also be influenced by shifts in pelagic fish species through changes in predation pressure.

Another example is the increased risk of oil spills under increasing exploration efforts of the Barents Sea oil resources anticipated under retreating Arctic ice cover. Oil compounds can have toxic effects on marine organisms, especially on sensitive species or early life stages, and these effects can exacerbate the physiological stress caused by ocean warming and/or acidification, as has been found in some fish larvae and zooplankton species (e.g. Ingvarsdóttir et al. 2012). While the ecological impacts of maritime industries on marine organisms are discussed in more detail in the following chapters of this book, it is important to mention that anthropogenic impacts can interact with climate change drivers and lead to additive or synergistic impacts on marine organisms, and that there is a significant and governance-relevant scientific knowledge gap with regard to these interactive effects.

3.4.2 Integrating Potential Ecosystem Shifts and Adaptation of User Groups

The depicted ecological changes in the Barents Sea exemplify the upcoming trends under progressing climate change in the Arctic. They illustrate the necessity for an ecosystem-level assessment, and adaptive governance regimes that can respond to unexpected changes. The non–linear behavior of marine systems, producing phenomena such as tipping points, regime shifts and multiple equilibria, is among the foremost challenges for science–based governance and societal adaptation to climate change in the marine realm (Perry et al. 2011; Rice et al. 2014a). These can also arise from interactions with human exploitation, and slow societal adaptation to fast environmental changes has led to overuse and collapse of living marine resources in the past (Hannesson and Herrick 2006; Pershing et al. 2015).

Considerable differences exist in adaptation capacities to climate change impacts among stakeholders in the Northern Norwegian and Barents Sea region. Some user groups, such as small–scale fishers and local tourism entrepreneurs, possess significantly fewer adaptation options to potential ecological shifts under climate fluctuations and climate change. These stakeholders groups in the Barents Sea region are interested in a better incorporation of environmental fluctuations, ecological interactions among fish species and changes in primary and secondary productivity into scientific projections and management processes (Koenigstein et al. 2016b; Tiller et al. 2016).

To assess impacts of climate change and use by societies on marine ecosystems, the importance of ecosystem-based management strategies is increasingly recognized, and ecological simulation models are important tools for this (Crowder et al. 2008; Plaganyi 2007; Essington and Punt 2011). The impacts of climate change and ocean acidification affect a wide range of organisms and organismal processes, as well as their trophic interactions and interactions with anthropogenic drivers such as overfishing and pollution. This complex task requires the development of advanced models, integrating effects on different processes in various organisms and incorporating experimental results (Blackford 2010; Le Quesne and Pinnegar 2012; Koenigstein et al. 2016a).

The variety of user groups potentially affected by climate-related ecological changes in the Barents Sea illustrate that future changes in Arctic marine ecosystems should not be addressed exclusively from a sectoral, e.g. fisheries management perspective. Important ecological and societal trade–offs exist among the different resources and societal user groups. Food-web mediated impacts on whales and seabirds or lower trophic levels would lead to governance-relevant trade-offs among fish provisioning and other ecosystem services. For instance, reductions in fisheries exploitation may help to secure mammal and seabird populations and the cultural and recreational services provided by them. On the other hand, in a situation with high fish stock levels as with the current Barents Sea cod stock, increased fishing could alleviate pressure on mammals and seabirds by reducing predation on forage fish species (Bogstad et al. 2015).

The Barents Sea thus exemplifies the promise and utility of ecosystem-based management approaches, and some aspects of species interactions and environmental drivers have already been incorporated in fisheries management regimes (Gjøsaeter et al. 2012). An important novel tool for this process are regional end-to-end models of marine systems, which can help to understand ecosystem processes, improve ecosystem surveys, and assess integrated management strategies (Michalsen et al. 2013). The end-to-end model ATLANTIS is one of the models currently available for the Barents Sea (Hansen et al. 2016; Hansen et al. 2019). The ecological trade-offs in use of marine living resources, the observed and upcoming changes in biological distribution and productivity, and the differences in adaptive capacity of user groups demonstrate the need for advancing adaptive, science-based governance frameworks for the Arctic marine areas that integrate environmental changes and ecosystem interactions for an adaptation to climate change impacts in the Arctic.

3.4.3 *The Barents Sea as an Example for Ecosystem-Based Governance of Arctic Living Marine Resources*

The identified ecological changes and affected user groups are subject to a range of governance mechanisms in the Barents Sea area. These can serve as an example for identifying management and adaptation options for changes in marine ecosystem services and affected societal groups relevant for other Arctic areas under future climate change. Fisheries in the Barents Sea are managed in international collaboration by the Joint Norwegian-Russian Fishery Commission, and illegal fishing has practically been eliminated (FAO 2013). Catch quota are based on scientific advice from the International Council for the Exploration of the Sea (ICES) and the Institute of Marine Research (Mikalsen and Jentoft 2001). National and regional fisheries associations participate in the Management Council, and fishers provide catch logs to inform stock management and are often surveyed by catch control systems (Johnsen 2013; Jentoft and Mikalsen 2014).

Over recent years, Norway has progressively incorporated many aspects of an ecosystem-based approach to fisheries management (Gullestad et al. 2017; Skern-Mauritzen et al. 2018). In 2016, the harvest control rule for the Barents Sea cod stock has been adjusted to allow increased catches at high stock levels, with the aim to reduce impacts on forage fish and possible trade-offs with mammals and seabirds populations (Skern-Mauritzen et al. 2018).

Since 2007, Norwegian Integrated Management plans for the Barents Sea and the area around Lofoten islands prescribe a regular assessment of the state of marine ecosystems, human uses and their ecological impacts, and put sensible areas such as cold-water coral reefs under a special protection status (Hoel et al. 2009; Harsem and Hoel 2012). These plans recognize the significant uncertainties and potentially large impacts of climate change on the Barents Sea and aim at an improved integration of ecosystem aspects and user groups (Hoel and Olsen 2012). Within the ICES Integrated Ecosystem Assessment framework, a working group has been established for the Barents Sea (ICES 2016). Overall, management of marine living resources and areas can be considered to be well-prepared to enable adaptation to future changes in fish stocks and marine ecosystems under climate change.

Nevertheless, some scientific uncertainties regarding responses of marine ecosystems to climate drivers remain to be better understood and necessitate precautionary governance approaches. For instance, as is worldwide practice in fisheries management, the existing assessment models use stock-recruitment functions that assume a stable dependence of recruitment on the spawning stock, and thus fail to fully resolve the influences of environmental and biotic drivers (Rice et al. 2014b; Pepin 2016). Advancement towards use of more detailed models and tools resolving multiple environmental drivers and spatial life histories will be necessary for fast-changing Arctic and Sub-Arctic regions, while at the same time acknowledging and incorporating uncertainties in decision-making processes.

Furthermore, the Barents Sea case study also illustrates that participation of affected stakeholder groups should be ensured on a regional level, and social consequences of adaptation measures, such as quota distributions with regard to small-scale fisheries in the high North, need to be considered in governance of future climate change impacts in the Arctic. Other user groups affected by climate change in the Barents Sea, apart from the fisheries sector, are not as well-organized and participate as affected stakeholders mainly via the Norwegian integrated coastal zone management procedures. This includes state representatives and user groups such as fish farmers, tourism entrepreneurs, environmental and outdoor activity organizations, transportation and military sectors, land owners and indigenous Sámi representatives (Buanes et al. 2004). The centralistic national management regimes in Norway tend to give well-organized and economically powerful groups, e.g. the industrial fisheries and the aquaculture sector, more political influence than smaller user groups and indigenous peoples, creating barriers e.g. for local co-management of fish stocks (Hoel and Olsen 2012; Jentoft and Mikalsen 2014). Recently, fisheries associations and environmental NGOs have successfully joined forces to fight off oil and gas exploration plans around the Lofoten Islands because of concerns for the recruitment of Barents Sea cod (Jentoft and Mikalsen 2014). A Strategic Council for Tourism with planned yearly meeting may become an additional platform for the climate-related concerns of the marine ecotourism sector (NMTI 2012).

The Norwegian management of living marine resources in the Barents Sea can thus serve as an example for governance of Arctic marine ecosystems under climate change. Management regimes that are adaptive, science- and ecosystem-based, internationally collaborative, and involve a wide variety of stakeholders, will have higher potential to make sustainable use of emerging economic opportunities, and at the same time ameliorate negative impacts of climate change in the Arctic seas (Skern-Mauritzen et al. 2018). A continued re-assessment of the environmental and ecological conditions for the currently positive trends in fish biomass, and adaptive governance mechanisms for responding to adverse environmental changes, should be exemplary for other Arctic regions under continuing climate change. This also underscores the importance of protecting areas which are important for sensitive life stages of fish and zooplankton, such as spawning areas, from additional stressors e.g. through oil pollution. International treaties for conservation of Arctic marine biodiversity and regulation of Arctic fisheries should provide important frameworks to achieve this for the Arctic Ocean (De Lucia et al. 2018).

3.5 Conclusions

In the Barents Sea, as in other marginal Arctic seas, multiple climate change drivers such as warming, sea ice reduction and future ocean acidification are observed and anticipated to affect biological productivity, marine fish stocks, marine mammals and seabirds. These changes are progressively impacting fisheries, tourism and other cultural and supporting marine ecosystem services. The impacts are highly interdependent, difficult to predict, and interact with anthropogenic drivers, which are expected to increase under further climate change. The combined impacts thus pose complicated challenges for governance of marine areas and resources in the Arctic.

This demonstrates the necessity for advancing ecosystem-based governance regimes for the Arctic areas. The Norwegian management regimes can serve as examples for options to address these challenges. A comprehensive assessment of future regional and local changes in marine ecosystems, improved precautionary consideration of scientific knowledge gaps, amplified participation of diverse stakeholder groups, and better recognition of the cultural significance of marine species by incorporating the public, would further improve the chances for governance of Arctic marine living resources and areas to meet the challenges posed by global change.

References

Aanesen, M., Armstrong, C., Czajkowski, M., Falk-Petersen, J., Hanley, N., & Navrud, S. (2015). Willingness to pay for unfamiliar public goods: Preserving cold-water coral in Norway. *Ecological Economics, 112*, 53–67. https://doi.org/10.1016/j.ecolecon.2015.02.007.

Allison, E. H., & Bassett, H. R. (2015). Climate change in the oceans: Human impacts and responses. *Science, 350*(6262), 778–782. https://doi.org/10.1126/science.aac8721.

Allison, E. H., Perry, A. L., Badjeck, M.-C., Neil Adger, W., Brown, K., Conway, D., Halls, A. S., et al. (2009). Vulnerability of national economies to the impacts of climate change on fisheries. *Fish and Fisheries, 10*(2), 173–196. https://doi.org/10.1111/j.1467-2979.2008.00310.x.

AMAP. (2013). *AMAP assessment 2013: Arctic ocean acidification*. Oslo: Arctic Monitoring and Assessment Program. https://www.amap.no/documents/doc/amap-assessment-2013-arctic-ocean-acidification/881

Armstrong, C. W., Holen, S., Navrud, S., & Seifert, I. (2012). *The economics of ocean acidification – A scoping study*. Oslo: NIVA.

Arrigo, K. R., van Dijken, G., & Pabi, S. (2008). Impact of a shrinking Arctic ice cover on marine primary production. *Geophysical Research Letters, 35*(19), 529. https://doi.org/10.1029/2008GL035028.

Aune, M., Aschan, M. M., Greenacre, M., Dolgov, A. V., Fossheim, M., & Primicerio, R. (2018). Functional roles and redundancy of demersal Barents Sea fish: Ecological implications of environmental change. edited by Athanassios C Tsikliras. *PLoS One, 13*(11), e0207451. https://doi.org/10.1371/journal.pone.0207451.

Barrett, R. T. (2007, November). Food web interactions in the southwestern Barents Sea: Black-legged kittiwakes *Rissa Tridactyla* respond negatively to an increase in Herring Clupea Harengus. *Marine Ecology Progress Series, 349*, 269–276. https://doi.org/10.3354/meps07116..

Beaumont, N. J., Austen, M. C., Atkins, J. P., Burdon, D., Degraer, S., Dentinho, T. P., Derous, S., et al. (2007). Identification, definition and quantification of goods and services provided by marine biodiversity: Implications for the ecosystem approach. *Marine Pollution Bulletin, 54*(3), 253–265. https://doi.org/10.1016/j.marpolbul.2006.12.003.

Berge, J., Heggland, K., Lønne, O. J., Cottier, F., & Hop, H. (2015). First records of Atlantic Mackerel (Scomber Scombrus) from the Svalbard Archipelago, Norway, with possible explanations for the extensions of its distribution. *Arctic, 54*, 68. https://doi.org/10.2307/24363888.

Blackford, J. C. (2010). Predicting the impacts of ocean acidification: Challenges from an ecosystem perspective. *Journal of Marine Systems, 81*(1–2), 12–18. https://doi.org/10.1016/j.jmarsys.2009.12.016.

Bluhm, B. A., Gebruk, A. V., Gradinger, R., & Hopcroft, R. R. (2011). Arctic marine biodiversity: An update of species richness and examples of biodiversity change. *Oceanography, 24*(3), *232–248*.

Bogstad, B., Gjøsæter, H., Haug, T., & Lindstrøm, U. (2015). A review of the battle for food in the Barents Sea: Cod vs. Marine Mammals. *Frontiers in Ecology and Evolution, 3*, 29. https://doi.org/10.3389/fevo.2015.00029.

Bopp, L., Resplandy, L., Orr, J. C., Doney, S. C., Dunne, J. P., Gehlen, M., Halloran, P., et al. (2013). Multiple stressors of ocean ecosystems in the 21st century: Projections with CMIP5 models. *Biogeosciences Discussions, 10*(2), 3627–3676. https://doi.org/10.5194/bgd-10-3627-2013.

Bouchard, C., Geoffroy, M., LeBlanc, M., Majewski, A., Gauthier, S., Walkusz, W., Reist, J. D., & Fortier, L. (2017). Climate warming enhances polar cod recruitment, at least transiently. *Progress in Oceanography, 156*, 121–129. https://doi.org/10.1016/j.pocean.2017.06.008.

Brander, K. (2012). Climate and current anthropogenic impacts on fisheries. *Climatic Change, 119*, 9–21. https://doi.org/10.1007/s10584-012-0541-2.

Browman, H. I., & Stergiou, K. I. (2005). Politics and socio-economics of ecosystem-based management of marine resources. *Marine Ecology Progress Series, 300*, 241–296. https://doi.org/10.3354/meps300241.

Buanes, A., Jentoft, S., Karlsen, G. R., & Maurstad, A. (2004). In whose interest? An exploratory analysis of stakeholders in Norwegian coastal zone planning. *Ocean and Coastal Management, 47*, 207–223. https://doi.org/10.1016/j.ocecoaman.2004.04.006.

Büscher, J. V., Form, A. U., & Riebesell, U. (2017). Interactive effects of ocean acidification and warming on growth, fitness and survival of the cold-water coral Lophelia Pertusa under different food availabilities. *Frontiers in Marine Science, 4*, 119. https://doi.org/10.3389/fmars.2017.00101.

CBD Secretariat. (2014). *An updated synthesis of the impacts of ocean acidification on marine biodiversity*. (CBD technical series no. 75). Secretariat of the convention on biological diversity. Eds: Hennige, S., Roberts, J. M., & Williamson, P.). Montreal, technical series no. 75, 99 pages.

Charles, A. (2012). People, oceans and scale: Governance, livelihoods and climate change adaptation in marine social–ecological systems. *Current Opinion in Environmental Sustainability, 4*(3), 351–357. https://doi.org/10.1016/j.cosust.2012.05.011.

Christiansen, J. S., Mecklenburg, C. W., & Karamushko, O. V. (2014). Arctic marine fishes and their fisheries in light of global change. *Global Change Biology, 20*, 352–359. https://doi.org/10.1111/gcb.12395.

Collins, M., Knutti, R., Arblaster, J., Dufresne, J. L., Fichefet, T., Friedlingstein, P., Gao, X., et al. (2013). Chapter 12 – long-term climate change: Projections, commitments and irreversibility. In *IPCC 5th assessment report*. Cambridge/New York: Cambridge University Press.

Cooley, S. R., & Doney, S. C. (2009). Anticipating ocean acidification's economic consequences for commercial fisheries. *Environmental Research Letters, 4*(2), 024007. https://doi.org/10.1088/1748-9326/4/2/024007.

Crowder, L. B., Hazen, E. L., Avissar, N., Bjorkland, R., Latanich, C., & Ogburn, M. B. (2008). The impacts of fisheries on marine ecosystems and the transition to ecosystem-based management. *Annual Reviews, 39*(1), 259–278. https://doi.org/10.1146/annurev.ecolsys.39.110707.173406.

Cury, P. M., Shin, Y. J., Planque, B., Durant, J. M., Fromentin, J.-M., Kramer-Schadt, S., Stenseth, N. C., Travers, M., & Grimm, V. (2008). Ecosystem oceanography for global change in fisheries. *Trends in Ecology & Evolution, 23*(6), 338–346. https://doi.org/10.1016/j.tree.2008.02.005.

Dahlke, F. T., Leo, E., Mark, F. C., & Pörtner, H.-O. (2017). Effects of ocean acidification increase embryonic sensitivity to thermal extremes in Atlantic Cod, *Gadus Morhua*. *Global Change Biology, 23*(4), 1499–1510. https://doi.org/10.1111/gcb.13527.

Dalpadado, P., Ingvaldsen, R. B., Stige, L. C., Bogstad, B., Knutsen, T., Ottersen, G., & Ellertsen, B. (2012). Climate effects on Barents Sea ecosystem dynamics. *ICES Journal of Marine Science, 69*(7), 1303–1316. https://doi.org/10.1093/icesjms/fss063.

Dannevig, H., & Hovelsrud, G. K. (2015). Understanding the need for adaptation in a natural resource dependent community in Northern Norway: Issue salience, knowledge and values. *Climatic Change, 135*(2), 261–275.

Darnis, G., Robert, D., Pomerleau, C., Link, H., Archambault, P., John Nelson, R., Geoffroy, M., et al. (2012). Current state and trends in Canadian Arctic marine ecosystems: II. Heterotrophic food web, Pelagic-Benthic coupling, and biodiversity. *Climatic Change, 115*(1), 179–205. https://doi.org/10.1007/s10584-012-0483-8. Springer Netherlands.

De Lucia, V., Prip, C., Dalaker Kraabel, K., & Primicerio, R. (2018). Arctic Marine biodiversity in the high seas between regional and global governance. *Arctic Review on Law and Politics, 9*(0), 264. https://doi.org/10.23865/arctic.v9.1470.

Denman, K., Christian, J. R., Steiner, N., Pörtner, H.-O., & Nojiri, Y. (2011). Potential impacts of future ocean acidification on marine ecosystems and fisheries: Current knowledge and recommendations for future research. *ICES Journal of Marine Science, 68*(6), 1019–1029. https://doi.org/10.1093/icesjms/fsr074.

Derocher, Andrew E, Nicholas J Lunn, and Ian Stirling. 2004. Polar bears in a warming climate. *Integrative and comparative biology* 44 (2). Oxford University Press:Oxford 163–176. doi:https://doi.org/10.1093/icb/44.2.163.

Descamps, S., Aars, J., Fuglei, E., Kovacs, K. M., Lydersen, C., Pavlova, O., Pedersen, Å. Ø., Ravolainen, V., & Strøm, H. (2017). Climate change impacts on wildlife in a high Arctic archipelago – Svalbard, Norway. *Global Change Biology, 23*(2), 490–502. https://doi.org/10.1111/gcb.13381.

deYoung, B., Barange, M., Beaugrand, G., Harris, R., Perry, R. I., Scheffer, M., & Werner, F. (2008). Regime shifts in marine ecosystems: Detection, prediction and management. *Trends in Ecology & Evolution, 23*(7), 402–409. https://doi.org/10.1016/j.tree.2008.03.008.

Doney, S. C., Ruckelshaus, M., Emmett Duffy, J., Barry, J. P., Chan, F., English, C. A., Galindo, H. M., et al. (2012). Climate change impacts on marine ecosystems. *Annual Review of Marine Science, 4*, 11–37.

Drinkwater, K. F. (2006). The regime shift of the 1920s and 1930s in the North Atlantic. *Progress in Oceanography, 68*(2–4), 134–151.

Drinkwater, K. F., Beaugrand, G., Kaeriyama, M., Kim, S., Ottersen, G., Perry, R. I., Pörtner, H.-O., Polovina, J. J., & Takasuka, A. (2010). On the processes linking climate to ecosystem changes. *Journal of Marine Systems, 79*(3–4), 374–388. https://doi.org/10.1016/j.jmarsys.2008.12.014.

Drinkwater, K. F., Miles, M., Medhaug, I., Otterå, O. H., Kristiansen, T., Sundby, S., & Gao, Y. (2014, May). The Atlantic multidecadal oscillation: Its manifestations and impacts with special emphasis on the Atlantic region north of 60°N. *Journal of Marine Systems 133*, 117–130. https://doi.org/10.1016/j.jmarsys.2013.11.001.

Duarte, C. M., Agustí, S., Wassmann, P., Arrieta, J. M., Alcaraz, M., Coello, A., Marbà, N., et al. (2012). Tipping elements in the Arctic marine ecosystem. *Ambio, 41*(1), 44–55. https://doi.org/10.1007/s13280-011-0224-7.

Durant, J. M., Hjermann, D. Ø., Frederiksen, M., Charrassin, J. B., Le Maho, Y., Sabarros, P. S., Crawford, R. J. M., & Chr Stenseth, N. (2009). Pros and cons of using seabirds as ecological indicators. *Climate Research, 39*(2), 115–129. https://doi.org/10.3354/cr00798.

Durner, G. M., Douglas, D. C., Nielson, R. M., Amstrup, S. C., McDonald, T. L., Stirling, I., Mauritzen, M., et al. (2009). Predicting 21st-century polar bear habitat distribution from global climate models. *Ecological Monographs, 79*(1), 25–58. https://doi.org/10.1890/07-2089.1.

Eide, A., Heen, K., Armstrong, C., Flaaten, O., & Vasiliev, A. (2013). Challenges and successes in the management of a shared fish stock – The case of the Russian–Norwegian Barents Sea Cod fishery. *Acta Borealia, 30*(1), 1–20. https://doi.org/10.1080/08003831.2012.678723.

Ellingsen, I. H., Dalpadado, P., Slagstad, D., & Loeng, H. (2007). Impact of climatic change on the biological production in the Barents Sea. *Climatic Change, 87*(1–2), 155–175. https://doi.org/10.1007/s10584-007-9369-6.

Eriksen, E., Skjoldal, H. R., Gjøsæter, H., & Primicerio, R. (2017, February). Spatial and temporal changes in the Barents Sea Pelagic compartment during the recent warming. *Progress in Oceanography, 151*, 206–226. https://doi.org/10.1016/j.pocean.2016.12.009

Essington, T. E., & Punt, A. E. (2011). Implementing ecosystem-based fisheries management: Advances, challenges and emerging tools. *Fish and Fisheries, 12*(2), 123–124. https://doi.org/10.1111/j.1467-2979.2011.00407.x.

FAO. (2012). *Technical guidelines for responsible fisheries 13: Recreational fisheries 194 pages*. Rome: United Nations Food and Agriculture Organization.

FAO. (2013). *Fishery and aquaculture country profiles – Norway*. Rome: United Nations Food and Agriculture Organization. http://www.fao.org/fishery/facp/NOR/en

Filin, A., Belikov, S., Drinkwater, K., Gavrilo, M., Jørgensen, L. L., Kovacs, K. M., Luybin, P., McBride, M. M., Reigstad, M., & Strøm, H. (2016). *Future climate change and its effects on the ecosystem and human activities*. Barents Portal. www.barentsportal.com/barentsportal/index.php/en/mor

Fiskeridirektoratet. (2018). *Economic and biological figures from Norwegian fisheries (Økonomiske Og Biologiske Nøkkeltal Frå Dei Norskefiskeria)*. Bergen: Norwegian Fisheries Directorate.
Fluhr, J., Strøm, H., Pradel, R., Duriez, O., Beaugrand, G., & Descamps, S. (2017). Weakening of the subpolar Gyre as a key driver of North Atlantic seabird demography: A case study with Brünnich's Guillemots in Svalbard. *Marine Ecology Progress Series, 563*, 1–11. https://doi.org/10.3354/meps11982.
Fogarty, M. J., Gamble, R., & Perretti, C. T. (2016). Dynamic complexity in exploited marine ecosystems. *Frontiers in Ecology and Evolution, 4*, 1187. https://doi.org/10.3389/fevo.2016.00068.
Fossheim, M., Primicerio, R., Johannesen, E., Ingvaldsen, R. B., Aschan, M. M., & Dolgov, A. V. (2015). Recent warming leads to a rapid Borealization of fish communities in the Arctic. *Nature Climate Change, 5*(7), 673–677. https://doi.org/10.1038/nclimate2647.
Frainer, A., Primicerio, R., Kortsch, S., Aune, M., Dolgov, A. V., Fossheim, M., & Aschan, M. M. (2017). Climate-driven changes in functional biogeography of Arctic marine fish communities. *Proceedings of the National Academy of Sciences of the United States of America, 114*(46), 12202–12207. https://doi.org/10.1073/pnas.1706080114.
Gao, K., & Häder, D.-P. (2017). Effects of ocean acidification and UV radiation on marine photosynthetic carbon fixation. *Systems Biology of Marine Ecosystems, 14*, 235–250. https://doi.org/10.1007/978-3-319-62094-7_12.
Gattuso, J.-P., Magnan, A., Bille, R., Cheung, W. W. L., Howes, E. L., Joos, F., Allemand, D., et al. (2015). Contrasting futures for ocean and society from different anthropogenic CO2 emissions scenarios. *Science, 349*(6243), aac4722–aac4722. https://doi.org/10.1126/science.aac4722.
Gjøsaeter, H., Tjelmeland, S., & Bogstad, B. (2012). Ecosystem-based management of fish species in the Barents Sea. In *Global progress in ecosystem-based fisheries management* (pp. 333–352). Alaska Sea Grant: University of Alaska Fairbanks. https://doi.org/10.4027/gpebfm.2012.017.
Gullestad, P., Abotnes, A. M., Bakke, G., Skern-Mauritzen, M., Nedreaas, K., & Søvik, G. (2017). Towards ecosystem-based fisheries management in Norway – practical tools for keeping track of relevant issues and prioritising management efforts. *Marine Policy, 77*, 104–110. https://doi.org/10.1016/j.marpol.2016.11.032.
Hamilton, C. D., Lydersen, C., Ims, R. A., & Kovacs, K. M. (2015). Predictions replaced by facts: A keystone species' behavioural responses to declining Arctic Sea-Ice. *Biology Letters, 11*(11), 20150803. https://doi.org/10.1098/rsbl.2015.0803.
Hamre, J. (1994). Biodiversity and exploitation of the main fish stocks in the Norwegian – Barents Sea ecosystem. *Biodiversity and Conservation, 3*(6), 473–492. https://doi.org/10.1007/BF00115154.
Hannesson, R., & Herrick, S. F. (2006). *Climate change and the economics of the world's fisheries: Examples of small pelagic stocks*. Cheltenham: Edward Elgar.
Hansen, Cecilie, Mette Skern-Mauritzen, Gro van der Meeren, Anne Jähkel, and Ken Drinkwater. 2016. Set-up of the Nordic and Barents Seas (NoBa) Atlantis model, Havforskningsinstituttet, Bergen. http://hdl.handle.net/11250/2408609.
Hansen, C., Drinkwater, K. F., Jähkel, A., Fulton, E. A., Gorton, R., & Skern-Mauritzen, M. (2019). Sensitivity of the Norwegian and Barents Sea Atlantis end-to-end ecosystem model to parameter perturbations of key species. *PLoS One, 14*(2), e0210419. https://doi.org/10.1371/journal.pone.0210419.
Harsem, Ø., & Hoel, A. H. (2012). Climate change and adaptive capacity in fisheries management: The case of Norway. *International Environmental Agreements: Politics, Law and Economics, 13*(1), 49–63. https://doi.org/10.1007/s10784-012-9199-5. Springer Netherlands.
Haug, T., Bogstad, B., Chierici, M., Gjøsæter, H., Hallfredsson, E. H., Høines, Å. S., Hoel, A. H., et al. (2017). Future harvest of living resources in the Arctic ocean North of the Nordic and Barents Seas: A review of possibilities and constraints. *Fisheries Research, 188*, 38–57. https://doi.org/10.1016/j.fishres.2016.12.002.
Haynie, A. C., & Pfeiffer, L. (2012). Why economics matters for understanding the effects of climate change on fisheries. *ICES Journal of Marine Science, 69*(7), 1160–1167. https://doi.org/10.1093/icesjms/fss021.

Hoegh-Guldberg, O., Rongshuo Cai, E. S., Poloczanska, P. G., Brewer, S., Sundby, K., Hilmi, V. J. F., & Jung, S. (2014). The ocean. In: *Climate change impacts, adaptation, and vulnerability. Part B Regional aspects. Contribution of Working Group II to the Fifth Assessment Report of the Intergovernmental Panel on Climate Change* (pp. 1–77). WGII Contribution to the IPCC Fifth Assessment Report (AR5): Chapter 30.

Hoel, A. H., & Olsen, E. (2012). Integrated ocean management as a strategy to meet rapid climate change: The Norwegian case. *Ambio, 41*(1), 85–95. https://doi.org/10.1007/s13280-011-0229-2.

Hoel, A. H., von Quillfeldt, C., & Olsen, E. (2009). Norway and integrated oceans management – The case of the Barents Sea. In A. H. Hoel (Ed.), *Best practices in ecosystem-based oceans management in the Arctic* (pp. 1–10). Tromsø: Norsk Polarinstitutt.

Hollowed, A. B., & Sundby, S. (2014). Change is coming to the Northern oceans. *Science, 344*(6188), 1084–1085. https://doi.org/10.1126/science.1251166.

Hop, H., & Gjøsæter, H. (2013). Polar Cod (Boreogadus Saida) and capelin (Mallotus Villosus) as Key species in marine food webs of the Arctic and the Barents Sea. *Marine Biology Research, 9*(9), 878–894. https://doi.org/10.1080/17451000.2013.775458.

Hoppe, C. J. M., Wolf, K. K. E., Schuback, N., Tortell, P. D., & Rost, B. (2018). Compensation of ocean acidification effects in Arctic phytoplankton assemblages. *Nature Climate Change, 8*(6), 529–533. https://doi.org/10.1038/s41558-018-0142-9.

Hunt, G. L., Jr, P. S., Walters, G., Sinclair, E., Brodeur, R. D., Napp, J. M., & Bond, N. A. (2002). Climate change and control of the southeastern Bering Sea pelagic ecosystem. *Deep Sea Research Part II: Topical Studies in Oceanography, 49*(26), 5821–5853. https://doi.org/10.1016/S0967-0645(02)00321-1.

Hunt, G. L., Blanchard, A. L., Boveng, P., & Dalpadado, P. (2013). The barents and chukchi seas: Comparison of two Arctic shelf ecosystems. *Journal of Marine Systems, 49*(26), 5821–5853.

ICES. (2016). *Final report of the Working Group on the Integrated Assessments of the Barents Sea (WGIBAR)*. Murmansk: ICES.

ICES AFWG. (2016). *Report of the Arctic Fisheries Working Group (AFWG)*. Copenhagen: ICES.

Ingvarsdóttir, A., Bjørkblom, C., Ravagnan, E., Godal, B. F., Arnberg, M., Joachim, D. L., & Sanni, S. (2012). *Journal of Marine Systems, 93*(C), 69–76. https://doi.org/10.1016/j.jmarsys.2011.10.014.

Jentoft, S., & Mikalsen, K. H. (2014). Do national resources have to be centrally managed? Vested interests and institutional reform in Norwegian fisheries governance. *Maritime Studies, 13*(1), 5. https://doi.org/10.1186/2212-9790-13-5.

Johannesen, E., Ingvaldsen, R. B., Bogstad, B., Dalpadado, P., Eriksen, E., Gjøsæter, H., Knutsen, T., Skern-Mauritzen, M., & Stiansen, J. E. (2012). Changes in Barents Sea ecosystem state, 1970–2009: Climate fluctuations, human impact, and trophic interactions. *ICES Journal of Marine Science, 69*(5), 880–889. https://doi.org/10.1093/icesjms/fss046.

Johnsen, J. P. (2013). Is fisheries governance possible? *Fish and Fisheries, 15*(3), 428–444. https://doi.org/10.1111/faf.12024.

Jorgensen, C., Enberg, K., Dunlop, E. S., Arlinghaus, R., Boukal, D. S., Brander, K., Ernande, B., et al. (2007). Ecology: Managing evolving fish stocks. *Science, 318*(5854), 1247–1248. https://doi.org/10.1126/science.1148089.

Jørgensen, L. L., Planque, B., Thangstad, T. H., & Certain, G. (2015, June). Vulnerability of Megabenthic species to trawling in the Barents Sea. *ICES Journal of Marine Science*, *73*, 84–97 https://doi.org/10.1093/icesjms/fsv107.

Kaiser, B. A., Kourantidou, M., & Fernandez, L. (2018, March). A case for the commons: The Snow Crab in the Barents. *Journal of Environmental Management, 210*, 338–348. https://doi.org/10.1016/j.jenvman.2018.01.007.

Katsanevakis, S., Stelzenmüller, V., South, A., Sørensen, T. K., Jones, P. J. S., Kerr, S., Badalamenti, F., et al. (2011). Ecosystem-based marine spatial management: Review of concepts, policies, tools, and critical issues. *Ocean and Coastal Management, 54*(11), 807–820. https://doi.org/10.1016/j.ocecoaman.2011.09.002.

Kjesbu, O. S., Bogstad, B., Devine, J. A., Gjøsæter, H., Howell, D., Ingvaldsen, R. B., Nash, R. D. M., & Skjæraasen, J. E. (2014). Synergies between climate and management for Atlantic Cod fisheries at high latitudes. *Proceedings of the National Academy of Sciences, 111*(9), 3478–3483. https://doi.org/10.1073/pnas.1316342111.

Klima- og Miljødepartementet. (2011). *First update of the integrated management plan for the marine environment of the Barents Sea–Lofoten area*. Regjeringen.no. March 11. http://www.regjeringen.no/nb/dep/md/dok/regpubl/stmeld/2010-2011/meld-st-10-20102011.html?id=682050

Koenigstein, S., & Goessling-Reisemann, S. (2014). *Ocean acidification and warming in the Norwegian and Barents Seas: Impacts on marine ecosystems and human uses*. University of Bremen, artec Sustainability Research Center. https://doi.org/10.5281/zenodo.8317.

Koenigstein, S., Mark, F. C., Gößling-Reisemann, S., Reuter, H., & Pörtner, H.-O. (2016a). Modelling climate change impacts on marine fish populations: Process-based integration of ocean warming, acidification and other environmental drivers. *Fish and Fisheries, 17*(4), 972–1004. https://doi.org/10.1111/faf.12155.

Koenigstein, S., Ruth, M., & Reisemann, S. G. (2016b). Stakeholder-informed ecosystem modeling of ocean warming and acidification impacts in the Barents Sea region. *Frontiers in Marine Science, 3*, 93. https://doi.org/10.3389/fmars.2016.00093.

Koenigstein, S., Dahlke, F. T., Stiasny, M. H., Storch, D., Clemmesen, C., & Pörtner, H.-O. (2018). Forecasting future recruitment success for Atlantic Cod in the warming and acidifying Barents Sea. *Global Change Biology, 24(1)*, 526–535. https://doi.org/10.1111/gcb.13848.

Kordas, R. L., Harley, C. D. G., & O'Connor, M. I. (2011). Community ecology in a warming world: The influence of temperature on interspecific interactions in marine systems. *Journal of Experimental Marine Biology and Ecology, 400*(1–2), 218–226. https://doi.org/10.1016/j.jembe.2011.02.029.

Kortsch, S., Primicerio, R., Fossheim, M., Dolgov, A. V., & Aschan, M. (2015). Climate change alters the structure of Arctic marine food webs due to poleward shifts of Boreal generalists. *Proceedings of the Royal Society B: Biological Sciences, 282*, 1814–20151546. https://doi.org/10.1098/rspb.2015.1546.

Kovacs, K. M., Lydersen, C., Overland, J. E., & Moore, S. E. (2010). Impacts of changing sea-ice conditions on Arctic marine mammals. *Marine Biodiversity, 41*(1), 181–194. https://doi.org/10.1007/s12526-010-0061-0.

Kroeker, K. J., Kordas, R. L., & Crim, R. (2013). Impacts of ocean acidification on marine organisms: Quantifying sensitivities and interaction with warming. *Global Change Biology, 19*, 1884–1896. https://doi.org/10.1111/gcb.12179.

Landa, C. S., Ottersen, G., Sundby, S., Dingsør, G. E., & Stiansen, J. E. (2014). Recruitment, distribution boundary and habitat temperature of an Arcto-boreal gadoid in a climatically changing environment: A case study on Northeast Arctic Haddock (*Melanogrammus Aeglefinus*). *Fisheries Oceanography, 23*(6), 506–520. https://doi.org/10.1111/fog.12085.

Langangen, Ø., Färber, L., Stige, L. C., Diekert, F. K., Barth, J. M. I., Matschiner, M., Berg, P. R., et al. (2019). Ticket to spawn: Combining economic and genetic data to evaluate the effect of climate and demographic structure on spawning distribution in Atlantic Cod. *Global Change Biology, 25*(1), 134–143. https://doi.org/10.1111/gcb.14474.

Le Quéré, C, & Metzl, N. (2004). *Natural processes regulating the ocean uptake of CO2*. In: Field, CB & Raupach, M.R. (eds.): Towards CO2 stabilization: Issues, Strategies, and Consequences Publisher. Island Press, Wahsington DC.

Le Quesne, W. J. F., & Pinnegar, J. K. (2012). The potential impacts of ocean acidification: Scaling from physiology to fisheries. *Fish and Fisheries, 13*(3), 333–344. https://doi.org/10.1111/j.1467-2979.2011.00423.x.

Lind, S., Ingvaldsen, R. B., & Furevik, T. (2018). Arctic warming hotspot in the northern Barents Sea linked to declining Sea-Ice import. *Nature Climate Change, 8*(7), 634–639. https://doi.org/10.1038/s41558-018-0205-y.

Loeng, H., & Drinkwater, K. (2007). An overview of the ecosystems of the Barents and Norwegian Seas and their response to climate variability. *Deep Sea Research Part II, 54*, 2478–2500. https://doi.org/10.1016/j.dsr2.2007.08.013.

Long, R. D., Charles, A., & Stephenson, R. L. (2015, July). Key principles of Marine ecosystem-based management. *Marine Policy, 57*, 53–60. https://doi.org/10.1016/j.marpol.2015.01.013.

Mangel, M., & Levin, P. S. (2005). Regime, phase and paradigm shifts: Making community ecology the basic science for fisheries. *Philosophical Transactions of the Royal Society of London. Series B, Biological Sciences, 360*(1453), 95–105. https://doi.org/10.1098/rstb.2004.1571.

Manizza, M., Follows, M. J., Dutkiewicz, S., Menemenlis, D., Hill, C. N., & Key, R. M. (2013, November). Changes in the Arctic ocean CO2 sink (1996–2007): A regional model analysis. *Global Biogeochemical Cycles, 27*, 1108–1118. https://doi.org/10.1002/2012GB004491

McNicholl, D. G., Walkusz, W., Davoren, G. K., Majewski, A. R., & Reist, J. D. (2015). Dietary characteristics of co-occurring polar Cod (Boreogadus Saida) and capelin (Mallotus Villosus) in the Canadian Arctic, Darnley Bay. *Polar Biology, 39*(6), 1099–1108. https://doi.org/10.1007/s00300-015-1834-5.

Metcalfe, J. D., Le Quesne, W. J. F., Cheung, W. W. L., & a Righton, D. (2012). Conservation physiology for applied management of marine fish: An overview with perspectives on the role and value of telemetry. *Philosophical Transactions of the Royal Society of London. Series B, Biological Sciences, 367*(1596), 1746–1756. https://doi.org/10.1098/rstb.2012.0017.

Michalsen, K., Dalpadado, P., Eriksen, E., Gjøsæter, H., Ingvaldsen, R. B., Johannesen, E., Jørgensen, L. L., Knutsen, T., Prozorkevich, D., & Skern-Mauritzen, M. (2013). Marine living resources of the Barents Sea–Ecosystem understanding and monitoring in a climate change perspective. *Marine Biology Research, 9*(9), 932–947. https://doi.org/10.1080/17451000.2013.775459.

Mikalsen, K. H., & Jentoft, S. (2001). From user-groups to stakeholders? The public interest in fisheries management. *Marine Policy, 25*(4), 281–292. https://doi.org/10.1016/S0308-597X(01)00015-X.

Myksvoll, M. S., Jung, K. M., Albretsen, J., & Sundby, S. (2013). Modelling dispersal of eggs and quantifying connectivity among Norwegian coastal cod subpopulations. *ICES Journal of Marine Science, February, 71*, 957–969. https://doi.org/10.1093/icesjms/fst022.

Nakken, O. (1998). Past, present and future exploitation and management of Marine resources in the Barents Sea and adjacent areas. *Fisheries Research, 37*(1–3), 23–35. https://doi.org/10.1016/S0165-7836(98)00124-6.

NMTI. (2012). *Destination Norway*. Norwegian Ministry of Trade and Industry. https://www.regjeringen.no/contentassets/1ce1d6cdcbac47739b3320a66817a2dd/lenke_til_strategien-engelsk.pdf

Nøttestad, L., Sivle, L. D., Krafft, B. A., Langård, L., Anthonypillai, V., Bernasconi, M., Langøy, H., & Axelsen, B. E. (2013). Ecological aspects of fin whale and humpback whale distribution during summer in the Norwegian Sea. *Marine Ecology, 35*(2), 221–232. https://doi.org/10.1111/maec.12075.

Øigård, T. A., Lindstrøm, U., Haug, T., Nilssen, K. T., & Smout, S. (2013). Functional relationship between harp seal body condition and available prey in the Barents Sea. *Marine Ecology Progress Series, 484*, 287–301. https://doi.org/10.3354/meps10272.

Olsen, E., Aanes, S., Mehl, S., Holst, J. C., Aglen, A., & Gjøsæter, H. (2010). Cod, Haddock, Saithe, Herring, and Capelin in the Barents Sea and adjacent waters: A review of the biological value of the area. *ICES Journal of Marine Science, 67*(1), 87–101. https://doi.org/10.1093/icesjms/fsp229.

Onarheim, I. H., Smedsrud, L. H., Ingvaldsen, R. B., & Nilsen, F. (2014, June). Loss of sea ice during winter North of Svalbard. *Tellus A: Dynamic Meteorology and Oceanography, 66*(1), 23933. https://doi.org/10.3402/tellusa.v66.23933.

Orlova, E. L., Boitsov, V. D., & Dolgov, A. V. (2005). The relationship between plankton, Capelin, and Cod under different temperature conditions. *ICES Journal of Marine Science, 62*, 1281–1292.

Ottersen, G., & Stenseth, N. C. (2001). Atlantic climate governs oceanographic and ecological variability in the Barents Sea. *Limnology and Oceanography, 46*(7), 1774–1780. https://doi.org/10.4319/lo.2001.46.7.1774.

Pepin, P. (2016). Reconsidering the impossible – Linking environmental drivers to growth, mortality, and recruitment of fish. *Canadian Journal of Fisheries and Aquatic Sciences, 73*(2), 205–215. https://doi.org/10.1139/cjfas-2015-0091.

Perry, R. I., Barange, M., & Ommer, R. E. (2010). Global changes in marine systems: A social–ecological approach. *Progress in Oceanography, 87*(1–4), 331–337. https://doi.org/10.1016/j.pocean.2010.09.010.

Perry, R. I., Ommer, R. E., Barange, M., Jentoft, S., Neis, B., & Sumaila, U. R. (2011). Marine social-ecological responses to environmental change and the impacts of globalization. *Fish and Fisheries, 12*(4), 427–450. https://doi.org/10.1111/j.1467-2979.2010.00402.x.

Pershing, A. J., Alexander, M. A., Hernandez, C. M., Kerr, L. A., Le Bris, A., Mills, K. E., Nye, J. A., et al. (2015). Slow adaptation in the face of rapid warming leads to collapse of the Gulf of Maine Cod fishery. *Science, 350*(6262), 809–812. https://doi.org/10.1126/science.aac9819.

Pinsky, M. L., Reygondeau, G., Caddell, R., Palacios-Abrantes, J., Spijkers, J., & Cheung, W. W. L. (2018). Preparing ocean governance for species on the move. *Science, 360*(6394), 1189–1191. https://doi.org/10.1126/science.aat2360.

Plaganyi, E. E. (2007). *Models for an ecosystem approach to fisheries* (FAO Fisheries Technical Paper. No. 477). Food and Agriculture Organization of the United Nations, Rome.

Poloczanska, E. S., Brown, C. J., Sydeman, W. J., Kiessling, W., Schoeman, D. S., Moore, P. J., Brander, K., et al. (2013). Global imprint of climate change on marine life. *Nature Climate Change, 3*, 919–925. https://doi.org/10.1038/nclimate1958.

Pörtner, H.-O., & Peck, M. A. (2010). Climate change effects on fishes and fisheries: Towards a cause-and-effect understanding. *Journal of Fish Biology, 77*(8), 1745–1779. https://doi.org/10.1111/j.1095-8649.2010.02783.x.

Pörtner, H.-O., Karl, D. M., Boyd, P. W., Cheung, W. W. L., Lluch-Cota, S. E., Nojiri, Y., Schmidt, D. N., & Zavialov, P. O. (2014). Ocean systems. In C. B. Field, V. R. Barros, D. J. Dokken, K. J. Mach, M. D. Mastrandrea, T. E. Bilir, M. Chatterjee, et al. (Eds.), *Climate change 2014: Impacts, adaptation, and vulnerability. Part A: Global and sectoral aspects. Contribution of Working Group II to the Fifth Assessment Report of the Intergovernmental Panel on Climate Change* (pp. 411–484). Cambridge/New York: Cambridge University Press.

Renaud, P. E., Berge, J., Varpe, Ø., Lønne, O. J., Nahrgang, J., Ottesen, C., & Hallanger, I. (2011). Is the poleward expansion by Atlantic Cod and Haddock threatening native polar Cod, Boreogadus Saida? *Polar Biology, 35*, 1–12. https://doi.org/10.1007/s00300-011-1085-z.

Rice, J., Jennings, S., & Charles, A. (2014a). Scientific foundation: Towards integration. In *Governance of Marine fisheries and biodiversity conservation* (Vol. 5, pp. 124–136). Chichester: Wiley. https://doi.org/10.1002/9781118392607.ch9.

Rice, J., Howard, I., & Browman. (2014b). Where has all the recruitment research gone, long time passing? *ICES Journal of Marine Science, 71*(8), 2293–2299. https://doi.org/10.1093/icesjms/fsu158.

Riebesell, U., & Gattuso, J.-p. (2015). Lessons learned from ocean acidification research. *Nature Climate Change, 5*(1), 12–14. https://doi.org/10.1038/nclimate2456.

Rijnsdorp, A. D., Peck, M. A., Engelhard, G. H., Mollmann, C., & Pinnegar, J. K. (2009). Resolving the effect of climate change on fish populations. *ICES Journal of Marine Science, 66*(7), 1570–1583. https://doi.org/10.1093/icesjms/fsp056.

Roberts, J. M. (2006). Reefs of the deep: The biology and geology of cold-water coral ecosystems. *Science, 312*(5773), 543–547. https://doi.org/10.1126/science.1119861.

Sakshaug, E., Bjørge, A., Gulliksen, B., Loeng, H., & Mehlum, F. (1994). Structure, biomass distribution, and energetics of the Pelagic ecosystem in the Barents Sea: A synopsis. *Polar Biology, 14*(6), 405–411. https://doi.org/10.1007/BF00240261.

Sandvik, H., Erikstad, K. E., Barrett, R. T., & Yoccoz, N. G. (2005). The effect of climate on adult survival in five species of North Atlantic Seabirds. *Journal of Animal Ecology, 74*(5), 817–831. https://doi.org/10.1111/j.1365-2656.2005.00981.x.

Simmonds, M. P., & Isaac, S. J. (2007). The impacts of climate change on Marine Mammals: Early signs of significant problems. *Oryx, 41*, 19–26.

Skern-Mauritzen, M., Olsen, E., & Huse, G. (2018). Opportunities for advancing ecosystem-based management in a rapidly changing, high latitude ecosystem. *ICES Journal of Marine Science, 75*(7), 2425–2433. https://doi.org/10.1093/icesjms/fsy150.

Skogen, M. D., Olsen, A., Børsheim, K. Y., Sandø, A. B., & Skjelvan, I.. (2014, March). Modelling ocean acidification in the Nordic and Barents Seas in present and future climate. *Journal of Marine Systems, 131*, 10–20. https://doi.org/10.1016/j.jmarsys.2013.10.005.

Slagstad, D., Ellingsen, I. H., & Wassmann, P. (2011). Evaluating primary and secondary production in an Arctic ocean void of summer sea ice: An experimental simulation approach. *Progress in Oceanography, 90*(1–4), 117–131. https://doi.org/10.1016/j.pocean.2011.02.009.

Sswat, M., Stiasny, M. H., Taucher, J., Algueró-Muñiz, M., Bach, L. T., Jutfelt, F., Riebesell, U., & Clemmesen, C. (2018). Food web changes under ocean acidification promote herring larvae survival. *Nature Ecology & Evolution, 2*(5), 836–840. https://doi.org/10.1038/s41559-018-0514-6.

Stenevik, E. K., & Sundby, S. (2007). Impacts of climate change on commercial fish stocks in Norwegian waters. *Marine Policy, 31*(1), 19–31. https://doi.org/10.1016/j.marpol.2006.05.001.

Stiasny, M. H., Mittermayer, F. H., Sswat, M., Voss, R., Jutfelt, F., Chierici, M., Puvanendran, V., Mortensen, A., Reusch, T. B. H., & Clemmesen, C. (2016). Ocean acidification effects on Atlantic Cod larval survival and recruitment to the fished population. *PLoS One, 11*(8), e0155448. https://doi.org/10.1371/journal.pone.0155448.

Stige, L. C., Ottersen, G., Brander, K., & Chan, K. S. (2006). Cod and climate: Effect of the North Atlantic oscillation on recruitment in the North Atlantic. *Marine Ecology, 325*, 227–241.

Sundby, S., & Nakken, O. (2008). Spatial shifts in spawning habitats of Arcto-Norwegian Cod related to multidecadal climate oscillations and climate change. *ICES Journal of Marine Science, 65*(6), 953–962. https://doi.org/10.1093/icesjms/fsn085.

Sundby, S., Drinkwater, K. F., & Kjesbu, O. S. (2016). The North Atlantic spring-bloom system—Where the changing climate meets the winter dark. *Frontiers in Marine Science, 3*, 14. https://doi.org/10.3389/fmars.2016.00028.

Sutton, R. T., & Hodson, D. L. R. (2005). Atlantic ocean forcing of North American and European summer climate. *Science, 309*(5731), 115–118. https://doi.org/10.1126/science.1109496.

Sydeman, W. J., Poloczanska, E., Reed, T. E., & Thompson, S. A. (2015). Climate change and Marine Vertebrates. *Science, 350*(6262), 772–777. https://doi.org/10.1126/science.aac9874.

The Royal Society. (2005). *Ocean acidification due to increasing atmospheric carbon dioxide*. London: Royal Society. https://royalsociety.org/~/media/Royal_Society_Content/policy/publications/2005/9634.pdf

Tiller, R, De Kok, J. L., Vermeiren, K., Richards, R., Van Ardelan, M., & Bailey, J. (2016, December). Stakeholder perceptions of links between environmental changes to their Socio-ecological system and their adaptive capacity in the region of Troms, Norway. *Frontiers in Marine Science, 3,* 444. https://doi.org/10.3389/fmars.2016.00267. Frontiers.

Turley, C. M., Roberts, J. M., & Guinotte, J. M. (2007). Corals in deep-water: Will the unseen hand of ocean acidification destroy cold-water ecosystems? *Coral Reefs, 26*(3), 445–448. https://doi.org/10.1007/s00338-007-0247-5.

Wassmann, P., & Reigstad, M. (2011). Future Arctic ocean seasonal ice zones and implications for Pelagic-Benthic coupling. *Oceanography, 24*(3), 220–231. https://doi.org/10.5670/oceanog.2011.74.

Wassmann, P., Reigstad, M., Haug, T., Rudels, B., Carroll, M. L., Hop, H., Gabrielsen, G. W., et al. (2006). Food webs and carbon flux in the Barents Sea. *Progress in Oceanography, 71*(2–4), 232–287. https://doi.org/10.1016/j.pocean.2006.10.003.

West, J. J., & Hovelsrud, G. K. (2010). Cross-scale adaptation challenges in the coastal fisheries: Findings from Lebesby, Northern Norway. *Arctic, 63*(3), 338–354. https://doi.org/10.2307/20799601.

Wittmann, A. C., & Pörtner, H.-O. (2013). Sensitivities of extant animal taxa to ocean acidification. *Nature Climate Change, 3*, 995–1001. https://doi.org/10.1038/nclimate1982.

Yaragina, N. A., & Dolgov, A. V. (2009). Ecosystem structure and resilience—a comparison between the Norwegian and the Barents Sea. *Deep-Sea Research Part II, 56*(21–22), 2141–2153. https://doi.org/10.1016/j.dsr2.2008.11.025.

Chapter 4
Oil Vulnerability Index, Impact on Arctic Bird Populations (Proposing a Method for Calculating an Oil Vulnerability Index for the Arctic Seabirds)

Nina J. O'Hanlon, Alexander L. Bond, Neil A. James, and Elizabeth A. Masden

Abstract In recent decades, political and commercial interest in the Arctic's resources has increased dramatically. With the projected increase in shipping activity and hydrocarbon extraction, there is an increased risk to marine habitats and organisms. This comes with concomitant threats to the fragile Arctic environment especially from oil, whether from shipping accidents, pipeline leaks, or sub-surface well blowouts. Seabirds are among the most threatened group of birds, and the main threats to these species at-sea are commercial fishing and pollution. Seabirds are vulnerable to oil pollution, which can result in mass mortality events. Species are affected to a differing extent, therefore it is important to objectively predict which species are most at risk from oil spills and where. Assessing the vulnerability of seabirds to oil is achieved through establishing an index for the sensitivity of seabirds to oil – Oil Vulnerability Index (OVI). This incorporates spatial information on the distribution and density of birds as well as on species specific behaviours and other life history characteristics. This chapter focuses on the threat of oil to seabirds, especially in the Arctic, and how an OVI can be used to highlight which species are most at risk and where within the Arctic region.

Keywords Marine · North Atlantic · Pollution

N. J. O'Hanlon (✉) · N. A. James · E. A. Masden
Environmental Research Institute, North Highland College UHI,
University of the Highlands and Islands, Thurso, UK
e-mail: nina.ohanlon@uhi.ac.uk; Neil.James@uhi.ac.uk; Elizabeth.Masden@uhi.ac.uk

A. L. Bond
Environmental Research Institute, North Highland College UHI,
University of the Highlands and Islands, Thurso, UK

Department of Life Sciences, The Natural History Museum, Tring, UK
e-mail: alex.bond@uhi.ac.uk

E. Pongrácz et al. (eds.), *Arctic Marine Sustainability*,
Springer Polar Sciences, https://doi.org/10.1007/978-3-030-28404-6_4

Over the last century, ice cover and thickness in the Arctic has decreased, resulting in increased opportunities for marine access. With the associated projected increase in shipping activity and hydrocarbon extraction in this region, there is an increased risk of negative ecological impacts on the marine environment. Seabirds are particularly vulnerable to oil pollution. Following a spill, seabirds frequently come in to contact with crude oil floating on the sea's surface, which can affect seabirds in many direct and indirect ways. Due to their ecology, some seabird species are more likely to be affected by oil than others. Therefore, we can estimate the vulnerability of different seabird species to oil by taking into account their behaviour and life history characteristics. This method allows us to create an index for the sensitivity of seabirds to oil – Oil Vulnerability Index (OVI). In this chapter we describe how we made small changes to the UK's Seabird Oil Sensitivity Index (SOSI) to assess seabird vulnerability to oil within the eastern North Atlantic to help identify which seabirds may be most sensitive to potential oil pollution associated with the predicted future increase in shipping and hydrocarbon exploration in the Arctic.

4.1 Introduction

In recent decades, political and commercial interest in the Arctic's resources have increased dramatically. Over the last century, Arctic ice cover and thickness have decreased, especially during the summer months, and is likely to decrease further as predicted global annual surface temperatures continue to rise with climate change (IPCC 2013). The increased melting of sea-ice in the Arctic increases the opportunities for marine access to this region, opening new shipping trade routes and access to unexploited oil and gas resources (Wilkinson et al. 2017). Trans-arctic shipping routes, such as the Northern Sea Route, provide a cheaper and quicker alternative than traditional, longer routes (Miller and Ruiz 2014). With the projected increase in shipping activity and hydrocarbon extraction in northern waters, there is an increased risk of negative ecological impacts on marine habitats and organisms. With this come the potential threats to the Arctic environment from those wanting to exploit these resources, especially oil, whether from shipping accidents, pipeline leaks or sub-surface well blowouts (Wilkinson et al. 2017).

Oil can enter the marine environment through natural seeps, however substantial quantities of oil enter from anthropogenic activities associated with sea-going vessels and oil exploration, extraction and accidental or deliberate discharge of oil during transportation (Clark 2001). The largest contribution of oil entering our oceans is from at-sea vessels (Committee on Oil in the Sea 2003). Large oil spills from tankers and routine hydrocarbon extraction operations are generally the most high-profile incidents, however sizeable quantities are discharged, mostly illegally, from washing vessel tanks and removing ballast (Committee on Oil in the Sea 2003).

4.2 Threat of Oil to Seabirds with Particular Focus on the Arctic

Seabirds are among the most threatened group of bird species, with 28% of the world's seabird species categorised as globally threatened (Birdlife 2012), and pollution being a key threat (Croxall et al. 2012). Seabirds are particularly vulnerable to oil pollution, which can cause mass mortality events, with even minor oil spills causing problems (Piatt and Ford 1996; Votier et al. 2005; Munilla et al. 2011). Seabirds generally encounter oil on the sea surface where it forms a thin film. Oil can affect seabirds directly through lethal and sub-lethal effects; it can suffocate individuals or destroy the insulating properties of feathers, as oil droplets adsorb to the feathers of birds because of their hydrophobic properties. The damage this causes to the feathers' microstructure reduces a bird's ability to insulate and waterproof itself resulting in hypothermia and reduced buoyancy, which can lead to starvation where individuals cannot fly or forage (Jenssen et al. 1985; Jenssen 1994; O'Hara and Morandin 2010).

If ingested, for example during preening when an individual attempts to clean its feathers, the oil and associated toxins can damage internal organs and affect metabolism, leading to dehydration and poisoning (Miller et al. 1978; Burger and Fry 1993; Paruk et al. 2016). Polycyclic aromatic hydrocarbons (PAHs) are a group of organic pollutants with mutagenic and carcinogenic properties released into the environment by the incomplete combustion of fossil fuels and the burning of organic matter. Paruk et al. (2016) found that exposure to PAHs found in oil resulted in lower body mass of adult and immature Common Loons *Gavia immer*. Seabirds can also be exposed to PAHs after major oil spills through ingesting contaminated prey (Alonso-Alvarez et al. 2007; Paruk et al. 2014). Even individuals that successfully preen themselves of oil have experienced negative consequences on their subsequent productivity (Corkhill 1973; Esler et al. 2000) and long-term survival (Esler et al. 2000; Peterson et al. 2003; Esler and Iverson 2010; Fraser and Racine 2016). Esler et al. (2000) found that 6–9 years after the Exxon Valdez oil spill there was a reduction in female Harlequin Duck *Histrionicus histrionicus* winter survival in oiled compared to unoiled areas; however, after 11–14 years no difference was observed indicating that it took over a decade for survival of Harlequin Ducks to recover (Esler and Iverson 2010).

Seabirds can also be impacted by oil spills indirectly through displacement from foraging habitats and reduced food availability where prey species are affected (Peterson et al. 2003; Velando et al. 2005). During the winter following the Sea Empress oil spill in Wales, Common Scoters *Melanitta nigra* were displaced from favoured foraging locations with individuals instead feeding in areas with energetically less profitable prey (Banks et al. 2008). Seabirds can also be affected negatively by clean-up efforts, through disturbance, and toxicity of dispersants (Jenssen 1994; Whitmer et al. 2018).

The productive, cooler temperature waters of the eastern North Atlantic Ocean hold large concentrations of seabirds (Wong et al. 2014), and internationally important numbers of many species (Birdlife 2019). However, these colder higher latitude waters can also increase seabirds' vulnerability to oil pollution (Fraser and Racine 2016). In cold water, oil can persist at the sea surface for extended periods as a more viscous, solidified form (Buist et al. 2000; Brandvik and Faksness 2008), whilst seabirds are more vulnerable due to potentially already being at a higher thermal stress (Ellis and Gabrielsen 2001). Consequently, only small amounts of oil may cause hypothermia and therefore increase mortality risk (Hartung 1967; Jenssen et al. 1985; Wiese and Ryan 2003). Wiese and Ryan (2003) found higher incidences of oiled birds in periods of colder ambient air temperatures, high winds and periods of increased onshore winds, which caused greater thermal stress for compromised birds.

Although large oil spills and disasters can affect and kill large numbers of individuals, chronic oil pollution is thought to have the greatest impact on seabirds because of its persistent nature over time (Wiese and Robertson 2004; O'Hara and Morgan 2006; Ronconi et al. 2015). Systematic beached bird surveys have occurred in the North Sea since the 1970s predominantly to monitor chronic oil pollution, which has revealed that seabird oiling rates (the number of oiled birds divided by the total number of birds found) have declined in this region, largely due to implemented legislation, such as the MARPOL protocol of 1973, and the amended protocol of 1978 (Jones 1980; Camphuysen 1998; Heubeck 2006; Stienen et al. 2017). In Canada, seabird oil rates also declined from 1984–2006, but remained high relative to other areas (Wilhelm et al. 2009).

4.2.1 Direct Monitoring of Seabird Vulnerability to Oil Via Beached Bird Surveys

To determine the extent to which, and mechanism of how oil affects seabirds we must be able to assess its impacts. One method for doing so is to search systematically for oiled individuals that have washed up on the shoreline (Camphuysen and Heubeck 2001; Wiese and Ryan 2003; Heubeck 2006). These "beached bird surveys" can provide details on the risk to different species of at-sea oiling, as well as monitor variation in the amount of oil at sea among years and locations (Furness and Camphuysen 1997; Heubeck et al. 2003).

From the data collected during beached bird surveys it is possible to derive a species-specific 'oiling rate', which can be defined as the number of birds found per km of coastline and the proportion of birds found oiled (Camphuysen 2007). These values can then be used to identify spatial and temporal variation in seabird vulnerability to oil, as survey methods are standardised (Wiese and Elmslie 2006). Beached bird surveys have shown that species with the highest oiling rates were those that would be expected from their behaviour of spending a large amount of time on the sea surface, such as auks (Alcidae) and eiders *Somateria* spp. (Camphuysen and Heubeck 2001; Wiese and Ryan 2003), whilst more aerial species and those that stay closer to the coast had lower rates of oiling (Wiese and Ryan 2003). However, it is possible that predatory and scavenging species, such as gulls or terns (Laridae), might have a higher risk if they are attracted to dead or incapacitated oiled individuals.

Beached bird surveys provide a relatively cost-effective way of monitoring oil pollution and the effect of spills on seabirds, with small changes in oiling rates successfully detected. This means that beached bird surveys can also be used to assess the effectiveness of measures to reduce oil pollution in the marine environment (Heubeck 2006). In addition to monitoring oiling rates, the surveys also increase public awareness of the issue. Beached bird surveys can, however, result in biases in the data collected. It is possible that birds that died from oil ingestion with no external signs of oiling are missed (Leighton 1995; Briggs et al. 1997), while there is also the possibility that birds become oiled after death. At times of reduced food availability or during severe winters, which increase seabird mortality generally, oiling rates can often be lower (Camphuysen and Heubeck 2001). Furthermore, not all oiled birds will reach land and a large proportion remain at sea. Many factors influence the number of clean and oiled birds that are found during beached bird surveys, including wind speed and direction, surface currents, sea surface temperatures (Camphuysen and Heubeck 2001), and other sources of mortality such as fishing effort or hunting pressure (Wilhelm et al. 2009). Finally, even if the birds do reach the coastline, they might still not be detected if they are scavenged before being found (Ford and Zafonte 2009) or if washed up in a remote area with lower survey effort. This final point is particularly relevant in the Arctic.

For the information gained from beached bird surveys to be useful to decision makers, an accurate number of all oiled birds needs to be established; however, this is not possible at present as only small areas are monitored, meaning assumptions need to be made if extrapolating to large, unsurveyed areas (Wiese and Elmslie 2006). It is also difficult to establish the effects of oil pollution on seabirds at the population level, as we generally do not have reliable estimates of seabird population numbers, and the effect of oil would need to be disentangled from other natural and anthropogenic mortality sources (Wiese and Elmslie 2006). It is very difficult to determine how the numbers of oiled birds that wash up oiled relate to those affected at sea and never recorded, and therefore current oiling rates have high levels of associated uncertainty.

4.2.2 Introduction to Oil Vulnerability Indices

Beached bird surveys provide information on the number of birds that have been affected by oil retrospectively. However, it is also useful to be able to predict the likely impacts of an oil spill on seabird populations and communities. Such predictions are useful, for example, in the planning of oil and gas developments, the opening of new shipping routes, and establishing environmental monitoring programmes.

Due to their ecology, some seabird species are likely to be affected to a greater extent than others. For example, pursuit diving seabirds such as seaducks (Anatidae: Mergiini), loons (Gaviidae) and auks are likely more susceptible to oiling than more aerial species such as gulls and terns because they spend a greater proportion of time on the sea surface and therefore are more likely to come into contact with oil (Camphuysen and Heubeck 2001; Heubeck 2006). Species that congregate together in large numbers, like many seaducks, those in polynyas, or those that undergo periods of flightlessness (or are naturally so) are also at high risk (Westphal and Rowan 1969; Burger 1993).

Assessing the vulnerability of seabirds to oil spill incidents incorporates information on the spatial abundance of seabirds as well as species-specific behaviours and other life history characteristics. To date, this has been achieved through modelling mortality based on behavioural characteristics, spatial distributions, and oil spill size (Fifield et al. 2009), or calculating an index for the sensitivity of seabirds to oil – Oil Vulnerability Index (OVI) (King and Sanger 1979; Williams et al. 1994), which we focus on here. Indices incorporate factors that affect the survival of a species in relation to oil spills using a scoring system, which varies depending on the study. The factors that contribute to the OVI are theoretically assessed and scored based on expert judgements, prior information, and knowledge about the behaviour of a species, foraging strategies, and demography. The OVI scores can range from low values indicating no or very little vulnerability to oil, to maximum scores, indicating high vulnerability (Camphuysen 2007). Furthermore, the OVI scores can be used to create a spatial OVI by combining them with species distribution and density data within areas of potential oil spill or oiling risk. For example, Renner and Kuletz (2015) created a spatial–seasonal analysis of the oiling risk from shipping traffic to seabirds in the Aleutian Archipelago, as did Wong et al. (2018) for the eastern Canadian Arctic. The OVI methodology has been further developed by Certain et al. (2015) to provide a mathematical argument for the combination of the factors used when producing a single sensitivity index. Earlier efforts (Williams et al. 1994) have also been refined for the United Kingdom continental shelf area after a thorough review of the contributing factors and spatial density data (Webb et al. 2016), resulting in the Seabird Oil Sensitivity Index (SOSI).

4.3 Case Study: Assessing the Feasibility of an Oil Sensitivity Index for the Eastern North Atlantic

Given the predicted increase in extractive hydrocarbon activity and traffic associated with trans-arctic shipping routes in northern Europe, and the Arctic more broadly, there is a need to assess the vulnerability of seabirds to oil in these areas using a unified and comparable approach. Although there are country-specific vulnerability indices for seabirds and oil within this region (Gavrilo et al. 1998; Clausen et al. 2016), not all jurisdictions have methods for assessing risks to seabirds from oil, and there is no region-wide assessment. Given the migratory nature of many seabirds in the eastern North Atlantic (Guilford et al. 2011; Frederiksen et al. 2012, 2016), an understanding of risk can only be done at the regional scale, where mortality often occurs away from breeding colonies (Harris and Wanless 1996; Tasker et al. 2000).

Here we discuss the rationale behind using the UK's SOSI approach as the foundation for a region-wide index of seabirds' oil sensitivity by examining how the SOSI can be modified to meet the requirements of the larger region, and its limitations. We then undertake a sensitivity analysis to establish which of the factors used to construct the current SOSI contribute the most to the final species-specific SOSI to guide the quality of data required, and data gaps that may limit our ability to apply the SOSI approach across the eastern North Atlantic region (Table 4.1).

Table 4.1 Seabird species commonly occurring as breeding or migrants in the eastern North Atlantic

Common name	Scientific name	Status	Birdlife red list category
Red-throated Loon	*Gavia stellata*	Breeding	Least concern
Arctic Loon	*Gavia arctica*	Breeding	Least concern
Common Loon	*Gavia immer*	Breeding	Least concern
Yellow-billed Loon	*Gavia adamsii*	Breeding	Near-threatened
Red-necked Grebe	*Podiceps grisegena*	Breeding	Least concern
Great-crested Grebe	*Podiceps cristatus*	Breeding	Least concern
Horned Grebe	*Podiceps auritus*	Breeding	Vulnerable
Black-necked Grebe	*Podiceps nigricollis*	Breeding	Least concern
Northern Fulmar	*Fulmarus glacialis*	Breeding	Least concern
Cory's Shearwater	*Calonectris borealis*	Migrant	Least concern
Great Shearwater	*Ardenna gravis*	Migrant	Least concern
Sooty Shearwater	*Ardenna grisea*	Migrant	Near-threatened
Manx Shearwater	*Puffinus puffinus*	Breeding	Least concern
Balearic Shearwater	*Puffinus mauretanicus*	Migrant	Critically endangered
European Storm-petrel	*Hydrobates pelagicus*	Breeding	Least concern
Leach's Storm-petrel	*Hydrobates leucorhous*	Breeding	Vulnerable
Northern Gannet	*Morus bassanus*	Breeding	Least concern
Great Cormorant	*Phalacrocorax carbo*	Breeding	Least concern
European Shag	*Phalacrocorax aristotelis*	Breeding	Least concern
Common Eider	*Somateria mollissima*	Breeding	Near-threatened
King Eider	*Somateria spectabilis*	Breeding	Least concern
Steller's Eider	*Polysticta stelleri*	Breeding	Vulnerable
Harlequin Duck	*Histrionicus histrionicus*	Breeding	Least concern
Long-tailed Duck	*Clangula hyemalis*	Breeding	Vulnerable
Common Scoter	*Melanitta nigra*	Breeding	Least concern
Velvet Scoter	*Melanitta fusca*	Breeding	Vulnerable
Goldeneye	*Bucephala clangula*	Breeding	Least concern

(continued)

Table 4.1 (continued)

Common name	Scientific name	Status	Birdlife red list category
Goosander	*Mergus merganser*	Breeding	Least concern
Red-breasted Merganser	*Mergus serrator*	Breeding	Least concern
Greater Scaup	*Aythya marila*	Breeding	Least concern
Red-necked Phalarope	*Phalaropus lobatus*	Breeding	Least concern
Red Phalarope	*Phalaropus fulicarius*	Breeding	Least concern
Pomarine Jaeger	*Stercorarius pomarinus*	Breeding	Least concern
Arctic Jaeger	*Stercorarius parasiticus*	Breeding	Least concern
Long-tailed Jaeger	*Stercorarius longicaudus*	Breeding	Least concern
Great Skua	*Catharacta skua*	Breeding	Least concern
Mediterranean Gull	*Larus melanocephalus*	Breeding	Least concern
Little Gull	*Hydrocoloeus minutus*	Breeding	Least concern
Sabine's Gull	*Xema sabini*	Migrant	Least concern
Black-headed Gull	*Larus ridibundus*	Breeding	Least concern
Mew Gull	*Larus canus*	Breeding	Least concern
Lesser black-backed Gull	*Larus fuscus*	Breeding	Least concern
European herring Gull	*Larus argentatus*	Breeding	Least concern
Yellow-legged Gull	*Larus michahellis*	Migrant	Least concern
Iceland Gull	*Larus glaucoides*	Breeding	Least concern
Glaucous Gull	*Larus hyperboreus*	Breeding	Least concern
Great black-backed Gull	*Larus marinus*	Breeding	Least concern
Ross's Gull	*Rhodostethia rosea*	Breeding	Least concern
Black-legged Kittiwake	*Rissa tridactyla*	Breeding	Vulnerable
Ivory Gull	*Pagophila eburnea*	Breeding	Near-threatened
Sandwich Tern	*Thalasseus sandvicensis*	Breeding	Least concern
Roseate Tern	*Sterna dougallii*	Breeding	Least concern
Common Tern	*Sterna hirundo*	Breeding	Least concern
Arctic Tern	*Sterna paradisaea*	Breeding	Least concern
Little Tern	*Sternula albifrons*	Breeding	Least concern
Black Tern	*Chlidonias niger*	Breeding	Least concern
Common Murre	*Uria aalge*	Breeding	Least concern
Thick-billed Murre	*Uria lomvia*	Breeding	Least concern
Razorbill	*Alca torda*	Breeding	Near-threatened
Black Guillemot	*Cepphus grylle*	Breeding	Least concern
Little Auk	*Alle alle*	Breeding	Least concern
Atlantic Puffin	*Fratercula arctica*	Breeding	Vulnerable

Common names and taxonomy follows Birdlife (2019)

4.3.1 Regional Sensitivity Indices

The production of the SOSI in the UK underwent a thorough review process to determine which factors would be most effective for oil pollution contingency planning and emergency response within UK waters (Webb et al. 2016). The SOSI has therefore been adopted by the UK's Joint Nature Conservation Committee (JNCC) to advise necessary actions when and where oil spills occur within the waters around the UK (JNCC 2017). We therefore used this most recent assessment of seabird vulnerability to oil to develop a regional approach for the eastern North Atlantic region.

The SOSI used in the UK incorporates eight factors that represent three principles to assess the sensitivity of seabird species to oil: (1) how likely individuals are to be affected by oil due to their behaviour (factors 1–3); (2) how vulnerable a population/species is (factors 4–6); and (3) how quickly a population or species might recover from an oil incident (factors 7–8). The eight factors are scored on a scale of 0.2–1.0 in increments of 0.2, from low to high sensitivity, and determined for each species (Webb et al. 2016) with total scores ranging from 1.6 to 8.0:

1. Proportion of time spent sitting on the water (using European Seabird at Sea data from 1995 to 2015). Species that more often sit on the ocean's surface are at greater risk of oiling and therefore have a higher score.
2. Percentage of tideline corpses contaminated with oil (based on Williams et al. 1994). Species with higher oiling rates (where a high percentage of tideline corpses are contaminated with oil) are assumed to be more sensitive to oil pollution and have a higher score.
3. Habitat flexibility (taken from Furness et al. 2013), defined as the range of habitats a species uses, scored from 0.2 (high habitat flexibility: tend to forage over large marine areas with little known association with particular marine features) to 1 (low habitat flexibility: tend to feed on very specific habitat features, such as shallow banks with bivalve communities, or kelp beds).
4. Percentage of biogeographical population within the UK continental shelf; a measure of how vulnerable a species is to mortality. Species with a high percentage of biogeographical population within the UK continental shelf are scored higher as this indicates the importance of the UK's population of a species globally.
5. Listing in Birds of Conservation Concern (BOCC) (scored depending on status levels from BOCC 2, BOCC 3 (Eaton et al. 2009) and BOCC 4 (Eaton et al. 2015). Species of higher conservation concern are scored higher.
6. Presence on EU Birds Directive Annexes; a third measure of how vulnerable a species is to mortality (with scores taken from Furness et al. 2012). Scored as 1 where species are listed in Annex 1 of the Directive (species of particular conservation concern, which have the highest level of protection), 0.6 where species are not listed on Annex 1 but are listed as a migratory species and 0.2 where species are not listed on Annex 1 or are not listed as migratory (European Commission 2009).
7. Potential annual productivity, score based on maximum and mean clutch size & age at first breeding, based on Williams et al. 1994 (a high score reflects a small maximum and mean clutch size with a high age of first breeding, whilst a low score reflects a large maximum and mean with a low age of first breeding). Species with high scores are expected to recover from an oil incident more slowly.

8. Adult annual survival rate, also a measure of how quickly a species may recover from an oil incident, with species with high scores (reflecting high annual survival rates) expected to take longer to recover from an oil incident.

There are several other factors that may influence the sensitivity of a species to oil, which are not included in the SOSI calculation, for example: ability to withstand oiling (Burger and Gochfield 2002), foraging/feeding behaviour (Schreiber and Burger 2002), aggregation at sea (Stone et al. 1995; Reid et al. 2001), coloniality at breeding sites (Schreiber and Burger 2002); and the extent to which species are attracted to vessels based on interactions with fisheries vessels (Wahl and Heinemann 1979; Skov and Durinck 2001). Increasing the number of factors used to calculate a sensitivity index may allow it to be more representative if it incorporates all aspects that might influence species' sensitivity to oil. However, obtaining adequate data to score these factors accurately is unlikely for many species, which will increase the uncertainty of index values. Having fewer and more broad factors may therefore prevent a false sense of precision in the index values. There is also an additional management cost of changing the current method and establishing whether alternative factors are appropriate. Effort may be better spent applying the current SOSI more widely, given its utility in a UK context. Furthermore, as the SOSI is an index, rather than an absolute value with a degree of certainty, applying it consistently, using the same method across regions, will make its caveats better understood by those using it. Due to the potential difficulty of obtaining sufficient data, both temporal and spatial, for species, especially in more remote areas such as across the Arctic, we used the SOSI used in the UK as a starting point for calculating an oil vulnerability index for species in the eastern North Atlantic Ocean, with a particular focus on the Northern Periphery of Europe and Arctic region: Denmark, England, the Faroe Islands, Finland, Greenland, Iceland, Ireland, Northern Ireland, Norway, Scotland, Svalbard (including Bjørnøya & Jan Mayan), Sweden, and Wales (Fig. 4.1).

Outside of the UK, the extent to which information on seabirds' vulnerability to oil exists across the eastern North Atlantic Ocean is currently variable (Camphuysen 2007). The Faroe Islands are relatively well covered, though there is limited winter seabird distribution data (Skov et al. 2002). The Baltic Sea has extensive data on seabird distribution and abundance at sea, however no oil sensitivity index has been established (Camphuysen 2007). The Norwegian Seas and those around Svalbard and Greenland are partly covered, but have incomplete data; for example, around Svalbard there are recent summer seabird at sea data for the Barents Sea to the south but not for northern waters (Camphuysen 2007). Data on at-sea distribution of seabirds in this region, particularly focused on the non-breeding season, have been collected through SEATRACK, an international seabird tracking programme covering the Barents, Norwegian and North Seas (http://www.seapop.no/en/seatrack/). Neither has looked at a comprehensive seabird vulnerability index for oil in their jurisdictions, but Greenland does have an oil vulnerability atlas for the west coast (Mosbech et al. 2004; Stjernholm et al. 2011; Clausen et al. 2016), with a similar atlas in preparation for east Greenland. Icelandic waters, as well as the high

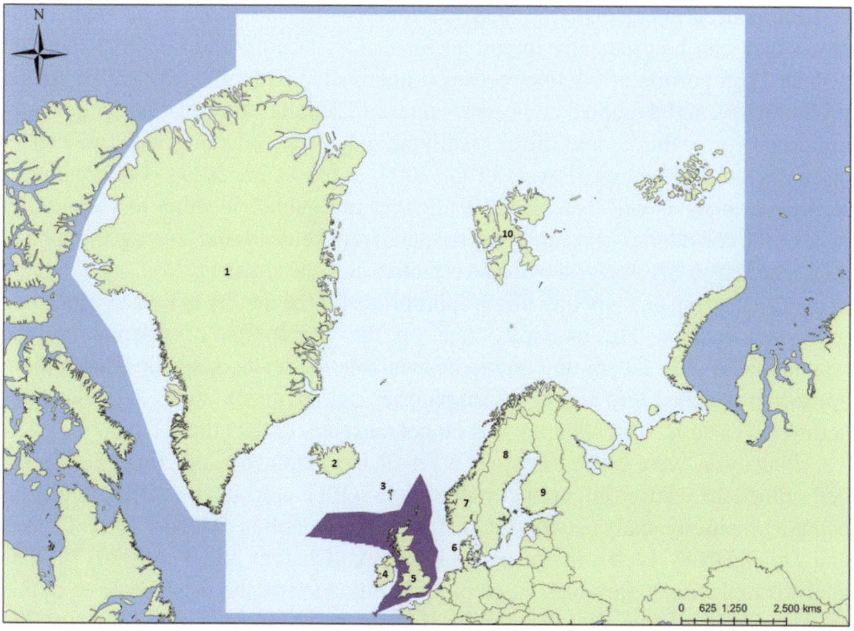

Fig. 4.1 The current spatial coverage of SOSI (Webb et al. 2016) in the United Kingdom Continental Shelf, shown in dark purple. We discuss expanding this to cover the Northern Periphery of Europe and Arctic region, shown in light blue, with a focus on *1* Greenland, *2* Iceland, *3* Faroe Islands, *4* Ireland, 5 United Kingdom (England, Northern Ireland, Scotland, Wales), *6* Denmark, *7* Norway, *8* Sweden, *9* Finland, and *10* Svalbard (including Bjornoya & Jan Mayan)

seas around Greenland, west of Ireland and around Svalbard are data deficient due to limited seabirds at sea data and having no assessment of seabird or habitat sensitivity to oil pollution.

4.3.2 Sensitivity Analysis of the SOSI Factors

Before expanding the SOSI to the eastern North Atlantic Ocean, we must assess the relative importance of the current SOSI factors. Developed in the UK, with a long history of seabird monitoring (Reid et al. 2001; Mitchell et al. 2004), data may not be necessarily available for all factors from all species across the entire region. A sensitivity analysis can therefore identify which factors are the most influential, and therefore where to focus future data collection, or use surrogate values from other species. As the SOSI adopts a binned approach, with individual parameter values scored from 0.2 to 1.0 for each factor, changing parameter values may not actually make any material difference to the overall result.

Factors used to calculate the SOSI do not consider variability in parameter values, which can be especially important for factors 1 (proportion of time spent on water), 2 (proportion of tideline corpses contaminated with oil), 7 (potential annual productivity), and 8 (annual adult survival rate). The latter two are demographically important for seabirds, and differ greatly among sites, reflecting local as well as more regional pressures (Lavers et al. 2009; Bond et al. 2011). Undertaking a sensitivity analysis will therefore also allow us to establish whether this variability needs to be considered when applying the parameter values to the larger geographical area. The sensitivity analysis will also establish how the current factors are weighted in the calculation, and whether this is appropriate for the quality of data available for individual factors. For example, data on the percentage of tideline corpses contaminated with oil are unlikely to be available for offshore species that seldom appear on beached bird surveys (Camphuysen and Heubeck 2001) and coastlines across the eastern North Atlantic that do not currently collect these data.

All analyses were carried out in R 3.5.1 (R Development Core Team 2018). We determined the relative importance of the eight SOSI factors on the SOSI calculation using a sensitivity analysis with Latin Hypercube Sampling (McKay 1992; Blower and Dowlatabadi 1994), in the R package *pse* (Chalom and Prado 2017). This analysis ranks input variables based on their influence on the model output, in this case the SOSI calculation, to identify the factors which have the most influence. We produced 200 random parameter combinations, with each factor value drawn from a uniform distribution between 0.2 and 1 at 0.2 increments, where all factors could vary. One parameter, at intervals of 0.2 between 0.2 and 1, was selected at random for each factor and combined with values from the other factors. We calculated partial rank correlation coefficients (PRCC) to determine the relative importance of each factor to the SOSI calculation (Blower and Dowlatabadi 1994). PRCC values reflect the estimated relative importance of each factor in the SOSI calculation, with PRCC values closer to 1 having the strongest influence.

The factors with the greatest impact on the final SOSI scores were factors 1 (proportion of time spent on water), 2 (proportion of tideline corpses contaminated with oil) and 3 (habitat flexibility), with all having high PRCC values >0.50 (Taylor 1990, Fig. 4.2). These factors with the most influence on the SOSI were those reflecting how likely individuals are to be affected by oil due to their behaviour. Factors 7 and 8, which reflect species demography (potential annual productivity, and adult annual survival rate), had the least influence on the final SOSI scores.

This highlights the importance of obtaining accurate parameter values and associated variability across the region of interest for factors 1–3 relating to behaviour given their relative importance in calculating species SOSI scores. The results also partially allay concerns that the SOSI does not consider spatial and temporal variation in seabird demography, with variability observed in the maximum and mean clutch size, age at first breeding and adult survival rate (Horswill and Robinson 2015). The relatively low influence of demographic factors suggests that it may not be essential to account for this variation, or uncertainty in these parameter values for understudied populations, or species, and that data from surrogate species, expert opinion, or local ecological knowledge may be an appropriate alter-

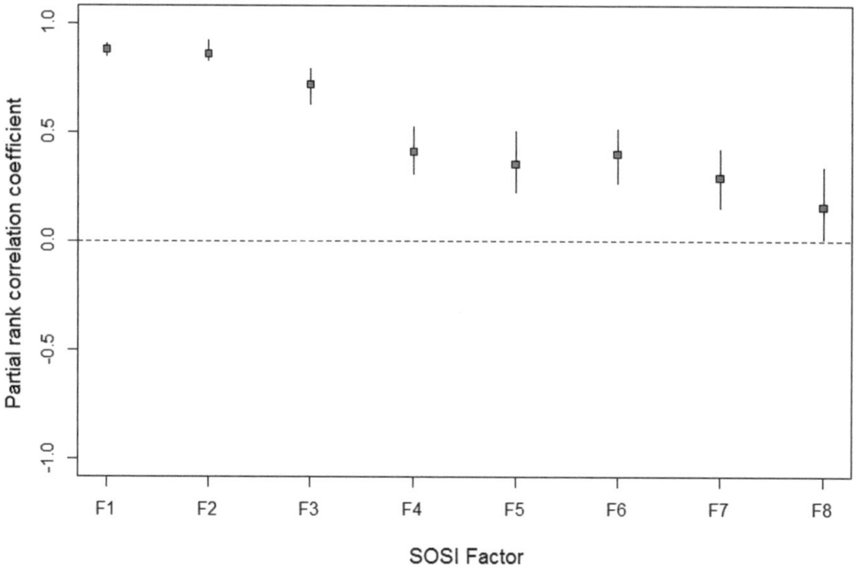

Fig. 4.2 The partial rank correlation coefficient (PRCC) for the eight SOSI factors (see text, Webb et al. 2016) with confidence intervals produced by bootstrapping. The PRCC values reflect the strength of the linear associations between the result and each of the input factors, after removing the linear effect of the other factors

native when data are lacking. Additionally, it means that it may not be necessary to account for the spatial and temporal variation observed in seabird demographic variables, especially as these parameter values are placed in bins with increments of 0.2 before inputting into the SOSI calculation.

4.3.3 Expanding the SOSI to the Eastern North Atlantic

If the application of the SOSI is to be extended into other areas, then we must first explore whether the eight factors relevant to assessing oil sensitivity in the UK are also relevant for the wider eastern North Atlantic region.

Factors 1 and 2 have the greatest influence on the overall SOSI score used in the UK, therefore it is important that the data on which these factor values are based are as accurate as possible. Where possible it would be beneficial to account for any variability in these factors across the region. The proportion of time individuals spend on the sea surface is likely consistent within species as this is largely driven by behaviour. To some extent, this consistency will also be the case for the proportion of tideline corpses contaminated with oil as this will also be related to seabirds' behaviour, with species that spend more time on the sea surface having higher oiling rates (Camphuysen 1998). Although the ranking of oiling rates for species are generally similar across different locations, the exact parameter values do differ (Camphuysen 1998). Current oiling rates are also skewed geographically with data from standardised beached bird surveys largely collated from around the North Sea, although there are also some beached bird surveys in Scandinavia (Camphuysen and Heubeck 2001), but these data are largely absent from more inaccessible coastlines (e.g., Greenland, Iceland or Svalbard). Factors 1 and 2 both reflect how likely individuals are to be affected by oil due to their behaviour and show a significant positive correlation ($r_s = 0.69$, $P < 0.001$, $N = 50$). Given this relationship and the challenges of accurately determining the number of tideline corpses, one option is to remove factor 2 and only include factor 1, where more robust data exist. Removing the proportion of oiled tideline corpses from the calculation increased the species' SOSI scores, apart from species that had the highest proportion of oil corpses (factor 2 score of 1), for which final scores stayed the same. However, the extent to which removing factor 2 increased a species' SOSI score varied, therefore the ranking of species' sensitivity did change.

Factor 3, reflecting how habitat flexibility influences the likelihood that individuals will be affected by oil due to their behaviour, also has an important influence on the overall SOSI score used in the UK. This factor considers the range of habitats a species uses and is therefore relatively straightforward to determine for most species. There may be some subjectivity in scoring this factor and therefore it is important that additional species in the eastern North Atlantic Ocean, not covered by the UK SOSI, are scored considering already published values (Garthe and Hüppop 2004, Furness et al. 2012) and expert judgement to ensure consistency.

To expand the SOSI used in the UK to a wider region factors 4–6 need to be replaced as these are specifically related to the UK (factors 4 and 5) or the European Union (factor 6), which does not encompass all jurisdictions in the eastern North Atlantic Ocean. These three factors refer to the conservation status of species and reflect how vulnerable a population or species is to threats such as oil pollution within the smaller region of the UK and European Union. One option could be to include, for each species, the percentage of biogeographical population within the eastern North Atlantic Ocean, however this will likely be difficult to determine as there is considerable uncertainty in regional and global population estimates of many species. Instead, it arguably makes more sense to use a species' global conservation status, as determined by the IUCN Red List (Birdlife 2019). Alternatively, as the countries in the study region are in Europe, the European Red List could be used (Birdlife 2015). By using the IUCN Red List categories, the index would be expandable to other regions, or even globally. This does have the disadvantage of removing any spatial or temporal variability in the factors associated with species' status and vulnerability, however.

We compared the actual winter (October – March) and summer (April – September) SOSI index scores for 50 UK seabird species against those calculated using scores for the IUCN Red list categories instead of factors 4, 5 and 6. In general, the resulting IUCN SOSI scores were lower than the original winter and summer scores. This is largely attributed to most species being listed on the IUCN Red List as Least Concern, which resulted in these species having the lowest score of 0.2 (species listed as Near-Threatened were scored as 0.4, Vulnerable as 0.6, Endangered as 0.8 and Critically Endangered as 1). The extent to which using these scores based on ICUN Red List categories changed a species' SOSI score varied, therefore the ranking of species' sensitivity did change.

In the UK's SOSI, factors 7 and 8 relate a species' demographic vital rates and reflect how quickly a population or species might recover from an oil incident. Relevant demographic data are available for the majority of the 62 seabird species that commonly occur within the eastern North Atlantic, although there are gaps for some species in the Arctic such as the Yellow-billed Loon *Gavia adamsii*, Ross's Gull *Rhodostethia rosea* and Iceland Gull *Larus glaucoides*, and demographic rates have high inter-annual variation. However, given their low influence in calculating the SOSI used in the UK, data from surrogate species may be helpful where this information is not available for specific species, but should be used cautiously.

Using ducks as an example, we calculated scores incorporating the above adaptations and compared these with the scores calculated by Webb et al. (2016), see Table 4.2. First, we calculated scores using the IUCN Red List status as a

Table 4.2 A comparison of the species-specific winter and summer SOSI scores for the duck (Anatidae) species from Webb et al. (2016) and the scores calculated using the proposed adapted SOSI approach

Species	Winter SOSI scores (rank)	Summer SOSI scores (rank)	Scores where factors 4, 5 & 6 are replaced by the IUCN Red List category	Scores where factors 4, 5 & 6 are replaced by the IUCN Red List category, and factor 2 is removed
Velvet scoter *Melanitta fusca*	0.657 (2)	0.657 (2)	0.657 (1)	0.727 (1)
Long-tailed duck *Clangula hyemalis*	0.570 (5)	0.570 (4)	0.570 (2)	0.694 (2)
Common Eider *Somateria mollissima*	0.651 (3)	0.651 (3)	0.542 (3)	0.659 (3)
Common scoter *Melanitta nigra*	0.712 (1)	0.667 (1)	0.336 (4)	0.366 (6)
Goldeneye *Bucephala clangula*	0.597 (4)	0.555 (5)	0.300 (5)	0.366 (6)
Greater Scaup *Aythya marila*	0.561 (6)	0.529 (6)	0.287 (6)	0.409 (4)
Red-breasted merganser *Mergus serrator*	0.396 (8)	0.396 (8)	0.270 (7)	0.409 (4)
Goosander *Mergus merganser*	0.427 (7)	0.427 (7)	0.260 (8)	0.317 (7)

Species are ranked by the scores where factors 4, 5 & 6 are replaced by the IUCN Red List category, reflecting most (1) to least (8) vulnerable to oil

replacement for factors 4, 5 and 6. Secondly, we calculated scores incorporating the IUCN Red List status as well as removing factor 2. Making these alterations to the SOSI used in the UK does alter the resulting scores and ranks among species, particularly for those with the highest scores. The IUCN Red List scores are generally lower than those calculated by Webb et al. (2016), as the global conservation status of these species is Least Concern, with the exception of the Long-tailed Duck *Clangula hyemalis* and Velvet Scoter *Melanitta fusca*, which have a global status of Vulnerable. The change in SOSI score and ranking for the Common Scoter is particularly large, as it is included on the Birds of Conservation Concern Red List in the UK (Eaton et al. 2015). Focusing on the rankings of the scores, the greatest change is for the Common Scoter and Long-tailed Duck, due to their differing conservation status on a local versus global scale.

Removing factor 2 from the calculation further changes the IUCN Red List scores, and more importantly the species rankings. All the included duck species have a maximum score of 1 for factor 1, as they are sensitive to oiling based on their at sea behaviour of spending a large proportion of time on the sea surface. However, the factor 2 scores, reflecting species-specific oiling rates, range from 0.4 (low oiling rates – Red-breasted Merganser *Mergus serrator* and Greater Scaup) to 0.8 (high oiling rates – Common and Velvet Scoter). Within the SOSI calculation, factor 2 is an aggravation factor (Certain et al. 2015; Webb et al. 2016), therefore removing factor 2 increases the scores for species where a low proportion of tideline corpses have been found contaminated with oil, as factor 1 is multiplied by factor 2 (Eq. 4.1, Webb et al. 2016). As removing factor 2 does change the ranking of how sensitive species are to oil pollution, it is important to consider the pros and cons of removing this factor as opposed to using data that are potentially biased to certain locations within the region of interest.

$$SOSI_i = \left(F_1 \times F_2\right)^{1-\frac{F_3}{F_3+0.5}} \times \left(\frac{F_4 + F_5 + F_6}{3}\right)^{1-\frac{\left(\frac{F_7+F_8}{2}\right)}{\left(\frac{F_7+F_8}{2}\right)+0.5}} \tag{4.1}$$

To create an oil vulnerability index suitable for the eastern North Atlantic Ocean based on the UK's SOSI method, changes will need to be made to the eight contributing factors. This should be achievable if firstly, factors 4, 5 and 6, concerning species' conservation status, are replaced by a single global assessment, specifically the IUCN Red List. Secondly, if there are no data on the proportion of oiled tideline corpses then there may be the option of removing this factor if it is thought that using parameter values for the North Sea would not be representative. However, it should be noted that making these changes did change the oil vulnerability scores, and more importantly the ranking of species, compared to using the existing SOSI method (median rank change for replacing factors 4, 5 and 6 with the IUCN Red List value = 3; median rank change for removing factor 2 from the proposed adapted SOSI calculation using the IUCN Red List values = 4). Nonetheless, this option is still likely the most comparative to the SOSI used in the UK compared to creating an entirely different method of assessing species sensitivity to oil.

4.4 Conclusion

Given that certain seabird species are more sensitive to oil pollution than others, understanding this variability, and where these species are concentrated is important to identify areas at sea where seabirds as a group are particularly vulnerable to threats such as oil pollution. This can then inform management decisions on the location of an extraction site, a sea-vessel transportation route, or mitigation and response measures for when spills do occur. Seabird vulnerability assessments can be accomplished by objectively predicting which seabird species are most at risk from oil pollution. Expanding the SOSI currently used in the UK, with some small modifications, should provide an applicable and useful oil vulnerability index for seabirds across the eastern North Atlantic region. Omitting the percentage of tideline corpses contaminated with oil (factor 2) from the calculation would remove the issues regarding biases in beached bird surveys, and poor spatial coverage of such surveys in the Arctic. Species' IUCN Red List status could easily substitute the three factors involving regional or national conservation status. By making these small changes we can use the SOSI method of assessing seabird vulnerability to oil within areas of the eastern North Atlantic in a way which should be comparable to the current SOSI for the UK, and which will improve our regional assessments of seabirds' vulnerability to oil spills. Incorporating this modified SOSI method with seabird at sea distribution and density data will allow important seabird concentrations to be identified within the eastern North Atlantic region. Producing a spatial SOSI will enable stakeholders to identify where seabirds may be most sensitive to potential oil pollution associated with the predicted future increase in shipping traffic and hydrocarbon exploration and extraction, especially if this approach is expanded across the Arctic.

Acknowledgements We would like to thank all the participants of the seabird and oil workshop held in Inverness, September 2017 where we discussed the benefits and limitations of expanding the UK's SOSI approach to the wider eastern North Atlantic region. The ERDF Interreg VB Northern Periphery and Arctic (NPA) Programme funded this activity through the APP4SEA project. Comments from anonymous reviewers improved this chapter.

References

Alonso-Alvarez, C., Munilla, I., López-Alonso, M., & Velando, A. (2007). Sublethal toxicity of the Prestige oil spill on yellow-legged gulls. *Environment International, 33*, 773–781.

Banks, A. N., Sanderson, W. G., Hughes, B., Cranswick, P. A., Smith, L. E., Whitehead, S., Musgrove, A. J., Haycock, B., & Fairney, N. P. (2008). The Sea Empress oil spill (Wales, UK): Effects on Common Scoter *Melanitta nigra* in Carmarthen Bay and status ten years later. *Marine Pollution Bulletin, 56*, 895–902.

BirdLife International. (2012). *Spotlight on seabirds. Presented as part of the BirdLife state of the world's birds website*. Available from: http://www.birdlife.org/datazone

BirdLife International. (2015). *European red list of birds*. Luxembourg: Office for Official Publications of the European Communities.
BirdLife International. (2019). *BirdLife data zone*. URL http://datazone.birdlife.org. Accessed 12 Feb 2019.
Blower, S., & Dowlatabadi, H. (1994). Sensitivity and uncertainty analysis of complex models of disease transmission: An HIV model, as an example. *International Statistical Review, 62*, 229–243.
Bond, A. L., Jones, I. L., Sydeman, W. J., Major, H. L., Minobe, S., Williams, J. C., & Byrd, G. V. (2011). Reproductive success of planktivorous seabirds in the North Pacific is related to ocean climate on decadal scales. *Marine Ecology Progress Series, 424*, 205–218.
Brandvik, P. J., & Faksness, L. G. (2008). Weathering processes in Arctic oil spills: Meso-scale experiments with different ice conditions. *Cold Regions Science and Technology, 55*, 160–166.
Briggs, K. T., Gershwin, M. E., & Anderson, D. W. (1997). Consequences of petrochemical ingestion and stress on the immune system of seabirds. *ICES Journal of Marine Science, 54*, 718–725.
Buist, I., Belore, R., Dickins, D., Hackenberg, D., Guarino, A., & Zhendi, W. (2000). Empirical weathering properties of oil in ice and snow. In *Proceedings of the thirty- second AMOP technical seminar on environmental contamination and response* (pp. 67–107). Vancouver: Environment Canada.
Burger, A. E. (1993). Estimating the mortality of seabirds following oil spills: Effects of spill volume. *Marine Pollution Bulletin, 26*(3), 140–143.
Burger, A. E., & Fry, D. M. (1993). Effects of oil pollution on seabirds in the Northeast Pacific. In *The status, ecology and conservation of marine birds of the North Pacific* (pp. 254–263). Ottawa: Canadian Wildlife Service.
Burger, J., & Gochfield, M. (2002). Effects of chemicals and pollution on seabirds. In J. Schreiber & E. A. Burger (Eds.), *Biology of marine birds* (pp. 485–525). New York: CRC Press.
Camphuysen, K. C. J. (1998). Beached bird surveys indicate decline in chronic oil pollution in the North Sea. *Marine Pollution Bulletin, 36*, 519–526.
Camphuysen, K. C. J. (2007). *Chronic oil pollution in Europe*. Brussels: IFAW.
Camphuysen, C. J., & Heubeck, M. (2001). Marine oil pollution and beached bird surveys: The development of a sensitive monitoring instrument. *Environmental Pollution, 112*, 443–461.
Certain, G., Jørgensen, L. L., Christel, I., Planque, B., & Bretagnolle, V. (2015). Mapping the vulnerability of animal community to pressure in marine systems: Disentangling pressure types and integrating their impact from the individual to the community level. *ICES Jornal of Marine Science, 72*, 1470–1482.
Chalom, A. & de Prado, P. I. K. L. (2017). *pse: Parameter space exploration with Latin Hypercubes. R package version 0.4.7.*
Clark, R. B. (2001). *Marine pollution* (5th ed.). Oxford: Oxford University Press.
Clausen, D. S., Mosbech, A., Boertmann, D., Johansen, K. L., Nymand, J., Potter, S., & Myryp, M. (2016). *Environmental oil spill sensitivity atlas for the Northwest Greenland (68°–72° N) coastal zone* (164 pp). DCE – Danish Centre for Environment and Energy.
Committee on Oil in the Sea. (2003). *Oil in the sea III: Inputs, fates, and effects* (278 pp). Washington, DC: National Research Council.
Corkhill, P. (1973). Oiled seabirds successfully cleaning their plumage. *British Birds, 66*, 535–537.
Croxall, J. P., Butchart, S. H. M., Lascelles, B., Stattersfield, A. J., Sullivan, B., Symes, A., & Taylor, P. (2012). Seabird conservation status, threats and priority actions: A global assessment. *Bird Conservation International, 22*, 1–34.
Eaton, M., Brown, A., Noble, D., Musgrove, A., Hearn, R., Aebischer, N., Gibbons, D., Evans, A., & Gregory, R. (2009). Birds of conservation concern 3: The population status of birds in the United Kingdom, Channel Islands and the Isle of Man. *British Birds, 102*, 296–341.
Eaton, M., Aebischer, N., Brown, A., Hearn, R., Lock, L., Musgrove, A., Noble, D., Stroud, D., & Gregory, R. (2015). Birds of conservation concern 4: The population status of birds in the United Kingdom, Channel Islands and Isle of Man. *British Birds, 108*, 708–746.

Ellis, H. I., & Gabrielsen, G. W. (2001). Energetics of free-ranging seabirds. In E. A. Schreiber & J. Burger (Eds.), *Biology of marine birds* (pp. 359–484). Boca Raton: CRC Press.

Esler, D., & Iverson, S. A. (2010). Female harlequin duck winter survival 11 to 14 years after the Exxon Valdez oil spill. *Journal of Wildlife Management, 74*, 471–478.

Esler, D., Schmutz, J. A., Jarvis, R. L., & Mulcahy, D. M. (2000). Winter survival of adult female harlequin ducks in relation to history of contamination by the "Exxon Valdez" oil spill. *The Journal of Wildlife Management, 64*, 839–847.

European Commission. (2009). *European union birds directive*. Available at: http://data.europa.eu/eli/dir/2009/147/oj

Fifield, D. A., Baker, K. D., Byrne, R., Robertson, G. J., Burke, C., Gilchrist, H. G., Hedd, A., Mallory, M. L., McFarlane Tranquilla, L., Regular, P. M., Smith, A., Gaston, A. J., Montevecchi, W. A., Elliot, K. H., Phillips, R. (2009). *Modelling seabird oil spill mortality using flight and swim behaviour environmental studies research funds report no. 186*. Dartmouth, 46 p.

Ford, R. G., & Zafonte, M. A. (2009). Scavenging of seabird carcasses at oil spill sites in California and Oregon. *Marine Ornithology, 37*, 205–211.

Fraser, G. S., & Racine, V. (2016). An evaluation of oil spill responses for offshore oil production projects in Newfoundland and Labrador, Canada: Implications for seabird conservation. *Marine Pollution Bulletin, 107*, 36–45.

Frederiksen, M., Moe, B., Daunt, F., Phillips, R. A., Barrett, R. T., Bogdanova, M. I., Boulinier, T., Chardine, J. W., Chastel, O., Chivers, L. S., Guilford, T., Jensen, G. H., & Gre, D. (2012). Multicolony tracking reveals the winter distribution of a pelagic seabird on an ocean basin scale. *Diversity and Distributions, 18*, 530–542.

Frederiksen, M., Descamps, S., Erikstad, K. E., Gaston, A. J., Gilchrist, H. G., Grémillet, D., Johansen, K. L., Kolbeinsson, Y., Linnebjerg, J. F., Mallory, M. L., McFarlane Tranquilla, L. A., Merkel, F. R., Montevecchi, W. A., Mosbech, A., Reiertsen, T. K., Robertson, G. J., Steen, H., Strøm, H., & Thórarinsson, T. L. (2016). Migration and wintering of a declining seabird, the thick-billed murre *Uria lomvia*, on an ocean basin scale: Conservation implications. *Biological Conservation, 200*, 26–35.

Furness, R. W., & Camphuysen, K. C. J. (1997). Seabirds as monitors of the marine environment. *ICES Journal of Marine Science, 54*, 726–737.

Furness, R. W., Wade, H. M., Robbins, A. M. C., & Masden, E. a. (2012). Assessing the sensitivity of seabird populations to adverse effects from tidal stream turbines and wave energy devices. *ICES Journal of Marine Science, 69*, 1466–1479.

Furness, R. W., Wade, H. M., & Masden, E. A. (2013). Assessing vulnerability of marine bird populations to offshore wind farms. *Journal of Environmental Management, 119*, 56–66.

Garthe, S., & Hüppop, O. (2004). Scaling possible adverse effects of marine wind farms on seabirds: Developing and applying a vulnerability index. *Journal of Applied Ecology, 41*, 724–734.

Gavrilo, M., Bakken, V., Firsova, L., Kalyakin, V., Morozov, V., Pokrovskaya, I. & Isaksen, K. (1998). *Oil vulnerability assessment for marine birds occuring along the northern sea route area* (INSROP Working Paper), 56 pp.

Guilford, T., Freeman, R., Boyle, D., Dean, B., Kirk, H., Phillips, R., & Perrins, C. (2011). A dispersive migration in the Atlantic Puffin and its implications for migratory navigation. *PLoS One, 6*(7), e21336.

Harris, M. P., & Wanless, S. (1996). Differential responses of Guillemot *Uria aalge* and Shag *Phalacrocorax aristotelis* to a late winter wreck. *Bird Study, 43*, 220–230.

Hartung, R. (1967). Energy metabolism in oil-covered ducks. *The Journal of Wildlife Management, 31*, 798–804.

Heubeck, M. (2006). The Shetland beached bird survey, 1979–2004. *Marine Ornithology, 34*, 123–127.

Heubeck, M., Camphuysen, K. C. J., Bao, R., Humple, D., Sandoval, A., Cadiou, B., Bräger, S., & Thomas, T. (2003). Assessing the impact of major oil spills on seabird populations. *Marine Pollution Bulletin, 46*, 900–902.

Horswill, C., & Robinson, R. A. (2015). Review of seabird demographic rates and density dependence. *JNCC Report, 552*, 1–126.
IPCC. (2013). Summary for policymakers. In T. F. Stocker, D. Qin, G. K. Plattner, M. Tignor, S. K. Allen, J. Boschung, A. Nauels, Y. Xia, & V. Bex (Eds.), *Climate change 2013: The physical science basis. Contribution of Working Group I to the Fifth Assessment Report of the Intergovernmental Panel on Climate Change* (pp. 2–23). Cambridge: Cambridge University Press.
Jenssen, B. M. (1994). Review article: Effects of oil pollution, chemically treated oil, and cleaning on thermal balance of birds. *Environmental Pollution, 86*, 207–215.
Jenssen, B. M., Ekker, M., & Bech, C. (1985). Thermoregulation in a naturally oil-contaminated black-billed Murre *Uria aalge. Bulletin of Environmental Contamination and Toxicology, 35*, 9–14.
JNCC. (2017). *Using the seabird oil sensitivity index to inform contingency planning* (7 pp). Peterborough: Joint Nature Conservation Committee.
Jones, P. H. (1980). Beached birds at selected Orkney beaches 1976-8. *Scottish Birds, 11*, 1–12.
King, J. G., & Sanger, G. A. (1979). Oil vulnerability index for marine oriented birds. In *Conservation of marine birds of Northern North America* (pp. 227–239). Washington, DC: U.S. Fish and Wildlife Service.
Lavers, J. L., Jones, I. L., Robertson, G. J., & Diamond, A. W. (2009). Contrasting population trends at two razorbill colonies in Atlantic Canada: Additive effects of fox predation and hunting mortality? *Avian Conservation and Ecology, 4*(2), 3.
Leighton, F. A. (1995). The toxicity of petroleum oils to birds: An overview. In L. Frink (Ed.), *Wildlife and oil spills* (pp. 10–22). Newark: Response, Research, Contingency Planning, Tri State Bird Rescue and Research.
McKay, M. D. (1992). Latin Hypercude sampling as a tool in uncertainty analysis of computer models. In J. J. Swain, D. Goldsman, R. C. Crain, & J. R. Wilson (Eds.), *Proceedings of the 24th conference on winter simulation* (pp. 557–564). New York: ACM.
Miller, A. W., & Ruiz, G. M. (2014). Arctic shipping and marine invaders. *Nature Climate Change, 4*, 413–416.
Miller, D. S., Peakall, D. B., & Kinter, W. B. (1978). Ingestion of crude oil: Sublethal effects in herring gull chicks. *Science, 199*, 315–317.
Mitchell, P. I., Newton, S. F., Ratcliffe, N., Eds, T. E. D., Dunn, T. E., Poyser, A. D., & May, L. (2004). *Seabird populations of Britain and Ireland: Results of the seabird 2000 census*. London: JNCC, Poyser.
Mosbech, A., Boertmann, D., Olsen, B. Ø., Olsvig, S., von Platen, F., Buch, E., Hansen, K. Q., Rasch, M., Nielsen, N., Møller, H. S., Potter, S., Andreasen, C., Berglund, J. & Myrup, M. (2004). *Environmental oil spill sensitivity atlas for the South Greenland coastal zone*. DCE – Danish Centre for Environment and Energy, 611 pp.
Munilla, I., Arcos, J. M., Oro, D., Álvarez, D., Leyenda, P. M., & Velando, A. (2011). Mass mortality of seabirds in the aftermath of the Prestige oil spill. *Ecosphere, 2*, art83.
O'Hara, P. D., & Morandin, L. A. (2010). Effects of sheens associated with offshore oil and gas development on the feather microstructure of pelagic seabirds. *Marine Pollution Bulletin, 60*, 672–678.
O'Hara, P. D., & Morgan, K. H. (2006). Do low rates of oiled carcass recovery in beached bird surveys indicate low rates of ship-source oil spills? *Marine Ornithology, 34*, 133–140.
Paruk, J. D., Long, D., IV, Perkins, C., East, A., Sigel, B. J., & Evers, D. C. (2014). Polycyclic aromatic hydrocarbons detected in common loons (*Gavia immer*) wintering off coastal Louisiana. *Waterbirds, 37*, 85–93.
Paruk, J. D., Adams, E. M., Uher-Koch, H., Kovach, K. A., Long, D., Perkins, C., Schoch, N., & Evers, D. C. (2016). Polycyclic aromatic hydrocarbons in blood related to lower body mass in common loons. *Science of the Total Environment, 565*, 360–368.
Peterson, C. H., Rice, S. D., Short, J. W., Esler, D., Bodkin, J. L., Ballachey, B. E., & Irons, D. B. (2003). Long-term ecosystem response to the Exxon Valdez oil spill. *Science, 302*, 2082–2086.

Piatt, J. F., & Glenn Ford, R. (1996). How many seabirds were killed by the Exxon Valdez oil spill. In S. D. Rice, R. B. Spies, & D. A. Wolfe (Eds.), *Proceedings of the Exxon Valdez oil spill symposium, Anchorage, Alaska, 2–5 February, 1993* (pp. 712–719). Bethesda: American Fish Society.

R Development Core Team. (2018). *R: A language and environment for statistical computing*.

Reid, J. B., Pollock, C. M., & Mavor, R. (2001). Seabirds of the Atlantic Frontier, north and west of Scotland. *Continental Shelf Research, 21*, 1029–1045.

Renner, M., & Kuletz, K. J. (2015). A spatial-seasonal analysis of the oiling risk from shipping traffic to seabirds in the Aleutian Archipelago. *Marine Pollution Bulletin, 101*, 127–136.

Ronconi, R. A., Allard, K. A., & Taylor, P. D. (2015). Bird interactions with offshore oil and gas platforms: Review of impacts and monitoring techniques. *Journal of Environmental Management, 147*, 34–45.

Schreiber, E. A., & Burger, J. (2002). *Biology of marine birds*. Florida: CRC Press.

Skov, H., & Durinck, J. (2001). Seabird attraction to fishing vessels is a local process. *Marine Ecology Progress Series, 214*, 289–298.

Skov, H., Upton, A. J., Reid, J. B., Webb, A., Taylor, S. J., & Durinck, D. (2002). *Dispersion and vulnerability of marine birds and cetaceans in Faroese waters* (106 pp). Peterborough: Joint Nature Conservation Committee.

Stienen, E. W. M., Courtens, W., Van de walle, M., Vanermen, N., & Verstraete, H. (2017). Long-term monitoring study of beached seabirds shows that chronic oil pollution in the southern North Sea has almost halted. *Marine Pollution Bulletin, 115*, 194–200.

Stjernholm, M., Boertmann, D., Mosbech, A., Nymand, J., Merkel, F., Myrup, M., Siegstad, H., Clausen, D., & Potter, S. (2011). *Environmental oil spill sensitivity atlas for the northern West Greenland (72°–75° N) coastal zone*. DCE – Danish Centre for Environment and Energy, 798 pp.

Stone, C. J., Webb, A., & Tasker, M. L. (1995). The distribution of auks and Procellariiformes in north-west European waters in relation to depth of sea. *Bird Study, 42*, 50–56.

Tasker, M. L., Camphuysen, C. J. K., Cooper, J., Garthe, S., Montevecchi, W. A., & Blaber, S. J. M. (2000). The impacts of fishing on marine birds. *ICES Journal of Marine Science, 57*, 531–547.

Taylor, R. (1990). Interpretation of the correlation coefficient: A basic review. *Journal of Diagnostic Medical Sonography, 1*(35), 35–39.

Velando, A., Munilla, I., & Leyenda, P. M. (2005). Short-term indirect effects of the Prestige oil spill on a marine top predator: Changes in prey availabilty for European shags. *Marine Ecology Progress Series, 302*, 1–22.

Votier, S. C., Hatchwell, B. J., Beckerman, A., McCleery, R. H., Hunter, F. M., Pellatt, J., Trinder, M., & Birkhead, T. R. (2005). Oil pollution and climate have wide-scale impacts on seabird demographics. *Ecology Letters, 8*, 1157–1164.

Wahl, T. R., & Heinemann, D. (1979). Seabirds and fishing vessels: Co-occurrence and attraction. *The Condor, 81*, 390–396.

Webb, A., Elgie, M., Hidef, C. I., Pollock, C., Barton, C., Burns, S., & Hawkins, K. (2016). *Sensitivity of offshore seabird concentrations to oil pollution around the United Kingdom: Report to oil & gas UK*. 103 pp.

Westphal, A., & Rowan, M. K. (1969). Some observations on the effects of oil pollution on the Jackass Penguin. *Ostrich, 40*, 521–526.

Whitmer, E. R., Elias, B. A., Harvey, D. J., & Ziccardi, M. H. (2018). An experimental study of the effects of chemically dispersed oil on feather structure and waterproofing in Common Murres (*Uria aalge*). *Journal of Wildlife Diseases, 54*(2), 315–328. 2017-01–016.

Wiese, F. K., & Elmslie, K. I. M. (2006). Underuse and misuse of data from beached bird surveys. *Marine Ornithology, 34*, 157–159.

Wiese, F. K., & Robertson, G. J. (2004). Assessing seabird mortality from chronic oil discharges at sea. *Journal of Wildlife Management, 68*, 627–638.

Wiese, F. K., & Ryan, P. C. (2003). The extent of chronic marine oil pollution in southeastern Newfoundland waters assessed through beached bird surveys 1984–1999. *Marine Pollution Bulletin, 46*, 1090–1101.

Wilhelm, S. I., Robertson, G. J., Ryan, P. C., Tobin, S. F., & Elliot, R. D. (2009). Re-evaluating the use of beached bird oiling rates to assess long-term trends in chronic oil pollution. *Marine Pollution Bulletin, 58*(2), 249–255.

Wilkinson, J., Beegle-Krause, C., Evers, K. U., Hughes, N., Lewis, A., Reed, M., & Wadhams, P. (2017). Oil spill response capabilities and technologies for ice-covered Arctic marine waters: A review of recent developments and established practices. *Ambio, 46*, 423–441.

Williams, J. M., Tasker, M. L., Carter, I. C., & Webb, A. (1994). A method of assessing seabird vulnerability to surface pollutants. *Ibis, 137*, 147–152.

Wong, S. N., Gjerdrum, C., Morgan, K. H., & Mallory, M. L. (2014). Hotspots in cold seas: The composition, distribution, and abundance of marine birds in the North American Arctic. *Journal of Geophysical Research: Oceans, 119*, 1–15.

Wong, S. N. P., Gjerdrum, C., Gilchrist, H. G., & Mallory, M. L. (2018). Seasonal vessel activity risk to seabirds in waters off Baffin Island, Canada. *Ocean and Coastal Management, 163*, 339–351.

Chapter 5
Conflicts Between Arctic Industries and Cetaceans

Charla J. Basran and Marianne H. Rasmussen

Abstract As in many oceans around the globe, there is extensive and increasing interactions between cetaceans and industries in the Arctic. The Arctic hosts 16 cetacean species (whales, dolphins, and porpoises), some of which are seasonal visitors while others are year-round inhabitants (CAFF 2017). This chapter focuses on six cetacean species on which extensive research has been conducted, exemplified by: baleen whales; blue whale (*Balaenoptera musculus*) and humpback whale (*Megaptera novaeangliae*), large toothed whales; sperm whale (*Physeter macrocephalus*) and orca (*Orcinus orca*), and small toothed whales; white-beaked dolphin (*Lagenorhynchus albirostris*) and harbour porpoise (*Phocoena phocoena*). These can be used as representatives of all Arctic cetacean species. The industries which have the most conflict with cetaceans in the Arctic include shipping, oil exploration, and commercial fisheries. This chapter will explore the interactions between the six example species and these industries, and the impacts these interactions can have on both. It will also touch on some further conflicts between other Arctic activities and cetaceans as industries expand and human presence in the Arctic Ocean increases.

Keywords Arctic · Cetaceans · Conflicts · Shipping · Oil-exploration · Entanglement

C. J. Basran (✉) · M. H. Rasmussen
University of Iceland, Reykjavík, Iceland
e-mail: cjb2@hi.is

E. Pongrácz et al. (eds.), *Arctic Marine Sustainability*,
Springer Polar Sciences, https://doi.org/10.1007/978-3-030-28404-6_5

5.1 Arctic Cetaceans

5.1.1 Baleen Whales: Blue Whale (Balaenoptera musculus) and Humpback Whale (Megaptera novaeangliae)

Baleen whales are generally characterized by the hundreds of baleen plates they have along their upper jaw and two blowholes on the top of their head. In addition, they are known to be primarily solitary animals that are communicating over long distances using low frequency sounds, and many are making long migrations between summer feeding grounds and winter breeding grounds. There is a total of seven baleen whale species which can be found in the Arctic seasonally or year-round: blue whale (*Balaenoptera musculus*), humpback whale (*Megaptera novaeangliae*), fin whale (*Balaenoptera physalus*), gray whale (*Eschrichtius robustus*), bowhead whale (*Balaena mysticetus*), sei whale (*Balaenoptera borealis*), and minke whale (*Balaenoptera acutorostrata*).

The blue whale (Fig. 5.1) is the largest baleen whale and largest animal on the planet, growing up to a maximum of 33 meters long and weighing up to 150 tonnes (Wilson and Mittermeier 2014). The blue whale communicates with the lowest frequency of any cetacean, with a range of 16–100 Hz (Cummings and Thompson 1971). The North Atlantic blue whale song consists of a single tonal unit around 19 Hz (Mellinger and Clark 2003). Source levels of blue whale moans measured by Akamatsu et al. (2014) in Icelandic waters were between 159–169 dB re 1μPa rms from tagged animals. The hearing of blue whales is unknown, but it is assumed that they can hear at least the same frequencies as their vocalizations. Some blue whale populations are seasonal migrators to Arctic waters, where they spend the summertime feeding primarily on krill. Their habitat range in the Arctic extends from northern Norway (up to Svalbard), north of Iceland, and east and west Greenland (Wilson and Mittermeier 2014).

The humpback whale (Fig. 5.2) is another member of the baleen whale group, growing up to a maximum of 17 meters long and weighing up to 34 tonnes (Wilson and Mittermeier 2014). They produce different kinds of sounds such as grunts and

Fig. 5.1 A blue whale (*Balaenoptera musculus*), the largest cetacean in the Arctic and on the planet

Fig. 5.2 A humpback whale (*Megaptera novaeangliae*) breaching out of the water

groans, mostly in a wide frequency range of 10 Hz to 10 kHz (Thompson et al. 1979; Cerchio and Dahlheim 2001). In Alaskan feeding grounds moans were recorded with dominant frequency at 300–500 Hz and source levels of 175–192 dB re 1 μPa @1 m (Thompson et al. 1986). Humpback whales are famous for their songs, first reported by Payne and McVay (1971), and they are reported singing in the tropical breeding grounds and recently also in sub-Arctic waters during the Icelandic winter (Magnúsdóttir et al. 2014, 2015). The hearing of humpback whales has been modelled based on anatomical data which indicates that humpback whales have maximum hearing sensitivity between 2 to 6 kHz, and good sensitivity between 700 Hz to10 kHz, resembling their vocalizations (Houser et al. 2001). They are also seasonal migrators, spending their summer feeding season both in sub-arctic and Arctic waters, where they range from northern Norway (including north of Svalbard), north of Iceland, east and west Greenland, and northern Canada and Alaska. Here, they are feeding on krill and small schooling fish species (Wilson and Mittermeier 2014).

5.1.2 Toothed Whales: Sperm Whale (Physester Macrocephalus), Orca (Orcinus orca), White-Beaked Dolphin (Lagenorhynchus albirostris) and Harbour Porpoises (Phocoena phocoena)

Toothed whales are generally characterized by having true teeth, a single blowhole, and using echolocation to hunt prey and for orientation. This group includes sperm whales, beaked whales, dolphins, and porpoises and therefore is comprised of a

Fig. 5.3 A sperm whale (*Physester macrocephalus*), the largest toothed whale in the Arctic and on the planet: fluking (lifting the tail) and resting at the surface

large range of species. Unlike the baleen whales, they are known to live in social groups and communicate with higher frequency whistles and calls. There is a total of nine toothed whale species found living in the Arctic: sperm whale (*Physester macrocephalus*), beluga whale (*Delphinapterus leucas*), narwhal (*Monodon monoceros*), northern bottlenose whale (*Hyperoodon ampullatus*), orca (also known as killer whale) (*Orcinus orca*), white-beaked dolphin (*Lagenorhynchus albirostris*), Atlantic white-sided dolphin (*Lagenorhynchus acutus*), long-finned pilot whale (*Globicephala melas*), and harbour porpoise (*Phocoena phocoena*).

Sperm whales (Fig. 5.3) are the largest of the toothed whales, with the males growing up to 19 m in length and weighing up to 70 tonnes (Wilson and Mittermeier 2014). They are deep diving specialists, with a maximum dive depth of 3000 m, feeding on squid and deep-sea fish. Sperm whales echolocate using clicks with a center frequency of 15 kHz (Møhl et al. 2003) and they produce the loudest sound of all cetaceans with source level of their clicks reaching up to 236 dB re 1 mPa peRMS (Møhl et al. 2003). The hearing of sperm whales is unknown, but as other species, it is assumed they can at least hear the same frequency range as the sounds they produce. Different from all other cetaceans found in the Arctic, only male sperm whales are seen in Arctic waters. Their known range in the Arctic spreads from northern Norway, north of Iceland, and the east coast of Greenland; all the way up to the southern ice edge (Wilson and Mittermeier 2014).

Orcas (Fig. 5.4) are the largest dolphin species in the world, with females growing up to 8 meters in length and weighing up to 5 tonnes and males growing up to 10 meters in length and weighing up to 7 tonnes (Wilson and Mittermeier 2014). The orca's echolocation clicks have a center frequency between 45–80 kHz (Au et al. 2004), and the average source level varies from 173–202 dB re 1 μPa peak-to-peak @ 1 m (Simon et al. 2007). Orcas communicate with calls and whistles, and different pods have different dialects of calls (Strager 1995). Whistles have been reported to include high frequencies with a fundamental frequency up to 40 kHz (Samarra et al. 2010). From studies of captive orcas, they are known to have best hearing between 18–42 kHz, but they reliably responded to 100 kHz tones at ca. 95 dB (Szymanskia et al. 1999). Orcas are found all around Arctic waters, extending

Fig. 5.4 Orcas (*Orcinus orca*), also known as killer whales

Fig. 5.5 A pod of white-beaked dolphins (*Lagenorhynchus albirostris*)

as far as the sea ice allows, and therefore with diminishing sea ice their range is expanding. Here they are feeding on a variety of fish and other marine mammal species (Wilson and Mittermeier 2014).

White-beaked dolphins (Fig. 5.5) grow up to 3 meters in length and weigh up to 350 kilograms (Wilson and Mittermeier 2014). They produce broadband echolocation clicks with very high frequencies: up to 250 kHz (Rasmussen and Miller 2002), with source levels up to 219 dB re 1 mPa (Rasmussen et al. 2002). White-beaked

Fig. 5.6 Harbour porpoises (*Phocoena phocoena*), the smallest cetacean in the Arctic

dolphins communicate with whistles that have a fundamental frequency up to 35 kHz (Rasmussen and Miller 2002) and harmonics up to 65 kHz (Rasmussen et al. 2006). Source levels of the whistles are up to 160 dB (Rasmussen et al. 2006). White-beaked dolphins hear in the frequency range from 1–150 kHz (Nachtigall et al. 2008). They are found only living in the temperate to arctic waters of the North Atlantic Ocean and up into the Arctic Ocean, where they range from north of Norway and Iceland, and along eastern Greenland. All year round they can be spotted in pods usually up to 50 individuals, feeding on a variety of small schooling fish (Wilson and Mittermeier 2014).

Harbour porpoises (Fig. 5.6) are the smallest Arctic cetacean species, growing up to 1.6 m in length and weighing up to 65 kg (Lockyer 2003). They produce narrow band clicks centered around 130 kHz (Au et al. 1999) with source levels up to 160–205 dB re 1 μPa pp @ 1 m (Au et al. 1999; Villadsgaard et al. 2007). Harbour porpoises do not whistle but communicate with high-rate clicks only (Clausen et al. 2011). It is known that harbour porpoises hear best between 16–140 kHz (Kastelein et al. 2002). They are widely distributed from temperate to Arctic waters, where they range from northeastern Russia, north of Norway and Iceland, along eastern and western Greenland, northeastern Canada and Alaska. Throughout their range they are most often found in shallow waters, though some populations are known to travel large distances to deep waters, feeding on a wide variety of fish and cephalopods depending on location and season (Wilson and Mittermeier 2014; Nielsen et al. 2018).

5.2 Conflicts with Arctic Industries

5.2.1 Ship Strikes

One of the major threats that the baleen whales face all along their migration route, including into the Arctic, is that of ship strikes from both large shipping vessels and cruise ships, as well as fishing vessels and whale watching vessels. These ship strikes are known to lead to serious injury and/or death of the whale, particularly if the ship is 80 meters or longer or when the ship involved is travelling at a speed of 14 knots or greater (Laist et al. 2001). The other major factor in ship strike occurrence is where shipping routes are overlapping with areas that have high densities of whales. This will likely be of greatest risk in areas which are narrow passages, where the ship traffic and whales in the area would be highly concentrated (Williams and O'Hara 2010).

Research on blue whales has found that they make little-to-no lateral movement to swim out of the way of an oncoming ship, but they performed slow-decent shallow avoidance dives in 55% of studied cases (McKenna et al. 2015). These factors make blue whales particularly vulnerable to ship strikes. Though ship-strikes may not be a large contributing factor in the slow recovery of blue whale populations in post-whaling times (Monnahan et al. 2015), it is still considered a serious concern in order to ensure their recovery.

Humpback whales are also considered as commonly struck by ships (Laist et al. 2001) and are actually one of the most reported species involved in a ship strike incident, though this is likely due in-part to the coastal habitat of humpback whales and therefore higher chance of detectability of ship-struck animals in coastal waters compared to other species that are struck further away from the coast (Jensen and Silber 2004). In addition to increasing ship traffic world-wide, humpback whale populations have recovered well in post-whaling times and therefore their increase in numbers around the world has also likely been a factor in increasing ship strike risk.

The question is, why do these whales not swim out of the way? Modelling suggests that baleen whales at the surface of the water may have difficulties acoustically detecting oncoming ships (Allen et al. 2012). Ship noise has been found to have lower source level sound at the surface of the water, and ships tend to have an "acoustic shadow" caused by the bow of the boat, lowering the detectable noise directly in front of the ship. These factors mean that baleen whales may not be able to detect the ship until it is too late for them to swim out of the way (Allen et al. 2012). Research also suggests that the likelihood of a ship strike may also be dependent on what the whale is doing at the time of approach. When whales are sleeping or feeding there is evidence they are much less responsive to the noise of an approaching vessel (Laist et al. 2001). Other factors may also increase some whales' vulnerability to ship strikes, such as parasites, illness, or entanglement in fish gear which may cause some animals to spend more time than usual at the surface.

From the other side of the issue, whale collisions can cause ship damage. This is unlikely to be an issue for the large cargo ships and cruise ships, but can occur for smaller vessels such as fishing boats or whale watching boats. In addition, there are cases where deceased whales have actually been brought to port on the front of large vessels that struck them. This is generally a negative situation that companies would like to avoid and therefore mitigation measures have been explored and adopted in some areas. Shipping lane changes have been proposed, and even some lanes have been rerouted, in order to reduce the risk of ship strike (Redfern et al. 2013). In addition, speed restrictions have been put in places in some areas with high whale densities near shipping lanes (van der Hoop et al. 2015). The use of forward-facing sonar has also been proposed as a way for ships to see large whales in their path in time to change course (Miller and Potter 2001).

The decrease in Arctic sea ice has opened up the possibly for increased shipping, new shipping routes, and an increase in Arctic tourism via cruise ships. Shipping time between Europe and Asia is estimated to be reduced by up to 50% by using the Northern Sea Route during the summertime now, with thinning ice conditions (Aksenov et al. 2017). It is predicted that by 2020 65 million tonnes of cargo will be shipped through the Bering Strait along the Northern Sea Route, compared to just 1.36 million tonnes in 2013 (Huntington et al. 2015). Marine spatial planning measures should be taken into account to minimize the impacts that increased traffic can have on Arctic cetaceans.

5.2.2 *Response to Oil Exploration*

A major source of loud sounds in the ocean that can cause communication masking and behavioural disruptions to cetaceans is that of seismic airgun surveys used for oil exploration (Di Iorio and Clark 2010). An airgun is towed behind a ship and emits high-powered blasts of air causing a seismic wave to penetrate deep into the seabed in order to find oil deposits. These loud blasts can travel great distances through the ocean, and in some cases are picked up by hydrophones over 3000 km from the source (Nieukirk 2004), and therefore may affect a wide range of ocean species.

Seismic survey blasts are believed to "mask" blue whale communication, making communication more difficult since the loud, low frequency signals from the airgun are produced in the same frequency range as blue whale calls (Di Iorio and Clark 2010). Research has found that blue whales significantly increase their calling rate during seismic survey days. This is hypothesized to be compensating for the increased noise in the ocean by increasing the chance that their message is received by other members of the species. One observation showed that a blue whale within 10 km of a ship with an active airgun array stopped calling for an hour and moved away from the vessel, showing potential avoidance behaviour (McDonald et al. 1995).

Research on humpback whale response to oil exploration has found different responses, with some showing avoidance, some showing approach, and some showing little reaction (Malme et al. 1985). Some humpback whales show a "startle response", meaning the animal quickly changes direction or behaviour in direct response to the sound. Trials conducted during humpback whale migration found that animals avoided the vessel with the airgun in operation when it was within 1.2–4.4 km away (McCauley 2000). It was also noted that some animals spend an unusually long time at the surface. This could be related to the fact that the sound level is lower at the surface of the water, in the same way that was described for ship noise in Sect. 5.2.1, and could be considered a "vertical avoidance response" (Weir 2008). This response could potentially make the animals more vulnerable to a ship strike. Interestingly, some humpback whales have been observed having the opposite response, actually approaching the vessel with an active airgun within 100–400 m. This is hypothesized to be male humpback whales that are responding to the airgun sound believing it is the sound of a competitor breaching (jumping out of the water) (McCauley 2000). This could potentially be dangerous for these whales, since it is considered plausible that baleen whales may experience at least temporary hearing impairment when exposed to the airgun at close range (Gedamke et al. 2011).

Just as is the case with baleen whales, some toothed whales are also known to respond to airguns used for oil exploration since they are believed to be audible to all species of cetacean due to the wide bandwidth of sound (Goold and Coates 2006). Interestingly, research has found that sperm whales do not detectibly or significantly react to or avoid airguns in use (Miller et al. 2009; Stone and Tasker 2006; Madsen et al. 2002), though there is some evidence that they may decrease their foraging during airgun activity due to longer periods spent at the surface. Orcas have been found to show "localized avoidance" to active airguns, where the animals distanced themselves from the source (Stone and Tasker 2006), also suggesting behavioural change.

There has currently been little scientific research on white-beaked dolphin response to airgun oil exploration specifically. This is problematic since they are an abundant sub-Arctic and Arctic species which is likely sensitive to this type of noise, based on the information known about other small cetacean species. One study found that white-beaked dolphins were observed moving away from a vessel with an operating airgun in 26.5% of sightings versus only 4.8% of sightings when the airgun was not operating (Stone and Tasker 2006), suggesting an avoidance response to the sound.

Harbour porpoises are thought to be one of the most sensitive cetaceans to airgun noise. Apparent avoidance of a vessel with an active airgun has been observed when the animals were upwards of 70 km away (Bain and Williams 2006). They are also considered to be the most sensitive to having temporary hearing impairment caused by seismic surveys than any other cetacean (Lucke et al. 2009). Other research showed that even without an apparent avoidance response, there appears to be other changes in behaviour. Harbour porpoises were found to produce 15% less "buzzing" sounds when within 25 km of a vessel with an active airgun (Pirotta et al. 2014). This buzzing is associated with either prey capture or communication (Verfuß et al. 2009; Clausen et al. 2011) meaning a reduction is indication of behavioural disruption.

From the other perspective, the oil industry faces increasing public and scientific pressure in regards to environmental impacts. For example, the United States has tried to pass legislation in order to be allowed to explore and drill oil within the current Arctic National Wildlife Refuge. This was met by opposition from 37 leading Arctic scientists, stating "…the cumulative impact of many seemingly small changes is significant. New development on the coastal plain of the Arctic Refuge, one of the nation's and planet's premier protected areas, will only contribute to these harmful impacts on wildlife. For all these reasons, we oppose oil and gas exploration, development and production on the coastal plain of the Arctic Refuge." (Bowyer et al. 2017). In Norway, a public opinion poll conducted in 2017 found that the majority of Norwegians would choose to minimize oil industry activity to protect the environment (P. Wijnen, August 4 2017). There is even opposition from other Arctic industries, such as the fishing and tourism sectors in Norway which have spoken out against negative impacts of oil exploration (G. Fouche, March 20 2009). Based on this type of opposition, the industry as a whole has a need to look for new technologies for less invasive surveys or turn their focus to alternatives to oil. One such technology is the marine vibrator. This has been developed by the industry to produce a lower sound source than an airgun, while working in much the same way. Preliminary tests of the impact of the marine vibrator on blue whales and humpback whales have been conducted in Iceland, and found that animals, which were tagged with acoustic behavioural recorders, were constantly feeding and seemingly not interrupted by the marine vibrator sound (Akamatsu et al. 2014; Schnitzler et al. 2018). These marine vibrators have yet to be tested on toothed whales.

Arctic oil exploration has been expanding and is likely to continue with the decreasing sea ice opening up new exploration possibilities. The Arctic continental shelf is now one of the world's largest oil prospect areas remaining (Gautier et al. 2009). Large oil companies in Russia have been exploring their Arctic shelf and believe 20–30% of Russia's oil production will come from that area by 2050 (T. Paraskova, October 19 2017). Norway is considered another major leader in the Arctic oil industry and there are 93 oil exploration blocks located in the Barents Sea (R. Milne, June 21 2017). The increasing interest in oil exploration in the Arctic could have serious impacts on Arctic cetaceans. Precautionary research needs to be conducted in order to minimize seismic survey impacts on these animals.

5.2.3 *Entanglement and Bycatch*

Entanglement (when a cetacean gets caught up in fishing lines or netting), and bycatch (when a cetacean is caught in fishing gear and drowns) are some of the leading causes of human-induced injury and mortality to whales. This is likely to be an issue anywhere that there is an overlap between fishing activities and cetaceans. Entanglement and bycatch are known to be increasing over the years as fishing industries develop and broaden, especially as the ice cover decreases in the Arctic (Meyer et al. 2011). For the large baleen whales, becoming entangled in fishing gear often does not lead to immediate drowning, and can last for hours, days, or even months, causing a wide range of potential impacts on the animal. Impacts on the animals in addition to the possibility of drowning include rope lacerations and infection risk and inability to feed/starvation (Cassoff et al. 2011). Furthermore, dragging entangling gear can lead to a disruption in the whale's energy budget which could affect their ability to migrate (Moore and van der Hoop 2012; van der Hoop et al. 2013). Entanglement is also a stressful event for an animal, leading to an increase in stress hormones being released, which has been linked to compromising the immune system and lowering reproductive success (Robbins and Mattila 2001; Rolland et al. 2017). Entanglement and bycatch are known to occur in all types of fishing gear to some extent, with set nets, long-lines, traps/pots, and purse seines, as well as "ghost gear" (i.e. fishing gear that has been lost at sea) being the main contributors (Butterworth et al. 2012).

Blue whales sometimes become entangled in fishing gear, though seemingly to a lesser extent than humpback whales. There is little information about blue whale entanglement other than news reports of disentanglement efforts off California, USA. This may be attributed to the blue whale's extremely large size and tendency to spend less time in shallow coastal waters, where the majority of entangling gear is used. It may also be due to their elusive nature and generally small populations, making them much less sighted and studied than the humpback whale.

Conversely, entanglement of humpback whales has been extensively studied in some populations, and they are known to be one of the most vulnerable species to this issue (Cole et al. 2006; Benjamins et al. 2012). Scarring studies estimating non-lethal entanglement rates found that 29–50% of humpback whales in North Pacific populations (Robbins et al. 2007; Neilson et al. 2009), 48–65% of humpback whales in the western North Atlantic (Robbins and Mattila 2004; Robbins 2009), and 25% of humpback whales around Iceland (Basran et al. 2019) have been previously entangled in fishing gear at least once in their lifetime. This is likely to be an underestimation and is not taking into account whales that died as a result of an entanglement. Due to their large size, humpback whales often free themselves from all or part of the entangling gear, but this is not always the case. In the western North Atlantic, 16% of reported entangled humpback whale cases were fatal to the animal (Benjamins et al. 2012).

Both sperm whales and orcas come in contact with fisheries, particularly long-line fisheries. They can become entangled in fishing line and occasionally net, just as baleen whales, though it is generally less of an issue for large toothed whales and there is little scientific information on this. There are a few reports of orcas being hooked in the back and bearing line scars consistent with previous entanglement (Visser 1998, 2000). Furthermore, there are some first-hand accounts of orcas being encircled in, and even sometimes drowning in herring purse seines (H. Hjallason, personal communication 2018). Studies on sperm whales found that six out of eight animals that were studied post-mortem after stranding in Ecuador had gillnet gear still attached to the body or bared clear entanglement injuries (Felix et al. 1997). Furthermore, sperm whales have been reported trapped and entangled in drift nets (Pace et al. 2008), and carrying long-line (Purves et al. 2004). In addition to becoming entangled in these nets, sperm whales are also known to ingest the nets by mistake (Jacobsen et al. 2010). One type of entanglement that seems to be unique to sperm whales is entanglement in deep sea cables. Since sperm whales are an extremely deep-diving species, they encounter these deep sea cables used for telecommunication, unlike most other cetacean species. This issue dates all the way back to 1953 when 14 sperm whales were reported entangled in cables (Heezen 1953). Though little is known about large toothed whale entanglement, it is an issue that likely goes hand-in-hand with the learned behaviour of associating fishing boats and gear with an easy meal (see Sect. 5.2.4.3 Depredation).

Due to the small size of dolphins and porpoises, getting caught-up in fishing gear and drowning is a major issue all over the world. White-beaked dolphins can sometimes be subject to bycatch, though it seems to be less of an issue for this species than it is for harbour porpoises. Despite this, they were assessed as being at "high-risk" of bycatch in setnet fisheries, just as the harbour porpoise (Brown et al. 2013). In Iceland white-beaked dolphins are the third most common bycatch species in the lumpfish (*Cyclopterus lumpus*) gillnet fishery, with 54 animals caught in 2013

(Palsson et al. 2015). Further bycatch evidence is reported from England and Wales where four white-beaked dolphins washed up on shore and the cause of death was determined as bycatch (Kirkwood et al. 1997). There are even small numbers reported caught in other types of gear, such as mid-water trawls, where they made up 1.5% of total dolphins caught (n = 71) around Ireland (Couperus 1997). White-beaked dolphin bycatch does not appear to be an issue that would be damaging populations in general, though it may be having an impact on localized individuals or small populations (MacLeod 2013).

Harbour porpoises are known to be extremely vulnerable to bycatch, particularly getting caught in gillnets (Brown et al. 2013) which are used extensively in sub-Arctic and Arctic waters to catch species such as cod (*Gadus morhua*), monkfish (*Lophius piscatorius*), and lumpfish. In Norwegian gillnet fisheries alone, it is estimated that 6900 harbour porpoises die as bycatch every year (Bjorge et al. 2013). In Iceland, just the lumpfish gillnet fishery accounts for an estimated annual bycatch of 551 harbour porpoises (Iceland Marine and Freshwater Institute 2018), and an estimated additional 1450–1650 end up as bycatch in the cod gillnet fishery (Palsson et al. 2015). Other sub-Arctic and Arctic fisheries, such as the salmon gillnet fishery in Alaska and many Greenlandic fisheries, are understudied in terms of bycatch. Harbour porpoise bycatch is likely high in understudied fisheries as well, though it is currently not properly documented (Moore et al. 2009). These high numbers of bycatch mortalities can mean that the deaths exceed the calculated "acceptable take" from the populations. In Norway for example, the aforementioned estimated annual bycatch would need to come from a population of over 400,000 individuals in order to be within sustainable limits, but it is considered unlikely that the population is that large (Bjorge et al. 2013).

Entanglement and bycatch of cetaceans does not only cause several issues for the animals, but for the fishers or fishing companies as well. Large whale collisions with fishing gear often lead to extensive gear damage, which in-turn leads to down-time for repairs, new investments, and loss of catch. In some cases, such as with set-net or traps/pots which are left in the water unattended for some time, the gear is just lost completely since it was carried away by the whale. This all equals a financial burden to the fishing companies. In the Canadian western North Atlantic alone, reported gear loss due to whales has been estimated to cost hundreds of thousands of dollars per year and could be upwards of one million dollars when including down-time losses[1] (Lien 1979; Lien and Aldrich 1982). From the fishers' perspective, bycatch of smaller cetaceans is generally less of an issue than that of the large whales. Even so, both harbour porpoise and white-beaked dolphin bycatch causes some net damage and potential downtime for repairs. Severely entangled animals, particularly dolphins, may need to be cut out of the net causing more extensive damage. Large, unrepairable holes cause the loss of approximately 2–3 lumpfish gillnets per season (March–April) in north Iceland alone (S. Karlsson, personal communication 2018). As one can imagine, due to these issues fishing companies then find it in their best interest to avoid cetacean entanglement or bycatch.

[1] Converted into current dollar estimates to account for inflation.

Several mitigation methods have been developed in attempt to minimize the conflict between cetaceans and fishing, and the impacts on both parties. There are several mitigation methods already in use in some places. One such measure is the use of "pingers": devices that make a noise in the best-known hearing range of the target cetacean species in order to alert them of fishing gear in the water (Harcourt et al. 2014). Porpoise/dolphin "pingers" have even been made mandatory in some fisheries around the world in attempt to reduce bycatch (Europa 2010; NOAA 2010) and are known to have some experimental and "real-world" success in reducing porpoise and dolphin bycatch (Cox et al. 2007). "Weak links" have also been developed for set-nets: links designed to break under the force of a large whale in order to avoid both entanglement and gear damage (Consortium for Wildlife Bycatch Reduction n.d.). In addition, in some places area closures have been put into effect, often limiting the time of year and type of fishing gear that is allowed to be used in certain areas that are considered important cetacean habitat, such a crucial feeding or calving areas (Vanderlaan et al. 2011).

With the changing Arctic environment and the warming of the Arctic Ocean, fish species are moving poleward, and therefore expanding the fishery possibilities. Currently the majority of commercial fishing takes place in sub-artic waters and "arctic corridors", and targets mainly sub-arctic/boreal fish species (Christiansen et al. 2014), but as these target species move further northwards into the Arctic, Arctic fishing is expected to expand. Modelling has predicted that the annual landed fish catch in the Arctic will increase by 39% by 2050 compared to the year 2000 (Lam et al. 2016). The expanding fishing industry in the Arctic potentially means greater risk to Arctic cetaceans with respect to entanglement and bycatch and therefore mitigation measures should be implemented to avoid increasing conflict.

5.2.4 *Other Conflicts Between Arctic Industries and Cetaceans*

5.2.4.1 Ship Noise

An increase in ship traffic in Arctic waters means more noise pollution in the ocean. Different types of vessels make different kinds of underwater noise, with most engines producing noise in low frequencies from 20–1000 Hz (Richardson et al. 1996). Therefore, animals communicating in this frequency range will be most affected in terms of masking of their communication. Masking is defined by Erbe et al. (2016) as: "The process by which the threshold of hearing for one sound is raised by the presence of another (masking) sound; and the amount by which the threshold of hearing for one sound is raised by the presence of another (masking) sound, expressed in dB". Communication masking is therefore depending on the hearing of the animals (their audiogram), the loudness of the signal, the frequency of the signal, the distance to the source (since sound attenuates with increasing distance), and the frequency of the signals the animals are using. Of the six example species, communication masking by ship noise is most relevant for blue whales, humpback whales, sperm whales, killer whales, and white-beaked dolphins, but it is even recently believed to interfere with harbour porpoise communication as well (Hermannsen et al. 2014). It is likely that ship noise can cause masking and communication difficulties for all 16 of the Arctic cetacean species, and therefore shipping routes through important habitat should be avoided where possible.

5.2.4.2 Whale Watching

Whale watching is an increasing industry around the world, including in sub-Arctic and Arctic waters. In general, whale watching can be considered a positive for the whales, increasing the general public's knowledge and the push to protect the whales and the ocean ecosystem by connecting people with them. With diminishing Arctic sea ice and shifts in species distribution, most likely due to climate change, whale watching in the sub-Arctic and Arctic is likely to continue to grow and evolve. Despite the positive image, there can be some negative consequences of this industry if not operated with care. Just as with increased shipping, increased whale watching also means more ship engine noise in the ocean and potential masking of whale communication for all types of whales, as well as potentially leading to more collisions between vessels and cetaceans (Parsons 2012). Behavioural changes around whale watching boats have also been noted, including changes in diving, breathing rate, swimming direction, swimming speed, feeding, and resting. In terms of the six example species, these behavioural changes have been specifically noted for humpback whales (Corkeron 1995), sperm whales (Richter et al. 2006), and orcas (Noren et al. 2009). To combat the impacts of whale watching, voluntary, or in some cases mandatory, whale watching code-of-conducts have been put in place, often limiting the distance a boat can get to the animals, the speed they can go, and the time they can spend there (Garrod and Fennell 2004).

5.2.4.3 Offshore Wind Development

Plans for offshore wind farms, to produce clean renewable energy by harvesting the Arctic's windy conditions, are underway in several sub-Arctic/Arctic countries including Russia, Norway, and Finland. These developments are being designed to withstand ice and harsh Arctic conditions. Of particular concern for marine mammals is the loud noise from pile-driving required to construct the wind turbines in some cases. This noise presents much of the same issues as seismic surveys: behavioural changes, potential communication masking (particularly of blue whale calls), and potential hearing impairment at close range (Bailey et al. 2010). Since windfarms are often constructed in shallow, coastal waters, the harbour porpoise is thought to be one of the more effected species. Research found they would likely show "minor disturbance" at a sound level of 90 dB re 1 μPa (or estimated to be 70 km away from the source), and "major disturbance" at a sound level of 155 dB re 1 μPa (or estimated to be 20 km from the source) (Dähne et al. 2013). As with other underwater noise issues, it is likely that most Arctic cetaceans could be affected by offshore wind farm construction depending on location, construction method, sound levels, and distance to the source. One way to eliminate the issues with construction is the creation of floating wind farm platforms that are just anchored to the sea floor (Roddier et al. 2010). This development also allows for turbines in deeper waters, to potentially choose sites with less impact on marine mammal habitats.

5.2.4.4 Depredation

Both sperm whales and orcas have a unique interaction with fisheries known as depredation. Depredation is when the whales specifically target the fishing boats as a food source, and essentially steal the fish off the lines or damage the fish that have been hooked. This is a learned behaviour that appears to be an issue all over the world, including in sub-Arctic and Arctic waters, as both species have a large distribution range. In Alaska, fish damage due to depredation by sperm whales has been found to occur in 46–65% of long-line sets when sperm whales are known to be present (Hill et al. 1999; Sigler et al. 2008). The target species of the long-line fishery is sablefish (black cod) (*Anoplopoma fimbria*). Similarly, orcas in Alaska and surrounding waters are depredating sablefish, as well as arrowtoothed flounder (*Atheresthes stomias*), Greenland turbot (*Reinhardtius hippoglossoides*) and several other long-line fished species.

Depredation has an effect on the catch of these long-line fisheries, and therefore there are cost implications to the companies. Where sperm whale depredation was recorded, there was an estimated 2% loss of sablefish catch in Alaskan waters (Sigler et al. 2008). Due to depredation from orcas, there is a predicted annual loss of 11–29% of sablefish, 10–22% of arrowtoothed flounder, and 22% of Greenland turbot in Alaska and surrounding waters (Peterson et al. 2013). The fact that sperm whale and killer whale depredation is overlapping in these areas means the accumulative impacts of this issue are even larger. These two species seem to be the only culprits of this in Arctic waters. In response to these losses, fisheries have changed their methods, moving away from long-lines towards the use of pot gear or trawls, or having altered their fishing season in order to try to mitigate depredation (Peterson et al. 2013). In addition to the loss of catch, depredation can also have an impact on fish stock assessments used to set fishing quotas, since depredation is causing unrecorded fish removal (Purves et al. 2004). Depredation is known to have been increasing in the past decade and should be considered an important issue in the aforementioned expanding Arctic fisheries.

Overall, there is a lot of interaction and potential conflicts between human industrial activity in the Arctic and cetacean species living there seasonally or year-round. As discussed, human presence in Arctic waters is likely to continue to expand to new locations, and generally increase. Continuous research, monitoring, and impact mitigation plans will be needed to maintain sustainable development in the Arctic that is taking into account the impacts industries can have on Arctic flora and fauna, with cetaceans being an example of just one group of animals that are known to be widely affected by industries and industrial development.

References

Akamatsu, T., Rasmussen, M. H., & Iversen, I. (2014). Acoustically invisible feeding blue whales in northern Icelandic waters. *The Journal of the Acoustical Society of America, 136*, 939–944.

Aksenov, Y., Popova, E. E., Yool, A., et al. (2017). On the future navigability of Arctic sea routes: High-resolution projections of the Arctic Ocean and sea ice. *Marine Policy, 75*, 300–317.

Allen, J. K., Peterson, M. L., Sharrard, G. V., et al. (2012). Radiated noise from commercial ships in the Gulf of Maine: Implications for whale/vessel collisions. *The Journal of the Acoustical Society of America, 132*(3), EL229–EL235.

Au, W. W. L., Ford, J. K. B., Horne, J. K., et al. (2004). Echolocation signals of free-ranging killer whales (*Orcinus orca*) and modeling of foraging for chinook salmon (*Oncorhynchus tshawytscha*). *The Journal of the Acoustical Society of America, 115*, 901–909.

Au, W. W. L., Kastelein, R. T., et al. (1999). Transmission beam pattern and echolocation signals of a harbor porpoise (*Phocoena phocoena*). *The Journal of the Acoustical Society of America, 106*, 3699–3705.

Bain, D. E., & Williams, R. (2006). *Long-range effects of airgun noise on marine mammals: Responses as a function of received sound level and distance*. Report by Sea Mammal Research Unit (SMRU), University of St Andrews, and University of Washington, p. 5.

Bailey, H., Senior, B., Simmons, D., et al. (2010). Assessing underwater noise levels during pile-driving at an offshore windfarm and its potential effect on marine mammals. *Marine Pollution Bulletin, 60*, 888–897.

Basran, C. J., Bertulli, C. G., Cecchetti, A., et al. (2019). First estimates of entanglement rate of humpback whales *Megaptera novaeangliae* observed in coastal Icelandic waters. *Endangered Species Research, 38*, 67–77.

Benjamins, S., Ledwell, W., & Davidson, A. R. (2012). Assessing changes in numbers and distribution of large whale entanglements in Newfoundland and Labrador, Canada. *Marine Mammal Science, 28*(3), 579–601.

Brown, S. L., Reid, D., & Rogan, E. (2013). A risk-based approach to rapidly screen vulnerability of cetaceans to impacts from fisheries bycatch. *Biological Conservation, 168*, 78–87.

Bjorge, A., Skern-Mauritzen, M., & Rossman, M. C. (2013). Estimated bycatch of harbour porpoise (*Phocoena phocoena*) in two coastal gillnet fisheries in Norway, 2006-2008. Mitigation and implications for conservation. *Biological Conservation, 161*, 164–173.

Bowyer, T. R., Boylan, M., Brodie, J., Brown, S., et al. (2017, November 9). *Letter to: Committee on energy and natural resources, United States Senate*. https://www.audubon.org/sites/default/files/arctic_refuge_science_letter_2017_11_09_final_00000003.pdf. Accessed 21 Sep 2018.

Butterworth, A., Clegg, I., & Bass, C. (2012). *Untangled – Marine debris: A global picture of the impact on animal welfare and of animal-focused solutions*. London: World Society for the Protection of Animals.

CAFF. (2017). *State of the arctic marine biodiversity report*. Conservation of Arctic Flora and Fauna International Secretariat, Akureyri, Iceland. 978-9935-431-63-9.

Cassoff, R. M., Moore, K. M., McLellan, W. A., et al. (2011). Lethal entanglement in baleen whales. *Diseases of Aquatic Organisms, 96*, 175–185.

Cerchio, S., & Dahlheim, M. (2001). Variation in feeding vocalizations of humpback whales, *Megaptera novaeangliae*, from Southeast Alaska. *Bioacoustics, 11*, 277–295. https://doi.org/10.1080/09524622.2001.9753468.

Christiansen, J. S., Mecklenburg, C. W., & Karamushko, O. V. (2014). Arctic marine fishes and their fisheries in light of global change. *Global Change Biology, 20*, 352–359.

Clausen, K. T., Wahlberg, M., Beedholm, K., et al. (2011). Click communication in harbour porpoises (*Phocoena phocoena*). *Bioacoustics, 20*(1), 1–28. https://doi.org/10.1080/09524622.2011.9753630.

Cole, T., Hartley, D., & Garron, M. (2006). *Mortality and serious injury determinations for baleen whale stocks along the eastern seaboard of the United States, 2000–2004*. U.S. Department of Commerce: NOAA National Marine Fisheries Service.

Consortium for Wildlife Bycatch Reduction. (n.d.). *Fishing technique modifications*. http://www.bycatch.org/research/consortium/fishing-gear-modi–fications. Accessed 15 Dec 2013.

Corkeron, P. J. (1995). Humpback whales (*Megaptera novaeangliae*) in Hervey Bay, Queensland: Behaviour and responses to whale-watching vessels. *Canadian Journal of Zoology, 73*(7), 1290–1299.

Couperus, A. S. (1997). Interactions between Dutch midwater trawl and Atlantic white-sided dolphins (*Lagenorhynchus acutus*) southwest of Ireland. *Journal of Northwest Atlantic fishery science, 22*, 209–218.

Cox, T. M., Lewison, R. L., Zydelis, R., et al. (2007). Comparing effectiveness of experimental and implemented bycatch reduction measures: The ideal and the real. *Conservation Biology, 21*(5), 1155–1164.

Cummings, W. C., & Thompson, P. O. (1971). Underwater sounds from the blue whale, *Balaenoptera musculus*. *The Journal of the Acoustical Society of America, 50*, 1193–1198.

Dähne, M., Gilles, A., & Lucke, K. (2013). Effects of pile-driving on harbour porpoises (*Phocoena phocoena*) at the fist offshore wind farm in Germany. *Environmental Research Letters, 8*(2013), 025002.

Di Iorio, L., & Clark, C. W. (2010). Exposure to seismic surveys alters blue whale acoustic communication. *Biology Letters, 6*, 51–54.

Erbe, C., Reichmuth, C., & Cunningham, K. (2016). Communication masking in marine mammals: A review and research strategy. *Marine Pollution Bulletin, 103*, 15–38.

Europa. (2010). *Summaries if EU legislation: Protecting cetaceans against incidental catch* (Council regulation (EC) No 812/2004) http://europa.eu/legislation_summaries/environment/nature_and_biodiversity/l66024_en.htm. Accessed 14 Dec 2013.

Felix, F., Haase, B., Davis, J. W., et al. (1997). A note on recent stranding and bycatches of sperm whales (Physeter macrocephalus) and humpback whales (Megaptera novaeangliae) in Ecuador. *Report for the Internation Whaling Commission, 47*, 917–919.

Fouche, G. (2009, March 20). Opposition grows to Norway's Arctic oil Search. *The Guardian*. https://www.theguardian.com/environment/2009/mar/20/oil-exploration-polar-circle-norway. Accessed 20 Sept 2018

Gautier, D. L., Bird, K. J., Charpentier, R. R., et al. (2009). Assessment of undiscovered oil and gas in the Arctic. *Science, 324*, 1175–1179.

Garrod, B., & Fennell, D. A. (2004). An analysis of whalewatching codes of conduct. *Annals of Tourism Research, 31*(2), 334–352.

Gedamke, J., Gales, N., & Frydman, S. (2011). Assessing risk of baleen whale hearing loss from seismin surveys: The effect of uncertainty and individual variation. *The Journal of the Acoustical Society of America, 129*(1), 496–506.

Goold, J. C., & Coates, R. F. W. (2006). *Near source, high frequency air-gun signatures*. A paper prepared for the IWC seismic workshop, St. Kitts, 24–25 May 2006.

Harcourt, R., Pirotta, V., Heller, G., et al. (2014). A whale alarm fails to deter migrating humpback whales: An empirical test. *Endangered Species Research, 25*, 35–42.

Heezen, B. C. (1953). Whales entangled in deep sea cables. *Deep Sea Research, 4*, 105–114.

Hermannsen, L., Beedholm, K., Tougaard, J., et al. (2014). High frequency components of ship noise in shallow water with a discussion of implications for harbor porpoises (*Phocoena phocoena*). *The Journal of the Acoustical Society of America, 136*, 1640–1653.

Hill, P. S., Laake, J. L., & Mitchell, E. (1999). *Results of a pilot program to document interactions between sperm whales and longline vessels in Alaska waters*. NOAA Technical Memorandum NMFS-AFSC-108.

Houser, D. S., Helweg, D. A., & Moore, P. B. W. (2001). A bandpass filter-bank model of auditory sensitivity in the humpback whale. *Aquatic Mammals, 27*, 82–91.

Huntington, H. P., Daniel, R., Hartsig, A., et al. (2015). Vessels, risks, and rules: Planning for safe shipping in the Bering Strait. *Marine Policy, 51*, 119–127.

Iceland Marine and Freshwater Institute. (2018). *Bycatch of seabirds and marine mammals in Lumpsucker Gillnets 2014–2017*. Report March 2018.

Jacobsen, J. K., Massey, L., & Gulland, F. (2010). Fatal ingestion of floating net debris by two sperm whales (*Physeter macrocephalus*). *Marine Pollution Bulletin, 60*, 765–767.
Jensen AS, Silber GK (2004). *Large whale ship strike database*. NOAA Technical Memorandum, NMFS-OPR-January 2014.
Kastelein, R. A., Bunskoek, P., Hagedoom, M., et al. (2002). Audiogram of a harbour porpoise (*Phocoena phocoena*) measured with narrow-band frequency modulated signals. *The Journal of the Acoustical Society of America, 112*, 334–344.
Kirkwood, J. K., Bennett, P. M., Jepson, P. D., et al. (1997). Entanglement in fishing gear and other causes of death in cetaceans stranded on the coasts of England and Wales. *The Veterinary Record, 141*, 94–98.
Laist, D. W., Knowlton, A. R., Mead, J. G., et al. (2001). Collisions between ships and whales. *Marine Mammal Science, 17*(1), 35–75.
Lam, V. W. Y., Cheung, W. W. L., & Sumaila, U. R. (2016). Marine capture fisheries in the Arctic: Winners or losers under climate change and ocean acidification. *Fish and Fisheries, 17*, 335–357.
Lien, J. (1979). *A study of entrapment in fishing gear: Causes and prevention*. Progress Report, March 1, 1979.
Lien, J., & Aldrich, D. (1982). *Damage to the inshore fishing gear in Newfoundland and Labrador by whales and sharks during 1981*. CAFSAC Marine Mammal Committee Meetings, St. John's NFL, May 18–19, 1982.
Lockyer, C. (2003). Harbour porpoise (*Phocoena phocoena*) in the North Atlantic: Biological parameters. *NAMMCO Science Publications, 5*, 71–90.
Lucke, K., Seibert, U., Lepper, P. A., et al. (2009). Temporary shift in masked hearing thresholds in a harbor porpoise (*Phocoena phocoena*) after exposure to seismic airgun stimuli. *The Journal of the Acoustical Society of America, 125*, 4060–4070.
MacLeod, C. D. (2013). *IV. White-beaked Dolphins in the Northeast Atlantic: A brief review of their ecology and potential threats to conservation status*. In Proceedings of the ECS/ASCOBANS/WDC workshop towards a conservation strategy for White-beaked Dolphins in the Northeast Atlantic. Document 4.1c.
Madsen, P. T., Mohl, B., Neilsen, B. K., et al. (2002). Male sperm whale behaviour during exposures to distant seismic survey pulses. *Aquatic Mammals, 28*(3), 231–240.
Magnúsdóttir, E. E., Rasmussen, M. H., Lammers, M. O., et al. (2014). Humpback whale songs during winter in subarctic waters. *Polar Biology, 37*, 427–433.
Magnúsdóttir, E. E., Rasmussen, M. H., Lammers, M. O., et al. (2015). Humpback whale (*Megaptera novaeangliae*) song unit and phrase repertoire progression on a subarctic feeding ground. *The Journal of the Acoustical Society of America, 138*, 3362–3374.
Malme, C. I., Miles, P. R., Tyack, P., et al. (1985). *Investigation of the potential effects of underwater noise from petroleum industry activities on feeding humpback whale behavior* (OSC Study MMS 85-0019). Prepared for U.S. Department of the Interior Minerals Management Service.
McCauley, R. D., Fewtrell, J., Duncan, A. J., et al. (2000). *Marine seismic surveys: Analysis and propagation of air-gun signals; and effects of air-gun exposure on humpback whales, sea turtles, fishes and squid*. Report R99-15. Prepared for Australian Petroleum Production Exploration Association.
McDonald, M. A., Hildebrand, J. A., & Webb, S. C. (1995). Blue and fin whales observed on a seafloor array in the Northeast Pacific. *The Journal of the Acoustical Society of America, 96*(2), Pt.1.
McKenna, M. F., Calambokidis, J., Oleson, E. M., et al. (2015). Simultaneous tracking of blue whales and large ships demonstrates limited behavioral responses for avoiding collision. *Endangered Species Research, 27*, 219–232.
Mellinger, D. K., & Clark, C. W. (2003). Blue whale (*Balaenoptera musculus*) sounds from the North Atlantic. *The Journal of the Acoustical Society of America, 114*(2), 1108–1119.
Meyer, M. A., Best, P. B., Anderson-Reade, M. D., et al. (2011). Trends and interventions in large whale entanglement along the South African coast. *African Journal of Marine Science, 33*(3), 429–439.

Miller, P. J. O., Johnson, M. P., & Madsen, P. T. (2009). Using at-sea experiments to study the effects of airguns on the foraging behavior of sperm whales in the Gulf of Mexico. *Deep-Sea Research Part I, 56*, 1168–1181.

Miller, J. H., Potter, D. C. (2001). *Active high frequency phased-array sonar for whale shipstrike avoidance: target strength measurements*. MTS/IEEE Oceans 2001. An Ocean Odyssey. Conference proceedings (IEEE Cat. No.01CH37295), Honolulu, HI, USA (4): 2104–2107. https://doi.org/10.1109/OCEANS.2001.968324

Milne, R. (2017, June 21). Norway opens up record 93 blocks for Arctic oil exploration. In *Financial Times*. https://www.ft.com/content/a120d578-567e-11e7-9fed-c19e2700005f. Accessed 21 Sept 2018.

Monnahan, C. C., Branch, T. A., & Punt, A. E. (2015). Do ships threaten the recovery of endangered eastern North Pacific blue whales? *Marine Mammal Science, 31*(1), 279–297.

Møhl, B., Wahlberg, M., Madsen, P. T., et al. (2003). The monopulsed nature of sperm whale clicks. *The Journal of the Acoustical Society of America, 114*, 1143–1154.

Moore, M. J., & van der Hoop, J. M. (2012). The painful side of trap and fixed net fisheries: Chronic entanglement of large whales. *Journal of Marine Biology, 2012*, 1–4.

Moore, J. E., Wallace, B. P., Lewison, R. L., et al. (2009). A review of marine mammal, sea turtle and seabird bycatch in USA fisheries and the role of policy in shaping management. *Marine Policy, 33*, 435–451.

Nachtigall, P. E., Mooney, T. A., & Taylor, K. A. (2008). Shipboard measurements of the hearing of white-beaked dolphins, *Lagenorhynchus albirostris*. *The Journal of Experimental Biology, 211*, 642–647.

Nieukirk, S. L., Stafford, K. M., Mellinger, D. K., et al. (2004). Low-frequency whale and seismic airgun sounds recorded in the mid-Atlantic. *The Journal of the Acoustical Society of America, 115*(4), 1832–1843.

Nielsen, N. H., Teilmann, J., Sveegaard, S., et al. (2018). Oceanic movements, site fidelity and deep diving in harbour porpoises from Greenland show limited similarities to animals from the North Sea. *Marine Ecology Progress Series, 597*, 259–272.

Neilson, J. L., Straley, J. M., Gabriele, C. M., et al. (2009). Non-lethal entanglement of humpback whales (Megaptera novaeangliae) in fishing gear in northern Southeast Alaska. *Journal of Biogeography, 36*, 452–464.

NOAA – Fisheries Service: Protected Resources Division. (2010). *Harbor porpoise take reduction plan*. http://www.nero.noaa.gov/prot_res/porptrp/ptci.html. Accessed 30 Nov 2013.

Noren, D. P., Johnson, A. H., Rehder, D., et al. (2009). Close approaches by vessels elicit surface active behaviors by southern resident killer whales. *Endangered Species Research, 8*(3), 179–192.

Pace, D. S., Miragliuolo, A., & Mussi, B. (2008). Behaviour of a social unit of sperm whales (Physeter macrocephalus) entangled in a driftnet off Capo Palinuro (Southern Tyrrhenian Sea, Italy). *Journal of Cetacean Research and Management, 10*(2), 131–135.

Palsson, O. K., Gunnlaugsson, Th., & Olafsdottir, D. (2015). Meðafli sjófugla og sjávarspendýra í fiskveiðum á Íslandsmiðum (By-catch of sea birds and marine mammals in Icelandic fisheries). Hafrannsóknir nr. 178.

Paraskova, T. (2017, October 19). *Russia goes all in on arctic oil development*. https://oilprice.com/Energy/Crude-Oil/Russia-Goes-All-In-On-Arctic-Oil-Development.html. Accessed 21 Sept 2018.

Parsons, E. C. M. (2012). The negative impacts of whale-watching. *Journal of Marine Biology, 2012*, 1–9.

Payne, R., & McVay, S. (1971). Songs of humpback whales. *Science, 173*, 585–597.

Peterson, M. J., Mueter, F., Hanselman, D., et al. (2013). Killer whale (*Orcinus orca*) depredation effects on catch rates of six groundfish species: Implications for commercial longline fisheries in Alaska. *ICES Journal of Marine Science, 70*(6), 1220–1232.

Pirotta, E., Brookes, K. L., & Graham, I. M. (2014). Variation in harbour porpoise activity in response to seismic survey noise. *Biology Letters, 10*(5), 20131090.

Purves, M. G., Agnew, D. J., Balguerias, E., et al. (2004). Killer whale (*Orcinus orca*) and sperm whale (*Physeter macrocephalus*) interactions with longline vessels in the Patagonian toothfish fishery at South Georgia, South Atlantic. *CCAMLR Science, 11*, 111–126.
Rasmussen, M. H., Lammers, M., Beedholm, K., et al. (2006). Source levels and harmonic content of whistles in white-beaked dolphins (*Lagenorhynchus albirostris*). *The Journal of the Acoustical Society of America, 120*, 510–517.
Rasmussen, M. H., & Miller, L. A. (2002). Whistles and clicks from white-beaked dolphins, *Lagenorhynchus albirostris* recorded in Faxaflói Bay. *Aquatic Mammals, 28*, 78–89.
Rasmussen, M. H., Miller, L. A., & Au, W. W. L. (2002). Source levels of clicks from free-ranging white beaked dolphins (*Lagenorhynchus albirostris* Gray 1846) recorded in Icelandic waters. *Journal of the Acoustical Society of America, 111*, 1122–1125.
Redfern, J. V., McKenna, M. F., Moore, T. J., et al. (2013). Assessing the risk of ships striking large whales in marine spatial planning. *Conservation Biology, 27*(2), 292–302.
Richardson, W. J., Greene, C. R., Jr., Malme, C. I., et al. (1996). *Marine mammals and noise*. San Diego: Academic. https://doi.org/10.1017/S0025315400030824.
Richter, C., Dawson, S., & Slooten, E. (2006). Impacts of commercial whale watching on male sperm whales at Kaikoura, New Zealand. *Marine Mammal Science, 22*(1), 46–63.
Robbins, J., Mattila, D. K. (2001). Monitoring entanglements of humpback whales (*Megaptera novaeangliae*) in the Gulf of Maine on the basis of caudal peduncle scarring (Document SC/53/NAH25). Unpublished report to the 53rd Scientific Committee Meeting of the International Whaling Commission. Hammersmith, London.
Robbins, J., & Mattila, D. (2004). *Estimating humpback whale (Megaptera novaaeangliae) entanglement rates on the basis of scar evidence*. Report to the Northeast Fisheries Science Center National Marine Fisheries Service.
Robbins, J. (2009). Entanglement scarring on North Pacific humpback whales. In J. Calambokidis (Ed.), *Symposium on the results of the SPLASH humpback whale study*. Final Report and Recommendations, Quebec City, Canada.
Robbins, J., Barlow, J., Burdin, A. M., et al. (2007). Preliminary minimum estimates of humpback whale entanglement frequency in the North Pacific Ocean based on scar evidence. (Unpublished Document SC/59/BC15).
Roddier, D., Cermelli, C., Aubault, A., et al. (2010). WindFloat: A floating foundation for offshore wind turbines. *Journal of Renewable and Sustainable Energy, 2*, 033104.
Rolland, R. M., McLellan, W. A., Moore, M. J., et al. (2017). Fecal glucocorticoids and anthropogenic injury and mortality in North Atlantic right whales *Eubalaena glacialis*. *Endangered Species Research, 34*, 417–429.
Samarra, F. I. P., Deecke, V. B., Vinding, K., et al. (2010). Killer whales (*Orcinus orca*) produce ultrasonic whistles. *The Journal of the Acoustical Society of America, 128*, EL205–EL210.
Schnitzler, J., Rasmussen, M. H., Lucke, K., et al. (2018). *Impact of vibroseismic underwater sound on the behaviour of baleen whales*. In ESOMM conference, Hague, the Netherlands.
Sigler, M. F., Lunsford, C. R., Straley, J. M., et al. (2008). Sperm whale depredation of sablefish longling gear in the Northeast Pacific Ocean. *Marine Mammal Science, 24*(1), 16–27.
Simon, M., Wahlberg, M., & Miller, L. (2007). Echolocation clicks from killer whales (*Orcinus orca*) feeding on herring (*Clupea harengus*) (L). *The Journal of the Acoustical Society of America, 121*, 749–752.
Stone, C. J., & Tasker, M. L. (2006). The effects of seismic airguns on cetaceans in UK waters. *Journal of Cetacean Research and Management, 8*(3), 255–263.
Strager, H. (1995). Pod-specific call repertoires and compound calls of killer whales, *Orcinus orca* Linnaeus, 1758, in the waters of northern Norway. *Canadian Journal of Zoology, 73*, 1037–1047.
Szymanskia, M. D., Bain, D. E., Wong, S., et al. (1999). Killer whale (*Orcinus orca*) hearing: Auditory brainstem response and behavioral audiograms. *The Journal of the Acoustical Society of America, 106*, 1134–1141.

Thompson, T. J., Winn, H. E., & Perkins, P. J. (1979). Mysticete sounds. In H. E. Winn & B. L. Olla (Eds.), *Behavior of marine animals, Vol. 3: Cetaceans*. New York: Plenum.

Thompson, P. O., Cummings, W. C., & Ha, S. J. (1986). Sounds, source levels, and associated behaviour of humpback whales, Southeast Alaska. *The Journal of the Acoustical Society of America, 80*(3), 735–740.

van der Hoop, J. M., Vanderlaan, A. S. M., & Cole, T. V. N. (2015). Vessel strikes to large whales before and after the 2008 ship strike rule. *Conservation Letters, 8*(1), 24–32.

van der Hoop, J., Moore, M., Fahlman, A., et al. (2013). Behavioral impacts of disentanglement of a right whale under sedation and the energetic costs of entanglement. *Marine Mammal Science, 30*, 282–307.

Vanderlaan, A. S. M., Smedbol, R. K., & Taggart, C. T. (2011). Fishing-gear threat to right whales (*Eubalaena glacialis*) in Canadian waters and the risk of lethal entanglement. *Canadian Journal of Fisheries and Aquatic Sciences, 68*, 2174–2193.

Verfuß, U. K., Miller, L. A., Pilz, P. K., et al. (2009). Echolocation by two foraging harbour porpoises (*Phocoena phocoena*). *The Journal of Experimental Biology, 212*, 823–834.

Villadsgaard, A., Wahlberg, M., & Tougaard, J. (2007). Echolocation signals of wild harbour porpoises, *Phocoena phocoena*. *The Journal of Experimental Biology, 210*, 56–64.

Visser, I. N. (2000). Killer whale (*Orcinus orca*) interactions with longline fisheries in New Zealand waters. *Aquatic Mammals, 26*(3), 241–252.

Visser, I. N. (1998). Prolific body scars and collapsing dorsal fins on killer whales (*Orcinus orca*) in New Zealand waters. *Aquatic Mammals, 24*(2), 71–81.

Weir, C. R. (2008). Overt responses of humpback whales (*Megaptera novaeangliae*), sperm whales (*Physeter macrocephalus*), and Atlantic spotted dolphins (*Stenella frontalis*) to seismic exploration off Angola. *Aquatic Mammals, 34*(1), 71–83.

Williams, R., & O'hara, P. (2010). Modelling ship strike risk to fin, humpback and killer whales in British Columbia, Canada. *Journal of Cetacean Research and Management, 11*(1), 1–8.

Wilson, D. E., & Mittermeier, R. A. (2014). Blue whale; humpback whale; sperm whale; Orca; white-beaked dolphin; harbour porpoise. In *Handbook of the mammal of the world. Vol 4. Sea mammals*. Barcelona: Lynx Edicions.

Wijnen, P. (2017, August 4). Survey: Majority choose climate before oil. In *Norway Today*. http://norwaytoday.info/news/climate-oil-survey-majority/. Accessed 20 Sept 2018.

Part II
Transport Infrastructure

Chapter 6
Social Responsibility Practice of the Evolving Nature in the Sustainable Development of Arctic Maritime Operations

Antonina Tsvetkova

Abstract This qualitative single-case study aims to explore how supply chain operations incorporate the social responsibility aspect in response to contextual influence and what possible effects does the implementing of social responsibility principles bring into SCM.

The study presents the implementation of social responsibility into SCM practices in the Russian Arctic context. Data from 22 semi-structured interviews, personal observations and archival materials are interpreted through the institutional logics approach.

The study reveals how social responsibility principles evolve in the existing SCM practice and enable supply chains to contribute to the needs of local communities in terms of the values of the society. The findings show that social responsibility initiatives in the existing SCM practice became possible after the satisfaction of economic and environmental concerns of cargo transportation in the Russian Arctic. Further, the case study indicates how contextual challenges make a company reconsider its core competencies and the role of supply chain practices to be more resilient and socially responsible. The findings reveal that social responsibility initiatives gain a more fertile ground for further development when they contribute also to strengthening financial performance in the supply chain.

More empirical studies on social responsibility in achieving SCM sustainability are suggested.

Keywords Sustainable supply chain management · Social sustainability · Institutional logics · Case study · Arctic shipping · Mono-towns

---

A. Tsvetkova (✉)
Faculty of Logistics, Molde University College, Molde, Norway
e-mail: antonina.tsvetkova@himolde.no

E. Pongrácz et al. (eds.), *Arctic Marine Sustainability*,
Springer Polar Sciences, https://doi.org/10.1007/978-3-030-28404-6_6

6.1 Introduction

Maritime transport plays an essential role in facilitating global trade. Approximately 80% of international trade by volume and over 70% by value is carried out by sea and is handled by seaports worldwide. One of the primary concerns of shipping management is to ensure cost-effective cargo movements between geographical locations. A relatively recent focus within the supply chain literature has addressed the potentiality of maritime operations towards sustainable development with ample room for understanding the corresponding challenges and opportunities.

The majority of the extant literature on sustainable or "green" supply chains has primarily focused on issues in making supply chain operations environmentally friendly through the minimization of environment impact and fuel consumption (Carter and Rogers 2008; Silvestre 2015; Lam and Lai 2015; Mansouri et al. 2015) to improve economic performance.

At the same time, the concept of sustainability encompasses not only the environmental and economic dimensions but also the social impact of supply chain operations. The balancing of the three equally dimensions – economic, environmental, and social – has been identified as crucial in making management decisions and companies' performance for achieving sustainability in supply chain operations. Although previous research on sustainable SCM has advanced the understanding of issues and possibilities of sustainable supply chains, there is, however, still to date, a significant shortfall in the literature regarding the social aspect of sustainability in SCM. While the intensity of research in the environmental dimension has recently more than tripled (Gurtu et al. 2015), the social aspect has received extremely limited attention that was stressed by many scholars (Seuring and Müller 2008; Wu and Pagell 2011; Beske 2012; Sarkis 2012; Ahi and Searcy 2015, Mani et al. 2016), creating difficulty in measuring advances in sustainable SCM practices (Davidson 2011).

Further, the social aspect has been considered as closely related to the environmental concern when there can happen negative social effects due to the water and coastal pollution affecting tourism and increased risk of diseases in humans through eating seafood from polluted areas (Lam and Lim 2016). Thus, previous research has implied the achievement of social sustainability as an indirect contribution by reducing environmental impact and thereby improving the quality of life (Mansouri et al. 2015). Just a few researchers have paid attention to this gap and emphasized the importance of embracing social responsibility considerations by logistics management (Carter and Jennings 2002; Murphy and Poist 2002). Social responsibility generally means that an organization's behaviour needs to be measured by more than its economic desirability but how it affects the public overall, customers, and more importantly local communities. However, previous research to date has mostly focused on the proposition of conceptual frameworks, which ties together standalone concepts like environment, ethics, working conditions, human rights, safety issues and philanthropy (Carter and Jennings 2002), as well as considering external pressures from regulations and stakeholders (Miao et al. 2012). Despite the advantages and contributions of these frameworks, there is still a lack of understanding of how organizations address the role of SCM as socially responsible and develop the social aspect in supply chain operations in real practice over time as a result of the interaction of supply chain operations, pressures of regulations and context.

Addressing the above-mentioned shortcomings in the literature on sustainability of supply chains, *this study aims to explore how supply chain operations incorporate the social responsibility aspect in response to contextual influence and what possible effects of implementing social responsibility principles into SCM.*

In doing so, this study presents an empirical case of the development of social responsibility principles in the existing SCM practice by a mining company in the Russian Arctic. The Russian Arctic provides a unique empirical setting, where mining companies are challenged by harsh climatic conditions, remoteness, sparse transportation links with other regions and global markets, and a limited amount of suppliers and logistics providers. To ensure regular deliveries of cargoes for their own needs and their customers, they have to apply specific supply chain strategies that enable to meet not only industrial activities but also the social needs of local communities in extremely remote Arctic areas. Maritime transport is mostly the only connecting artery for the survival of both the industry and local communities but is challenged by many regulatory norms.

This study employs the institutional logics approach, which is an alternative theoretical lens to previous research that has focused on sustainable SCM practices and social responsibility. This approach is helpful in revealing conditions to which social responsibility begins to permeate supply chain operations, how it can be incorporated and why certain practices are chosen without an obvious economic return (DiMaggio and Powell 1983). It also provides insights into the role of different actors in the development of sustainable supply chains.

This study is organized as follows. The next section presents knowledge of social sustainability in supply chains in the extant literature. This is followed by the theoretical framework taken from the institutional logics approach. The fourth section describes the research method. Then the context and the case study are presented. The findings are discussed in the following section. The study concludes with implications for theory and practice, as well as several guidelines for future research opportunities.

6.2 Social Sustainability in Supply Chains: Literature Review

The concept of sustainability is defined as meeting today's needs without compromising the future generation needs (WCED 1987) that integrates economic, environmental and social responsibilities. Sustainability requires the balancing of these three equally important dimensions.

Emphasizing the stakeholders' interests, Seuring and Müller (2008, p. 1700) define sustainable supply chain management (SSCM) as

> the management of material, information and capital flows as well as cooperation among companies along the supply chain while taking goals from all three dimensions of sustainable development, i.e., economic, environmental, and social, into account which are derived from customer and stakeholder requirements.

A more recent definition by Hassini et al. (2012, p. 70) combines the desired performance with the stakeholders' interests as

> The management of supply chain operations, resources, information, and funds in order to maximize the supply chain profitability while at the same time minimizing the environmental impacts and maximizing the social well-being.

To turn a supply chain into a sustainable supply chain, cooperation and integration are needed at all stages: from raw materials purchase to end customers' consumption (Seuring and Müller 2008) that can be feasible through partnerships based on trust. Additionally, several scholars have emphasized that sustainable supply chains can occur only through learning, change and innovative solutions (Pagell and Wu 2009; Tsvetkova 2011; Silvestre 2015).

Over the past decade, an increasing number of research has addressed challenges and issues in the more sustainable management of the supply chains. Recent literature reviews have pointed to the intensity of research in the environmental dimension of sustainability to minimize environmental impact and fuel consumption, thereby giving rise to "green" SCM (Seuring and Müller 2008; Quarshie et al. 2016). Most of this research has narrowed down the social sustainability to the environmental concern regarding possible negative effects from the environmental pollution on human health, safety and quality of life. However, as emphasized by many scholars, the social aspect itself has been rarely addressed in the SCM studies (Seuring and Müller 2008; Wu and Pagell 2011; Beske 2012; Sarkis 2012; Ahi and Searcy 2015, Mani et al. 2016).

The transition from mostly environmental concerns towards a more sustainable perspective implies the inclusion of the social aspect into consideration. This can create greater managerial challenges due to the complexity of involving a wide range of stakeholders with various motives, goals, and demands that may interpret the same situation differently (Matos and Hall 2007) with the appearance of so-called "stakeholder ambiguity" (Hall and Vredenburg 2005).

Scholars (Sharma and Ruud 2003) suggest viewing social sustainability as an ethical code of conduct for human survival and future development that needs to be accomplished *"in a mutually inclusive and prudent way."* Further, in the supply chain literature, the social aspect of sustainability has been defined from a corporate social responsibility (CSR) perspective (Carter and Jennings 2002; Ciliberti et al. 2008). Most studies on social responsibility in SCM have focused on purchasing decisions and social issues due to incorrect behaviour of suppliers, including ethics and safety conditions, emphasizing thereby supplier relationships as a challenge towards social sustainability (Carter and Jennings 2001; Boyd et al. 2007; Ciliberti et al. 2008). Another issue of social sustainability has been addressed to human rights, working conditions welfare and labour safety (Quarshie et al. 2016).

Further, Seuring and Müller (2008) argue that for establishing a sustainable supply chain, an organization should seek a strategy that is beneficial in terms of social, environmental and economic issues, dealing with trade-offs between these three dimensions and establishing minimum requirements for suppliers. However, little attention is paid to the potential influence of contextual factors such as firm location, regulatory norms, local community and network ties, and all other taken-for-granted values. There is still an issue within the sustainable supply chain discourse that is under what conditions they emerge and when they can succeed. At the same time, strategy implementation defines the nature of the interaction between the supply chain and context or the external environment in which strategies on sustainability are developed. Changes in context may also make it necessary to change strategy implementation or formulation that may, in turn, change the links between all supply chain players, including local communities (Tsvetkova and Gammelgaard 2018). Further, Lawrence and Suddaby (2006) argue that research often neglects that, in certain cases, companies are able to react to contextual pressures by incorporating new SCM practices and own behavioural norms.

Social responsiveness or responsibility of organizations can be based on the obligation to

> pursue those policies, to make those decisions or to follow those lines of actions that are desirable in terms of the objectives and values of our society (Bowen 1953, p. 6).

In supply chains, which link together various organizations from different fields of business with different goals and ways of managing, common understanding of social responsibility becomes a major challenge for all supply chain players, including suppliers, manufacturers, customers, and society or local communities. Thus, the social aspect concerns fair opportunities, involvement inside and outside the community, not only the boundaries of a single organization. At the same time, despite both topics – SCM and CSR – have been increasingly observed in the literature, little attention is paid to the understanding of how SCM can contribute to the needs of society or local communities through incorporating social responsibility considerations. While sustainability research emphasizes now the importance of considering interactions of all the three sustainable dimensions, little is still known about possible effects of integrating SCM and social responsiveness on local communities (the society) within a supply chain. That makes our knowledge of understanding of sustainability insufficient to create truly sustainable supply chains (Pagell and Shevchenko 2014).

To investigate social responsibility in SCM practices, this study applies the lenses of an institutional logics perspective as outlined in the section below.

6.3 Theoretical Framework: Social Responsibility in SCM through Institutional Logics

One of the main features of sustainable SCM is that it is based on the inter-organizational field that affects and is affected by the interaction and integration between different organizations across the supply chain (Sarkis et al. 2011). This study argues that a greater understanding of these interactions between multiple tiers of supply chain actors can increase the effective implementation of sustainable SCM practices. In turn, this creates heterogeneity of goals, motives demands and ways of managing within the supply chain. This study uses the concept of "institutional logics" to understand the reasons for this heterogeneity in behaviour of different actors. Thornton and Ocasio (1999, p. 804) have defined institutional logics as.

> "the socially constructed, historical patterns of material practices, assumptions, values, beliefs, and rules by which individuals produce and reproduce their material subsistence, organize time, and space, and provide meaning to their social reality". Therefore, "rather than positing homogeneity and isomorphism in organizational fields, the institutional logics approach views any context as potentially influenced by contending logics of different societal sectors".

Thus, institutions create expectations that define legitimate and appropriate actions for organizations, i.e. what is acceptable behaviour (DiMaggio and Powell 1991), and also form the logics by which laws, rules, and taken-for-granted behavioural expectations appear natural and permanent (Zucker 1987). It then affects how organizations make decisions. The previous studies have shown the possible dynamics of institutional logics in terms of their evolution over time (Thornton and Ocasio 1999) and also in terms of the contradictions and competition between the different logics at any one point in time (Greenwood et al. 2011).

Within the supply chain, organizations have to deal with different societal levels like markets, professions, corporations, communities and so on, which are associated with distinctive institutional logics (Scott 2014). Each logic defines a specific set of behavioural models for motivating social entities and organizing institutional and business environments (Thornton et al. 2012).

In the supply chain context, it can be argued that encouraging organizations to think more sustainably creates prerequisites for new logic, which tries to replace, compete with or complement other dominant logics like market and financial logics. Thus, the institutional logic approach is helpful to investigate the role and behaviour of various stakeholders in facilitating or hindering the implementation of sustainability into SCM; however, to date, it is mostly neglected in the SCM literature (Sayed et al. 2017).

6.4 Method

6.4.1 Research Design

A qualitative single-case was chosen to explore a social sustainability practice in SCM within Russian Arctic shipping in order to achieve a deep analysis, to gain experience with issues related to this context, and to develop knowledge by connecting these issues with the existing literature. This social sustainability practice was incorporated into SCM and maritime operations by a Russian mining and metallurgic company, whose core activities are extraction and processing of mineral resources, as well as production, marketing, and sales of non-ferrous and precious metals (hereafter, the focal company). Within Russian Arctic shipping chains, this company was the main actor in the elaboration and implementation of social sustainability into SCM.

The exploratory case study research method was employed to capture the real-life contextual settings of the remote Russian Arctic areas where local industries and communities are isolated from other areas with developed infrastructure and global market, and maritime transport is the only connecting artery. It helped in identifying the potential influence of the context on the development of social sustainability in the focal company's SCM and its further effects on local communities. Additionally, this context was argued to have greater potential for understanding the interaction of different institutional logics of supply chain players' behaviour involved in the development of social responsibility initiatives into maritime operations by the focal company.

One of the misunderstandings about the single case study is that its findings obtained from specific contextual factors can hardly contribute to theoretical knowledge, as one cannot generalize (Flyvbjerg 2006). However, a single-case is useful for tracing and in-depth understanding the process by which a phenomenon takes place (Maxwell 1996; Bennett and Elman 2010) from bounded real-world settings and interpreting them *"in terms of the meanings people bring to them"* (Denzin and Lincoln 2005). Further, this single case study approach enabled the collection of rich data from multiple organizations within a supply chain in Russian Arctic shipping and rigorous analysis of evidence on sequences and conjunctures of events within the selected case.

6.4.2 Data Collection

This study was accomplished by using multiple data sources. First, a literature review was conducted to understand the state of previous research related to the topic of this study. Then, 22 semi-structured and in-depth face-to-face interviews were conducted to gain more insights into the development of social responsibility initiatives within the focal company's maritime operations in the Russian Arctic.

The interviews were targeted at the management personnel of the focal company, port terminal operators, suppliers, and vessel administration, as well as local people, and regional authorities. The interviewees were selected on the basis of their involvement in the focal company's supply chain operations related to social sustainability as well as accessibility. The interviews emphasized tracing events and actions associated with the development of social responsibility initiatives in SCM, additionally to identify the role of different supply chain actors in these events and actions. The interviews were arranged in the port of Murmansk, St. Petersburg, the port of Dudinka during five periods: May 2014, November 2014, October 2015, December 2015 and then May 2016. The interviews were conducted in Russian and then translated into English. All the interviews were hand-written and recorded after receiving the consent of each interviewee to be transcribed later. The interviewees (excepting for two fishermen who were interviewed on an ice floe at the mouth of the Yenisei River) were provided with a draft of interview protocols and transcripts via email to ensure the validity of descriptions and interpretations. If it was necessary and feasible, follow-up interviews with additional questions were conducted via email, telephone or in person to obtain more in-depth data.

Also, various secondary sources like trade journals, the focal company's annual reports, internal archival materials, press releases and official websites were consulted for collecting data.

Further, the findings were supported by personal observations during a trip on board the container vessel, ice-class Arc7, owned by the focal company, on its regular voyage from Murmansk port to Dudinka port between April 28 and May 06, 2016. Data collected during this voyage on board helped reveal different and often-contradictory institutional logics of behaviour among all stakeholders involved in developing social sustainability within maritime operations. Further, personal observations were made during several full-day visits to the focal company's port terminal and the interviewees' own offices at the focal company's site.

The multiple data sources and several periods of interviewing allowed revealing the process of how social responsibility was incorporated into the focal company's SCM practice by different initiatives and operational activities.

6.4.3 Data Analysis

Data collected from multiple sources appeared as fragments of a whole story that were combined together to understand how social supply chain responsibility evolved in this particular context.

The data analysis was guided by identifying the main institutional logics embedded in the data and evaluating their effects on the incorporation of social responsibility considerations into the focal company's supply chain. It allowed revealing the interplay between the focal company's actions, state regulatory pressures, as well as social challenges and needs of local communities in remote Russian Arctic regions. When constructing the case story, the meaning of interview contents was essential for interpreting organizational actions and behaviour of all supply chain players involved in the process of social sustainability in SCM. The case presentation was based on many qualitative pieces of data, including individual opinions, perceptions and experiences of the interviewees. Interview citations were used to support the claims. The data analysis was constructed in a chronological way focusing on prerequisites, sequences and conjunctures of the events within the selected case.

The case story starts with the description of contextual settings of remote Russian Arctic regions, including the main features and challenges of local industries and communities. Then the process of incorporating social responsibility into the focal company's SCM was presented.

6.5 Case Presentation

6.5.1 Contextual Settings of the Russian Arctic – Interplay of Local Industries and Communities

The focal company's manufacturing is located in an extremely remote northern Russian area close to the prolific regional mineral resources. It makes the resource extraction and development more efficient to organize supply chains for processed products than for raw materials given that they are of a lower quantity and higher value (Plaizier et al. 2012). The focal company's customers are steel producers, hydropower utilities, and machine-building plants in Europe, Asia and North America, which use the focal company's mineral and metal production as semi-product inputs for their industrial processes.

One of the main features of the northern enterprises like the focal company is that they created mono-industry towns (mono-towns) around themselves during the Soviet period. Mono-towns in the Russian Arctic region represent specialized urban settlements where the socio-economic development mostly or fully depends on the performance of one or a few town-forming enterprises. This phenomenon obtained attention after the Soviet Union collapse, which has resulted in a worsening of the socio-economic situation in mono-towns. The transition towards a market economy broke the existing linkages that provided the functioning of the dominant enterprises and revealed their weaknesses. Enterprises were not able to face the tough open-economy rivalry due to their uncompetitive production, obsolete facilities and infrastructure, the state non-participation and improper management (World Bank Report 2010). Unlike in the Soviet Union where town-forming enterprises were providing jobs and social services to local residents, then in 2000s many companies stopped performing such social function (Institute of Regional Policy 2008). However, the focal company took social responsibility for the local residents living in the quite large mono-industry town around the focal company's business activities, including social infrastructure and delivery of essential goods and even food products.

There has been only one transport connection through the Arctic waters, namely the sea segment between the ports of Dudinka, Murmansk and Arkhangelsk (see Appendix). The focal company's primary concern was that its supply chain could ensure regular year-round deliveries of products to customers as well as of cargoes from suppliers to meet its manufacturing needs. Thus, ensuring the regular cargo deliveries matter not only to the focal company's manufacturing operations but also to the survival of local people.

6.5.2 Contextual and Regulatory Challenges as Prerequisites for Implementing New Supply Chain Practices

The location in the northern latitudes implies specific contextual challenges like remoteness, long distances, severe arctic climate, sparse transportation networks, and lack of transport infrastructure. There is an utterly limited choice of suppliers and transport links with other regions, mostly via sea routes, complicated navigation through ever-changing ice conditions. Additionally, only ice-strengthened vessels and icebreaker assistance are required for operating in the icy waters among hummocks. These challenges, the so-called "lock-ins", made the focal company's cargo deliveries vulnerable to possible disruptions and increased delivery time, isolating the focal company from its customers and global markets, as well as local residents from suppliers of food and essential commodities.

Another challenge was related to state regulation and state involvement in cargo shipping in Russian Arctic waters. The focal company's mineral extraction and manufacturing activities are major components of the state and regional economic development because of the large-scale exploitation of natural resources. Its cargo flows, including the finished metal products to the global market and materials/equipment to meet the manufacturing, made up the bulk of the total cargo shipped on that sea segment; thereby affecting the development of cargo shipping in the western part of the Russian Arctic. Thus, the state is a principal stakeholder of the focal company's activities. It is worth noting that the state used to play two roles – first as the regulator of cargo shipping in the Russian Arctic waters and second as the sole supplier of transport infrastructure, including ice-strengthened vessels and icebreaker assistance.

Over several decades up to the beginning of the 2000s, a state-owned shipping company based in Murmansk was the only sea carrier providing ice-class vessels and nuclear assistance during winter navigation from October to May. The focal company was the largest cargo owner in the region, and its freight traffic between the ports of Dudinka and Murmansk comprised about 45% of the profits of this sole carrier. However, the focal company experienced strong domination from this state shipping company due to dependence on its actions and behaviour.

The economic crisis and decline in production by many industrial enterprises during the 1990s caused a significant decrease in shipping activities within Russian

Arctic waters. Freight traffic often became interrupted and irregular. To compensate the economic losses, the state constantly increased tariffs on icebreaker services. At the beginning of the 2000s, cargo deliveries became unprofitable for the focal company due to this continuous increase in transportation costs. Further, the cargo transportation situation was compounded by the fact that the fleet of ice-class vessels and icebreakers became obsolete and required renovation.

6.5.3 Development of New Supply Chain Practices towards Sustainability

The limited choice of suppliers/sea carriers, a shortage of nuclear icebreakers and ice-class vessels, constantly increasing tariffs for icebreaker assistance caused uncertainty about the reliability of cargo delivery. A senior manager of the focal company emphasized:

> Any disruptions of cargo delivery could violate the manufacturing process and could result in significant economic losses for our company.

6.5.3.1 Developing the Focal company's Own Transport Infrastructure

Between 2006 and 2008, the focal company put into operation five ice-class (Arc-7) container vessels with a freight-carrying capacity of 16,000 tons. The technical capabilities of these vessels allowed them to overcome the over 1.5 m-thick ice shield creating an ice-free area around them. A senior manager of the focal company pointed to this action as the most important step in developing the supply chain:

> Having our own fleet guaranteed cargo transportation without icebreaker assistance all year round. Its availability enabled solving one of our most urgent strategic objectives: to ensure transport independence from the constant use of icebreaker assistance and government policies in Arctic shipping, in order not to pay the obligatory icebreaker fees when sailing along the Northern Sea Route.

In 2014, the focal company finished the construction of own trans-shipment terminal in Murmansk port. The terminal can process all kinds of own cargoes: nickel matte from the Norilsk region to the Murmansk region; metal finished products from the port of Dudinka and Murmansk region for export to European ports; and commercial cargoes from suppliers for delivering for social needs of the remote industrial monotown. Up to 700,000 tons of cargo is trans-shipped through Murmansk port annually. Before the construction of own terminal, there was only one provider of stevedore services – the state organization "Murmansk Sea Trading Port". This provider increased tariffs for its services by almost 50% during several years between 2010 and 2013. In addition, only one berth was available for handling the cargoes of different clients, including the focal company. Other berths in the port of Murmansk were designed to handle coal, which was the main port cargo. The risk of cargo delays was considerable. Moreover, the proximity of coal transshipment could result in contamination of containers by coal dust.

The construction of its own trans-shipment terminal released the focal company from heavy dependence on the state organization's actions and behaviour. As emphasized by a senior manager:

> The processing of cargoes at our terminal has significantly reduced our costs and improved the company's stability. Further, using our own terminal enabled us to develop a new activity like transportation of commercial cargoes to meet not only the manufacturing processes but also social needs and municipal facilities in extremely remote areas of the Russian Arctic. Trucks with commercial containers drive up directly to the loading operations. This saves resources and handling time.

6.5.3.2 Ensuring Environmental Efficiency

The focal company's vessels are equipped with an innovative electric device – Azipod propulsion – that consisted of the electric motor driving the fixed pitch propeller mounted on a gondola. The technical characteristics of these vessels made them more environmentally friendly and up to 20% more energy efficient with reduced fuel consumption. As highlighted a senior manager of the focal company:

> This type of vessels require a minimum need for lubricants; this reduces potential leaks. Lower fuel consumption reduces emissions. Azipod propulsion also allows the use of biodegradable lubricants. At the same time, these vessels are highly harbour maneuvered with decimeter accuracy. It is very important in icy waters where there is very narrow access to northern harbours like the port of Dudinka in the winter navigation to enter quickly and safely.

Thus, the new Arctic vessels created tangible fuel savings for the focal company and provided strong compliance with the national regulations and the International Convention for the Prevention of Pollution from Ships.

6.5.3.3 Ensuring Cost Efficiency

The building and subsequent commissioning of its own fleet demanded heavy investments by the focal company in what were considered non-core assets. As noted by a senior manager of the focal company:

> The cost of building the first vessel was 71.7 million euros, and each of the following four vessels came to about 82 million euros [the figures were taken as of 2010]. The price of building rose due to the substantial increase in world prices for metal commodities at that time. Despite so huge investments, the company covered its costs for a short time. Icebreaker fees were so high that the commissioning of its own fleet recommenced the company's costs related to icebreaker assistance and freight by third-party vessels.

Thus, the commissioning of its own fleet enabled the focal company to cut transportation costs and, thereby, to improve the efficiency of its supply chain considerably.

6.5.3.4 Reformulating the Existing SCM Practices

The development of own transport infrastructure encouraged the focal company to implement new supply chain practices, which had hardly ever been used before in Arctic navigation. These new practices – such as containerization, the "open water" principle, and the cargo circulation principle, to avoid empty vessel voyages – constituted specific operations, focused on ensuring maritime safety and improving the performance of cargo transportation. As noted by a senior manager of the focal company:

> These new solutions for Arctic navigation to sail without icebreaker assistance along the whole length of the Northern Sea Route shortened delivery time and cut transportation costs.

Containerization assured safety and security of products at all the stages of transportation as well as reduced product delivery periods by decreasing the time of cargo trans-shipment and handling operations at ports. As highlighted by a senior manager of the focal company:

> Using own trans-shipment terminal, containerization reduced transportation costs for 1 ton of cargo on average by 15% through eliminating the costs of bulk cargo handling at ports.

The "open water" principle meant that, by monitoring ice conditions, the master chose the best route to navigate the vessel through open-water channels or light ice, using a sophisticated online information system. It allowed adherence to the tight schedule of vessel traffic, avoiding any disruption and reducing delivery time and fuel consumption.

The principal of cargo circulation implied not only commercial benefits for the focal company but also social responsibility initiative for social needs as outlined in the subsection below.

6.5.4 Incorporating Social Responsibility into SCM and Meeting the Needs of Local Communities

When sailing in Arctic waters, shipping companies often encounter a cost issue that vessels might have to travel empty in one direction due to insufficient cargo accumulation in the northern regions. In order to avoid empty vessel voyages from the port of Murmansk to Dudinka, the focal company introduced the principle of cargo circulation by taking onboard commercial cargoes. As emphasized by a senior manager of the focal company:

> The principle of cargo circulation implies that our vessels deliver not only industrial cargoes to meet the needs of our processing and manufacturing facilities, but also take onboard commercial cargoes for local residents. The volume of commercial cargoes by our own vessels aggregates 1/5 of the total amount of cargoes. Thus, avoiding empty vessel voyages secures an additional income.

6.5.4.1 Commercial Cargo for Meeting Social Needs

After having launched its own fleet, the focal company became the sole maritime carrier on the sea segment between the ports of Murmansk, Arkhangelsk and Dudinka. It implied that the focal company became the only carrier that had the necessary resources to provide regular cargo transportation for local residents of the remote monotown built around the manufacturing facilities. A senior manager of the focal company told how the commercial cargo activity started:

> Earlier we focused mostly on our own industrial and general cargoes to meet the manufacturing needs in the remote region. However, our company is a town-forming enterprise, and we had to think about the social needs in addition. We reluctantly started developing the transportation of commercial cargoes. In the beginning, only two managers were involved in this activity.

The commercial cargo consisted of containers of food, essential goods, medicines, public transport like buses and cars, and other types of goods for social needs.

The involvement of social responsibility into SCM was an extra burden on the focal company's managers that complicated the fleet operations in the tight schedule of vessel traffic. As emphasized by a senior manager of the focal company:

> When scheduling the cargo and vessel traffic, we distributed industrial and commercial cargoes in order of importance. Figuratively speaking, when the tangerine time comes before the New Year holidays, we delivered tangerines instead of building industrial materials. We understood that the manufacturing sustainment primarily depended on the local people working in it.

There happened sometimes that commercial cargoes got stuck at the Arkhangelsk port terminal for several months to be loaded onboard the focal company's vessels towards the remote Arctic region because its priority was cargoes related to the manufacturing needs.

However, suppliers' requests for cargo transportation services to the Dudinka port intended for local residents and municipal needs of the remote monotown were increasing from year to year. The focal company's managers had to revise the practice of port operations to reduce the time of loading/unloading vessels. This made it possible to speed the turnover of the vessels between the ports of Murmansk, Dudinka, and Arkhangelsk, optimize the vessel schedule, and thereby increase the cargo traffic, including commercial cargo activity. As noted by a senior manager of the focal company:

> Our commercial department has expanded significantly. Unlike earlier when commercial container deliveries were not interesting to us, nowadays we focus, to some extent, on attracting this kind of cargo. Commercial deliveries have enabled a good extra income.

The delivery of commercial cargoes was carried out under contracts with different suppliers. The focal company was responsible for ensuring regular deliveries, but not for the safety of the products, e.g. in reefer containers. Inspection of commercial containers before loading onto the vessel was not carried out because this procedure required a special employee who would follow the cargo during its journey and check its quality and condition at the port terminals. Such kind of trust to suppliers caused actually frequent claims from suppliers themselves. As highlighted an interviewee of the vessel administration:

> Last winter one incident happened with a container full of onions for local food stores in the monotown when onion got frozen on its way. This fact was known only when it arrived in the port of destination. Perhaps onion was frozen before accepting onboard at the port of departure. The reefer container could be without electricity for a while during its delivery by a truck. When events like this occur, claims and complaints are addressed primarily to the vessel captain.

Despite these claims happened from time to time, the focal company did not change the order of work with suppliers and considered them unfounded. Besides, there was no other sea carrier that could carry out this kind of transportation on the sea segment between the ports of Murmansk, Dudinka and Arkhangelsk. As added an interviewee of the vessel administration to this situation:

> On their own initiative, the vessel mechanics inspect twice a day all containers onboard, including commercial cargo, whether they have electrical power. Although this is not provided for by contracts with suppliers. If necessary, the mechanics repair minor problems with the containers, mostly for safety reasons.

6.5.4.2 Transport of Local Indigenous People

In addition to transporting commercial cargoes of food products and essential commodities, social responsibility in the focal company's supply chain operations was addressed the transport of local people from extremely remote settlements along the High North coast in the mouth of the Yenisei River towards the nearest port of Dudinka. This social responsibility aspect in maritime operations was agreed with the regional authorities. Local indigenous people from extremely remote small settlements, located along the banks of the mouth of the Yenisei River and Taymyr, could receive access to medical services, hospital facilities and the nearest airport, and even without the fare. Local passenger traffic was carried out during winter navigation from September to June when the northern river navigation stopped until the summer period, and these settlements became completely isolated. As noted by a representative of the vessel administration:

> Actually, transportation of local indigenous people is rather hectic and disturbing for the vessel crew. The vessel stops in the ice in a certain place and the locals approach it on a snowmobile or on foot. The vessel's capacity is about 6-8 extra people. But sailors are afraid of being infected by local indigenous people because they are from too isolated places where local viruses are not dangerous for them but can cause great harm to the crew members. Thus, local passengers eat not with the crew, use disposable dishes and are told not to leave the cabins unnecessarily.

6.5.4.3 Trade Exchange with Local Indigenous People

Another social responsibility aspect initiated by seamen was the development and maintenance of trade relations with local indigenous people. Trade was rather a barter exchange of essential commodities, food and medicine for valuable species of local fish. As highlighted by a representative of the focal company:

> The exchange of fish and commodities also takes place on the ice quite far from the coast. Managers in offices have their eyes closed to this activity since the main task of the vessel crew is to arrive at the port of destination in time without delay.

Therefore, the implementation of social responsibility initiatives into SCM mitigated the drawbacks of the extremely remote location in the Russian Arctic for the focal company and local indigenous people, decreased a significant risk due to the sparse transportation network, the probability of supply chain disruptions, and a highly limited choice of suppliers and carriers.

6.5.5 Summary

The incorporation and development of social responsibility principles into the existing SCM practice in the Russian Arctic by the focal company roughly encompasses three phases (see Table 6.1).

The process of developing supply chain operations to be socially responsible illustrates a movement from the situation of uncertainty and high risk of supply chain disruption towards ensuring regular cargo deliveries and sustainability of supply chain operations in all the three aspects – economic, environmental and social.

Further, the findings have indicated various supply chain participants who are involved in the development of supply chain social responsibility, gain some benefits and experience some drawbacks of this process (see Table 6.2).

6.6 Discussion

The development of social responsibility in the existing SCM practice in the Russian Arctic emerged as a consequence of several strategic actions by the focal company. A number of contextual and regulatory challenges that became crucial for the economic activities in the Russian Arctic encouraged the focal company to develop own transport infrastructure and introduce a new SCM practice of sailing without icebreaker assistance. These strategic actions made it possible to improve cargo transportation efficiency, keep a high degree of vessel circulation between the ports despite the ice conditions, as well as reduce the drawbacks of the focal company's remote location. New innovative SCM practices like the principle of vessel circulation encouraged the focal company to develop the transportation of commercial

Table 6.1 The development of new supply chain practices towards social responsibility

Period	Prerequisites for implementing new SCM practices	Development of own transport infrastructure and new SCM practices	Incorporating social responsibility into SCM
Process	***Contextual challenges:*** Remoteness, the only transport link; Domination of the state sea carrier; Lack of icebreakers; Obsolescence of ice-strengthened vessels; The only one berth available for container cargoes in Murmansk port; The only state provider of stevedore services; ***Regulatory constraints:*** constantly increasing tariffs; unstable, coercive state policy	Commissioning own Arctic fleet with new technologies; Constructing own trans-shipment terminal in Murmansk port; New SCM practices (containerization, the "open water" principle, cargo circulation principle)	***Economic aspect of sustainability:*** Reducing transportation costs; Reducing delivery time; Additional income from the transportation of commercial cargoes; Reducing costs of cargo handling in Murmansk port ***Environmental aspect of sustainability:*** Reducing fuel consumption and emissions by the own vessels; Reducing the number of lubricants and potential leaks; Strong compliance with the national and international regulations ***Social responsibility aspect:*** Transporting commercial cargoes to meet social needs in extremely remote areas; Transporting local indigenous people to provide access to medical services, hospital facilities and the nearest airport; Trade exchange with local indigenous people; Reducing drawbacks of the remote location
Consequences	Uncertainty over supply chain disruption; Decision to build own Arctic fleet and develop own transport infrastructure	Improving cargo transportation efficiency and flexibility; Ensuring reliability of the supply chain on the sea segment; Implementing transportation of commercial cargoes; Mitigating drawbacks of the remote location; Building supply chain Operations to be more sustainable	
Institutional logics	Logic of stability: Focus on ensuring regular cargo deliveries; Logic of transport cost reduction: Focus on lower transport costs	Logic of stability: Focus on ensuring regular cargo deliveries; Financial logic: Focus on profitability	Logic of stability: Focus on ensuring regular cargo deliveries; Financial logic: Focus on profitability; Sustainability logic: Concern regarding social, environmental and economic aspects

Table 6.2 Benefits and drawbacks of supply chain participants in the aspect of supply chain social responsibility

Stakeholders	Benefits	Drawbacks
Focal company's managers	Profitability of commercial cargo transportation; No empty voyages	Customers' claims on a damaged cargo (mostly food products) at arriving in the port of destination – no contractual terms to check cargo quality before its loading onto the board at the port terminal
Suppliers	Commercial interest; Relationships with the sea carrier based on trust – No checking of cargo before loading	Cargoes due to several trans-shipments and long distances may be damaged at the port of destination
Seamen onboard the vessels owned by the focal company	Trade exchange of valuable species of local fish and commodities from the mainland with local indigenous people (during the winter navigation)	Extra trouble due to local passengers on board – Considering them as alien; Fear of contracting local viruses
Local people from extremely remote settlements that are isolated in the winter period from September to June	Trade exchange with seamen; Receiving access to medical services, hospital treatment and the nearest airport (without the fare)	–
Local community	Regular deliveries of food products, medicaments, essential products, public transport like buses and cars, and other types of goods for social needs to remote regions; Survival and support of the local monotown's infrastructure	–

cargoes, in addition to its own industrial cargoes, in order to meet the social needs of local residents of the remote monotown and settlements. Thus, the social aspect emerged in the supply chain due to the availability of new resources, new SCM practices, as well as arrangements with local authorities by the focal company.

The institutional logics approach has been suggested to focus on the implementation of social responsibility principles in the supply chain. The findings reveal the populations of several institutional logics, which have developed at various stages of this process (see Table 6.1). Originally, the logic of ensuring regular cargo deliveries and the logic of transport cost reduction were dominant in the supply chain. Then, supply chain strategic initiatives that seemed to be illusive in so harsh climatic conditions but, nevertheless, under the certain circumstances could change the existing SCM practices and develop a new institutional logic of sustainable supply chain practices.

After achieving all the three dimensions – economic, environmental and social – in the SCM practice, the financial logic still overrides the sustainability logic. However, it cannot be viewed as a trade-off of developing social responsibility principles because, in this case, it works as a driver. The financial logic is dominant and well established because the focal company continues to invest in the development of transport infrastructure, and thereby strengthen the sustainability logic. Further, this is due to the fact that senior managers often experience that dealing with social issues and needs has proven difficult and, to some extent, troublesome. Public expectations evolve; relationships between existing supply chain practices, costs and competitive benefits may disentangle or can be difficult to define. The findings, however, reveal that social responsibility initiatives gain a more fertile ground for further development when they contribute also to strengthening financial performance in the supply chain. Thereby, the findings illustrate that under certain conditions there is no need to pursue in making the sustainable logic dominant if the current dominant logic, e.g. the financial logic in this case study, contributes to its further development.

The extant literature on SCM and CSR concentrates on external pressure and incentives as drivers for sustainable SCM set mainly by two groups – customers that accept services of supply chain operations and all modes of governmental control (Seuring and Müller 2008). In our case, however, transportation of commercial cargoes encouraged the incorporation of social responsibility principles into SCM and became a strategic response by the focal company to contextual challenges e.g. harsh winter conditions and remoteness, as well as the lack of infrastructure and governmental support. Thereby, the findings support the supposition that institutional factors may play a role in how supply chain strategic initiatives evolve and that such initiatives are not only objective, rational processes of goal setting and activity planning. Further, this study illustrates how social responsibility can provide the economic aspect when commercial benefits are under the highest priority.

The findings outline that the focal company took responsibility for ensuring regular deliveries when it was able to influence contextual challenges and institutional constraints through implementing its supply chain strategies. This, in turn, resulted in specific social outcomes like the transport of indigenous people from extremely remote settlements. This finding is in line with Klassen and Vereecke's results (2012) concluding that responsibility can emerge as an outcome of both arrangements with local authorities and ethical overtones of the focal company. Therefore, social responsiveness encompasses that the focal company is responsible for its own strategic actions and decision-making, which can influence the people and social surroundings of this company. It is also conferred when there are available capabilities, including the company's willingness and resources, to link responsibility and necessary actions for social issues and needs.

Further, this study reveals one more distinctive feature of sustainable SCM. When the focal company is pressured by customers' claims on damages of commercial cargoes at arriving in the port of destination, it usually passes this pressure on to suppliers. After becoming the sole maritime carrier, the focal company has to take a longer part of the supply chain into account than needed earlier for meeting only manufacturing needs and pure economic reasons. While previous research looks more at the factors that are external to the supply chain, this study also emphasizes the importance of exploring which internal factors act as barriers for implementing sustainable supply chains and which factors support such developments. At the same time, the findings identify that all supply chain participants, including suppliers, manufacturers, customers and society understand social responsibility differently due to what benefits and drawbacks they gain in the process of social responsibility incorporation (see Table 6.2). This study provides deeper insights into how supply chain strategies can contribute to the needs of local communities in terms of the values of the society in remote Russian Arctic areas by incorporating the social responsibility principles in the existing practice of cargo transportation. Therefore, the study emphasizes the importance of understanding not only how social issues are managed but also how and whom the solutions of those social issues are targeted to.

6.7 Conclusion and Implications for Theory and Practice

Through the lenses of the institutional logics approach, this study presents the implementation of social responsibility principles into the existing SCM practice by exploring how they emerge under certain circumstances and evolve into actions, producing a contribution to the needs of local communities. In the case studied here, social responsibility initiatives were not planned beforehand either for strategic advantage or for legitimacy; they became possible after the satisfaction of economic and environmental concerns of cargo transportation in the Russian Arctic. This study emphasizes that a supply chain becomes sustainable when it creates value not only for the focal company but also directly for local economies through supporting and building capabilities in local communities. This takes a particular value in remote areas with sparse transportation networks like the Arctic regions. Through the analysis of building the supply chain to be more resilient and socially responsible, this study provides deeper insights into how companies can involve all the three dimensions – economic, environmental, and social – in achieving further sustainability of supply chain operations. That is still rare in the extant literature (Seuring and Müller 2008).

Responding to calls for conducting more case study-based research within the SCM field (Näslund 2002; Seuring 2005; Pagell and Wu 2009; Stock et al. 2010; Quarshie et al. 2016), the findings provide an in-depth understanding of social sustainability in particular empirical settings. The single case study illustrates the development of the existing SCM practices where different contextual factors converge and, thereby, making resistance to state coercive pressures a viable option towards further sustainability. This could happen quite often in real practice than normally anticipated by the extant literature. Further, the findings highlight that the building of supply chain sustainability can no longer be ignored in the Arctic regions, as it is one of the most important ways for further industrial and social development.

Reflection on the contextual challenges and circumstances before developing new SCM practices like the social aspect of sustainability may be crucial in choosing a set of necessary strategic actions. This may be explained by the fact that there are institutional forces and logics, which being often invisible are able nonetheless to exert considerable impact on strategic initiatives (Tsvetkova and Gammelgaard 2018) like sustainable SCM practices. By considering the interaction of context, regulatory pressures or drivers and the potential impacts of new strategic initiatives on the existing SCM practice, managers will gain a better understanding of how to develop SCM to be more socially responsible, resilient and further sustainable. This is particularly important when extending business to countries or regions with sparse transportation networks where local regulations may present contextual challenges.

This study is helpful for the logistics managers and policymakers to understand different patterns of social sustainability and this can act as a key tool for decision-making in sustainability in remote Arctic areas, particularly concerning mining companies and offshore oil/gas development projects.

The findings may be also valuable for managers to indicate the importance of supplier quality improvement by developing monitoring and collaborative capabilities when integrating social responsibility into SCM. This will help in addressing not only social issues but also gaining competitive benefits for both the carriers and suppliers.

6.8 Limitations and Further Research

This study expands previous research on sustainability in supply chains by providing deep insights into the real-life situation of the incorporation of social responsibility into the existing SCM practice. However, the Russian Arctic is a specific context of increased complexity, not at least because of its harsh natural conditions and extreme remoteness. Further research should include case studies on social responsibility in SCM in other contexts, which could include other alternative institutional influences on how organizations develop their supply chain operations to be socially responsible.

Further, this study focuses on the transportation of general commercial cargoes for social needs. Supply chains with some specialized products that have a more direct or specific impact on the environment (e.g. chemical products suppliers) may provide further insight into institutional influences within sustainable SCM practices.

The findings suggest that individual initiatives by seamen who are involved directly in social responsibility practices in SCM and, to some extent, personally committed to social responsibility activities can drive these initiatives even in the absence of a supportive organizational culture. Further research is necessary to explore the functional interplay between the logistics managers and other supply chain participants (e.g. at the grassroots level).

There are still fundamental issues that future research need to address to explore how to create supply chains more responsible and sustainable, and how the sustainable logic can become the dominant logic in the supply chain.

Appendix: Mapping the Supply Chain of the Focal company's Cargo Transportation

Figures 6.1 and 6.2 illustrate the focal company's geographical location and its physical flows of cargoes. Both figures present also the supply chain configuration, which remained the same after strategy implementation, but the SCM practice changed.

Figure 6.1 maps the supply chain when delivering metal finished products from the manufacturing location to domestic and global markets: by railway to Dudinka port (Link 1), then by sea to the ports of Murmansk and/or Arkhangelsk (Link 3), then by railways to domestic customers (Link 4) or to Rotterdam/Hamburg ports to the international market by own vessels or third-party vessels (Link 5). Strategy implementation made it possible to use the eastern part of the Northern Sea Route to reach Asian ports (Link 6).

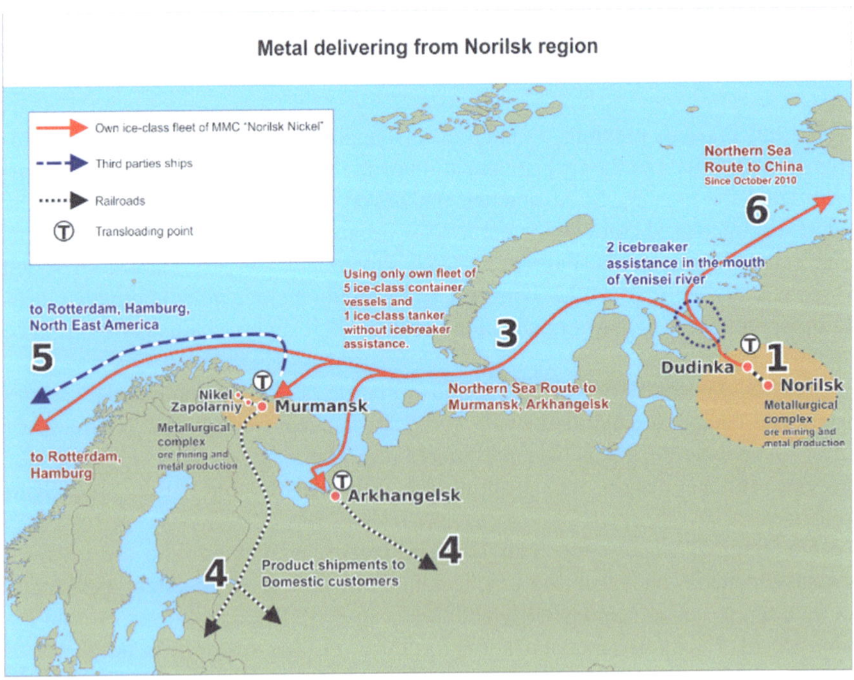

Fig. 6.1 Delivery of finished products from the location to the market

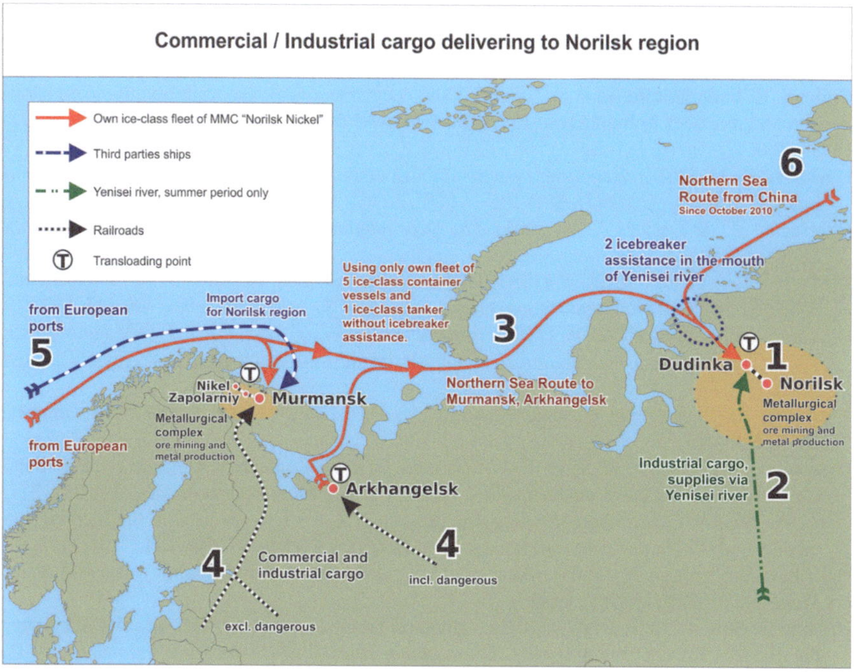

Fig. 6.2 Delivery of cargoes from suppliers towards the location

Figure 6.2 maps the supply chain when delivering industrial and commercial cargoes to the manufacturing location: by sea from the ports of Murmansk and Arkhangelsk (Link 3), along the Yenisei River only during the summer navigation by own barges (Link 2).

References

Ahi, P., & Searcy, C. (2015). Measuring social issues in sustainable supply chains. *Measuring Business Excellence, 19*(1), 33–45.

Bennett, A., & Elman, C. (2010). Case study methods. In C. Reus-Smit & D. Snidal (Eds.), *The Oxford handbook of international relations* (p. 29). Oxford: Oxford University Press.

Beske, P. (2012). Dynamic capabilities and sustainable supply chain management. *International Journal of Physical Distribution and Logistics Management, 42*(4), 372–387.

Bowen, H. R. (1953). *Social responsibilities of the businessman*. New York: Harper & Row.

Boyd, D. E., Spekman, R. E., Kamauff, J. W., & Werhane, P. (2007). Corporate social responsibility in global supply chains: A procedural justice perspective. *Long Range Planning, 40*(3), 341–356.

Carter, C. R., & Jennings, M. M. (2001). Social responsibility and supply chain relationships. *Transportation Research. Part E, 38*, 37–52.

Carter, C. R., & Jennings, M. M. (2002). Logistics social responsibility: An integrative framework. *Journal of Business Logistics, 23*(1), 145–180.

Carter, C. R., & Rogers, D. S. (2008). A framework of sustainable supply chain management: Moving toward new theory. *International Journal of Physical Distribution and Logistics Management, 38*(5), 360–387.

Ciliberti, F., Pontrandolfo, P., & Scozzi, B. (2008). Logistics social responsibility: Standard adoption and practices in Italian companies. *International Journal of Production Economics, 113*, 88–106.

Davidson, K. M. (2011). Reporting systems for sustainability: What are they measuring? *Social Indicators Research, 100*(2), 351–365.

Denzin, N. K., & Lincoln, Y. S. (2005). *The Sage handbook of qualitative research* (2nd ed.). Thousand Oaks: SAGE.

DiMaggio, P. J., & Powell, W. W. (1983). The iron cage revisited: Institutional isomorphism and collective rationality in organizational fields. *American Sociological Review, 48*, 147–160.

DiMaggio, P. J., & Powell, W. W. (1991). Introduction. In W. W. Powell & P. J. DiMaggio (Eds.), *The new institutionalism in organizational analysis* (pp. 1–38). Chicago: University of Chicago Press.

Flyvbjerg, B. (2006). Five misunderstandings about case-study research. *Qualitative Inquiry, 12*(2), 219–245.

Greenwood, R., Raynard, M., Kodeih, F., Micelotta, E. R., & Lounsbury, M. (2011). Institutional complexity and organizational responses. *The Academy of Management Annals, 5*(1), 317–371.

Gurtu, A., Searcy, C., & Jaber, M. Y. (2015). An analysis of keywords used in the literature on green supply chain management. *Management Research Review, 38*, 166–194.

Hall, J., & Vredenburg, H. (2005). Managing the dynamics of stakeholder ambiguity. *MIT Sloan Management Review, 47*(1), 11–13.

Hassini, E., Surti, C., & Searcy, C. (2012). A literature review and a case study of sustainable supply chains with a focus on metrics. *International Journal of Production Economics, 140*(1), 69–82.

Institute of Regional Policy. (2008). *Monotowns of Russia: how to survive the crisis?* English edition: Моногорода России: как пережить кризис?" (trans), p. 81.

Klassen, R. D., & Vereecke, A. (2012). Social issues in supply chains: Capabilities link responsibility, risk (opportunity), and performance. *International Journal of Production Economics, 140*, 103–115.

Lam, J. S. L., & Lai, K. (2015). Developing environmental sustainability by ANP-QFD approach: The case of shipping operations. *Journal of Cleaner Production, 105*, 275–284.

Lam, J. S. L., & Lim, J. M. (2016). Incorporating corporate social responsibility in strategic planning: Case of ship-operating companies. *International Journal of Shipping and Transport Logistics, 8*(3), 273–293.

Lawrence, T. B., & Suddaby, R. (2006). Institutions and institutional work. In S. R. Clegg, C. Hardy, W. R. Nord, & T. Lawrence (Eds.), *The handbook of organization studies* (2nd ed.). London: Sage.

Mani, V., Agarwal, R., Gunasekaran, A., Papadopoulos, T., Dubey, R., & Childe, S. J. (2016). Social sustainability in the supply chain: Construct development and measurement validation. *Ecological Indicators, 71*, 270–279.

Mansouri, S. A., Lee, H., & Aluko, O. (2015). Multi-objective decision support to enhance environmental sustainability in maritime shipping: A review and future directions. *Transportation Research Part E, 78*, 3–18.

Matos, S., & Hall, J. (2007). Integrating sustainable development in the extended value chain: The case of life cycle assessment in the oil & gas and agricultural biotechnology industries. *Journal of Operations Management, 25*, 1083–1102.

Maxwell, J. A. (1996). *Qualitative research design: An interactive approach*. Thousand Oaks: SAGE.

Miao, Z., Cai, S., & Xu, D. (2012). Exploring the antecedents of logistics social responsibility: A focus on Chinese firms. *International Journal of Production Economics, 140*(1), 18–27.

Murphy, P. R., & Poist, R. F. (2002). Socially responsible logistics: An exploratory study. *Transportation Journal, 41*(4), 23–35.

Näslund, D. (2002). Logistics needs qualitative research – Especially action research. *International Journal of Physical Distribution and Logistics Management, 32*(5), 321–338.

Pagell, M., & Shevchenko, A. (2014). Why research in sustainable supply chain management should have no future. *Journal of Supply Chain Management, 50*(1), 44–55.

Pagell, M., & Wu, Z. (2009). Building a more complete theory of sustainable supply chain management using case studies of 10 exemplars. *Journal of Supply Chain Management, 45*(2), 37–56.

Plaizier, W., McCool, M., & Cretenot, G. (2012). The risks in remote locations. *Supply Chain Management Review, 16*(1), 52–53.

Quarshie, A. M., Salmi, A., & Leuschner, R. (2016). Sustainability and corporate social responsibility in supply chains: The state of research in supply chain management and business ethics journals. *Journal of Purchasing and Supply Management, 22*, 82–97.

Sarkis, J. (2012). Models for compassionate operations. *International Journal of Production Economics, 139*(2), 359–365.

Sarkis, J., Zhu, Q., & Lai, K. H. (2011). An organizational theoretic review of green supply chain management literature. *International Journal of Production Economics, 130*(1), 1–15.

Sayed, M., Hendry, L. C., & Bell, M. Z. (2017). Institutional complexity and sustainable supply chain management practices. *Supply Chain Management: An International Journal, 22*(6), 542–563.

Scott, W. R. (2014). *Institutions and organizations: Ideas, interests and identities* (4th ed.). Thousand Oaks: SAGE.

Seuring, S. (2005). Case study research in supply chains – An outline and three examples. In H. Kotzab, S. Seuring, M. Müller, & G. Reiner (Eds.), *Research methodologies in supply chain management* (pp. 75–90). Heidelberg: Physica.

Seuring, S., & Müller, M. (2008). From a literature review to a conceptual framework for sustainable supply chain management. *Journal of Cleaner Production, 16*, 1699–1710.

Sharma, S., & Ruud, A. (2003). On the path to sustainability: Integrating social dimensions into the research and practice of environmental management. *Business Strategy and the Environment, 12*, 205–214.

Silvestre, B. S. (2015). Sustainable supply chain management in emerging economies: Environmental turbulence, institutional voids and sustainability trajectories. *International Journal of Production Economics, 167*(C), 156–169.

Stock, J. S. L., Boyer, S., & Harmon, T. (2010). Research opportunities in supply chain management. *Journal of the Academy of Marketing Science, 38*(1), 32–41.

Thornton, P. H., & Ocasio, W. (1999). Institutional logics and the historical contingency of power in organizations: Executive succession in the higher education publishing industry, 1958–1990. *The American Journal of Sociology, 105*(3), 801–843.

Thornton, P. H., Ocasio, W., & Lounsbury, M. (2012). *The institutional logics perspective: A new approach to culture, structure, and process*. Oxford: Oxford University Press.

Tsvetkova, A. (2011). Innovations for sustainability of maritime operations in the Arctic. In B. Th, C. Jahn, & W. Kersten (Eds.), *Maritime logistics in the global economy: Current trends and approaches* (Vol. 5). Lohmar-Koln: Josef Eul Verlag Gmbh.

Tsvetkova, A., & Gammelgaard, B. (2018). The idea of transport independence in the Russian Arctic: A Scandinavian institutional approach to understanding supply chain strategy. *International Journal of Physical Distribution and Logistics Management, 48*(9), 913–930.

WCED. (1987). *Report of the world commission on environment and development: "Our common future"*. Oxford: Oxford University Press.

World Bank Report. (2010). The world bank in Russia. *Russian Economic Report, 22*, 29. http://siteresources.worldbank.org/INTRUSSIANFEDERATION/Resources/305499-1245838520910/rer_22_eng.pdf. Accessed 29 Nov 2018.

Wu, Z., & Pagell, M. (2011). Balancing priorities: Decision-making in sustainable supply chain management. *Journal of Operations Management, 29*, 577–590.

Zucker, L. (1987). Institutional theories of organizations. *Annual Review of Sociology, 13*(1), 443–464.

Chapter 7
Miles and Meters Matter: Political Effects on the Shipping Routes of Measurement Techniques in the Arctic

Eda Ayaydın

Abstract States sought sovereignty over the Arctic Region by discovering the vastness of this uncharted territory. Coastal states developed measurement techniques to take the biggest share of the region. One of these measurement techniques is the Sector Principle that the Canadian senator Pascal Poirier introduced in 1907. Other Arctic states, such as the United States of America and Norway, objected to this technique. However, Russia also adopted and started to use this principle in order to draw Arctic borders in 1926. Before the Sector Principle was introduced, the Median Line Principle had been used and is still in use. Therefore, this new technique created political disputes on the controversial areas in the Arctic. Thus, another problem occurred apart from the unsolved disputed regions; the states also argued their way of measuring and calculating while preparing their Arctic claims to the UNCLOS. The Law of the Sea (1982) brought rules for gaining sovereignty for the 5 coastal states in the Arctic. On the other hand, currently, climate change threat increases the immense geopolitical importance of the region regarding petroleum, oil & gas and especially new shipping routes opportunities. Therefore, sovereignty rights in the region became much more significant for littoral states. Accordingly, this paper will try to see how technical systems have impacted on political claims – especially on shipping routes – and will analyse the history of acquisition of the sovereignty in the Arctic by two measuring techniques. The focus will be on Sector Principle within the sovereignty concept and geopolitical framework.

Keywords Sovereignty · Measurement techniques · Governance · UNCLOS · Shipping routes

---

E. Ayaydın (✉)
University of Versailles, Science Po Bordeaux, Paris, France
e-mail: eda.ayaydin@uvsq.fr

E. Pongrácz et al. (eds.), *Arctic Marine Sustainability*,
Springer Polar Sciences, https://doi.org/10.1007/978-3-030-28404-6_7

Abbreviations

CHNL	Centre for High North Logistics
CLCS	Commission on the Limits of the Continental Shelves
EEZ	Exclusive Economic Zones
LNG	Liquid Natural Gas
NSR	Northern Sea Route and
NWP	Northwest Passage
TPP	Trans-Polar Passage
TSR	Trans-Arctic Sea Route
UNCLCS	United Nations Commission on the Limits of the Continental Shelf
UNCLOS	United Nations Law of the Sea Convention

7.1 Introduction

"There is no word more ambiguous in its meaning than *possession*" indicated the U.S Supreme Court.[1] *Possessio* was used by ancient Romans to refer to occupation or control of a land or a property.[2] This is precisely the subject of this chapter within the frame of the Berlin Conference of 1884–1885, which brought a new concept to international arena: *Animus possidendi (effective occupation).* Berlin Conference was organized in order to seal the fate of Africa. However, it inherited effective occupation that obliges colonizers to be *psychically* in the occupied region for the establishment of the authority to sustain their sovereignty. The concept of sovereignty is the major backbone of international law and politics that can be described as "to have controlling and practicing power on possessed entity or territory". The discussions on the concept of sovereignty have lasted for centuries, from "use of sovereignty" to "sovereign types" in historical, judicial and political contexts. The legitimization of sovereignty and possession by the international arena is a core matter. The Russia's possession of Alaska during eighteenth century is an example of effective occupation in terms of psychical presence by establishing a colonial administration: Russian-Alaskan Company. Russia remained in the region until 1867 and was known as the sovereign of Alaska until the Alaska Treaty that ceded Alaska to the USA.

Today, in 2019, Arctic is not unclaimed any more. There is not specific Arctic Charter or Arctic Treaty. United Nations Law of the Sea Convention (UNCLOS), that rules the Arctic governance issues in terms of possession and sovereignty claims, came into force in 1982. Five Arctic states (Canada, Denmark, Norway, Russian Federation, and USA) have their own Arctic strategies and sovereignty

[1] *National Safe Deposit Co. v. Stead*, 232 U.S. 58, 34 S. Ct. 209, 58 L. Ed. 504 [1914], quoted by Klein Christine A., *Property: Cases, Problems and Skills*, Wolters Kluwer, New York, 2016, p. 62.

[2] Radin Max, "Fundamental Concepts of the Roman Law", *California Law Review*, Vol: 13, Issue: 3, Berkeley, pp. 207–228.

claims. However, they need to measure their claimed areas and prove it to the Commission on the Limits of the Continental Shelves (CLCS). The commission examines each of the scientific claims and if the measurements are accurate, the states would gain legitimate sovereignty on the claimed territories and they can exploit the high Arctic and its resources.

The aim of this chapter is to analyse the measurement techniques in the Arctic as tools of maximization of power for states as rational actors of the international system. Specifically, the chapter questions how measurement techniques impact on political claims and at which extend they are implemented.

7.2 Measurement Techniques

Arctic region has remained a mystery, until today for some aspects. States, especially Arctic littoral states, conducted several explorations to discover the region in order to possess, control and exploit. This section analyses measurement techniques as consequences of the adjacent states' efforts to legitimate their power.

7.2.1 Sector Principle and Median Line Principle

While making claims on the ice deserts, governments seek to establish their limits, and, consequently seek to find new methods.[3] The Sector Principle was one of the controversial methods that used meridians as a convenient way of delimitating territorial claims. Firstly, the principle was used as a quasi-official method for the publication of the map prepared by the Department of the Interior of Canada in 1904, which showed the 141° and 60° meridians as boundaries. Eventually, it gained prominence by Pascal Poirier's efforts. The Canadian senator defined for the first time in 1907 the Sector Principle at the Canadian Senate. According to Poirier, the area between the 60° and 141° meridians up to the 90° North latitude belong to Canada. These longitudinal lines look as a pie shape emanating from the North Pole.

According to Donath Pharand, there are three features of the Sector Principle: boundary, contiguity and customary law.[4] Generally, the treaties of 1825 and 1867 are given as examples in boundary treaties using the same criteria, i.e. longitudes without mentioning the term of "sector".[5] However, Christopher Joyner finds this point controversial, underlining the fact that both treaties were made for the land

[3] Crawford James, *Brownlies's Principles of Public International Law*, Oxford University Press, Oxford, 2012, p. 241.

[4] Pharand Donath, *Legal Status of the Arctic Regions*, Edmond Mongomery Publications, Toronto, 2000, p. 424–425.

[5] 1825 is the Anglo-Russian Convention on the boundary and 1867 is Alaska's ceding treaty from Russia to the United States.

areas, not for the ice.[6] Legal status of the ice is not clear but the consensus is that it ought to be considered as land because of the thickness of the ice.

States tried to draw their lines that are contiguous to each other, by the Sector Principle. Contiguity requires state practice as a feature of the Sector Principle. However, Norway, one of the principal states in the Arctic, rejects the Sector Principle strongly. Therefore, Norwegian rejection of the Sector Principle makes it inaccurate for the international law.

There is a negotiated boundary between Canada and the United States at the Beaufort Sea, located between the Yukon Territory and Alaska. Traditionally, the United States suggested equidistance line. However, Canada presented the 141° longitude measured by the Sector Principle as delimitation, based on the 1825 Anglo-Russian Convention. Canada considered the 141° as delimitation because the equidistance line favours the United States duc to being slightly convex coast of Alaska and concave coast of the Yukon.

The western sector line of Canada was drawn by the Convention of 1825 between Russia and Canada signed in St. Petersburg.[7] The convention does not indicate clearly that the boundary between Alaska and Canada was a sector line that extended till the North Pole. However, it stated:

> …dans son prolongement jusqu'à, (sic.) la mer Glaciale…[8]

At this point, the *lingua-franca* of the time comes to the surface. Byers argues that the 1825 Anglo-Russian treaty was written in French, which caused some translation problems. When it is translated to English, the preposition "jusqu'à" in the sentence "dans son prolongement jusqu'à, la mer Glaciale" might include the frozen sea.[9] Therefore, the 1825 treaty can be interpreted as the beginning of sectorial ideas before Poirier, in 1907.

According to the 1867 convention, the western line was at the 65° 35′ North latitude that corresponded to a point on the Bering Strait, *i.e.* there was not a distinction for land obligation. Accordingly, the boundary might be on ice or sea. At this point, it would be fair to say that the Sector Principle was used partly before its official promotion in the Canadian Parliament in 1907.

Canada's claim on the Arctic started in 1907 by the introduction of the Sector Principle by Pascal Poirier. Its official sector lines were drawn in 1925.[10] The

---

[6] Joyner Christopher, *Antarctica and the Law of the Sea*, Martinus Nijhoff Publishers, Dordrecht, 1992, pp. 54–59.

[7] Witschel Georg, Winkelmann Ingo, Tiroch Katrin, *New Chances and New Responsibilities in the Arctic Region, Papers from the International Conference at the German Federal Foreign Office in Cooperation with the Ministries of Foreign Affairs of Denmark and Norway 11–13 March 2009*, Berliner Wissenschafts-Verlag, Berlin, 2010, p. 215.

[8] Hyde Charles, *International Law Chiefly; Interpreted and Applied by the United States*, Little, Brown and Company Press, Boston, 1947, pp. 60–61.

[9] Byers Michael, *International Law and the Arctic*, Cambridge University Press, Cambridge, 2013, p. 72.

[10] Pharand Donath, *op. cit.* p. 425.

Canada's position was unclear before 1925 on whether the claim was covering only the land or also including the sea. However, in 1925, the government passed an order-in-council, which emphasized preservation of wildlife within the sector lines. This legislation claimed the Arctic sector line up to North Pole which showed Canadian intend on the sovereignty both on the seas and the land. In 1939, Thomas Alexander Crear, the Minister of Mines and Resources, declared the Sector Principle as official at the House of Commons.

Canada did not solely add the territory within the sector lines, but also claimed the frozen sea. Ottawa desired sovereignty over the Northwest Passage according to the "historic title statute" that indigenous people of Canada should have brought. The Inuit used to live at the Arctic part of the Canada and their economical territory, their *biotopos*, was at the Arctic for thousand years.[11] However the United States and the other Arctic states (except Russia) saw the Northwest Passage as an international strait.

Russia does not oppose to Canadian sovereignty claim on the Northwest Passage because Moscow also rules the Northern Sea Route as internal waters, therefore both countries share similar interests.

Even if Canada invented the Sector Principle, it was the Soviet Union that used it first. St. Petersburg adopted the principle in 1916 and Moscow implemented officially in 1926 (Moscow became the Soviet capital in 1918). It was the first formal declaration of a sector claim made by the Soviet Union through the Central Executive Committee of the USSR.[12] Moscow calculated its sector lines on the 168° of longitude west and the 32° longitude east. It was published at the *Izvestia* newspaper as:

> All known lands and islands lying north of the Soviet Union which were not already acknowledged to belong to some other power, as well as that which might be discovered in the future, were part of the USSR.[13]

According to this approach, the Soviet Union claimed also the land in the Arctic as well as the frozen sea, such as East Siberia, Chuckchi and Laptev Sea, regardless if these lands and frozen seas are "discovered or undiscovered." In addition, Kremlin added several islands in Siberia into its claim by using this new measurement technique.

It was not easy to make measurements in this kind of remote areas -massive, not discovered at all- as the Arctic, bearing in mind the technology of the early 1900s. This is where explorers gain importance with their expeditions and reports to their respective governments during the 1900s. For instance, Canadian explorer Viljhamur Stefansson went to Arctic discovery after the Soviet Union adopted the Sector

[11] Byers Michael, *Who Owns the Arctic: Understanding Sovereignty Disputes in the North*, Douglas & McIntyre Publishing, Vancouver, 2009, p. 50.

[12] Joyner Christopher, *Antarctica and the Law of the Sea*, Martinus Nijhoff Publishers, Dordrecht, 1992, p. 56.

[13] Lajeunesse Adam, *Lock, Stock and Icebergs: A History of Canada's Arctic Maritime Sovereignty*, UBC Press, Vancouver, 2016, p. 27.

Principle in 1926. Stefansson claimed Wrangel Island (Russian Territory) as Canadian land and asked permission to the government for the occupation of the region, however the Soviet Union sent troops to the region immediately.[14] Explorers might sometimes have caused political tensions in the region, which created mistrust between the Arctic states due to the lack of technological advancement.

The opponent of Sector Principle is Median Line Principle. This principle leads to a line that every point is equidistant to the nearest points of baselines.[15] International lawyers started to examine median line method during 1940s.[16] Geneva Convention of 1958 on the Continental Shelf highlighted median line. Furthermore, the regulations of the Convention require that the borders should be drawn according to equidistance to nearest points on the coast from a median line. Eighty-nine percent of opposite coastal states of the world used the Median Line Principle with respect to delimitation. This method divides offshore through the land border between parties.

These measurements are important in terms of fishing, natural resources, international straits and security. So, baselines are vital for the scope of the claims and the rights over the high amount of fishing and natural resources.

Baselines are very significant to coastal states' claims because they are the starting lines of the claimed zone,[17] and they are the outer limits of the states. Therefore, baselines are highly important for delimitation of Exclusive Economic Zones (EEZ). Territorial sea, contiguous zone and exclusive economic zone were defined in the UNCLOS, respectively as 12, 24 and 200 nautical miles.

There are two types of baselines; normal baseline and straight baseline. Normal baseline is a low-water line intersects the coast. Normal baseline should be a low water line along the coast. Coastlines in the Arctic often change by erosion therefore low water line also shifts.

Straight baseline can be applied strictly. According to article 7 of the UNCLOS, the straight baseline is permitted if the coastal line is indented or in presence of a fringe of islands along the coast. The International Court of Justice states that straight line might be used as an exception to normal baseline in specific circumstances. However, in this anthropogenic epoch, this principle of ICJ does not seem physically applicable.

[14] *Ibid*, p. 27.

[15] Sjöberg Lars, "The Three-Point Problem of the Median Line Turning Point: on the Solution for the Sphere and Ellisoid", *International Hydrographic Review*, Vol: 3, No: 1, 2002, pp. 81–87.

[16] Howard Roger, *The Arctic Gold Rush: The New Race for Tomorrow's Natural Resources*, Bloomsbury Academic Press, London, 2009, p. 59.

[17] Schofield Clive, "Departures from the Coast: Trends in the Application of Territorial Sea Baselines under the Law of the Sea Convention", *the International Journal of Marine and Coastal Law,* Vol. 27, 2012, pp. 723–732.

7.2.2 Clash of Measurement Techniques

Arctic shorelines are predominantly ice covered. However, the situation changes with the warming and baselines are getting more vulnerable to erosion. Baselines of the rock coasts are more stable as it is the case in Svalbard. The highest erosion is visible at the East Siberia, Laptev and Beaufort coast. For instance, erosion at the Beaufort Sea is 10–30 m per year.[18] Ice is melting, permafrost thaws in the coastal areas, moreover massive storms, waves and sea level rise make the coasts more vulnerable. As a result, this affects technically and politically the basepoints on the baselines. If a claim of any state is proven that its basepoint is on the ice, this situation will bring controversies with it as the global warming is increasing in the North and the probability of melting of the ice basepoints is increasing, too. In such case, the current international law system is incapable of responding. That is why certain amendments are required for the UNCLOS. Otherwise, in the absence of a legal statute, further boundary struggles would occur in the Arctic.

Russia became a party to the UNCLOS in 1997 and formulated a submission to the continental shelf commission in 2001 in which outer limits included the North Pole. Moscow submitted its claim by using the sector principal in 2001 to the United Nations Commission on the Limits of the Continental Shelf (UNCLCS). By contrast, the United States, Norway and Denmark were strongly against the Sector Principle even if they could gain more territory by this method. On the other hand, Canada would have gained more sovereignty by the Median Line Principle due to its broad coastline.[19]

Russia adopted a new Arctic strategy in 2008, emphasizing the region's importance and major source of revenue for Russian economy.[20] Direct effect on the Russian economy might create struggle and conflict due to the question of rights over the Arctic oil. Naturally, for more oil, there is a need of more "territory" in the North Pole. At this point, the geopolitical and economic significance of the Northern Sea Route and Northwest Passage can be understood.

Russia and Canada supposed that the sector method was more favourable for acquisition of sovereignty in the Arctic. By withdrawing Russian western line to the east, the use of the sector method was causing Russia to have less territorial sovereignty. In 2010, Russia concluded that the Sector Principle was not for its advantage. Therefore, Russia abandoned this method by signing the "Maritime Delimitation and Cooperation in the Barents Sea and the Arctic Ocean Treaty." At the beginning, Russia and Norway, as parties of this treaty, applied the Sector

[18] Jones Benjamin, Puckett Catherine, Vinas Jose Marias, "Erosion Doubles along Part of Alaska's Arctic Coast: Cultural and Historical Sites Lost", *U.S. Department of the Interior*, U.S. Geological Survey, 18 February 2009. (https://soundwaves.usgs.gov/2009/05/research2.html) accessed in 20 May 2017.

[19] *Ibid*, p. 61.

[20] Johansson Taffsir, Donner Patrick, *The Shipping Industry, Ocean Governance and Environmental Law in the Paradigm Shift in Search of a Pragmatic Shift for the Arctic*, Springer, New York, 2015, p. 60.

Principle as a result of the Russian demand. However, they could not compromise and finally, they divided the sea into two equal parts. Therefore, out of 176, 000 km^2 of the area, Russia got 88, 000 km^2 and Norway got 88.000 km^2.[21]

During Barents talks, Russia and Norway gave indication of the possibility of cooperation between the two countries. For example, in 2008, as leading energy suppliers of Europe, Russia and Norway signed a partnership in the exploration and production of oil and gas. Russia's Gazprom, Norway's Statoil and France's Total signed an agreement on development of Shtokman gas field in 2011.[22] Moreover, following the Barents agreement in 2012, Russian Rosneft and Statoil signed another partnership on the Barents Sea and the Sea of Okhotsk.[23]

One can say that after the collapse of the Soviet Union, Arctic policies of Russian Federation were not negatively affected, however changes occurred, such as the abandon of Sector Principle. The Soviet Union was one of the biggest supporters of the Sector Principle; Moscow abandoned the theory as a result of Barents Talks that ended in 2010 after 40 years of negotiations.

The Beaufort Sea is the main disputed area between Canada and the United States. Here, the United States and Canada have 6250 square nautical miles of overlapping claimed territorial waters. Canada drew its line from the 141° meridian according to Sector Principle relying on the Anglo-Russian boundary treaty of 1825. However, the U.S supports to draw a median line as a maritime boundary.

Due to their different delineation methods, Canada and USA cannot draw their continental shelves. Therefore, they cannot conduct resource exploitation in that coast of Beaufort Sea. On the other hand, both countries' scientists are working together to collect data and to map the region.

Another problem between Canada and the United States is the Northwest Passage issue that has been going on for the last hundred years. Canada never renounced its claim on this mythical passage. The United States has been insistent on declaring the Northwest Passage as an international strait. The Passage brings advantages to the states on shipping as it shortens the way in half between Atlantic and Pacific Oceans. By the warming, the use of this route started to be possible; therefore, the United States does not want to yield the Passage's sovereignty to Canada.

The United States has always considered the Arctic as an open region to the international community. The American administration deems the Northwest Passage to be international, as well. It does not want to receive a passing permission and pay a fee. This is one of the reasons why the U.S. does not ratify the UNCLOS.

[21] Dahl-Jorgensen Arnfinn, Eger Magnus Karl, Floistad Brit, Larsen-Mejlaender Morten, Lothe Lars, Wergeland Tor, Ostreng Willy, *Shipping in Arctic Waters, A Comparison of the Northeast, Northwest and Trans-Polar Passages*, Praxis Publishing, Chichester, 2013, p. 262.

[22] "Heads of Gazprom, Statoil and Total discuss implementation of Shtokman project"', *Gazprom* official web page, 8 July 2011, (http://www.gazprom.com/press/news/2011/july/article115065/), Accessed in August, 2018.

[23] Honneland Geir, *Russia and the Arctic: Environment, Identity and Foreign Policy*, I. B Tauris & Co., London, 2016, pp. 104–105.

Arctic has been a stable region until today with the absence of any conflict. The region is physically changing and the leaders are changing, too. So, this alteration would affect the future of the new blue ocean.

By the invitation of Danish Minister of Foreign Affairs and the Prime Minister of Greenland, five Arctic coastal states came together in 27 May 2008 in Illulisat, Greenland. All the adopted decisions were binding. The conference emphasized that the consensus on the disputes on the boundaries should be reached by bilateral or multilateral agreements as in the Convention on the Law of the Sea. Hans Island case is one of the relevant examples.

The name of the Hans Island comes from the Arctic explorer Hans Hendrik.[24] The Island is situated at the Nares Strait, which connects Baffin Bay to the Lincoln Sea and separates Ellesmere Island and Greenland.[25] Hans Island is between the sovereignty of Canada and Denmark. 1973 Canada-Denmark continental shelf agreement did not include the Lincoln Sea. Both countries compromised to use the median line method to divide their disputed territorial sea. But they still have a dispute on the basepoints.

7.3 Troubled Ice, Troubled Waters

Arctic has always been an issue for the states, before the climate change era or irrespective of effects of the climate change. Psychically irreparable environmental and ecological changes are leading up to new shipping routes as well as oil and gas access, fishing and tourism in the High North. This section of the chapter examines the process of acquisition of sovereignty in the Arctic and the geopolitical emphasis of the promising shipping routes in the regard of the impacts of measurement techniques.

7.3.1 Acquisition of Sovereignty in the Arctic

Svalbard is an Arctic archipelago lying towards the Barents Sea. Novaya Zemlaya and Franz Josef Land are situated at the East of Svalbard and Greenland is at the west coast of Svalbard. Dutch explorer Willem Barents discovers Svalbard in 1596 while he was trying to find Northern Sea Route.[26] Afterwards, Svalbard became an

---

[24] Hund Jon Andrew, *Antarctica and the Arctic Circle: A Geographic Encyclopedia of the Earth's Polar Regions*, ABC-CLIO Publishing, California, 2014, p. 107.

[25] Smith Robert W., Bradford Thomas, "Island Disputes and the Law of the Sea: An Examination of Sovereignty and Delimitation Disputes", in Schofield Clive, Harris Andrew (eds), *Maritime Briefing*, Vol:2, No:4, International Boundaries Research Unit, Durham, 1998, pp. 24–25.

[26] Conway Martin, *No Man's land: A History of Spitsbergen from its Discovery in 1596 to the beginning of the Scientific Exploration of the Country*, Cambridge University Press, Cambridge, 1906, pp. 33–38.

interesting place for explorers to discover new regions therefore numerous explorers made expeditions to this archipelago. For instance, Benjamin Leigh Smith was one of the nineteenth century American explorers who made five expeditions to the island. He named 31 points of the Svalbard Island such as Brochoyo, Leighbreen, and Kapp Leigh Smith.[27] After crashing his ship to ice, he had to survive 10 months at Franz Josef Land.[28] Even though the discoveries and commercial activities had started during the sixteenth century, ships continued to collapse in nineteenth century despite technological improvements. According to the reports of Willem Barents, Svalbard was rich in hunting resources. In a short period, states such as the Danish-Norwegian Kingdom, Britain, Netherlands, France and Spain started to explore and to exhaust resources. The states found a basis for exploitation of Svalbard under the "mare liberum[29]" term of Hugo Grotius.[30]

The legal statute of Svalbard was *terra nullius* during the nineteenth century.[31] In the twentieth century, Oslo asked to cancel the *terra nullius* statute of Svalbard after the separation of Norway from Sweden in 1905. During the Paris Peace Conference in 1919, Norway requested the revision of this no man land's statute. The conference established the Spitsbergen Commission that recognized the Norwegian sovereignty over Svalbard in 1920. Additionally, 46 parties were signatories to the treaty and the commission gave the equal fishing and mining rights to the signatories.[32] The same treaty, on the other hand, put a ban on any presence of naval bases or military exercises on the island.

Norway and Russia had 40 years dispute on the archipelago. The 1920 Treaty gave the sovereignty of the archipelago to Norway. Besides the fact that the treaty gave mining and fishing rights to all the signatory parties, mostly Norway and Russia use this right, at present.

As mentioned above, Svalbard cannot be used for military aims and the parties cannot install military on the archipelago. But, when in 2017, Norway-Russia relations have been tense about the Svalbard Island, the NATO parliamentary

---

[27] Credland Arthur, "Benjamin Leigh Smith: A Forgotten Pioneer", *Cambridge Core*, Vol: 20, No: 125, 1980, 127–145.

[28] *Ibid*, pp. 127–145.

[29] Churchill Robin, Ulfstein Geirs, "The Disputed Maritime Zones around Svalbard", *Changes in the Arctic Environment and the Law of the Sea, Panel IX*, Martinus Nijhoff Publishers, Leiden, 2010, pp. 551–593.

[30] Conway Martin, *Op. Cit*., pp. 33–38.

[31] Ulfstein Geir, *The Svalbard Treaty; from Terra Nullius to Norwegian Sovereignty*, Scandinavia University Press, Oslo, 1995, pp. 34–38.

[32] Afghanistan, Albania, Argentina, Australia, Austria, Belgium, Bulgaria, Canada, Chile, China, Czech Republic, Denmark, Dominican Republic, Egypt, Estonia, Finland, France, Germany, Greece, Hungary, Iceland, India, Ireland, Italy, Japan, Latvia, Lithuania, Monaco, Netherlands, New Zealand, North Korea, Norway, Poland, Portugal, Romania, Russia, Saudi Arabia, Slovakia, South Africa, South Korea, Spain, Sweden, Switzerland, United Kingdom, United States, Venezuela.

assembly planned to meet in Longyearbyen, Svalbard. Moscow feeling threatened by the NATO reacted sharply, by calling the Svalbard treaty:

> The NATO meeting is at odds with the spirit of the 1920 Svalbard treaty and that is part of a provocative policy.[33]

Russia is not a member of NATO, which is the main reason of the Russian protest.

Barents Sea is also highly significant in terms of fishing. The sea includes high amount of fish types such as cod, haddock, shrimp and herring. Norway and Russia are the main fishing countries in Barents but also Greenland, Iceland and the EU have rights of use for fishing.

On the other hand, Barents have serious amount of petroleum resources in its depth. Russia and Norway already started exercises in Snohvit. And Norway started drilling in Barents in February 2017 under the big oppositions of environmentalists.[34] Therefore, the Svalbard Treaty is important because of Svalbard archipelago's geopolitical and geostrategic location.

The international legal statute of coastal states to exclusive rights over resources in their continental shelf goes back to the 1945 Truman Proclamation on the continental shelf.[35] 1958 Geneva Convention established the essential legal statute of the continental shelf.[36] The Convention states:

- *the nature of a coastal State's rights over the continental shelf is 'sovereign rights for the purpose of exploring it and exploiting its natural resources', which meant 'the mineral and other non-living resources' of the shelf together with' sedentary species,*
- *The 'sovereign rights' are exclusive to that state,*
- *a coastal state rights over its adjacent continental shelf do not depend upon or an express proclamation.*[37]

As seen in the statement above, according to the Convention, the outer limits of a coastal state are ambiguous. It provided the criteria for the outer limit as the sea bed within the tranche of waters of a depth of 200 meters or beyond that limit, as long as the adjacent water sea bed allows to exploitation of the natural resources.[38]

---

[33] Staalesen Atle, "Russian Svalbard Protest Totally without Merit", *The Independent Barents Observer*, 21 April 2017 (https://thebarentsobserver.com/en/arctic/2017/04/russian-svalbard-protest-totally-without-merit) Accessed in April 2017.

[34] Staalesen Atle, "Going All in Norway Proposes Massive Opening of Artic Shelf", *The Independent Barents Observer*, 13 march 2017 (https://thebarentsobserver.com/en/industry-and-energy/2017/03/going-all-norway-opens-its-arctic-shelf-wide-open-oil-drilling), Accessed in May 2017.

[35] United States, *Proclamation 2667*, "Policy of the United States with Respect to the Natural Resources of the Subsoil and Sea Bed of the Continental Shelf", 28 September 1945.

[36] Wasum-Rainer Susanne, Winkelmann Ingo, Tiroch Katrir, *Arctic Science, International Law and Climate Change: Legal Aspects of Marine Science in the Arctic Ocean*, Springer-Verlag, Berlin, 2012, p. 122.

[37] *Geneva Convention on the Continental Shelf*, 29 April 1958, Article 2.

[38] Wasum-Rainer Susanne, Winkelmann Ingo, Tiroch Katrir, *Op. Cit.*, p. 123

Because of the ambiguity of the limits, it was supposed that there was a political and economic pressure on lawmakers for offshore hydrocarbon exploitability criteria.

If we apply this ambiguous outer limit criterion to the Arctic, it results for sure in political crises because of the economic interests on promising underground features and shipping routes of the region. On the other hand, as a most damaged region by the global warming, the Arctic keeps its physical vulnerability to any harmful activity. Therefore, this kind of open-ended criteria leads up more oil and gas exploitations and additional cooperation with external states on the Arctic resources. As a result, it might end up further destruction on the region's environment under the on-going warming.

The International Court of Justice introduced the "natural prolongation" term in the context of continental shelf in 1969. Therefore, in the case of proven natural prolongation of any continental shelf, the claimant state would have right over prolonging area. According to this concept, Canada, Denmark and Russia claim their natural prolongation of coastal shelves extend until the Lomonosov Ridge. This ridge is important because of the density of hydrocarbon resources that fall within the deep waters around the North Pole. The coasts of the Arctic littoral states are shallow. Therefore, Arctic states make efforts for vast continental shelves to extend their sovereignty.

The four parts of the Geneva Convention have been included in UNCLOS in 1982. The UNCLOS is a legal regime that rules the World's ocean space regarding jurisdiction, territorial sea, continental shelf, exclusive economic zone and international deep-sea beds.

The UNCLOS also approved the regulation of the Geneva Convention of 200 nautical miles limit of the coastal shelves. According to the UNCLOS, if any claim is proven by the CLCS, the claimant state can extend its EEZ to 350 nautical miles. The addressee of the Arctic adjacent states is the Commission on the Limits of the Continental Shelf. This commission might give recommendations to the claimant parts. The constituted limits in pursuit of the CLCS are legally binding.[39] After the final decision of the CLCS, the littoral states acquire rights on exploitation exercises in the specified limits; they could also run co-operations with other non-Arctic states with their owned sovereignty.

Extractive industries in the region are one of the important economic features of the Arctic and underground resources are not the motivation for states to make submissions for the outer limits of continental shelf. States have to submit claims as the requirement comes from UNCLOS. When states have made scientific explorations to map the continental shelf close to their coast, they have room for finding evidence for the CLCS that is beneficial from their viewpoint. That is why they try to convince CLCS that their continental shelf is as large as it can be.

There is not any consensus on what will happen to the zone beyond the exclusive economic zones of the Arctic Five states. These five Arctic states are not only of concern here, but also non-Arctic states are stakeholders as well, in terms of

[39] "United Nation Convention on the Law of the Sea", 10 December 1982, Article 76.

shipping, fishing and extraction. Apart from that, very important and famous feature of Arctic, the underground resources are explored and exploited by extractive industry. Private or state owned companies need to correspond the national regulations of the state in which the extractive company will operate. At the same time, the companies have to get the right of the ocean floor which part they will conduct drilling. As, the underwater borders, are not clear yet in the Arctic, the measurement techniques will play an important role. If the area is measured by median line principle, 350 NM will be the limit to EEZ in best conditions, but if sector principle is used, the area will contain the 90th degree however will be narrower.

It is clear that the Article 234[40]on ice-covered areas of the UNCLOS needs amendments in order to respond to the current situation of the region. Even though the region's governance is subject to the Law of the Sea convention, the current content of the convention is not capable of serving a reply to the probable emergent political and legal structure as ensued from the change in the region.

7.3.2 *Why Icy Arctic Waters?*

One of the most important economic bringing of the melting sea ice of the Arctic Ocean is the appearance of new shipping routes. Not only is the Arctic Ocean itself a crucial back aisle connecting Pacific and Atlantic, but also Northern Sea Route (NSR) and Northwest Passage (NWP) gain currency in the geopolitical consideration.

New shipping routes are always needed as long as trade continues between states. Shipping in the Arctic Ocean might be seen in three different ways:

- intra-arctic shipping: sailing from one point to another destination within the Arctic,
- Arctic routes which are used for going to a point inside or outside of the region,
- Trans-Arctic Sea Route (TSR) or Trans-Polar Passage (TPP) imply the use of Arctic Ocean as a transit route between Pacific and Atlantic.

TSR's sea ice condition is much denser and remoter than NSR and NWP. Moreover, it offers a very little time period in a year to navigate through. Of course, crossing through Trans-Arctic Sea Route is a good opportunity in terms of not being needed

[40] *UNCLOS, SECTION 8, ICE-COVERED AREAS, Article 234*, "Ice-covered areas Coastal States have the right to adopt and enforce non-discriminatory laws and regulations for the prevention, reduction and control of marine pollution from vessels in ice-covered areas within the limits of the exclusive economic zone, where particularly severe climatic conditions and the presence of ice covering such areas for most of the year create obstructions or exceptional hazards to navigation, and pollution of the marine environment could cause major harm to or irreversible disturbance of the ecological balance. Such laws and regulations shall have due regard to navigation and the protection and preservation of the marine environment based on the best available scientific evidence."

to Russian Northern Sea Route or Canadian Northwest Passage and not paying fees. However, the physical terms are not convenient for regular use.

The Northern Sea Route is under the legislation of Russian Federation. The official page of NSR shows that in 2015, 125 of 207 transit destinations were to Russia and 82 of them are to other countries, mostly (64%) China and South Korea.[41] 94% of type of ships were tankers which were carrying liquids as oil and gas. The ships also need to apply by submitting a detailed list of information for permission from the Administration of the NSR before 4 months from the envisioned time of transit. On the other hand, the fee for the NSR is highly controversial. In 2003, Russian Federation promulgated a decree called "About Changing Rates for Ice-Breaking Fleet Services on the NSR" and the fees increased significantly. In 2000, ships were paying 7.5 USD per ton, this price increased to 23 USD per ton in 2003 and the fee for NSR reached to 40 USD per ton in 2009. The German shipping company *Beluga*'s two ships accompanied by two ice-breakers passed the NSR with 3500 tons in 2009. According to the interview conducted with the President and CEO of Beluga Shipping Niels Stolberg, Beluga's gain by using NSR was 20.000 per day.[42] This shows that NSR administration did not apply the costly tariffs and German Beluga made a profitable transition. Thus, shipping tariffs are not stable and the fee of the NSR is open to negotiation.

Another issue about the NSR is the type of ships. The type of the ships sailing through the NSR must be in accordance with the requirements of the Russian Maritime Register of Shipping. The ships are classified according to the level of ice strengthening. The vessels classifications are Ice1 to Ice 3 for non-Arctic ships, Arc4 to Arc9 for Arctic ships, and the ice-breakers operating in the NSR must be classified from Ice-breaker6 to Ice-breaker9. The admittance criteria of navigation is classified such as; the vessels without ice reinforcements Ice 1-Ice 3 can sail between the period July to November, the vessels Arc4-Arc 9 categories with ice reinforcements can sail between November–July, and Ice-breakers6-Icebrakers9 can operate from January to June.[43]

Northern Sea Route shortens the way between North Western Europe and Asia approximately in half. However, NSR is a seasonal way, same as Trans-Arctic Sea Route and Northwest Passage. Even with the climate change, the region keeps its physically hard conditions and NSR is in service around 135 days depending on the weather conditions.[44]

Northwest Passage is different than NSR in terms of application of fees. There is no fee for transition from the NWP. The Canadian government monitors Passage's navigation, but the monitoring process is certainly not sufficient to prevent unauthorized transition or hazardous accidents. German and Canadian scientists started

[41] CHNL (Centre for High North Logistics), official information page for NSR. (http://www.arctic-lio.com/docs/nsr/transits/Transits_in_2015.pdf) Accessed in August, 2018.

[42] Chernova Svetlona, Volkov Anton, *Economic feasibility of the Northern Sea Route container shipping development,* Masteroppgaver i bedriftsøkonomi, Bodo University Publishing, 2010.

[43] *Ibid*, (http://www.arctic-lio.com/nsr_iceclasscriteria), Accessed in August, 2018.

[44] *Ibid*, (http://www.arctic-lio.com/nsr_generalareadescription), Accessed in August, 2018.

to conduct a project –*Passages* -, installing maritime surveillance system in order to reduce the hazardous risks and have a safe navigation throughout a year at the NWP. Nevertheless, Canadian Arctic remains weak compared to the Northern Sea Route in terms of search and rescue capabilities.[45] Compared to the route between Western Europe and Asia via Suez Canal, Northwest Passage offers 5000 Nautical Miles shorter route, which means substantial time, fuel and employee savings. However, the Northwest Passage is not secure because of the lack of monitoring systems and the weakness of search and rescue activities. There is lack of data and infrastructure concerning sensor and communication technology.[46] Therefore, unauthorized fishing activities and transitions that occur in the region might also create conflicts between countries.

As it is largely supposed, the Suez and Panama Canals will be replaced by Northern Sea Route and Northwest Passage for transportation between Western Europe and Asia as long as the extreme climate conditions of the Arctic keep warming. While Suez Canal is taking 11, 400 Nautical Miles and 32 days between North West Europe and Far East, NSR takes 7, 200 Nautical Miles and 18 days.[47] One can say that the Arctic shipping routes are more advantageous for Northern and Asian countries.

The oil and gas extractions have already started in the Arctic, and serious percentage of this oil and gas is exported from the Arctic to giant economies such as China, India and Europe. On the other hand, the intensity of iron, copper, nickel, phosphates and bauxite in the Arctic must not be ignored. For the transportation of fossil energy resources and minerals, pipelines can be constructed between Russia and China (e.g. for the transportation of gas from Chayandinskoye- Yakutia to China, "Power of Siberia" gas pipeline is already constructed[48]), but the shipping routes will play an important role for the transportation from the Arctic to Europe or India. By the upcoming project between Russia and Germany, the Nord Stream 2 project raised the question of Russian leverage over Europe. Therefore, the role of the shipping routes are not only geopolitically significant, but also substitutes for the controversial transportation tools.

The Russian natural gas giant Novatek and French energy company Total produced 3,5 million tons of natural gas in Yamal Peninsula in the first 8 months of

---

[45] Byers Michael, "Canada's Arctic nightmare just came true: The Northwest Passage is commercial", *The Globe and Mail*, 28 November 2017. (https://www.theglobeandmail.com/opinion/canadas-arctic-nightmare-just-came-true-the-northwest-passage-is-commercial/article14432440/) Accessed in August, 2018.

[46] German-Canadian Research Project, "Research News", *Fraunhofer*, November 2016, pp. 2–3. (https://www.fraunhofer.de/content/dam/zv/en/press-media/2016/November/ResearchNews/rn11_2016_FKIE_Safe%20navigation%20through%20the%20Northwest%20Passage.pdf) Accessed in August, 2018.

[47] Brathen Svein, Schoyen Halvor, "The Northern Sea Route versus the Suez Canal: cases from bulk shipping", *Journal of Transport Geography*, Vol: 19, Issue: 4, 2011, pp. 977–983.

[48] "Power of Siberia gas pipeline completed by 90.5 per cent, 1954 kilometers built", Official web page of *Gazprom*. (http://www.gazprom.com/press/news/2018/july/article446731/)' Accessed in August, 2018.

2018.[49] An ice-class Arc-4 vessel delivered the natural gas to Europe (France and Netherlands) and Arc-7 classed vessel transported the natural gas without any support of ice-breaker to the markets in Asia in 19 days instead of 35 days by Suez Canal. The LNG (Liquid Natural Gas) ice class vessels built by Novatek completed 47 round trips in NSR. Novatek and Total have already commenced the second train of production of LNG on 12 July, 2018 which means the shipping along NSR will be continued throughout the year 2018.

Due to the harsh conditions of the region, Arctic does not pose a danger like piracy and chokepoints as in Somali, Malacca Strait, South China Sea and Gulf of Aden. On the other hand, from the Mahanian point of view,[50] the USA tries to sustain its hegemonic maritime power as the leader of the seas. Domination of the USA on the sea-lanes creates strategic competition with other states. However, the USA is not the only one, other states also noticed the role of the sea power. China, the second larger energy consumer claimed to be a shipbuilding nation by 2015. Compared to its giant economy, China is a weak country regarding sea power. However, Arctic shipping routes offer a free area without USA domination and also safe sea-lanes.

The last but not least significance of the Arctic shipping routes is to become a cruise ship route between Europe and America or Europe and Asia. The increasing number of touristic activities to the North Polar Region must not be ignored. Furthermore, by the change of climate, some of the fish species have also changed their rotas towards much northern waters for the need of cold. This would also increase the probability to see much more fishing vessels in the seasonally open Arctic waters.

7.4 Conclusion

"The summer is in name only in the Arctic"; lines from a TV series on John Franklin's lost expedition in the search of the Northwest Passage in Canadian waters during 1845–1848. It is becoming less true day by day, provoking new appetites! Canada was the first country came up with the idea of Sector Principle, an alternative to median line principle. By Sector Principle, Canada would annex the Northwest Passage to its sovereignty. As the first supporter of the Sector Principle, Canada, is the last littoral state that made its Arctic claim to the CLCS in 2019. The claim will show the determination of the Arctic's second largest state on the Northwest Passage and it will show which methods were used for measurements.

[49] Humpert Malte, "Novatek's Yamal LNG Doubles Production Capacity Ahead of Schedule", *High North News*, 13 August 2018. (http://www.highnorthnews.com/novateks-yamal-lng-doubles-production-capacity-ahead-of-schedule/)' Accessed in August, 2018.

[50] Alfred Thayer Mahan associates the power of the state with the sea force. His piece *The Influence of Sea Power upon History, 1660–1783* details the advantages of the sea power.

While shipping continues well along Northern Sea Route under Russian legislation, there is an on-going discussion on the legitimacy of the NSR rule of Russia according to international law. USSR adopted Sector Principle in 1926 and according to a decree in 1990, the NSR was defined as "situated within its inland seas, territorial waters or exclusive economic zone adjacent to the USSR Northern Coast and include sea ways suitable for leading ships in ice" as suits with the description of sector concept where the islands, sea and land belong to states' sovereignty between chosen longitudes.[51]

The Soviets persisted on the sector principle not only for the territorial gain but also for the self-protection for its northern part from the enemies, ensuring the national economic interests, preventing air-bases and guarding Northern Sea Route. This endeavour of the Arctic's largest country was questioned by the USA by sending a vessel during 1962–1968 (Cuban Missile Crises era) to conduct a hydrological research in the Northern Sea Route whether the sector principle was suitable with international law. The USA sees the Northern Sea Route as an international strait as it does with the Northwest Passage where Washington DC has a dispute with Ottawa.

Miles and meters have always been an issue when it comes to sovereignty regarding economical activities such as oil and gas extraction, fishing and shipping. Alteration of impossible icy waters to bergy waters[52] during summer time paved the way for shipping through the new passages. Neither ice nor seasons are the only difficulties of shipping in the Arctic waters, moreover technological advancement is highly required to be capable of navigation which costs a lot as for building suitable ships as required, unstable costly fees, assurance, valid polar ship certificate and waiting for transition permission.

Bibliography

Brathen, S., & Schoyen, H. (2011). The Northern Sea route versus the Suez Canal: Cases from bulk shipping. *Journal of Transport Geography, 19*(4), 977–983.

Byers, M. (2009). *Who owns the Arctic: Understanding sovereignty disputes in the North*. Vancouver: Douglas & McIntyre Publishing.

Byers, M. (2013). *International law and the Arctic*. Cambridge: Cambridge University Press.

Byers, M. (2017). Canada's Arctic nightmare just came true: The Northwest Passage is commercial. *The Globe and Mail*, 28 November 2017. https://www.theglobeandmail.com/opinion/canadas-arctic-nightmare-just-came-true-the-northwest-passage-is-commercial/article14432440/. Accessed in Aug 2018.

Chernova, S., & Volkov, A. (2010). *Economic feasibility of the Northern Sea Route container shipping development*. Masteroppgaver i bedriftsøkonomi: Bodo University Publishing.

Churchill, R., & Ulfstein, G. (2010). The disputed maritime zones around Svalbard. In *Changes in the Arctic environment and the law of the sea, panel IX* (pp. 551–593). Leiden: Martinus Nijhoff Publishers.

[51] Dahl-Jorgensen Arnfinn, *et. alii, op. cit.* p. 257.

[52] Bergy waters: freely navigable water in terms of ice concentration. *Polar Code*, 2017, Annex 10, Chapter 1, p. 10.

Conway, M. (1906). *No man's land: A history of Spitsbergen from its discovery in 1596 to the beginning of the scientific exploration of the country*. Cambridge: Cambridge University Press.

Crawford, J. (2012). *Brownlies's principles of public international law*. Oxford: Oxford University Press.

Credland, A. (1980). Benjamin Leigh Smith: A forgotten pioneer. *Cambridge Core, 20*(125), 127–145.

Dahl-Jorgensen, A., Eger, M. K., Floistad, B., Larsen-Mejlaender, M., Lothe, L., Wergeland, T., & Ostreng, W. (2013). *Shipping in Arctic waters, a comparison of the northeast, northwest and trans-polar passages*. Chichester: Praxis Publishing.

Honneland, G. (2016). *Russia and the Arctic: Environment, identity and foreign policy*. London: I. B Tauris.

Howard, R. (2009). *The Arctic gold rush: The new race for Tomorrow's natural resources*. London: Bloomsbury Academic Press.

Humpert, M. (2018). Novatek's Yamal LNG doubles production capacity ahead of schedule. *High North News*, 13 August 2018. http://www.highnorthnews.com/novateks-yamal-lng-doubles-production-capacity-ahead-of-schedule/. Accessed in Aug 2018.

Hund, J. A. (2014). *Antarctica and the Arctic circle: A geographic encyclopedia of the earth's polar regions*. California: ABC-CLIO Publishing.

Hyde, C. (1947). *International law chiefly; interpreted and applied by the United States*. Little, Brown and Company Press: Boston.

Johansson, T., & Donner, P. (2015). *The shipping industry, ocean governance and environmental law in the paradigm shift in search of a pragmatic shift for the Arctic*. New York: Springer.

Jones, B., Puckett, C., & Vinas, J. M. (2009). Erosion Doubles along Part of Alaska's Arctic Coast: Cultural and Historical Sites Lost. *U.S. Department of the Interior*. U.S. Geological Survey, 18 February 2009. https://soundwaves.usgs.gov/2009/05/research2.html. Accessed in 20 May 2017.

Joyner, C. (1992). *Antarctica and the law of the sea*. Dordrecht: Martinus Nijhoff Publishers.

Klein, C. A. (2016). *Property: Cases, problems and skills*. New York: Wolters Kluwer.

Lajeunesse, A. (2016). *Lock, stock and icebergs: A history of Canada's Arctic maritime sovereignty*. Vancouver: UBC Press.

Mahan, A. T... *The influence of sea power upon history, 1660–1783*. Internetim yok tam referans bul

Pharand, D. (2000). *Legal status of the Arctic regions*. Toronto: Edmond Mongomery Publications.

Radin, M. Fundamental concepts of the Roman law. *California Law Review, 13*(3), 207–228. Berkeley.

Schofield, C. (2012). Departures from the coast: Trends in the application of territorial sea baselines under the law of the sea convention. *The International Journal of Marine and Coastal Law, 27*, 723–732.

Sjöberg, L. (2002). The three-point problem of the median line turning point: On the solution for the sphere and ellisoid. *International Hydrographic Review, 3*(1), 81–87.

Smith, R. W., & Bradford, T. (1998). Island disputes and the law of the sea: An examination of sovereignty and delimitation disputes. In S. Clive & H. Andrew (Eds.), *Maritime briefing, Vol: 2, No: 4* (pp. 24–25). Durham: International Boundaries Research Unit.

Staalesen, A. (2017a). Going all in Norway proposes massive opening of artic shelf. *The Independent Barents Observer*, 13 March 2017. https://thebarentsobserver.com/en/industry-and-energy/2017/03/going-all-norway-opens-its-arctic-shelf-wide-open-oil-drilling. Accessed in May 2017.

Staalesen, A (2017b). Russian Svalbard Protest Totally without Merit. *The Independent Barents Observer*, 21 April 2017. https://thebarentsobserver.com/en/arctic/2017/04/russian-svalbard-protest-totally-without-merit. Accessed in Apr 2017

Ulfstein, G. (1995). *The Svalbard Treaty; from Terra Nullius to Norwegian Sovereignty*. Oslo: Scandinavia University Press.

Wasum-Rainer, S., Winkelmann, I., & Tiroch, K. (2012). *Arctic science, international law and climate change: Legal aspects of marine science in the Arctic Ocean*. Berlin: Springer.
Witschel, G., Winkelmann, I., & Tiroch, K. (2010). *New chances and new responsibilities in the Arctic region*. Papers from the international conference at the German Federal Foreign Office in Cooperation with the Ministries of Foreign Affairs of Denmark and Norway 11–13 March 2009. Berlin: Berliner Wissenschafts-Verlag.

Webography

http://www.arctic-lio.com
http://www.gazprom.com
http://www.highnorthnews.com
https://soundwaves.usgs.gov
https://thebarentsobserver.com
https://www.fraunhofer.de
https://www.theglobeandmail.com

Chapter 8
Black Carbon, Maritime Traffic and the Arctic

Olli-Pekka Brunila, Tommi Inkinen, Vappu Kunnaala-Hyrkki, Esa Hämäläinen, and Katariina Ala-Rämi

Abstract Maritime transportation covers approximately 90% of the global traffic volumes. The global fleet consists of approximately 100,000 diesel ships, around 250 LNG ships, and a smaller number of methanol or even electric ferries. When it comes to maritime transportation, the Arctic sea route is becoming more and more interesting for the shipping industry as it has been estimated that the Northeast Passage can shorten the travelling distance significantly compared to Suez Canal.

Black Carbon (BC) is the second largest contributor to climate change emissions after carbon dioxide (CO2). BC particles spread out from different sources and the majority of BC emissions are transmitted to the Polar Regions from other parts of the globe. The share of global BC emission from international shipping is estimated to be up to 3% of the global total.

The Northern Sea Route can shorten the travelling distance, but it is important to find out, will the increase of maritime traffic effect the BC emissions in the Arctic. This paper considers how BC from ships' fuel affects the Arctic. This paper also discusses alternative fuels and emission abatement technologies, which can decrease the emissions from ships and may also affect the BC emissions in the Arctic in the future.

Keywords Black carbon · Emission abatement · Arctic · Ship traffic

---

O.-P. Brunila (✉) · T. Inkinen · E. Hämäläinen
University of Turku, Turku, Finland
e-mail: olli-pekka.brunila@utu.fi

V. Kunnaala-Hyrkki
South-Eastern Finland University of Applied Sciences, Mikkeli, Finland

K. Ala-Rämi
University of Oulu, Oulu, Finland

E. Pongrácz et al. (eds.), *Arctic Marine Sustainability*,
Springer Polar Sciences, https://doi.org/10.1007/978-3-030-28404-6_8

8.1 Introduction

Over recent decades, temperatures in the Arctic have increased. The climate of the Arctic Region is known to be warming at almost twice the rate of the rest of the world. Reduction of global CO2 emissions is required to slow this warming, but there is also a need to reduce short-lived climate forcers. It has been estimated that the majority of influence on the radiative forcing in the Arctic is from external emissions of greenhouse gases and particulate matter (PM), with possibly half of Arctic temperature rise linked to black carbon (BC) (IMO 2015). Even though the atmospheric concentrations of BC in remote areas, such as the Arctic Region, are generally low, their effects on the regional climate may be substantial (e.g. Flanner 2013; Winiger et al. 2016; Sand et al. 2016; AMAP 2011; Corbett et al. 2007).

Sources, whose emissions are rich in BC, can be grouped into a small number of categories: diesel engines, industry, residential solid fuel, and open burning. The largest global sources are open burning of forests and savannas, solid fuels burned for cooking and heating, and on-road and off-road diesel engines. Dominant emitters of BC from other types of combustion depend on the location. Industrial activities are also significant sources, but e.g. shipping emissions provide only a minor contribution to BC emissions at the global scale. What makes shipping emissions noteworthy is that shipping emits into regions that otherwise have low concentrations of emissions. Currently, there are few sources of pollution within the Arctic itself so almost all BC is transported there from other regions (Bond et al. 2013).

Sea transportation covers approximately 90% of the global traffic volumes and it also contribute global climate change and causes health impacts through emission of greenhouse gases (GHGs) and other pollutants, including CO2, NOx, SOx and various species of particulate matter (PM) including BC. BC emissions from the shipping industry are thought to contribute about 2–3% of global BC (IMO 2015; Corbett et al. 2010).

Decreasing ice volumes in the Arctic sea has increase the interest in efforts to establish new trade passages. The Arctic sea ice retreat is opening up both the Northwest Passage (across Canadian Arctic waters) and the Northeast Passage (also known as the Northern Sea Route passing along the Siberian north coast), during parts of the year (Yumashev et al. 2017: Kiiski 2017). Journeys between Asia and Europe and Eastern US and Asia through the Arctic could cut travel distances by 25% and 50% respectively in the Northwest Passage and Northern Sea Route compared with the current sea routes. In addition, the routes are generally considered to be financially viable as they would bring savings to both time and fuel (Corbett et al., 2010; IMO 2015). Although increased Arctic shipping may provide commercial opportunities, the associated environmental issues should also be considered (IMO 2015; Corbett et al. 2010; Yumashev et al. 2017).

The purpose of this chapter is to consider the Arctic sea traffic and BC relations. This is done through a literature review and by presenting the latest BC emission level data. The regulatory framework is discussed as the backbone that controls commercial activities, and thus, the responsible business potentials in the Arctic maritime transport. The majority of the paper is conceptual and it applies latest measurement data in order to consider the problematics of the Arctic maritime transport. The topic is challenging due to constantly developing technologies for emission reduction and control and other uncertainties related to the shipping industry, such as the asymmetric growth in activity among ship types. These factors make it difficult to estimate the future effects of shipping in the Arctic Region (Corbett et al. 2010; VITO 2013).

8.2 Black Carbon

BC emissions come from the combustion process when fossil fuel or biomass is burned. Carbonaceous material is formed near flames during the combustion process (Bond et al. 2013). Fossil fuels are very widely used in transport sector, industry and the household sector, and forest fire produces a lot of BC emissions in world wide. After the open burning of forests and savannas, the largest emission sources include solid fuels burned for cooking and heating, and on-road and off-road diesel engines. Dominant emitters of BC from other types of combustion depend on the location. Industrial activities are also significant sources, but e.g. shipping emissions provide only a minor contribution to BC emissions at the global scale (Bond et al. 2013).

BC is the second major contributor to climate change after CO2 (Petzlod et al. 2013; Aplin 2015; Bond et al. 2013). BC is a short-lived climate forcer or a short-lived climate pollutant. In practice, BC has a greater effect on the Polar Regions' climate than CO2. BC emissions have notable local climatic effects in the Arctic, as the BC particles absorb solar heat very effectively. When such particles are deposited to a reflective surface, such as ice, they may significantly alter the albedo of the surface and increase the amount of absorbed solar heat, which in turn leads to warming of the surface. This leads to increase in the melting of ice and snow coverage and contributes directly to the climate change (Vihanninjoki 2014; Flanner et al. 2007). What makes shipping emissions noteworthy is that shipping emits into regions that otherwise have low concentrations of emissions. Currently, there are few sources of pollution within the Arctic itself so almost all BC is transported there from other regions (Bond et al. 2013).

As yet, there is no universal definition of BC and it has been problematic to reach a consensus on the matter. In order for measurement and emissions control technologies and policies to be able to operate as intended and in a cost-efficient fashion, it is vital reach an understanding on the matter. IMO's Maritime Environmental Committee (MEPS) has in its meeting approved a definition for BC: BC is a solid, carbon-based substance formed as carbon-based fuel's burning process is incomplete. As it enters the atmosphere it can powerfully absorb all lengths of visible light. Over 80% of BC's weight is pure carbon, of which majority have dual-bonds (sp2). In the atmosphere, the particles form into a sphere, which have an aerodynamic diameter of around 20–50 nm. Fresh BC can absorb 550 λ per

5 m^2 gram (Bond et al. 2013; IMO 2015; Mukunda et al. 2015). The ability of BC particles to absorb light depends on its consistency, shape, size distribution and particle mixing state.

BC's climate effects are either direct or indirect. The effects of different emissions on the climate are divided into three categories: (1) as it floats in the air, BC absorbs sunlight and thus heats up the atmosphere (direct effect); (2) BC affects cloud characteristics (indirect effect); (3) on top of ice and snow, BC absorbs lights and heats it up, thus expediting the melting process (snow effect). In the Polar Regions, the snow effect has the biggest effect on the climate. Globally, the snow effect is estimated to amount to up to 1%. On the other hand, for example, Finland's direct and indirect BC's climate effect only amounts to around one per mil. Estimates are based on climate simulation calculations. The biggest source of BC in Finland is small-scale burning of wood (Twigg 2009, AMAP 2011; AMAP 2015: Bond et al. 2013).

BC is a so-called primary particle, which means that it is in a solid form as it enters the atmosphere. As it forms, BC is also hydrophobic, meaning that it is water resistant. This phase only lasts hours (AMAP 2015). The formation of secondary BC particles takes place only as it reaches the atmosphere. As the carbon nucleus enlarges, they act as a surface for the carbon nitrogen from the gas stream to stick to. The particle grows but also becomes more dense as liquid solidifies and evaporates (AMAP 2015; Twigg 2009).

8.3 Black Carbon from Different Fuel Types

Sea transportation covers approximately 90% of the global traffic volumes. Shipping contribute significantly to global climate change and health impacts through emission of GHGs and other pollutants, including CO2, NOx, SOx and various species of particulate matter (PM) including BC. BC emissions from the shipping industry are thought to contribute about 2–3% of global BC (IMO 2015; Corbett et al. 2010), which makes sea transportation a minor BC producer compared to road transportation.

The emissions from maritime transport do not only depend on the total traffic but also on the characteristics of the fleet, which are at least equally important. Significant factors include the average engine power, engine type and fuel type (VITO 2013). The global fleet consists of approximately 100,000 diesel ships, around 250 LNG ships, and a smaller number of methanol or even electric ferries. According to Winther et al. (2014) & Timonen et al. (2017) in 2012, the largest share of Arctic ships' BC emissions originated from fishing ships (45%) followed by passenger ships (20%), tankers (9%), general cargo (8%) and container ships (5%).

Approximately 10–20% of global BC emission are from road transportation (Bond et al. 2013; Lund et al. 2014). From this, BC emissions from diesel engines are responsible of about 90%. In some countries, diesel engines produces 70% of total BC emissions (Lund et al. 2014). The International Council of Combustion Engines (CIMAC) came to the same results, as they measured BC amounts from different fuels by using Filter Smoke Number (FSN) method, which is standardized, in ISO 10054 and in ISO 8178. (CIMAC 2012; IMO 2015) FSN is an optical measurement based on filter darkening or photoacoustic method. In photoacoustic method, particles are heated by laser and the sound and lights absorbing are measured from particles. Typically a different measurement technology gives different results. In the small car diesel engines with automotive diesel fuel, the elementary carbon (EC) or BC can be over 70% in diesel particulate matter. Figure 8.1 presents typical diesel particulate matter composition by using ISO 8178 Measurement method, which is NSF (CIMAC 2012).

Figure 8.2 presents the share of EC/BC from four-stroke diesel engine with maritime type distillate fuel or light fuel oil (LFO). These kind of diesel engines are medium speed and running relatively heavy load. In these measurements, the

Fig. 8.1 Typical diesel PM from vehicle diesel engine and automotive diesel CIMAC (2012)

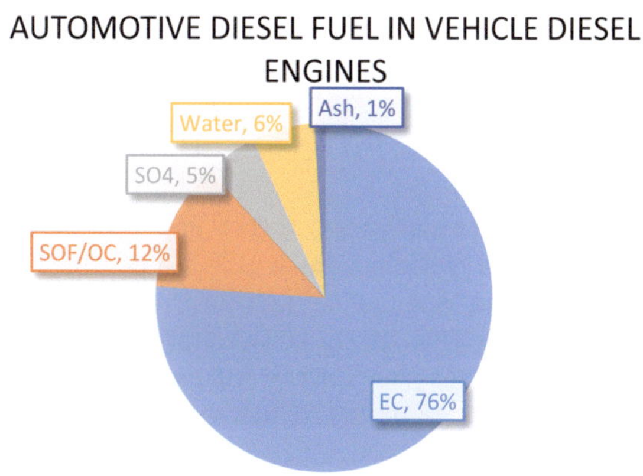

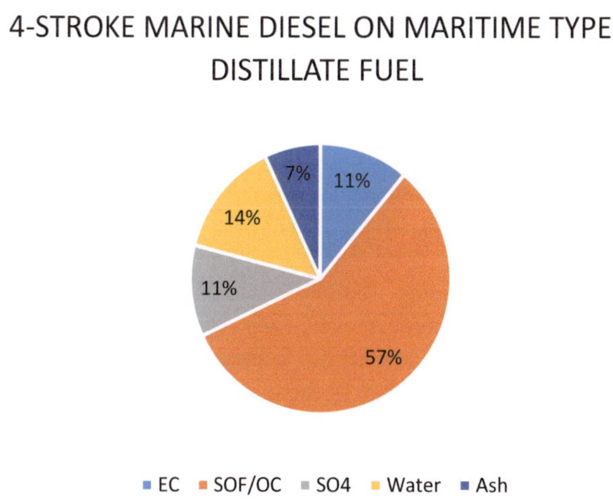

Fig. 8.2 Distillate fuel PM from 4 stroke marine diesel engine CIMAC (2012)

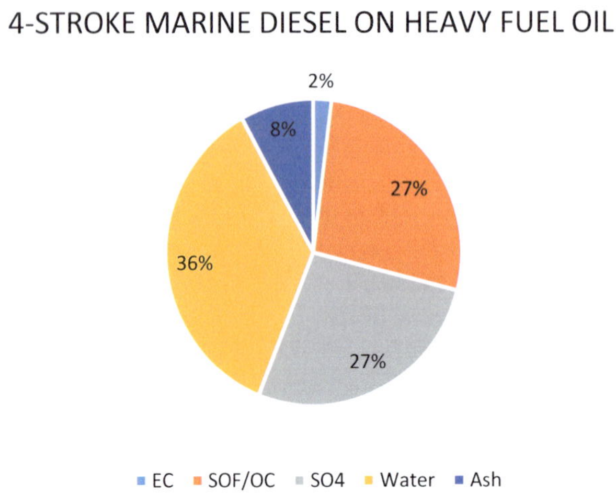

Fig. 8.3 Heavy fuel oil PM from 4-stroke marine diesel engine CIMAC (2012)

amount of EC in fuel PM was 11%. According to CIMAC (2012), variation is typically around 10–15%.

Figure 8.3 presents heavy fuel oil PM shares. When heavy fuel oil (HFO) is measured, the PM typically consist of 2–5% EC/BC. In the perspective of BC's warming effect, in automotive fuel in small diesel engines, PM consistency is approximately 75% BC. In HFO, in the PM, the share is only 2–5%. Compared to BC, Sulphates, Organic Carbon (OC) and mineral dust have a cooling effect on the environment (CIMAC 2012; Fuglestvedt et al. 2009).

New fuel types can significantly reduce the amount of emissions from shipping. Nevertheless, given the long turnover rate of maritime vessels, the effects of introducing alternative fuels will not be significant in the near-term (VITO 2013).

8.4 BC Emission Abatement Technologies

BC emissions have been measured and calculated in several studies. The amount of BC varies based on the fuel type and method of determination. According to CIMAC (2012), the BC emission vary from 0.1 to 1 g/kg per fuel burned. CIMAC (2012) also concluded that, for shipping, the emissions of BC seems to be highly overestimated. According to different studies, there are several ways to mitigate BC emissions (Azzara et al. 2015; CIMAC 2012; IMO 2015). In addition, IMO regulations for decreasing BC emissions are in the making. Table 8.1 below presents, how IMO categorizes BC abatement technologies.

Previous studies suggest that the use of diesel particulate filters, LNG, scrubbers and low-sulphur fuels (LSF) can reduce shipping BC emissions by up to 70% (Azzara et al. 2015). Based on the studies of Aakko-Saksa et al. (2017) and Timonen

Table 8.1 IMO abatement technologies for BC

(1) Fuel efficiency – vessel design (excludes engine, fuel options)
(2) Fuel efficiency – monitoring options
(3) Fuel efficiency – engine options
(4) Slow steaming
(5) Fuel treatments:
Colloidal catalysts
Water in Fuel Emulsion (WiFE)
(6) Fuel quality (traditional fuels)
(7) Heavy fuel oil – distillate
(8) Alternative fuels:
Biodiesel
LNG
Methanol – Dimethyl Ether (DME)
Nuclear
(9) Exhaust treatment:
Electrostatic precipitators (ESP)
Diesel particulate filter (PDF)
Diesel oxidation catalysts (DOCs)
Selective catalytic reduction (SCR)
Exhaust gas recirculation (EGR)
Exhaust gas scrubber (EGS)

et al. (2017) done in two different measurement campaign laboratories and on board, the laboratory measurements showed that BC amount was higher in 0.5% sulphur fuel than in 2.5% sulphur fuel at 25% engine load, but not at 75% engine load. Low BC amount was observed in the 0.1% Sulphur fuel and Bio30 fuels, with a particularly low BC and polycyclic aromatic hydrocarbons amounts for the Bio30 fuel. The measurements on board a modern ship showed that the new engine emission control technologies (SCR + scrubber) and lower sulphur (~0.7% sulphur) fuels dramatically reduced the BC and PM concentrations of ship exhaust when compared to those of an old marine engine at 25% engine load. The engine load also had less influence on BC for a newly built ship when compared to an old marine engine.

As stated above in Table 8.1, IMO has defined several abatement technologies, including scrubbers and slow steaming. In addition, emissions can be controlled with other environmental policy instruments. These include different kinds of negative and positive incentives and legislative actions, such as the introduction of emission control areas. Despite the fact that IMO has not yet regulated BC, it is already indirectly regulated by IMO's MARPOL Annex VI, which sets limits for nitrogen oxides and the sulphur content of fuel. BC is also indirectly regulated in the Emission Controlled Areas (ECAs), where a 0.1% limit on sulphur emissions is already in force. The Baltic Sea is one of these special emission controlled areas. In these areas, vessels need to use either low sulphur fuel or exhaust cleaning technologies such as scrubbers. In the future, IMO will drive further research on the impacts of BC, potentially bringing about future BC emission regulations (Aplin 2015; IMO 2015).

According to Brunila et al. (2017) there is no single solution that would decrease BC emissions. Different abatement technologies; scrubbers, fuel type selection, slow steaming, better maintenance for engines and better burning of fuel are all together factors that can decrease emissions and BC. Currently the focus is more on how to measure BC emission than in how to decrease BC emission because there are no regulations or limitations for BC at the moment.

8.5 Ship Traffic in the Arctic

The Northern Sea Route has been in the interest of seafarers for centuries. The decline in Arctic sea ice has reignited interest in efforts to establish new Arctic trade passages (Mjelde et al. 2014; Winther et al. 2014; Mukunda et al. 2018; Kiiski 2017; Yumashev et al. 2017). Polar Regions are warming twice as fast as other regions on Earth. Reason for this include the escalated shrinking of the snow and ice cover. As snow and ice disappear, more radiation from the sun is absorbed in the dark surfaces unearthed (Rubbel 2015; Vihanninjoki 2014). Shipping in the Arctic Region is expected to grow by around 1.8–5% by 2050 (AMAP 2015). The growth rate depends on the capacity of the already established shipping routes and the costs of the new route in comparison to the current routes (AMAP 2015, Vihanninjoki 2014).

The Arctic sea ice retreat is opening up both the Northwest Passage (across Canadian Arctic waters) and the Northeast Passage (also known as the Northern Sea Route, NSR, passing along the Siberian north coast), during parts of the year (Yumashev et al. 2017; Kiiski 2017). Journeys between Asia and Europe, and Eastern US and Asia through the Arctic could cut travel distances by 25% and 50% respectively in the Northwest Passage and Northern Sea Route, compared with the current sea routes. Both routes are considered to be financially viable as they would bring savings to both time and fuel (Corbett et al. 2010; IMO 2015). It has been estimated that around 5% of the world's trade could be shipped through the Northern Sea Route in the Arctic alone under year-round and unhampered navigability (Yumashev et al. 2017).

Although increased Arctic shipping may provide commercial opportunities, the associated environmental issues should also be considered. Increases in Arctic shipping will introduce direct near-surface emissions of pollution, including BC (IMO 2015; Corbett et al. 2010; Yumashev et al. 2017). Due to the increasing interest in Arctic ship routes, the IMO preparing limits in BC emissions of ships (Timonen et al. 2017; Brunila et al. 2017).

Currently, Arctic shipping does not contribute a significant amount to the region's emissions, but the emissions occur further north and thus, they have a stronger regional impact (Quinn et al. 2008). Direct emissions are significant contributors as Arctic warming is most sensitive to emissions within the region compared to the current emissions where most must survive long-range transport from its source before directly impacting the region (IMO 2015). Due to the remoteness and poor accessibility, the current emission levels in the Arctic are relatively low compared to the global averages. In such an environment, even small absolute increases are likely to lead to significant relative increases (Vihanninjoki 2014).

Most significant for the Arctic is the additional source of short-lived climate forcing agents, such as BC, from ships in proximal transport distance (Corbett et al. 2010). BC emissions might have significant local climatically effects in the Arctic, because BC particles absorb solar heat very effectively. When such particles are deposited to a reflective surface they may significantly alter the albedo of the surface and increase the amount of absorbed solar heat, which in turn leads to warming of the surface and contributes directly to the climate change (Flanner 2013; Vihanninjoki 2014).

In order for the shipping industry's interest on Arctic trade passages to grow, several requirements need to be met. First of all, the duration of potential ice-free periods has to increase or at least remain the same. Other factors that need to be taken into consideration are, for example, appropriate fuels, bunker prices, possible icebreaker assistance and cost of icebreaking and transit fees, investments in equipment and personnel training and insurance costs. The administration, coastal infrastructure improving the safety along with suitable rescue equipment also needs to be improved along the route. (AMAP 2015, Vihanninjoki 2014).

In general, the regulation of operation in the Arctic is still insufficient. Due to the lack of regulation, inappropriate equipment and preparations may lead to adverse consequences in the Arctic region. Vihanninjoki (2014) has listed several contributing factors related to shipping in the Arctic:

- Insufficiently equipped and insufficiently ice-strengthened vessels
- Crews that are not trained to handle the difficult navigational and operational challenges in the Arctic waters
- Lack of shoreside infrastructure
- Arctic waters are not very well charted
- Search and rescue infrastructure is limited and regionally varying

According to Mjelde et al. (2014), in the year 2012 almost 1350 vessels operated in the Arctic Region. These ships sailed a total of 5.8 million nautical miles and consumed about 0.166 million tons of distillate fuels and 0.135 million tons of HFO. Calculations show that these ships produced 105 tons BC annually. It can be said that, on average, the majority of large cargo vessels (wet and dry bulk) use HFO. HFO is most notably used in container vessels. The fishing fleet in the Arctic uses mainly distillate fuels. In the Arctic shipping, biggest BC emission peaks occur in the summer time, during which the weather and ice conditions are most suitable for the ships to operate. Geographically, the BC emission distributions are close to the Behring Sea, the Barents Sea, and the Labrador Sea, in which most of the popular sea routes are located.

Several studies have been conducted on how vessel traffic and emission load will developed in the future in the Arctic (e.g. Corbett et al. 2010; Dalsoren et al. 2010; Dalsoren 2013). Depending on the development scenario (high, business as usual, or low), it has been estimated that the global shipping will increase 1–3% in the Arctic areas. At the same time, shipping outside the Arctic areas will increase 2–3% depending on the scenario.

Increased shipping and increased emissions can have a huge effect in the polar latitudes. Emission peaks especially during the summer time will grow. Yet, it is unlikely that the Arctic will become a viable alternative for transit shipping in the near future. For example, the Northern Sea Route has only a fraction of the vessel traffic compared to the traffic quantity in the Suez Canal or other busier sea routes. Most of the traffic in the Arctic area consists of intra-Northern Sea Route journeys that are related to the Russian Yamal and Gydan Peninsula LNG projects. Currently, Russian energy companies have several energy projects in the Arctic areas that can increase the amount of ship traffic during a certain period of time, usually only for a month or two.

Only 19 vessels and 214,513 tons of cargo transited through the Northern Sea Route in 2016. In comparison, 16,800 vessels and 974 million tons of cargo transited the Suez Canal that year. The peak year in the Northern Sea Route was in 2013, when 71 vessels and 1.36 million tons of cargo transited through the route. In comparison, 16,600 vessels and 915 million tons of cargo transited through the Suez Canal in 2013 (Northern Sea Route Administration 2017). Operating in the Arctic will pose challenges especially to maritime safety and environmental issues. Polar Code entered into force in 1 January 2017 and it will improve the safety issues and standards (Yliskylä-Peuralahti et al. 2016).

8.6 Conclusions

In order to conclude, the following interpretations may be drawn concerning BC emissions and Arctic areas. A clear starting fact is that during the recent decades, temperatures in the Arctic have increased. Reduction of global CO2 emissions is required to slow the warming, but there is also a need to reduce short-lived climate forcers, especially BC, which is considered to be specifically harmful in the arctic environment (e.g. IMO 2015; Winiger et al. 2016; Sand et al. 2016; AMAP 2011). The temperature increase in the Arctic has also lead to the decline in Arctic sea ice. This, in turn, has reignited interest in efforts to establish new trade passages. Although increased Arctic shipping may provide commercial opportunities, the associated environmental issues should also be considered, since increases in Arctic shipping will introduce direct near-surface emissions of pollution, such as BC (IMO 2015; Corbett et al. 2010; Yumashev et al. 2017). Due to the remoteness and poor accessibility, the current emission levels in the Arctic are relatively low compared to the global averages. In such an environment, even small absolute increases are likely to lead to significant relative increases. BC emissions are likely to have notable local climatic effects in the Arctic, as the BC particles absorb solar heat very effectively. When such particles are deposited to a reflective surface they may significantly alter the albedo of the surface and increase the amount of absorbed solar heat, which in turn leads to warming of the surface and contributes directly to the climate change (Flanner 2013; Vihanninjoki 2014).

Second, the emissions from international maritime transport do not only depend on the total traffic but also on the characteristics of the fleet, which are at least equally important. Significant factors include the average engine power, engine type and fuel type. Yet, since the turnover rate of vessels is usually quite slow, the effects of alternative fuels are not that rapid (VITO 2013). There are several means for reducing emissions from shipping in the Arctic. IMO has defined several abatement technologies, including scrubbers and slow steaming. In addition, emissions can be controlled with other environmental policy instruments. These include different kinds of negative and positive incentives and legislative action, such as the introduction of emission control areas (Makkonen and Inkinen 2018). On the other hand, before there are clearly defined limits and regulations for BC and other emission in the Arctic areas, there will be a lack of investments for new cleaner technologies and cleaner fuels (Brunila et al. 2017). Currently, IMO has not defined the area that would become the 'Artic Emission Control Area'. In addition, the IMO has not defined whether BC regulations and limits should concern shipping merely in the Artic areas or should BC limits be introduced in international shipping more widely and concern also other areas, such as the Baltic Sea and North Sea?

Finally, it is difficult to estimate the future effects of shipping in the Arctic region. This is due to constantly developing technologies (for emission reduction), control regulations (IMO), and other uncertainties (such as asymmetric growth in activity between different types of ships) that are associated with the shipping industry (Corbett et al. 2010; VITO 2013). In addition, the introduction of new legislation related to e.g. emission control areas may lead to a reduction of interest of the shipping industry in the Arctic passages. In order for the shipping industry's interest on Arctic trade passages to grow, several requirements need to be met. First of all, the duration of potential ice-free periods has to increase. Other factors that need to be taken into consideration are, for example, bunker prices, possible icebreaker assistance and transit fees, investments in equipment and personnel training and insurance costs (Vihanninjoki 2014). In general, the regulation of operation in the Arctic is still insufficient. Due to the lack of regulation, inappropriate equipment and preparations may lead to adverse consequences in the Arctic region.

References

Aakko-Saksa, P., Murtonen, T., Vesala, H., Koponen, P., Timonen, H., Teinilä, K., Aurela, M., Karjalainen, P., Kuittinen, N., Puustinen, H., Piimäkorpi, P., Nyyssönen, S., Martikainen, J., Kuusisto, J., Niinistö, M., Pellikka, T., Saarikoski, S., Jokela, J., Simonen, P., Mylläri, F., Wihersaari, H., Rönkkö, T., Tutuianu, M., Pirjola, L., & Malinen, A. (2017). *Black carbon emissions from a ship engine in laboratory (SEA-EFFECTS BC WP1)* (Report VTT-R-02075-17).

AMAP. (2011). *The impact of black carbon on Arctic climate* (P. K. Quinn, A. Stohl, A. Arneth, T. Berntsen, J. F. Burkhart, J. Christensen, M. Flanner, K. Kupiainen, H. Lihavainen, M. Shepherd, V. Shevchenko, H. Skov, & V. Vestreng, Eds.). Oslo: Arctic Monitoring and Assessment Programme (AMAP).

AMAP. (2015). *Black carbon and ozone as Arctic climate forcers*. Oslo: Arctic Monitoring and Assessment Programme (AMAP).

Aplin, J. (2015). *Five things the shipping industry needs to know about black carbon*. Available at URL: know-about-black-carbon.

Azzara, A., Minjares, R. and Rutherford, D. (2015). *Needs and opportunities to reduce black carbon emissions from maritime shipping*. The International Council on Clean Transport (ICCT). Working Paper 2015-2. Available at URL: https://theicct.org/sites/default/files/publications/ICCT_black-carbon-maritimeshipping_20150324.pdf

Bond, T. C., Doherty, S. J., Fahey, D. W., Forster, P. M., Berntsen, T., DeAngelo, B. J., Flanner, M. G., Ghan, S., Kärcher, B., Koch, D., Kinne, S., Kondo, Y., Quinn, P. K., Sarofim, M. C., Schultz, M. G., Schulz, M., Venkataraman, C., Zhang, H., Zhang, S., Bellouin, N., Guttikunda, S. K., Hopke, P. K., Jacobson, M. Z., Kaiser, J. W., Klimont, Z., Lohmann, U., Schwarz, J. P., Shindell, D., Storelvmo, T., Warren, S. G., & Zender, C. S. (2013). Bounding the role of black carbon in the climate system: A scientific assessment. *Journal of Geophysical Research-Atmospheres, 118*(11), 5380–5552.

Brunila, O.-P., Ala-Rämi, K., Inkinen, T. & Hämäläinen, E. (2017). *Black carbon measurement in the Arctic – Is there a business potential?* Final report of the work package 3 in the sea effects black carbon project. Publications of Centre for Maritime Studies, University of Turku. A 73, 2017 Turku, Finland. ISBN 978–951–29–6983–8 (PDF).

CIMAC. (2012). Background information on black carbon emissions from large marine and stationary diesel engines – definition, measurement methods, emission factors and abatement technologies. The International Council on Combustion Engines.

Corbett, J. J., Wang, C., Winebrake, J. J., & Green, E. (2007). *Allocation and forecasting of global ship emissions*. Prepared for the clean air task force. Boston.

Corbett, J. J., Lack, D. A., Winebrake, J. J., Harder, S., Silberman, J. A., & Gold, M. (2010). Arctic shipping emissions inventories and future scenarios. *Atmospheric Chemistry and Physics, 10*, 9689–9704.

Dalsøren, S., Eide, M., Myhre, G., Endresen, O., Isaksen, I., & Fuglestvedt, J. (2010). Impacts of the large increase in international ship traffic 2000–2007 on tropospheric ozone and methane. *Environmental Science & Technology, 44*, 2482–2489.

Dalsøren, S. B., Samset, B. H., Myhre, G., Corbett, J. J., MInjares, R., Lack, D., & Fuglestvedt, J. S. (2013). Environment impacts of shipping in 2030 with a particular focus on the Arctic region. *Atmospheric Chemistry and Physics, 13*, 1941–1955.

Flanner, M. G. (2013). Arctic climate sensitivity to local black carbon. *Journal of Geophysical Research-Atmospheres, 118*, 1840–1851.

Flanner, M. G., Zender, C. S., Randerson, J. T., & Rasch, P. J. (2007). Present-day climate forcing and response from black carbonin snow. *Journal of Geophysical Research-Atmospheres, 112*, 2007.

Fuglestvedt, J., Berntsen, T., Eyring, V., Isaksen, I., Lee, D. S., & Sausen, R. (2009). Shipping emissions: From cooling to warming of climate and reducing impacts on health. *Environmental Science & Technology, 43*, 9057–9062.

IMO. (2015). *Investigation of appropriate control measures (abatement technologies) to reduce black carbon emissions from international shipping* (p. 2015). London: International Maritime Organization.

Kiiski, T. (2017). *Feasibility of commercial cargo shipping along the Northern Sea Route* (p. E12). Turku: Publications of University of Turku.

Lund, M. T., Berntsen, T. K., Heyes, C., Klimot, Z., & Samser, B. H. (2014). Gloal and regional climate impacts of black carbon and co-emitted species from on-road diesel sector. *Atmospheric Environment, 98*, 50–58.

Makkonen, T., & Inkinen, T. (2018). Sectoral and technological systems of environmental innovation: The case of marine scrubber systems. *Journal of Cleaner Production, 200*, 110–121.

Mjelde, A., Martinsen, K., Eide, M., & Endresen, O. (2014). Environmental accounting for Arctic shipping – A framework building on ships tracking data from satellites. *Marine Pollution Bulletin, 87*(1–2), 22–28.

Mukanda, M. G. M., Babu, S., Moorthy, K., Thakur, R., Chaubey, J., & Nair, V. (2015). Aerosol black carbon over svalbard regions of arctic. *Polar Science, 10*, 1–11.

Mukunda, M. G., Babu, S. S., Pandey, S. K., Nair, V. S., Vaishya, A., Girach, I. A., & Koushik, G. N. (2018). Scavening ratio of black carbon in the Arctic and the Antarctic. *Polar Science, 16*, 10–22.

Northern Sea Route Administration. (2017). *Vessel activity*. Northern Sea Route Administration. Available at URL: < http://www.nsra.ru/en/home.html>

Petzlod, A., Orgen, J. A., Fiebig, M., Li, S. M., Baltensperger, U., Holtzer-Popp, T., Kinne, S., Pappalardo, G., Sugimoto, N., Whrli, C., Wiedensohler, A., & Zhang, X. Y. (2013). Recommendation for reporting "Black Carbon" measurements. *Atmospheric Chemistry and Physics, 13*, 8365–8379.

Quinn, P. K., Bates, T. S., Baum, E., Doubleday, N., Fiore, A. M., Flanner, M., Fridlind, A., Garrett, T. J., Koch, D., Menon, S., Shindell, D., Stohl, A., & Warren, S. G. (2008). Short-lived pollutants in the Arctic: Their climate impact and possible mitigation strategies. *Atmospheric Chemistry and Physics, 8*(6), 1723–1735.

Rubbel M. M. (2015). *Black carbon deposition in the European Arctic from the preindustrial to the pre-sent*. Dissertationes Schola Doctoralis Scientiae Circumiectalis, Alimentariae, Biologicae. ISSN 2342–5423 (print).

Sand, M., Berntsen, T. K., von Salzen, K., Flanner, M. G., Langner, J., & Victor, D. G. (2016). Response of Arctic temperature to changes in emissions of short-lived climate forcers. *Nature Climate Change, 62*(016), 286–289.

Timonen, H., Aakko-Saksa, P., Kuittinen, N., Karjalainen, P., Murtonen, T., Lehtoranta, K., Vesala, H., Bloss, M., Saarikoski, S., Koponen, P., Piimäkorpi, P. & Rönkkö, T. (2017). *Black carbon measurement validation onboard (SEA-EFFECTS BC WP2)* (Report VTT-R-04493-1).

Twigg, M. (2009). Cleaning the air we breathe – controlling diesel particulate emissions from passenger cars. *Platinum Metals Review, 53*, 27–34.

Vihanninjoki, V. (2014). *Arctic shipping emissions in the changing climate*. Reports of the Finnish Environment Institute 41/2014.

VITO. (2013). Specific evaluation of emissions from shipping including assessment for the establishment of possible new emission control areas in European Seas. Paul Campling, P., Janssen, L., Vanherle, K., Cofala, J., Heyes, C. & Sander, R.Vision on Technology (VITO), Belgium.

Winiger, P., Andersson, A., Eckhardt, S., Stohl, A., & Gustafsson, Ö. (2016). The sources of atmospheric black carbon at a European gateway to the Arctic. *Nature Communications, 7*, 12776.

Winther, M., Christensen, J. H., Plejdrup, M. S., Ravn, E. S., Eriksson, Ó. F., & Kristenssen, H. O. (2014). Emission inventories for ships in the Arctic based on satellite sampled AIS data. *Atmospheric Environment, 91*, 1–14.

Yliskylä-Peuralahti, J., Ala-Rämi, K., Rova, R., Kolli, T., & Pongracz, E. (2016). *Matching safety and environmental regulations regarding the Inter-national maritime organization's polar code in Finland (POLARCODE)*. Publications of the govenrment's analysis, assessment and research activities 11/2016. Prime Minister's office, March 14th, 2016.

Yumashev, D., van Hussen, K., Gille, J., & Whiteman, G. (2017). Towards a balanced view of Arctic shipping: estimating economic impacts of emissions from increased traffic on the Northern Sea Route. *Climatic Change, 143*, 143–155.

Chapter 9
Impact of the Potential Implementation of Unmanned Aerial Vehicles on the Northern Sea Route Safety Monitoring

Nikita Kuprikov, Mikhail Kuprikov, and Maxim Shishaev

Abstract The chapter presents the peculiarities of creating an infrastructure, which will make it possible to provide a real time control over the infrastructure projects in the Arctic by way of the communication between the aerospace systems and the autonomous land and sea monitoring stations. The application of new information technologies and that of the hardware and software ensuring the competitiveness will enable one to safely and effectively use various high technologies in navigation, aviation, and space.

Keywords UAV · Northern Sea route · Arctic · Safety monitoring · Competitiveness

The economic and geopolitical significance of the Arctic is largely determined by the potentially largest gas and oil reserves. Hydrocarbon reserves in the Arctic are one of the main reasons for the keen interest of both the Arctic and non-Arctic states in this region. Large-scale Arctic projects, including international ones, such as Yamal LNG implemented together by Russia and China, will appear more and more often with the disappearance of eternal ice that previously chained the region.

In this situation, it is critical to create the most accurate prediction systems of ice cover behavior in areas that could potentially become sites for the development of new exploitable resources in the future. The unpredictability of ice behavior can become a threat both for cargo and passenger ships, and for drilling platforms and installations. Their operational efficiency may be undermined as a result of incorrect forecasting of ice cover behavior. With this in mind, it is necessary to develop integrated monitoring systems that would take into account all climatic and tempo-

N. Kuprikov (✉) · M. Kuprikov
Moscow Aviation Institute, Moscow, Russia

M. Shishaev
Kola Scientific Center, Apatity, Russia

E. Pongrácz et al. (eds.), *Arctic Marine Sustainability*,
Springer Polar Sciences, https://doi.org/10.1007/978-3-030-28404-6_9

Table 9.1 Geological estimates of undiscovered oil and gas in the Arctic

Petroleum province	Crude oil (billion barrels)	Natural gas (trillion cubic feet)	Natural gas liquids (billion barrels)	Total (oil equivalent in billions of barrels)
West Siberian Basin	3.66	651.50	20.33	132.57
Arctic Alaska	29.96	221.40	5.90	72.77
East Barents Basin	7.41	317.56	1.42	61.76
East Greenland Rift Basin	8.90	86.18	8.12	31.39
Yenisey-Khatanga Basin	5.58	99.96	2.68	24.92
Amerasian Basin	9.72	56.89	0.54	19.75
West Greenland-East Canada	7.27	51.82	1.15	17.06

Source: *Circum-Arctic Resource Appraisal: Estimates of Undiscovered Oil and Gas North of the Arctic Circle*: Kenneth J. Bird and others, United States Geological Survey, Fact Sheet 2008–3049. July 2008

ral factors. Such systems may include not only special Arctic satellite systems, but the prospective use of unmanned aerial vehicles (UAVs).

The resource potential of the Arctic is the main factor that forces the largest oil and gas companies to develop new systems for drilling and gas production, the safety of which must be guaranteed. The resource capacity of the Arctic accounts for 25% of all unexplored hydrocarbons in the world (Table 9.1) (NPC 2015).

At the same time, not least because of the melting of an unprecedented amount of Arctic ice (Arctic sea ice is now declining at a rate of 12.8% per decade, relative to the 1981 to 2010 average – NASA[1]), the life cycle of transport corridors in the Arctic Ocean is changing, and shipping in new conditions requires new approaches and working coordination systems (Fig. 9.1).

Climate change in now influencing the dynamics of atmospheric conditions, which introduce a whole range of new external influencing factors on the equipment and specialists controlling these systems (Novikov 2015). The development of the transport network including its associated infrastructure elements (lighthouses, berths, piers, wiring, information systems, etc.) requires continuous monitoring of a number of parameters for analyzing and planning, as well as for the sustainable development of the natural resources use and development (McDonald 2018). In this area, in addition to satellite measurements by the prospective Arktika satellite system, measurements directly on site are required, and this is where new types of autonomous measuring equipment can demonstrate its advantages.

The Arktika satellite network can perform a variety of remote-sensing tasks, such as monitoring of environmental conditions, and also provide reliable communications and navigation across the region. A pair of Arktika-M satellites fully

[1] Arctic Sea Ice Minimum by NASA. – Access: https://climate.nasa.gov/vital-signs/arctic-sea-ice/

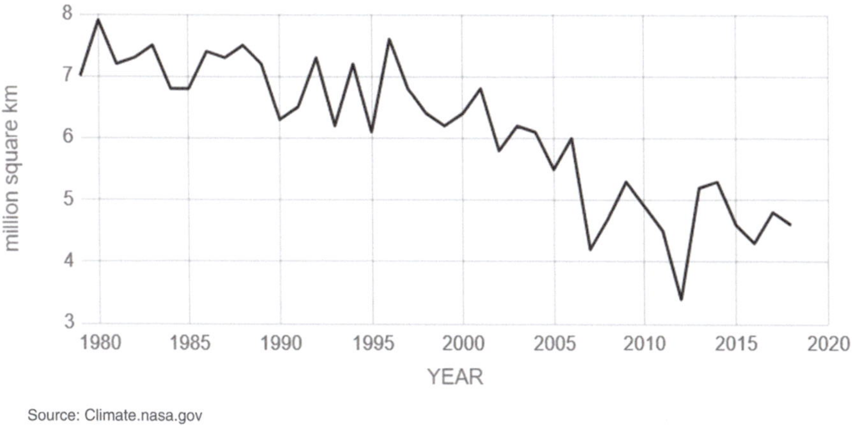

Fig. 9.1 Arctic Sea ice minimum

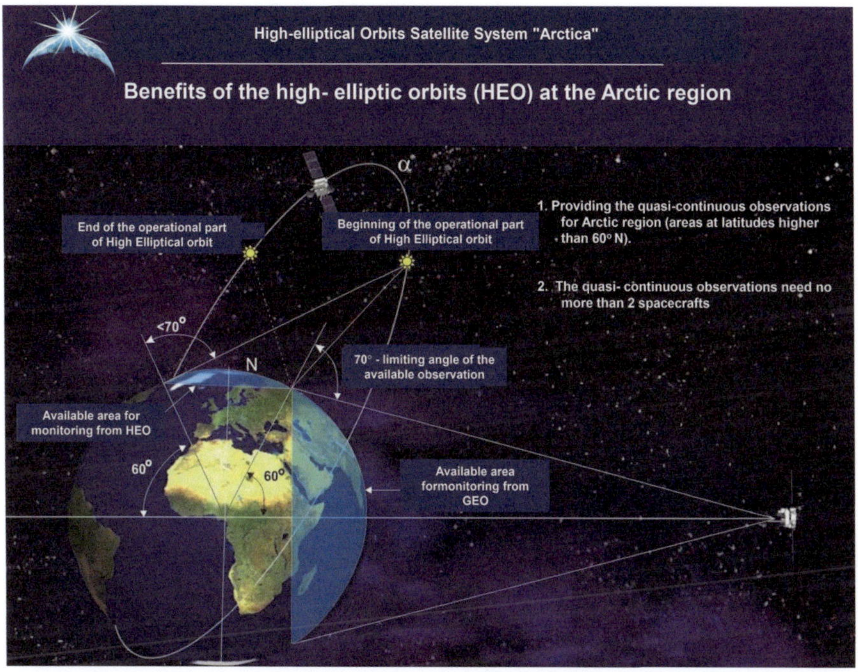

Fig. 9.2 Satellite system Arctica. (Source: Russian Federal Space Agency)

funded from the Russian space budget focused on meteorology and emergency communications. Each spacecraft carries a multi-spectral imager, known as MSU-GSM, along with transmitters for meteorological and rescue systems. An apogee (highest point) of their orbit is 40,000 km above the Earth surface and a perigee 1000 km. Such orbital parameters enable frequent overflies of the polar regions with practically uninterrupted view of the northern hemisphere (Fig. 9.2).

Oceanographic and meteorological observations along the route of the Northern Sea Route (NSR), on Arctic islands and near archipelagoes were regularly made before the 1990s of the last century. The islands are all situated within the Arctic Circle and are scattered through the marginal seas of the Arctic Ocean, namely, the Barents Sea, the Kara Sea, the Laptev Sea, the East Siberian Sea, the Chukchi Sea and the Bering Sea. The area extends 7000 km (4300 miles) from Karelia in the west to the Chukchi Peninsula in the east.

In subsequent years, however, the number of regular observation points was significantly reduced due to the overall decline in the state of the Russian science. Today in order to monitor and prevent particularly extreme weather events, a fundamentally new observation system that will allow to quickly monitor the conditions for the occurrence of such extreme weather events is required. In this connection, it is necessary to deploy additional monitoring systems for geophysical, hydrometeorological, ecological and oceanographic conditions on the route of the NSR and in remote and potentially exploitable areas of the Arctic shelf.

Global competition for world markets and main transport arteries leads to the constant scientific and technological development and improvement of satellite controlling systems that can be used in security monitoring and tracking/Russian companies with interests in the Arctic and the Far East strive for maximum economic efficiency in the design and operation of infrastructure along the NSR and adjacent territories. Some large energy companies including Norilsk Nickel, Lukoil and a number of other oil companies have acquired modern telecommunication, transport and information facilities and equipment to eliminate the deficit of information and to reduce risks that may adversely affect the cargo flows on the NSR and, as a consequence, their positions on the global market.

Infrastructure development in the Russian Arctic today is shaped by the tasks and capabilities of the icebreaker and cargo fleet, the Arctic ports and airfields, whose activities are largely limited by the navigation periods and constantly changing climatic and ice conditions on the NSR routes (from the multimodel simulations, the expected duration of the navigation period by the late twenty-first century will be approximately 3 to 6 months for the NSR[2]). These factors objectively set the requirements for the systematization of management mechanisms, production capacity of ports and improvement of working safety conditions.

Modern technological capabilities allow for efficient operational monitoring of the natural environment of the Arctic regions, which are highly vulnerable to external influences. There are many factors that can significantly change the shape the future of the region, including anthropogenic factors, monitoring of the climate changes over different times, monitoring of potential natural and man-made emergencies. They allow to systematically receive hydrometeorological and heliogeophysical information on a planetary scale. Given the infrastructural and climatic limitations, one of the first and foremost comprehensive measures to increase the competitiveness of the polar regions of Russia is to constantly monitor the infra-

[2]Arctic climate changes and possible conditions of Arctic navigation in the twenty-first century. – Access: https://link.springer.com/article/10.1134/S0001433810010032

structure of local ports, crude oil production, processing facilities and marine cargo vessels.

Measurement of seasonal changes in ice cover and the mass distribution of coastal ice sheets provides the necessary data for creation of a new-type ice forecasting system. The most important parameters are the movement of large ice sheets, their speed and lines of contact with the surface of the continent, as well as the thickness of the ice and the rate of melting of its surface. Advances in satellite remote sensing to study changes in ice mass provide many of these parameters except for ice thickness. However, means of measuring the thickness of ice from orbit, in addition to laser systems that measure changes in the height of the ice surface, have not yet been fully developed and are costly for local use.

The solution to this problem may lie in the use of UAVs (unmanned aerial vehicles) in the Arctic. Equipped with the equipment to collect the data, they can provide for information and research functions, including improvement of the reliability of transport planning for the NSR, operational control and dispatching of the transport process and telecommunications support for search and rescue. UAVs are key to monitoring the movement of icebergs in shallow water, which can cause damage to communications cables and pipelines placed on the coastal seabed. They can also transform the seabed topography, leading to possible changes in navigation conditions. At the same time, one of important practical tasks is to possess and maintain relevance of accurate nautical charts that would include all the changes taking place in this area. Now the Arctic territories of the Russian Federation are still considered as insufficiently developed in terms of the reliability of navigation, positioning, geodesy, cartography and telecommunications.

There are objective technical difficulties that can be resolved by the prospective local use of UAVs, for example, the covering and maintenance of control and correction stations along the NSR (Gorokhov et al. 2018). Unmanned aerial vehicles used to record and measure ice conditions should, in particular, be able to monitor the state of sea ice, coastal ice sheets, glaciers and ice caps, large icebergs. Given that the total length of the NSR is about 5600 km, and the coastal zone is about 370 km (200 nautical miles), the UAV must be equipped with laser altimeters, high-definition optics with photometric and other equipment, the weight of which cannot be large – no more than 50–150 kg. At the same time, the reserve charge for one mission of the UAV should be enough to cover 200 square miles of territory. In addition, the organization of ground take-off and landing sites for such UAVs with a runaway length of no more than 400 m, which do not require serious preparation, is necessary. The use of such airborne platforms, especially UAVs, can offer significant improvements in the ice cover measurements, as well as suggest new approaches to systematizing research and monitoring activities in the Arctic zone of the Russian Federation.

Moreover, it is necessary to note some uncertainty regarding the regulation of the international use of UAVs in the waters of the NSR. Today certain states with significant economic and political influence (including permanent members of the UN Security Council) can contribute to the development of existing international legal norms governing the use of their own UAVs to support their transport operations in the NSR, as well as to regulate and standardize polar research and ice measurements on the territory of foreign states (Zaikov 2015). These rules governing the use of foreign UAVs over the territories of the Arctic states may become a product of the interstate dialogue and can also take the form of public documents or joint standards that reflect the common goals and concerns of states about the future shape of their UAV management system in the Arctic.

Sustainable development of the Arctic, according to the UN expert community,[3] will be difficult to reach unless comprehensive control systems are strengthened and the control functions of the scientific community when conducting polar research are centralized. Under the conditions of global competition in the Arctic, the use of advanced technologies for successful monitoring of the NSR water area, the most promising of which are UAVs, can be the "springboard" in which the Arctic and subarctic states can work out the mechanisms of interaction and cooperation. Such an approach will not only allow year-round monitoring of the state of the ice cover, which is especially vulnerable under conditions of a global warming, but will also become the basis for creating the most accurate method for predicting ice movement, which, in turn, will minimize risks along the NSR as it increasingly becomes a regularized transport corridor.

References

Arctic Potential: Realizing the Promise of U.S. Arctic Oil and Gas Resources. National Petroleum Council (NPC), 2015. Part One. www.npcarcticpotentialreport.org (06.09.2018).

Gorokhov, A. M., Zaikov, K. S., Kondratov, N. A., Kuprikov, M. Y., Kuprikov, N. M., & Tamickij, A. M. (2018). Analysis of scientific and educational space of the Arctic zone of the Russian Federation and its contribution to social and economic development. *European Journal of Contemporary Education, 7*(3), 485–497. https://doi.org/10.13187/ejced.2018.3.485.

McDonald, J. (2018). *Drones and the European Union. Prospects for a common future*. London: The Royal Institute of International Affairs. 20 p.

Transport & Logistics in the Arctic. Almanah 2015. Issue 1. Edited by S.V. Novikov, Moscow: TEHNOSPHERA, 2015. pp. 68–73.

Zaikov, K. S. (2015). The "Arctic competition" problem and the marine transport hubs: Is it a clash of business interests or the knockout game? *Arctic and North, 19*, 35–55. https://doi.org/10.17238/issn2221-2698.2015.19.35.

[3] Polar Code by International Maritime Organization. – Access: http://www.imo.org/en/mediacentre/hottopics/polar/pages/default.aspx

Part III
Oil and Gas

Chapter 10
Handling the Preparedness Challenges for Maritime and Offshore Operations in Arctic Waters

Kay Fjørtoft and Tor Einar Berg

Abstract The purpose of this chapter is to provide an understanding of preparedness considerations and operations linked to sustainable maritime activities in polar waters. In this context, Arctic Ocean operations include fishing, aquaculture, offshore petroleum operations, ocean mining, tourist/explorer cruises and merchant shipping. Our mission is to share knowledge and understanding regarding preparedness, to support a sustainable development and minimise the consequences to the environment of maritime Arctic Ocean operations. The rising level of activity in Arctic waters requires improved emergency response capabilities beyond the capacity of governments. Preparedness is protection of the environment, property, and human beings, where the saving of lives and reduction of losses due to an accident or operation are the highest priorities. Oil and gas companies are required to develop their own emergency response organisations in order to obtain licences to operate. Their organisations must collaborate with public ones and must also be capable of operating alone. Where other sectors such as commercial shipping and fisheries are concerned, emergency response operations will depend on the availability of governmental resources and vessels of opportunity when an accident occurs. Preparedness is about preventing future accidents and minimising consequences when they do occur.

Keywords Arctic cruise · Arctic preparedness · Emergency towing · Oil spill · MarSafe · Polar Code

---

K. Fjørtoft (✉) · T. E. Berg
SINTEF, Trondheim, Norway
e-mail: kay.fjortoft@sintef.no; tor.e.berg@sintef.no

E. Pongrácz et al. (eds.), *Arctic Marine Sustainability*,
Springer Polar Sciences, https://doi.org/10.1007/978-3-030-28404-6_10

Abbreviations

AIS-Sat 1/2	Norwegian Automatic Identification System Satellite
AMSA	Arctic Marine Shipping Assessment
ARCSAR	Arctic and North Atlantic Security and Emergency Preparedness Network
BSEE	Bureau of Safety and Environmental
EMERCON	Ministry of Civil Defense, Emergencies and Disaster Relief
EPA	Emergency Preparedness Analysis
EPPR	Emergency, prevention, preparedness and response
GEO	Geostationary Earth Orbit
GMDSS	Global Maritime Distress Safety System
GNSS	Global Navigation Satellite System
HEO	High Elliptical Orbit
HIBLEO 2	Iridium satellite constellation
HSC	International Code of Safety for High-Speed Craft
IHO	International Hydrographic Office
ISPS Code	The International Ship and Port Facilities Security Code
ITC	Information and Communication Technologies
IMO	International Maritime Organization
INMARSAT	The International Maritime Satellite Organization
IO	Integrated Operations
JIP	Joint Industry Programme
JRCC	Joint Rescue Coordination Centre
LRS	Local Rescue Centre
MARPART	Maritime Preparedness and International Partnership in the High North
MARPOL	International Convention for the Prevention of Pollution from Ships
MarSafe	The Maritime Safety Research Program
MOSPA	Marine Oil Pollution Preparedness and Response in the Arctic
MTO	Man, Technology and Organisation
NORDLAB	Emergency Preparednes Laboratory
NORSOK	A standard used within the Norwegian petroleum sector
NOU	Norwegian Official Report
NOR VTS	Vardø Vessel Traffic Centre
OILPOL	Prevention of Pollution of the Sea by Oil
OSPD	Oil Spill Preparedness Division
PAME	Arctic Marine Environment Working Group
PSA	Petroleum Safety Authority
RRFP	Regional Reception Facilities Plan
SAR	Search and Rescue
SARiNOR	Search and rescue in the High North
SOP	Standard Operating Procedure
SOLAS	International Convention for the Safety of Life at Sea
VSAT	Very small aperture terminal
VTC	Vessel Traffic Centre

10.1 Introduction

When the Titanic collided with an iceberg off the coast of Newfoundland in 1912, information about drifting icebergs had not reached the ship's officers and navigators, and it took a long time before nearby vessels received a request for assistance. As recently as 1989, the Maxim Gorkiy cruise vessel, which sailed into an ice belt southwest of Svalbard experienced a similar lack of information (Kvamstad et al. 2009). The hull was damaged, and the passengers, crew and ship were safely rescued thanks to extremely good weather conditions and courageous on-scene commanders. In both cases, if the navigation officers had received information in advance they would have been able to take another and safer route, and the accidents could have been avoided.

Since the times of the Titanic until today, the extent of Arctic preparedness has grown rapidly. Technology for navigation is steadily improving, as is search and rescue (SAR)-capacity, international and maritime industry collaboration has improved, as has the ability to forecast undesirable conditions and call for navigational support. Preparedness is not only about SAR issues, but also involves having sufficient information to be able to plan and execute operations in a safe way. If the weather forecast is at an alert level, an operation might have to be postponed, or new sailing routes selected. If a forecast on space weather says that navigation or positioning satellites are interrupted, an operation might need to be aborted, or at least fail-safe procedures put in place before the operation starts (Fig. 10.1).

The future looks better regards preparedness (Schmied et al. 2016). Although we are heading for more digitalisation and we do not know the consequences, new technologies look promising as means to avoiding accidents and reducing uncertainty. Greater safety is generally anticipated, and many studies suggest that the number of maritime accidents that involve "human error" ranges from some 60% to 90% (Allianz Global 2018). We know that operational practices and activities in

Fig. 10.1 Arctic preparedness. (Source: SINTEF)

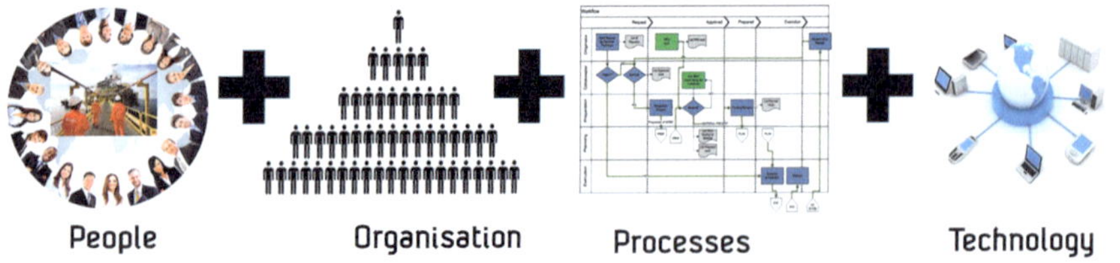

Fig. 10.2 IO is the integration of man/people, organisation, processes, technology. (Illustration: SINTEF)

several sectors will be changed, and that Information and Communication Technologies (ICT), digitalisation, and innovation technologies such as autonomous vessels will be important driving factors for sustainability and raised awareness. Autonomous and digitalisation are not the goals, but the means to make it safer. In parallel with the development of innovations, it is important to ensure that the human element will be taken care of, including the training and education of workers or operators. The introduction of new technology should therefore include the "*human in the loop*", when preparedness activities are under the microscope. The focus on integration between Man, Technology and Organisation (MTO) is central (Fig. 10.2).

Integrated Operations (IO) is the integration of people, organisations, work processes and information technology to be capable to make smarter decisions. It is enabled by global access to real-time information, collaborative technology and the integration of multiple expertise across disciplines, organisations and geographical locations, with the purpose of achieving better and more coordinated planning of operations and preventing events from occurring. IO is a tool for raised awareness which again leads to better preparedness.

This chapter describes some central Arctic Maritime Preparedness systems, how they operate, and provide examples of existing and future trends that influence the preparedness landscape in the Arctic environment.

10.2 The Starting Point for Maritime Preparedness

As the Titanic story tells it started back in the early nineteenth century (Titanic history 2009), where following quotes from the History.com website:

> The luxury steamship RMS Titanic sank in the early hours of April 15, 1912, off the coast of Newfoundland in the North Atlantic after sideswiping an iceberg during its maiden voyage. Of the 2240 passengers and crew on board, more than 1500 lost their lives in the disaster. According to some hypotheses, Titanic was doomed from the start by a design that many lauded as state-of-the-art. The Olympic-class ships featured a double bottom and 15 watertight bulkhead compartments equipped with electric watertight doors that could be operated individually or simultaneously by a switch on the bridge. It was these watertight bulkheads that inspired Shipbuilder magazine, in a special issue devoted to the Olympic

liners, to deem them "practically unsinkable." But the watertight compartment design contained a flaw that was a critical factor in Titanic's sinking: While the individual bulkheads were indeed watertight, the walls separating the bulkheads extended only a few feet above the water line, so water could pour from one compartment into another, especially if the ship began to list or pitch forward.

A little more than an hour after contact with the iceberg, a largely disorganized and haphazard evacuation began with the lowering of the first lifeboat. The craft was designed to hold 65 people; it left with only 28 aboard. Tragically, this was to be the norm: During the confusion and chaos during the precious hours before Titanic plunged into the sea, nearly every lifeboat would be launched woefully under-filled, some with only a handful of passengers.

The Titanic news was greeted with worldwide shock and outrage at the huge loss of life, and the regulatory and operational failures that led to it. The investigations of the accident led to major improvements in maritime safety. One of their most important legacies was the establishment in 1914 of the International Convention for the Safety of Life at Sea (SOLAS), which still governs maritime safety today. The Titanic disaster is regarded as the event that triggered the organised maritime preparedness system (Fig. 10.3).

For the offshore petroleum sector, the Alexander L. Kielland flotel disaster (NOU 1981) was an equivalent event that led Norwegian authorities to improve design and safety regulations for floating structures. It capsized in March 1980 near the Edda platform in the Ekofisk area of the North Sea; 123 died and 89 survived. A fracture led to the loss of one of the flotel's five columns. It then suffered a severe list which resulted in water reaching the topside before the rig capsized. A combination of strong winds, high waves and thick fog hampered rescue efforts. This accident had

Fig. 10.3 Der Untergang der „Titanic". (Source: Willy Stöver, Magazine Die Gartenlaube)

great significance for safety developments on the Norwegian Continental Shelf – including new regulations, a drastically improved regulatory regime and a new allocation of regulatory responsibilities.

Some outcomes of the accident investigation were:

- New buoyancy standards for offshore facilities
- Research projects were launched, including efforts to improve lifeboats
- New and improved survival suits were introduced for offshore use

A major step in international efforts to prevent oil pollution was the signing of the International Convention for the Prevention of Pollution of the Sea by Oil (OILPOL) in 1954, which in 1973 was subsumed into the International Convention for the Prevention of Pollution from Ships (MARPOL). Special problems are related to oil spill in Arctic waters. Eger (2010) has provided a brief overview of the effects of oil spills in Arctic waters, while the Arctic Council's working group on emergency, prevention, preparedness and response (EPPR) continues its efforts to improve responses to environmental and other emergencies (https://www.eppr.org).

10.3 IMO – International Maritime Organization

The International Maritime Organization (IMO), was established following an agreement that was formulated at a United Nations conference held in Geneva in 1948; its first meeting took place 10 years later. The IMO is headquartered in London, and it currently has 174 Member States and 3 Associate Members. The main duty of IMO is to be responsible for regulating international shipping, which is does by maintaining a comprehensive regulatory framework for shipping that includes safety, environmental concerns, legal matters, technical cooperation, maritime security and shipping efficiency.

The International Conference on Safety of Life at Sea took place in 1914 in response to the Titanic disaster. Its purpose was to bring the regulation of maritime safety under an international framework. The SOLAS Convention requires flag states to ensure that ships flagged by them comply with the minimum safety standards in the construction, equipment and operation of merchant ships. The treaty includes articles that set out general obligations followed by an annex divided into twelve chapters. Two new chapters were added in 2016 and 2017, in which the Polar Code is referred to as Chapter XIV.

- **Chapter I – General Provisions:** Includes regulations concerning the survey of the various types of ships.
- **Chapter II-1 – Construction – Subdivision and stability, machinery and electrical installations:** The subdivision of passenger ships into watertight compartments must be such that after assumed damage to the ship's hull the vessel will remain afloat and stable.

- **Chapter II-2 – Fire protection, fire detection and fire extinction:** Includes detailed fire safety provisions for all ships and specific measures for passenger ships, cargo ships and tankers.
- **Chapter IV – Radiocommunications:** This incorporates the Global Maritime Distress and Safety System (GMDSS).
- **Chapter V – Safety of navigation:** Identifies certain navigation safety services which should be provided by Contracting Governments and sets forth provisions of an operational nature applicable in general to all ships on all voyages.
- **Chapter VI – Carriage of Cargoes:** Covers all types of cargo (except liquids and gases in bulk) "which, owing to their particular hazards to ships or persons on board, may require special precautions".
- **Chapter VII – Carriage of dangerous goods**: Carriage of dangerous goods in packaged form; Construction and equipment of ships carrying dangerous liquid chemicals in bulk; Construction and equipment of ships carrying liquefied gases in bulk and gas; Special requirements for the carriage of packaged irradiated nuclear fuel, plutonium and high-level radioactive wastes on board ships.
- **Chapter VIII – Nuclear ships:** Outlines basic requirements for nuclear-powered ships and is particularly concerned with radiation hazards
- **Chapter X – Safety measures for high-speed craft**: The Chapter makes mandatory the International Code of Safety for High-Speed Craft (HSC Code).
- **Chapter XI-1 – Special measures to enhance maritime safety**: Clarifies requirements relating to the authorisation of recognised organisations.
- **Chapter XI-2 – Special measures to enhance maritime security:** Clarifies the International Ship and Port Facilities Security Code (ISPS Code)
- **Chapter XII – Additional safety measures for bulk carriers:** Includes structural requirements for bulk carriers over 150 metres in length.
- **Chapter XIII – Verification of compliance:** A systematic, independent and documented process for obtaining audit evidence and evaluating it objectively to determine the extent to which audit criteria are fulfilled
- **Chapter XIV – Safety measures for ships operating in polar waters:** The chapter makes mandatory, from 1 January 2017, the Introduction and part I–A of the International Code for Ships Operating in Polar Waters (the Polar Code).

Many countries have turned these international requirements into national laws so that the owners of any vessel at sea that is in breach of SOLAS V requirements may find themselves subject to legal proceedings. Regarding the focus on the Arctic, the Polar Code came into force in 2017. The Code sets out regulations for shipping in the Polar regions, principally relating to Ice navigation and ship design. The Polar Code, Chapter XIV, covers the full range of design, construction, equipment, operational, training, search and rescue and environmental protection matters relevant to ships operating in the inhospitable waters surrounding the two poles. The Code will not be in force for ships of less than 500 GT, fishing vessels or those entitled to sovereign immunity.

10.4 Arctic User Groups

Norway's general interests in the Arctic (Norwegian Ministries 2017) and those of most other countries with Arctic interests, include energy, transport, aeronautical, environment, fisheries, tourism, as well as security and the lives of local inhabitants. Energy and maritime are significant parts of the list regarding the number of users operating in the area.

The table below defines maritime activities, describing type of operation, type of vessel or installation, operating season(s) of the year, and region. The table was drawn up from a Norwegian perspective but is relevant to other countries and Arctic regions (Table 10.1).

Arctic maritime areas tend to be demanding to navigate in, according to previous studies such as the MarSafe project (Fjørtoft et al. 2010, 2012, 2013). In the north,

Table 10.1 Typical maritime activities in the Norwegian part of the Arctic (Source: SARiNOR)

Maritime activities in the Arctic	Types of vessel	Operational period	Region
Coastal traffic	Passenger and cargo vessels inspection vessels.	All year	Close to the coast and between Svalbard and the mainland.
Intercontinental transport	Transport of dry cargo, containers and tank products	All year depending on ice conditions	Sail in corridors. Ice-breakers are normally used on the northern sea routes (Eastern and Western passages).
Fisheries	Coastal and ocean-going fishing vessels	All year	The Norwegian coast for smaller vessels. Norwegian Sea, Barents Sea and Svalbard Region for larger vessels, close to the ice edge north of Svalbard
Petroleum activity	Exploration, construction and operation, fields that employ a wide range of offshore service vessels, floating and fixed installations and helicopters	All year, but most activity in the summer. Exploration is normally done in the summer period.	Close to the mainland for existing production. Expands to fields further offshore and closer to ice-fields
Maritime tourism	Recreational craft, small tourist boats and large cruise vessels.	Mostly summer but increasing in the autumn and spring season. Winter tourism is growing.	Both close to the mainland in open sea, and towards the islands around Svalbard. Arctic cruises are growing.
Research and other public-sector activities	Land- and sea-based research teams, military vessels and aircraft	All year	Both close to the mainland, open sea and in ice-infested waters

the winter season presents special challenges related to polar lows, fog, snow and ice that also challenge the safety and emergency response system's ability to carry out SAR operations. The farther north you go, the worse is access to good charts, good communication and positioning signals, while the distance to ports and necessary infrastructure increases. Furthermore, the relatively underdeveloped infrastructure for maritime shipping in the northern regions poses further challenges. An example is limited and unstable radio/satellite communication (Fjørtoft and Tu 2017) for information exchange, navigation and positioning. This can increase the risk of accidents, contingency response time, and increases risks to SAR personnel as well as other users in the area.

Farther offshore, there is a high level of activity of ocean-going vessels, including vessels with more than 50 people aboard (Borch et al. 2017). A smaller number of large seagoing vessels operate in more remote areas in the Norwegian Sea, the Barents Sea and the Svalbard area. Cod, herring and shrimp trawlers operate as far north as 83 degrees, which is north of Svalbard. SAR operations in the northern parts of the Barents Sea are very challenging due to the long distance for helicopter support, from both Longyearbyen and the mainland. The Svalbard zone is a challenging area also due to sea ice and ice, harsh weather, lack of port infrastructure, and the potentially long distance to helicopters and coast guard resources. Helicopters from Longyearbyen will have to refuel at depots to be able to reach the most remote areas. Fishing in the northern Barents Sea and Svalbard zone takes place most of the year, with particularly high activity from August to December. The typical size of the boats is between 25 and 60 m, with crews of 10–50 people, according to measurements of the traffic obtained from AIS signals (Automatic Identification System, tracking of vessels).

The level of transport in coastal regions around mainland Norway is relatively stable. Some of the coastal traffic is transport routes between Russia and mainland Norway, and the mainland and Svalbard. Today almost 300 loaded tankers sail annually to and from Russia along the Norwegian coast, and the number is expected to increase in the next few years, see Fig. 10.7, the largest ports are Murmansk and Arkhangelsk on the Russian side of the border. Many of the vessels are regarded as "risk vessels", which are defined by the Norwegian Coastal Administration as (1) tankers and vessels with hazardous and/or toxic cargo, (2) all vessels over 5000 tonnes, and (3) vessels carrying radioactive material.

Intercontinental transport in the Norwegian Sea and the Barents Sea is primarily related to oil and gas activity in northwest Russia. The main trend for intercontinental transport in the Arctic is the increase in tanker traffic (McCormick 2012). The number of tankers transiting the Norwegian coast fluctuates between 15 and 35 a month. The number of commercial ships calling at Norwegian ports in transit through the Norwegian Sea is expected to increase, due to rising commercial activity in northern Russia with the result that more traffic is sailing through the Northeast Passage.

The number of tourist vessels sailing in the Arctic regions has increased recent years. Arctic cruises along the coast of northern Norway in the winter season are becoming more popular and are a new offer from the cruise industry. About 100 cruise ships annually visit the Norwegian city of Tromsø during an eight-month

Fig. 10.4 MSC Meraviglia. (Photo: Robert Colquhoun, Shutterstock.com)

season, where we can list about 50 different cruise ships of different sizes and ages. In 2015, the largest ship that visited Tromsø was MSC Splendida, with 3900 passengers and a crew of about 1300, which means a total of 5200 people on board. In 2018, MSC Meraviglia arrived at Longyearbyen with 4500 passengers. In 2017 Tromsø had about 75,000 inhabitants, while Longyearbyen had a population of about 2300, which means the pressure on the infrastructure is high during a visit. Longyearbyen's hospital capacity at is low, is incapable of servicing such a huge number of people, which is critical in case of a health or accident emergency on board a cruise vessel. The main season for Arctic cruises is from May to September, but cruise ship traffic during the dark season (October–November to February–March) is rising, according to data from Port of Tromsø. One of the selling points for a dark season cruise is the northern lights (aurora borealis) in the region (Fig. 10.4).

Another tourist offer is expedition cruises. Many offers departing Svalbard start from Longyearbyen. Expedition cruises with small and medium-sized passenger ships with typically less than 200 passengers and normally have a duration of 3–14 days. They operate in remote areas around Svalbard and the east coast of the archipelago. Studies by Borch et al. (2017) show that interest in routes eastwards into Russian waters such as Franz Josef Land and the Northeast Passage as well as westwards between Svalbard and Greenland is growing. This creates further challenges to the preparedness system, due to the large distances, lack of infrastructure and limited SAR capacities involved.

The cruise industry represents a major risk due to the large number of passengers and crew, because of the size of the vessels and potential pollution in the event of an accident. Factors such as increasing number of operations in remote areas, and the variable qualifications of ships and crew of operating in the area are concerns. The major risk categories for the cruise ship industry include the following issues, based on studies by Borch et al. (2017) and Elgsaas and Offerdal (2018):

- Severe illness/injury and evacuation from remote areas
- Fire/explosion on board with the need for medical personnel and firefighters

- Grounding and shipwreck with the need for comprehensive rescue operations
- Accidents during transfer of passengers to smaller boats
- Violent acts on board with the need for integrated SAR operations.

Oil and gas activities in the Norwegian Arctic include exploration, development/construction, and production in coastal areas from the Norwegian Sea to the eastern Barents Sea. The trends are that the activities are moving northward in the Norwegian Sea and further east and west of the Barents Sea. Currently, several vessels are involved in seismic operations, and drilling rigs are engaged in exploration, accompanied by supply vessels. The exploration region also includes Jan Mayen to the north of Iceland. The region has limited infrastructure for environmental protection, preparedness or transport. Oil companies will be responsible for providing their own obligatory SAR capacity, but large operations will be performed in cooperation with public resources. In both eastern and western areas, cooperation between nations on the utilisation of SAR resources and risk-mitigation activities will be crucial in major crises or accidents.

There is a relatively high level of research activities in the Arctic. Areas of research include environmental monitoring and biological studies, with representatives from several nations. Some of the research vessels involved are commercial, and some vessels take care of both civilian and military needs. There is a rising trend the level of scientific activity, with more vessels and new areas of research, in ever more remote parts of the Arctic.

In September 2018, the Venta Maersk made the first ever transit of the Arctic Ocean by a container ship. It had an icebreaker escort in parts of the Northern Sea Route. This was a milestone for the East-West transport corridor, which is expected to grow in importance in the near future. One reason for growth is that large parts of the Arctic Ocean now seems to be free of ice during much of the summer. This means new opportunities and possibilities of new shipping corridors opening, which will lead to an increase in the number of users in these areas, Overland and Wang (2013). For all stakeholders operating in the Arctic region, safe and efficient operations are challenging and require good planning. Operations must be planned based on available and trusted data that make it possible to provide qualified forecasts and good prognoses. If possible, plans and information should be shared between companies and organisations in case of an accident. The "help each other's" principle is vital in the Arctic. There are many extra challenges in the region compared to the situation further south. One example is that weather information is frequently unavailable and may be very unreliable. Another example is the lack of adequate navigational infrastructure, combined with unreliable sea charts, partly due to lack of seabed mapping in Arctic waters. Arctic preparedness is essential for preventing undesirable situations and reducing the impact if something does happen. The only course to sustainable Arctic development is to prioritise safety and emergency preparedness and to develop the infrastructure needed for rising levels of activity (Fig. 10.5).

Fig. 10.5 Container transport in icy waters, a Maersk vessel. (Photo Jean Landry, Shutterstock.com)

10.5 Digital Infrastructure

A major problem in the Arctic is that almost all today's satellite communications systems are based on satellites in geostationary orbit round the Earth (GEO) on the equatorial plane. This also applies to Inmarsat, which is part of the GMDSS (Global Maritime Distress Safety System). In principle, Inmarsat C and VSAT communication systems have problems when used at latitudes above 75° north. Some systems give good results in certain areas outside this range, but this is not the general rule. Iridium can be used, but the current system has some quality of service problems. In addition, it provides only low data capacity and voice services. To maintain continuous coverage of the Arctic, satellites in HEO orbits (High Elliptical Orbit) will make a difference. An initiative led by Space Norway is currently in the design stage of launching two HEO satellites. Figure 10.6 illustrates these satellite configurations and their coverages showed in red and with the orbit directions.

A user terminal normally needs line-of-sight to a satellite. At high latitudes, GEO satellites can only be seen very low on the horizon. When the satellite and the user terminal are not at the same longitude, as is usually the case, it will be invisible from the user at significantly lower altitudes. The signal path length through the atmosphere between a user terminal and a GEO satellite – the 'slant range' – is highly dependent on the terminal's elevation angle, increasing dramatically at higher latitudes, thus escalating the signal deterioration due to the effects referred to above. The Earth's atmosphere has significant impact on radio wave propagation, but different phenomena are predominant depending on the frequency. In general:

- Lower frequencies, e.g. 2–3 GHz or L-band (1–2 GHz), are mainly influenced by the ionosphere, the effects decreasing with increasing frequency. Ionosphere effects are highly vulnerable to solar activity.

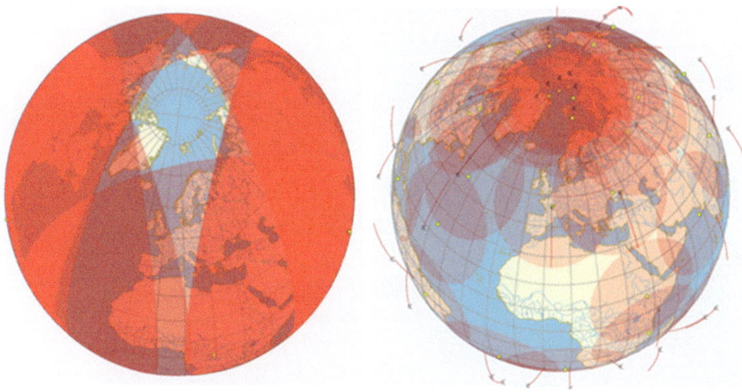

Fig. 10.6 Footprint of the satellites Inmarsat (Left) and Iridium (Right). (Source: SINTEF)

- Higher frequencies, e.g. Ku-band (12–18 GHz) and Ka-band (27–40 GHz), are mainly impaired by the troposphere (gases, clouds, rain, hail, snow, sleet, dust etc.), the effects increasing with increasing frequency.

One new initiative for a better communication infrastructure future in the Arctic is coming from Space Norway. They have been working with a project to launch two new satellites for communication covering the high north (Norwegian Government 2018). The satellites will offer users more bandwidth. The plans are to launch them into HEO orbit (High Elliptic Orbit). The Norwegian Government resolved in June 2018 that the Ministry of Industry and Fisheries may make a conditional commitment to Space Norway to realise the project to establish satellite-based broadband communications capacity in the High North. The introduction of two new satellites will improve the communication situation significantly in the region and will be a major step forward to addressing Arctic preparedness issues.

Another important safety system is called GMDSS, Global Maritime Distress and Safety System. It is an international system that aims to increase the probability that an alert will be sent when a vessel is in distress, which in turn will improve the likelihood that the alert will be received, and thus the improve our ability to locate survivors. It will also improve rescue communications and coordination, as well as providing mariners with vital maritime safety information. GMDSS uses terrestrial and satellite technology and ship-board radio systems. The only satellite provider of GMDSS is INMARSAT, but Iridium/HIBLEO 2 is in the process of applying to become a GMDSS service provider. Thuraya, a global satellite operator based in the United Arab Emirates has also indicated that it may be interested in becoming a GMDSS service provider in the future.

Another source of information is the Norwegian AIS-Sat 1 and 2 satellite system, which consists of two satellites that monitor ship traffic. AIS-Sat 1 and 2 is a collaboration project between the Norwegian Coastal Administration, the Norwegian Space Centre, the Norwegian Defence Research Establishment, and Kongsberg Seatex. AIS-Sat 1 came into operation in June 2010, while AIS-Sat 2 was launched

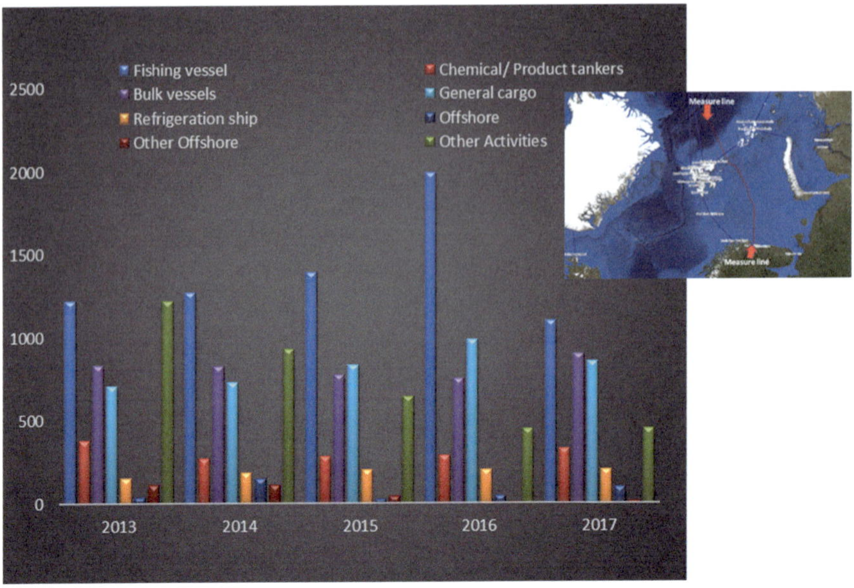

Fig. 10.7 Vessels crossing the border line between Norway and Russia; both directions. (Source: havbase.no)

in 2014. An example of statistical data based on AIS is shown in Fig. 10.7, the data is published in http://www.Havbase.no and based on the AIS data mentioned (Norwegian Coastal Administration 2018). In this example, the data are collected from the maritime boundary line shown on the map from the mainland, passing to the east of Spitsbergen and following the line between Norway and Russia. The data are from a five-year period from 2013 until 2017, and count ship crossings in both sailing directions, from east to west and from west to east. The bars in the figure show the vessel categories. Total number of crossings in the marked line was for the period 22, 031 vessels, with about 4400 vessels a year crossing the line. In 2013, 4660 ships crossed the line, there was a peak in 2016 with 4724 crossings. The statistics further indicate that fishing vessels are the largest category of vessels with a total of 6971 passings in the five-year period, with an average of 1394 crossings annually.

Navigational systems also face challenges in the area. Operation in the Arctic means a situation of 24 h bright daylight in the summer as well as pitch-darkness for 24 h in the winter. Problems with visibility due to fog, sleet or snow are common. In addition, navigation in ice can cause collisions with "bergy bits" that are very difficult to see as they only rise a few metres above the ice sheet or water. This means that new approaches to bridge design are required, as well as better sensors and presentation systems. Another issue is that if charts are available at all, they are often of low quality. This applies to details in the map, geometry as well as the

Fig. 10.8 The fishing vessel MV Remøy at 82 degrees north. (Photo Henning Flusund)

geodetic datum they use. The International Hydrographic Office (IHO) estimates that about 95% of the polar regions is uncharted or covered by charts with poor accuracy.

The oil and gas sector need not only a reliable infrastructure for data transmission, but also qualified navigational data for operational purposes. For example, highly accurate latitude and longitude positioning coordinates are important in drilling operations. Users require reliable Global Navigation Satellite System – GNSS signals that can be used for dynamic positioning. Deviations of more than a few centimetres can lead to critical situations in drilling activities or in vessel navigation. The accuracy of GNSS in the far north is less than the precision further south due to lack of reference stations as one example (Vigen and Ørpen 2013) (Fig. 10.8).

10.6 Norwegian SAR Preparedness Activities

The Joint Rescue Coordination Centres of Norway have overall operational responsibility during search and rescue operations in the area we have focused on in this chapter. Norway has two Joint Rescue Coordination Centres (JRCC), one in Bodø and one in Sola near Stavanger. The area covered by the JRCCs is from 57 degrees south till 90 degrees north. The centres have the overall responsibility for coordinating all maritime, air and land-based SAR services in the area. Coordination of accident responses ashore is delegated in most cases to one of the local rescue centres (LRS), in accordance with instructions from the JRCC Centres. JRCC North covers the area from 65 degrees north to the North Pole. In the west, the limit of its remit crosses the 0-meridian and in the east from the border Norway-Russia and an area

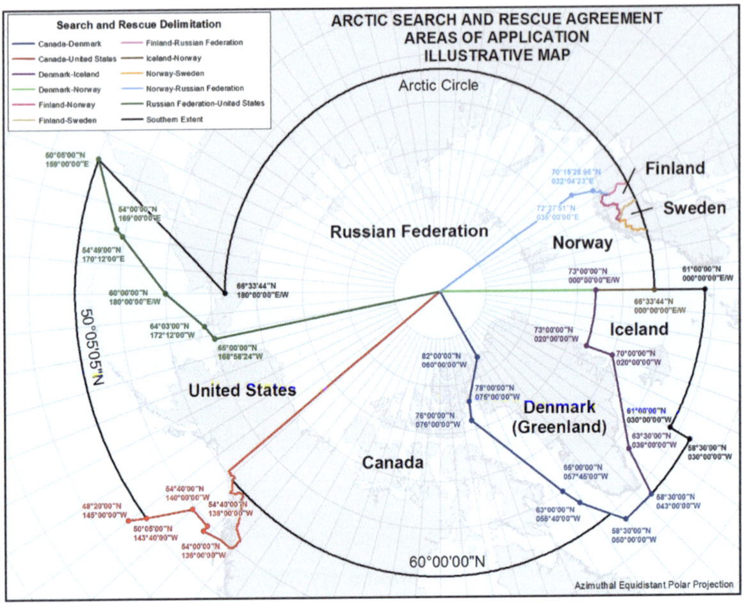

Fig. 10.9 Arctic SAR areas. (Illustration: Ministry of Foreign Affairs)

north north-east until it hits 35 degrees east, up to the North Pole. JRCC also has overall responsibility for all rescue services at Svalbard. The responsibility of JRCC is primarily to save people in acute distress (Fig. 10.9).

Regards Arctic SAR Norway has set out the following ambition:

- *"Norway shall be world leader in planning, coordinating and implementing maritime SAR operations at sea in the North."*

Search and rescue operations in the Arctic are demanding, comprising several challenges related to long distances, limited infrastructure and demanding climate conditions. This places tough demands on equipment, expertise and, not least, cooperation between the actors involved in emergency preparedness. For major actions, a wide range of emergency preparedness actors, stakeholders and institutions will be involved in SAR operations, including private and public actors, civil and military, and volunteer and professional aid organisations. In individual cases, there may also be close cooperation among the rescue services of several countries' associated with so-called host-support.

As one of the steps to reach that SAR goal the Norwegian project SARiNOR came up in 2013. The objectives of project SARiNOR, SARiNOR (2018), were to realizing the vision of being a world leader in maritime SAR operations. SARiNOR stands for Search and Rescue in the High North. The project was supported by the Norwegian government and Norwegian industry, with the objective of performing a feasibility study on possible improvements in operations in the Arctic, with a focus on SAR. The project analysed gaps between different SAR activities. The SARiNOR work packages covered technological gaps, organisational procedures and human factors such as needs for training and education to strengthen Arctic SAR capabilities:

- Alarm and notification
- Search
- Rescue and survival in cold climates
- Shared situational awareness
- Training and competence building

Experience gained from events and exercises during the SARiNOR project reveals the need for competence building for improved SAR preparedness in the Arctic. Good exchanges of knowledge among all the actors involved is important for building a common understanding of the situation, better knowledge and training, and strengthening the interaction and thereby realizing a powerful SAR capacity. The report of Work Package 7 (Borch et al. 2017), stated that current training and competence building can be improved. An analytical centre can make a significant contribution to learning and have pointed to run week-long practical sessions on utilising data from exercises and accidents as the basis of new training programmes.

The SARiNOR project has produced many findings that are relevant to Arctic preparedness. As mentioned above, one important issue is the potential to strengthen SAR collaboration between the Arctic countries. The elements identified include:

- Currently limited systematic review of events, especially in the north.
- A regional resource register is needed. We have a good national overview, but little coordination with other countries.
- Expertise requirements and status should be mapped, documented, and shared with other countries.
- Little focus on "lessons learned" from previous events, on training exercises and multinational exercises.
- An international training program on Arctic SAR should be developed.
- The importance of the SAR elements in the Polar Code should be mapped, practiced and adjusted as more knowledge is gained.

10.7 Environmental Protection Preparedness

Protection of the Arctic Marine Environment Working Group (PAME) is the most important player in environmental protection preparedness in Arctic waters. The Arctic Marine Shipping Assessment report from 2009 provides a good state-of-the-art summary (AMSA 2009). Environmental considerations and impacts on the four Arctic regions (Aleutian Islands/Great Circle Route, Barents and Kara Seas, Bering

Strait and Canadian Arctic) are described in the AMSA report, which presents findings for these regions as well as an overview of existing Arctic marine infrastructure. Annual reports on status on implementation of the AMSA 2009 report are prepared for the Ministers of the Arctic Council (Arctic Council 2017). The Arctic Council has also organized the work that has led to "Agreement on Cooperation on Marine Oil Pollution Preparedness and Response in the Arctic" (MOSPA) back in 2013. The Arctic Coast Guard Forum has defined 10 strategic goals for Coast Guard activity. One of these is to work collaboratively to advance the protection of the marine environment. Through joint exercises, they attempt to improve information exchange, sharing of expertise and handling major incidents (Arctic Council 2018). Examples are the Arctic Guardian exercise in 2017 and the planned Polaris 2019 (which will focus on Mass Rescue Operations in connection with to maritime incidents). Finally, the Norwegian SARiNOR 2 project (SARiNOR 2018) has carried out a study of challenges for oil spill response and salvage in Arctic waters.

10.8 Spill Preparedness

Oil-spill preparedness is essential for all commercial activities in Arctic waters. PAME's EPPR working group updated operational guidelines for marine oil pollution in 2017 (EPPR 2017). Different types of spills are due to different activities. In this section, we focus on oil spills from the petroleum and shipping sectors. Other types of spills such as toxic or nuclear waste will only be mentioned briefly, as will plastic waste in the sea.

10.8.1 Oil Spill

In connection with the growing interest in petroleum production in Arctic waters, oil and gas companies have begun to investigate additional challenges due to oil spills (Arctic Response Technology 2018). Nine oil companies participated in the Joint Industry Programme Arctic Response Technology programme (JIP Arctic Response Technology 2017). Challenges include low sea temperature, wave and weather conditions and sea ice. Governments demand safe operations and the lowest possible impact on the environment. One example is the standard operating procedures (SOPs) developed by the Oil Spill Preparedness Division (OSPD) of the Department of the Interior, Bureau of Safety and Environmental Enforcement (BSEE) (BSEE 2017). These procedures (SOPs) are known as the "OSPD Manual". Another example is the Norwegian guidelines prepared by the Petroleum Safety Authority (PSA), (PSA 2017). Chapter XIII has subsections on:

- Section 78 – Collaboration on preparedness against acute pollution
- Section 79 – Action against acute pollution

Basharat (2012) discusses the anticipated emergency preparedness challenges for arctic and sub-arctic areas based on an analysis of recorded accidents and how they were dealt with, using the NORSOK Z-013 standard (NORSOK 2010) as a baseline for the investigation. After an introduction to risk assessment, the standard has sections on:

- Section 9 – General requirements for emergency preparedness assessment
- Section 10 – Evaluation of emergency preparedness in the concept selection phase
- Section 11 – Emergency preparedness analysis (EPA) in concept definition, optimisation and detailed engineering phases
- Section 12 – Emergency preparedness analysis (EPA) in operational phase

The operability of oil-spill resources such as skimmers and oil booms, chemical dispersion and *in-situ* burning can be significantly reduced during the winter season. The large distances and lack of shore-based depots with winterized equipment reduce current oil spill preparedness in Arctic waters. A review of methods, capacity and technology for Arctic oil spill removal can be found in Ambio (2017). The recent SARiNOR2 report (SARiNOR2 2017) describes available governmental and oil company oil spill resources. The report adopts a tiered approach:

- Tier 1 – Local resources
- Tier 2 – Regional/national resources
- Tier 3 – International resources

One of the main objectives of the SARiNOR 2 project was to identify additional challenges facing operations in the High North. How to handle them is an important part of the project. The focus is on specifying consequence-reduction actions such as using available vessel and personnel resources, for instance fishing vessels, through dedicated crew training and the installation of relevant equipment on board. The report also discusses new actions for operations north of the border for the IMO Polar Code. One outcome of the project is a proposal for an action plan that comprise 19 actions grouped into four topics (in order of priority):

- Arctic preparedness base at Svalbard
- Improved competence, co-operation and collaboration on Arctic preparedness
- Command and control for surveillance and response in the High North
- Vessel programme for preparedness in the High North

A number of research and educational institutions are discussing initiatives to establish collaboration activities to address these action points. One example is Nord University in Bodø, which is developing new MSc level studies, starting research programmes and establishing NORDLAB – a new concept for preparedness and emergency management training and testing.

The increase in commercial shipping in Arctic waters increases the risk of accidental oil spills. Governments responsible for areas with increased traffic need to improve regional oil spill preparedness and to strengthen international collaboration. One example of regional improvements is EMERCON's set up of new rescue centres along the Northern Sea Route. Another one is the Norwegian Coastal Administration's procurement of three ice-strengthened oil boom systems. One of these systems will be installed on board the Norwegian Coast Guard vessel "KV Svalbard" and another in Varanger Fjord (close to the Russian boarder). The Arctic Coast Guard Forum could act as a platform for increased international collaboration in building up competence and operational experience for handling oil spills in Arctic waters as it will be linked to the seventh of the Forum's strategic goals "Work collaboratively to advance the protection of the marine environment".

10.8.2 Toxic Spills, Nuclear Waste and Plastic in the Sea

So far, these topics have been given little attention as specific problems in Arctic waters. PAME has started a working group on shipping waste challenges. In 2017 PAME prepared a document on regional waste management strategies for Arctic shipping (PAME 2017), in which it discussed the concept of a Regional Reception Facilities Plan (RRFP). The document was written as a proposal for consideration by IMO.

Increased shipping activities could increase the risk of toxic spills from dangerous cargos lost at sea (containers) or from grounded vessels. The spread of various types of toxic waste their influence on local habitats are topics for further research. This is a specific topic for the fish farming industry. Here the use of some delousing methods has been shown to have a negative effect on shrimp stocks close to fish farms.

Nuclear waste can be a result of accidents involving nuclear-powered naval vessels. Furthermore, the planned operation of Russia's first floating nuclear power plant, the Academician Lomonosov, (which will come into operation near Pevek by 2019), could be affected by operational disturbances which in the worst case would dump nuclear waste into the sea. An emergency centre, to be established in proximity to the floating nuclear plant, should include the ability to handle such a case. Finally, concern has recently been expressed regarding pollution from melting ice at former test-sites for nuclear missiles.

Plastic in the sea is becoming a global problem. Prolonged degradation time for plastics in cold water is a major challenge. Ways of "harvesting" floating plastics need to be investigated, and modifications of tools developed for warmer waters should be given high priority.

10.9 Property Protection Preparedness

Vessels and offshore installations represent large values. In the event of incidents/accidents it is important to prevent total loss of the units involved. This section focuses on saving drifting vessels, and resources for emergency towing. Other causes of total loss such as explosions and fires have been are not dealt with here. How to handle such cases for offshore floating units are described in each unit's emergency response plan.

10.9.1 Emergency Towing

A vessel in distress due to lack of power or damage to its steering system requires external help to prevent it from grounding or capsizing. In some situations, there will be a vessel of opportunity in the neighbourhood. Assistance from such a vessel will be limited unless it is a vessel with equipment and trained crew capable of making an initial emergency towing connection to prevent a drifting vessel from entering coastal waters, thus reducing the risk of grounding. Other vessels of opportunity may pick up survivors if they are abandoning the distressed vessel, and act as a communication centre with rescue organisations, waiting for a commercial seagoing tug or dedicated emergency towing vessel to arrive. Unfortunately, there are not many commercial seagoing ice-reinforced tugs on the market. In most cases, a coast guard vessel will be sent to assist the drifting vessel. Experience related to emergency towing is developed at national level. A Norwegian example is the workshops on emergency and ocean towing arranged annually by the Norwegian Coastal Administration in collaboration with SINTEF Ocean. Participants in these workshops come from tug companies, Norwegian Coast Guard vessels, the Norwegian Coastal Administration, Joint Rescue Coordination Centre, universities and research institutes. National experience will also be shared in two international coast guard forums – the North Atlantic Coast Guard Forum and the Arctic Coast Guard Forum.

Due to an increase in the number of drifting vessels and grounding incidents in Norwegian waters in the 1990s, the Norwegian Coastal Administration decided to set up a national emergency towing service. In the beginning, they allocated three emergency towing vessels. These vessels were stationed in waters with high environmental risks in the event of a maritime incident, and where there was a lack of commercial tugs. Through their Vessel Traffic Centres, Norwegian Coastal Administration oversees vessel movements in Norwegian waters. The VTC in Vardø (NOR VTS) is responsible for Northern Norway and Norwegian Arctic waters. If suspicious vessel tracks are observed, NOR VTS contacts the actual vessel and asks if assistance is needed. NOR VTS has the authority to dispatch a governmental emergency towing vessel to assist the vessel in distress. At present two vessels are located in northern Norway, Strilborg and Far Sabre. Both vessels are former offshore vessels with high towing power. In addition, Norwegian Coast Guard vessels can be allocated to emergency towing operations. As part of reorganisation of the emergency towing service, it has been decided that Norwegian Coast Guard vessels will take over this service. The first vessel will be working in parallel with one of the present offshore vessels from 2019. The second Coast Guard vessel will enter the service in 2020.

An example of a challenging towing operation occurred in northern Norway waters in October 2012. The fishing vessel “Kamaro” suffered an engine failure and started drifting under harsh weather conditions. Another fishing vessel from the same company came to assist and established a towing connection. The weather deteriorated and the towline broke. NOR VTS then asked a Norwegian Coast Guard vessel to assist the drifting “Kamaro”. “KV Harstad” arrived, established a new towing connection and started towing. In view of the actual weather conditions, it was decided to evacuate the vessel’s crew. The Joint Rescue Coordination Centre in Bodø was asked to assist. The initial plan was to lift the crew off by helicopter, but the weather conditions (full storm and 10 m high waves), the plan was revised. Crew members had to jump into the water to be airlifted by the rescue helicopter. It was feared that a second towline failure might occur, which it did during the crew rescue operation. A second Coast Guard vessel “KV Barentshav” arrived at the scene and succeeded in establishing a new towing connection and towing the vessel to port. This rescue operation is used as study case in emergency towing workshops arranged by the Norwegian Coastal Administration, the Coast Guard and SINTEF Ocean.

10.10 Arctic Preparedness Outlook Summary

The focus on ensuring the safety of people and the environment related to increased activity in Arctic waters takes high priority, both nationally and in international organisations such as the Arctic Council, IMO and the Arctic Coast Guard Forum. The Arctic Maritime Safety Cooperation project (Finnish Border Guard 2017) is a 3 years project aiming to develop practical level cooperation connected to Arctic maritime safety between Arctic states Coast Guards. Arctic Coast Guard Forum will run their second Live Exercise in Gulf of Bothnia early April 2019. Focus will be on Mass Rescue Operation connected to maritime incidents (Arctic Coast Guard Forum 2018). Another example is the European Commission funded ARCSAR project (Finne 2018). SARex Svalbard is a Norwegian project based on full-scale tests in Svalbard waters in 2019 and 2020, to develop new knowledge and improve survival kits.

In EPPR’s work plan for 2017–2019 two new expert groups will start work, one on Marine Environmental Response (linked to the MOSPA Agreement) and the other on Search and Rescue (linked to the SAR Agreement).

Russia’s Arctic SAR centres will increased from five to nine before end of 2020. The four new ones will be located in Tiksi, Anadyr, Pevek and Sabetta. In addition to serving traffic using Northern Sea Route, the centres will improve emergency preparedness for the local population.

In Norway, Nord University has an ambition to be an international competence centre on Arctic preparedness. Its Business School has been developing the new experienced-based master’s degree program in Preparedness and Emergency Management. Planned activities include:

- Establishment of a joint MSc programme in crisis management, emergency prevention, preparedness and response.
- Develop an integrated Arctic Safety and Security simulator network

Nord University also manages the MARPART (2018) initiative and the ARCSAR project. MARPART covers several topics, focusing on emergency response management and organisation:

- Patterns of risk in various sea areas
- Emergency response capacities of four Arctic countries (Greenland, Iceland, Norway, Russia)
- Organization of the emergency response system.

There is a need to improve oil spill handling technology for operations in ice infested waters. Recently Norway bought three systems designed for operation in Arctic waters. One of them is on board the ice-strengthened Coast Guard vessel "KV Svalbard". Exercises are needed to train operational procedures, observe efficiency and propose design changes to improve efficiency of oil spill handing in ice-infested waters.

Finally, the communication infrastructure is getting better. The Space Norway is currently in the final run to decide the contracts of two new satellites that will give a significant improvement to users in the high north. The satellites will be placed in an orbit covering the high north in an excellent way, at the same time as there will be enough bandwidth capacity to serve the user demands.

To sum up our thoughts for Arctic preparedness outlook we would like to go back to the introduction of this chapter. Arctic preparedness must cover the integration of people, organisations, work processes and information technology to be capable to make smarter decisions. The future is tomorrow, but we must start today the process of educating and training the people to operate in the high north, to understand and organise the preparedness system that builds on the available resources from multiple countries, and to develop the technological infrastructure that serves the digital demands. One nation cannot do the whole work alone, this must be done with a collaborative approach between the users in the area, between the different rescue coordination centres, and between private and public stakeholders. Preparedness is about knowing each other's capacities and to close identified gaps in the system by investment in new technology and infrastructure. As example, the communication is one of the most important key factors for a sustainable development of the Arctic environment. Luckily, many of mentioned issues are on the agenda and part of the outlook program, but we cannot stop and wait, we should continue to prioritise the investments and develop the best possible program to achieve improved Arctic Maritime Preparedness systems. This includes investments to improve human performance and survivability under arctic conditions, operational understanding and international cooperation related to handling of incidents, and in the technology. We must build the preparedness landscape in the Arctic environment to serve the users and the environment in a long-term perspective.

10.11 Authors

Since 1984, Tor E. Berg has held various positions in MARINTEK until his retirement in 2012 (MARINTEK changed its name to SINTEF Ocean in 2017). During the same period, he had a part-time chair in Marine Hydrodynamics at the Norwegian Institute of Technology (since 1996, the Norwegian University of Science and Technology – NTNU). Since 2012, Berg has held a part-time position as Scientific Advisor at SINTEF Ocean. In addition to project activities related to ship manoeuvring performance/simulator-based training of ship officers, he has participated in projects on safety of maritime operations in Arctic waters (especially the SARiNOR pre-project) and the development of competence systems for senior officers on Norwegian emergency towing vessels.

Since 1995, Kay E. Fjørtoft has been working as a researcher and research manager at MARINTEK, which changed its name to SINTEF Ocean in 2017. Fjørtoft has been involved in several research projects, mainly in of software architecture and development, integrated operations, maritime safety operations, operations in the Arctic environment, freight transport, port community systems and communications (telecom). He was managing one of the work-packages in the SARiNOR project and participated in several others. Fjørtoft has published book contributions, papers and articles, with focus on software architecture and maritime operational challenges, Arctic, as well as maritime communication. Many of them are primarily concerned with the Arctic environment. He also holds an ESA AP Ambassador role for Norway within the Business Applications programme.

References

Allianz Global Corporate & Specialty. (2018). *Safety and shipping review.*

Ambio. (2017). Oil spill response capabilities and technologies for ice-covered Arctic marine waters: A review of recent developments and established practices. *Kungliga Vetenskaps-akademien, Ambio, 46*(Suppl 3), S423–S441.

AMSA. (2009). *Arctic marine shipping assessment 2009 report.*

Arctic Council. (2009). https://www.pame.is/images/03_Projects/AMSA/AMSA_2009_report/AMSA_2009_Report_2nd_print.pdf

Arctic Council. (2017, May). *Status on implementation of the AMSA 2009 report recommendations*. PAME.

Arctic Council. (2018). https://www.arcticcoastguardforum.com/

Arctic Response Technology. (2017, May). *Arctic oil spill response technology – Joint industry programme summary report.*

Arctic Response Technology. (2018). *Oil spill preparedness.* http://www.arcticresponsetechnology.org/research-projects/

Basharat, S. (2012). *Proactive emergency preparedness in the Barents Sea.* MSc thesis, Nord University, Bodø, Norway.

Borch, O. J., et al., (2017). SARiNOR WP7, RAPPORT: «Behov for trening, øving og annen kompetanseutvikling innenfor søk og redning i nordområdene»-(in Norwegian).

Eger. (2010). *Effects of oil spills in Arctic waters.* Arctis Knowledge Hub

Elgsaas, I., & Offerdal, K. (2018). *Maritime preparedness systems in the Arctic – Institutional arrangements and potential for collaboration*. MARPART project report 3. Nord University, Bodø.

EPPR. (2017). *Operational guidelines – Agreement on cooperation on marine oil pollution preparedness & response in the Arctic*.

Finne, A. (2018). *SAR network goes cruising with EU funding*. http://www.highnorthnews.com/sar-network-goes-cruising-with-eu-funding/

Fjørtoft, K., & Tu, D. H. (2017) Arctic approach, the Northern area program. In *Communications and navigation challenges in the High North* (pp. 81–94). Fagbokforlaget. ISBN: 978-82-450-2112-7

Fjørtoft, K., Kvamstad, B., Bekkadal, F., et al. (2010). *SINTEF Ocean: Maritime safety management in the High North (MarSafe): Analysis of maritime safety management in the High North*. Trondheim: SINTEF.

Fjørtoft, K., Kvamstad, B., Bekkadal, F., et al. (2012). *SINTEF Ocean: Maritime safety management in the High North (MarSafe): Future needs and visions for maritime safety management in the High North*. Trondheim: SINTEF.

Fjørtoft, K., Bekkadal, F., & Kabacik, P. (2013). *Maritime operations in Arctic waters*. MTS/IEEE OCEANS – Bergen.

International Convention for the Safety of Life at Sea (SOLAS). (1974). http://www.imo.org/en/About/Conventions/ListOfConventions/Pages/International-Convention-for-the-Safety-of-Life-at-Sea-(SOLAS),-1974.aspx

Kvamstad, B., Fjørtoft, K., Bekkadal, F., Marchenko, A. V., & Ervik, J. L. (2009). A case study from an emergency operation in the Arctic seas. *TransNav International Journal on Marine Navigation and Safety of Sea Transportation, 3*, 455.

MARPART. (2018). www.MARPART.no

McCormick, D. (2012). *Arctic-ocean shipping doubles as melting ice opens sea lanes*. IEEE Spectrum.

NORSOK. (2010). *Risk and emergency preparedness assessment*. NORSOK standard Z-013, Edition 3.

Norwegian Coastal Administration. (2018). Statistic data from http://Havbase.no

Norwegian Government. (2018). *Prop. 55 S, Satellite communication in the High North*. https://www.regjeringen.no/contentassets/f4eaa67003e749c7949a31b59daa7daa/no/pdfs/prp-201720180055000dddpdfs.pdf

Norwegian Ministries. (2017). *Norway's Arctic strategy – Between geopolitics and social development, Oslo NOU*. https://www.regjeringen.no/contentassets/fad46f0404e14b2a9b-551ca7359c1000/arctic-strategy.pdf

Norwegian Technology Centre. (2001, September). *NORSOK Z-013 risk and emergency preparedness analysis*. Oslo: Norwegian Technology Centre.

NOU. (1981). "Alexander L. Kielland" ulykken, Norges Offentlige Utredninger. NOU 1981:11. Statens Trykningssentral, 1981 (in Norwegian).

Ocean Conservancy. (2017). *Navigating the north – An assessment of the environmental risks of Arctic vessel traffic*. Anchorage. Retrieved from https://oceanconservancy.org/wp-content/uploads/2017/06/ArcticVessel-Traffic-Report-WEB-2.pdf

Overland, J. E., & Wang, M. (2013). When will the summer Arctic be nearly sea ice free? *Geophysical Research Letters, 40*(10), 2097–2101. https://doi.org/10.1002/grl.50316.

PAME. (2017, May). *Regional waste management strategies for Arctic shipping; regional reception facilities plan (RRFP)*. PAME EPPR.

Petroleum Safety Authority Norway. (2017). *Guidelines regarding the activities regulations* (Last updated 18 december 2017). Petroleum Safety Authority Norway

SARiNOR. (2018). https://www.sarinor.no

SARiNOR2. (2017). *Salvage and preparedness for acute spills in Northern areas* (in Norwegian: Berging og beredskap mot akutt forurensning i nordområdene). Maritimt forum Nord (project leader K. E. Solberg)

Schmied, J., Borch, O. J., Roud, E. K. P., Berg, T. E., Fjørtoft, K., Selvik, Ø., & Parsons, J. T. (2016). Search andrescue operations contingencies in polar waters. In K. Latola & H. Savela (Eds.), *The interconnected Arctic — UArctic congress 2016*. Cham: Springer. https://doi.org/10.1007/978-3-319-57532-2_25.

The Titanic. (2009). https://www.history.com/topics/early-20th-century-us/titanic

Vigen, E., & Ørpen, O. (2013). *Distribution of GNSS augmentation service at high latitudes using the Iridium communication system*. ENC 2013 (s. 5). Vienna, ENC.

Chapter 11
Arctic Marine Oil Spill Response Methods: Environmental Challenges and Technological Limitations

Victor Pavlov

Abstract The most important issue, besides financial viability, which stops oil and gas industry from operating in full scale in Arctic seas, is the harsh environmental conditions, which can undermine oil spill response actions in case of oil spill accidents. The methods and technologies that are currently in use, cannot deliver efficient and fast oil recovery from the sea surface under icy, stormy, low visibility and extremely cold conditions. In 2015, Emergency Prevention, Preparedness and Response Working Group of the Arctic Council (EPPR), an international body that strengthens oil spill response (OSR) in the Arctic region, discussed the need of improvements in this field. In 2018, after comprehensive studies of the Arctic waters, EPPR concluded that natural climate conditions are too challenging for the present level of OSR preparedness in many regions. There are still many technological limitations and no optimized strategies for oil spill abatement. Thus, OSR issues are and will stay on the agenda in future, too.

This chapter will focus on main available OSR methods and give an overview of challenges and limitations set by the demanding Arctic environment. It will analyze the different methods, in situ burning, dispersants use, mechanical and physical response, and their efficiency, applicability, and multiple environmental, economic and technical parameters.

Keywords Arctic oil spill response

Abbreviations

API	American Petroleum Institute
BOPD	Barrels of Oil per Day
EPPR	Emergency Prevention, Preparedness and Response
HFO	Heavy Fuel Oil

V. Pavlov (✉)
Water, Energy and Environmental Engineering, University of Oulu, Oulu, Finland
e-mail: victor.pavlov@oulu.fi

E. Pongrácz et al. (eds.), *Arctic Marine Sustainability*,
Springer Polar Sciences, https://doi.org/10.1007/978-3-030-28404-6_11

ISB	In Situ Burning
OSR	Oil Spill Response
PACs	Polycyclic Aromatic Compounds
SIMA	Spill Impact Mitigation Assessment
UV	Ultraviolet
WOP	Window of Opportunity

11.1 Introduction

Arctic oil spill accident can be a challenging task for responders, as demonstrated by the description of the "Exxon Valdez" incident in the box below. For combating an oil spill, response authorities have several methods in their tool box: in situ burning, mechanical recovery with oleophilic skimmers and booms, dispersants use, and sorbents use. By its nature, these can be categorized as thermal, mechanical, chemical and physical. Core mechanisms behind each are respectively combustion, adhesion, dispersion and sorption. (ITOPF 2012; Liu et al. 2016) Due to its increasing relevancy, marine oil spill response measures in the Arctic have been explored in detail. Because of this, the Arctic region is not anymore one big entity with similar unbearable environment applicable to all its seas, the prevalent ambient factors of the sub-regions have been defined. Those factors limit oil spill response operations and the Kara Sea is among the most problematic sub-regions with its extremely low temperatures of −45 °C, long 9-month winters, gale-force winds and rapid glaciation (Vorobiev et al. 2005). Oil spill response in these marine areas remain ineffective due to their harsh weather patterns and extreme temperatures. (EPPR 2017a) Arctic OSR toolbox contains solutions, which are described in a number of reports, for instance, within Arctic OSR Technology – Joint Industry Programme (JIP 2014). However, most literature sources do not provide a concise description of operational conditions of the methods and their limiting metric parameters (Wilkinson et al. 2017) (Box 11.1).

Box 11.1: Example of near Arctic Oil Spill Event, Alaska, USA
The largest oil spill in the near Arctic latitudes so far took place in 1989 nearby Alaska, USA. The "Exxon Valdez" oil tanker was grounded on Bligh Reef within Prince's William Sound. Eight out of 11 oil tanks were damaged and leaked 41,000 m^3 of oil into the Gulf of Alaska. The spilt oil transformed into the "chocolate mousse" and reached the shoreline. It took 3 years to recover the contaminated beach. The abatement actions were delayed due to lack of prepared equipment. Only after 35 h of the accident, the tanker was surrounded by booms. While the weather was favorable, the chosen methods were ISB and to a very limited extent – use of dispersants. Thus, about 100 t were burned with 2 t residue generation, being 98% efficient. Very small part

(continued)

Box 11.1 (continued)
was dispersed from aircraft. The third day started a heavy storm, which lasted for 2 days. After the mousse formation, the only option to remove oil from water was via the mechanical methods. In total, about 11,000 people, 85 units of air fleet, and more than 1400 of various ships and boats were involved in the operation. High tides are common in the area, which spread the oil to vast areas of shoreline. The statistics of this ecological disaster record that 370,000 birds, 200 seals and countless number of fish were killed by the spill. Industrial catch of fish in the area was halted. (Vorobiev et al. 2005; WWF 2011)

This chapter provides a general overview of the oil spill response methods that can be used in the Arctic marine environment. The main contribution is to group the relevant data into one single structured manuscript: description, key technical parameters and environmental limitations of oil spill methods. Discussion of all included sections is based on the latest publications in the field. The presented data is provided to show a range of applicability of major oil spill response solutions at sea rather than recommending one or another method for certain marine conditions of the Arctic Ocean. Bioremediation along with supportive OSR tools for monitoring, remote sensing and contingency planning are not included in the discussion due to the scope of the work.

The rest of this chapter is organized as follows. The first three sections focus on primary oil spill response methods. Section 11.2 presents in situ burning (ISB). Section 11.3 examines mechanical oil spill response. Section 11.4 describes use of dispersants. Each section starts with a method overview. Then environmental challenges and technological limitation are demonstrated. In Sect. 11.5, secondary physical method is briefly mentioned, as a subsequent clean-up measure of primary OSR.

11.2 In Situ Burning Oil Spill Response

11.2.1 Overview of In Situ Burning

Brief Description The main thermal method, known worldwide since 1970s, is in situ burning (ISB) (Gelderen et al. 2015; McLeod and McLeod 1972). The basic concept is that oil is burnt locally under professional control and supervision, at the site where it was accidentally spilled. Almost no further transportation of recovered oil is organized. The exception is only residue materials after the controlled burn, which are to be collected if feasible. ISB is primarily considered for situations when oil is surrounded by ice. However, the method is only applicable if the oil has enough thickness to be ignited. (Michel et al. 2005; Buist et al. 2013)

Examples of Equipment In order to control the process and ensure that oil has reached the required thickness for ISB, oil slick is contained in one location. This is achieved either by means of ice sheets or fire-resistant booms: steel, ceramic, fiber-based, water-cooled or with stainless steel hemispheres. (Al-Majed et al. 2012) Oftentimes, response can be organized only with aerial means of transport, sent remotely from the shore. This is the case, when oil is naturally contained in ice. (API 2016, 2017; ART 2012; Buist et al. 2013; Fritt-Rasmussen et al. 2012) The response is conducted with igniters (e.g. handheld type or "helitorch") and ignition systems. When oil slick requires artificial containment, fireproof booms or herding agents are applied. Sometimes, promoters, such as diesel, gelled gasoline, gelled kerosene cubes, and reactive chemical compounds are used to improve ignition of the slick and flame spreading. For installing the booms two towing vessels can be employed; for spotting the spill visually, igniting and burn monitoring, aircrafts may be also utilized. Oil removal is done by igniting it with a torch from helicopter, sea vessel or other suitable way. The flame height is usually 1.5 times of the spill diameter. (API 2015; Buist et al. 2013; IPIECA 2016; Michel et al. 2005)

Efficiency of Oil Removal and Required Time The effectiveness of ISB is a function of different parameters, such as climate, oil thickness, type, degree of emulsification, slick size, diameter, and other factors. However, the efficiency and elimination rates are usually maintained high (API 2017). About 0.5 to 4.0 mm of oil slick is burnt per minute. In ideal conditions, which are by EPPR (2017a) with 75% of the time unlikely to happen, up to 300 tons (or 2000 m^3) of oil can be burnt in an hour, and the removal efficiency is up to 98%. In operational conditions, these numbers are unachievable. (Allen 1988; API 2017; Buist et al. 2013; IPIECA 2016; Lampela 2011; Li et al. 2016; Walton and Jason 1999) Most in situ burns are completed within minutes to hours (API 2015, 2016). The lighter the oil, the better it burns. In terms of its burning capacity, the order is: gasoline, diesel, jet fuel, heavy fuel oils, in a descending order (Table 11.1) (API 2016).

Sustainability of the Method

Environmental Aspect The ISB method removes oil from water and minimizes the impact to the marine ecosystems, as all toxic petroleum-originated evaporating chemicals are eliminated during the combustion process. (API 2017; EPA 1999; IPIECA 2014, 2016) Emitted heat is mostly – 97%, directed towards the sky. Only minor warmth is supplied back to the slick surface to create more hydrocarbon vapors and sustain the flame. The temperature of burning oil on water varies from 900 to 1200 °C in the flame, whereas the slick temperature on its upper oil layers is about 350 to 500 °C. The oil beneath it is close to the temperature of water and the environment: water at water-oil interface is never more than 100 °C, heat exchange between the liquids hardly happens due to insulating properties of the oil layer. (API 2015; Walton and Jason 1999) The only substance that is left after the operations is burn residue (API 2016). The residue cannot be burnt or gasified and is transported by sea currents and tides at sea. This left-over corresponds to 2 to 15% of oil slick total volume. For lighter oils, it is 1 mm thick slick spread on surface, for heavier oils – up to 5 mm thick. For a set of 100 crude oils, only half of oils a leave residue, which would float on the surface; for the other half – the residue would sink. (API 2016) It is a very viscous, dense and biologically unavailable by-product of ISB. It is composed of semi-burned oil, without volatiles and with presence of precipitated soot (IPIECA 2016). The residue recovery is a challenging operation in the Arctic conditions. As a negative impact to marine life, it affects benthic and coastal ecosystems by its physicochemical contamination. This can cause harmful effects to coastal wildlife due to direct physical contact, resulting possibly in coating and ingestion. The residue is likely to persist in the environment for long periods of time and end up stranding on shore. However, all toxic impacts are rather localized, sparse and have small surface damage (Al-Majed et al. 2012; ART 2012; API 2016) Burn residue can be collected from water via use of vacuum suction systems, subsea pumps, skimmers, sorbents or manual means – shovels, buckets, nets. (API 2015; IPIECA 2016)

Otherwise, residue collection depends on ice situation. If oil is naturally trapped in ice, less waste is generated. If oil is being contained by a fire boom, the boom gets dirty and further treatment is required. When ISB is conducted in ice free waters with booms, an OSR vessel or two are also in direct contact with the slick. So, further cleaning of the ship body is recommended. (IPIECA 2014) Overall, waste management aspect is less in scale of an issue in comparison with the mechanical method. In mechanical OSR, the scale of employed logistics and waste strategies can be overwhelming due to liquid oily waste to be collected, stored and transported. (ART 2012; IPIECA 2014)

Social Aspect The effects to the environment and human health are proved to be minor and local. Moreover, they are carefully monitored. In all ISB operations weather patterns, proximity to wildlife and human population centers have to be taken into account (API 2017). Health concerns are possible only, when there is a direct contact between people and the smoke plume. This can only occur, when people are not wearing personal protective gears and the plume lies at sea level in calm weather conditions. (IPIECA 2016) Emissions level is connected with the type of oil or petroleum product, which is being ignited. The smoke produced is predominantly composed of simple molecules of carbon dioxide (73%) and water (12%). Remaining 15% are mostly particulates of soot, with minor content of carbon monoxide, sulfur dioxide, nitrogen oxides, polyaromatic hydrocarbons, and others. (API 2015; API 2017; Buist et al. 2013; IPIECA 2014)

Economic Aspect ISB is a cheap OSR method as it does not require sophisticated equipment. The only requirement is for specialized booms and chemical herders. Personnel involvement can be low, and ISB implementation is rather fast. (ART 2012; Buist et al. 2013; Fritt-Rasmussen et al. 2012; IPIECA 2016) It is usually the cheapest method followed by dispersant use and mechanical recovery respectively (Doshi et al. 2018). An estimated cost of organizing ISB response of 1 ton of oil offshore is about 2700 euros (Li et al. 2016).

11.2.2 Environmental Challenges for In Situ Burning

Ambient Temperature The International Association of Oil and Gas Producers states that the lower limits that ISB can be applied are 0 °C of water and 11 °C of air. Extremely low temperatures add expenses, as adjustments may need to be made to the duration of work shifts, and heating systems may also be required. (IPIECA 2016)

Ice in Water Most of existing OSR methods are not very efficient in ice conditions, since they were initially developed for open water (Vorobiev et al. 2005). In Arctic conditions, ice coverage over water surface varies from 0 to 100% and can consist of differents types of ice, from seasonal to multi-year ice, with additional types. For ISB, ice effects are demonstrated in Fig. 11.1 (ART 2012).

In scenario 1 is the coverage is from 0 to 40%. These conditions favor OSR works, since ice can be managed by responders: e.g. fire-proof booms and herders can be used in operations. The ISB efficiency is rather high, and oil removal is full on. In scenario 2 is ice coverage of 40–60%, and from 90 to 100%. Responding to oil spill becomes harder and the efficiency is lower than theoretical – down to 60%. This is caused by ice physical obstacles, possible boom failures and oil weathering processes. In Scenario 3 water surface is covered 60–90%, or natural containment conditions. In this case ice plays a supportive role in OSR operations by containing the oil in its trap and dampening wave influence. No booms are required. Ice contains the oil and the burning process is responding with unique changes in its geometrical appearance due to mass and hear transfer (Farmahini Farahani et al. 2019). The efficiency is as high as in open water. (ART 2012; API 2015; IPIECA 2016)

As seen from above discussions, ice can both inhibit and support oil spill response. Natural containment conditions, 60–90% ice cover, prevent oil from spreading or slow it down considerably. When the oil slick is in one place, oil thickness sustains itself in a sufficient condition much better. The oil is preserved well,

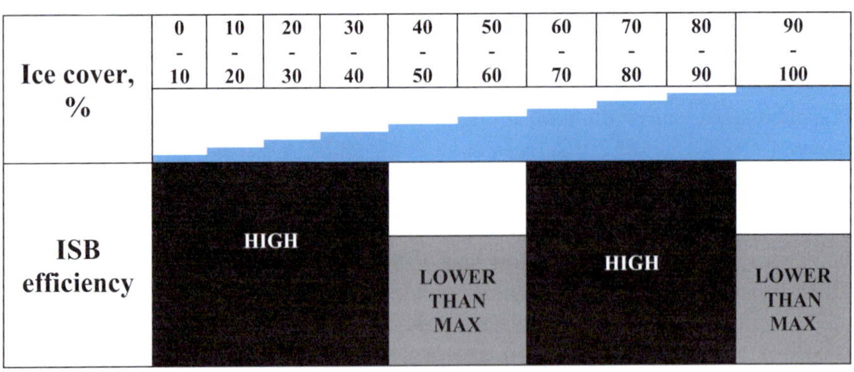

Fig. 11.1 Effects of ice cover to ISB

while other weathering processes are also inhibited: for instance, emulsification, evaporation and biodegradation. The last two are delayed due to influence of low temperatures together with lack of solar radiation in winter (Vergeynst et al. 2018). The speed of oil breakdown processes is slower. There are negligible effects from tidal current activities. These external conditions suit as an environment for ISB. Window of opportunity stays open much longer than in warmer latitudes. (Al-Majed et al. 2012; API 2016; Fritt-Rasmussen and Brandvik 2011) ISB is applicable in ice-infested waters in almost all conditions, where other OSR methods, such as mechanical and chemical are limited (Buist et al. 2013; Fritt-Rasmussen et al. 2012).

Mixed with snow on top of ice blocks, oil first needs to be collected in a cone pile. There is a need for a free pool of oil. Once organized this way, it may be ignited and burnt. If little oil volume is spilt on the snow, comprising up 3 to 4% of its volume, then fire starters, including diesel fuel or gelled gasoline, should be added to ignite the slick successfully and support combustion, removing up to 90% of the oil. The window of opportunity for such incidents extends to 2 weeks. (ART 2012; API 2017)

Wind Influence Calm water is an important condition for successful application of ISB. If the wind is stronger than 10–12 m/s, then ignition is not possible even with 10 mm oil slick thickness and absence of waves (Buist et al. 2013). This occurs due to lack of eligible inflammable vapour concentration, absence of which cannot support the combustion. The upper limit of bearable evaporation loss stays at the level of about 30%, depending however on oil types (Al-Majed et al. 2012; ART 2012; Buist et al. 2013; Michel et al. 2005; Walton and Jason 1999) When ignited, the burning process can be sustained with winds below 18 m/s (API 2016). However, winds of this strength can have negative influence on fire safety for ISB operators, taking into account the fact that the fire and the smoke spread downwind along the slick. The wind speed is also important for the smoke plume. It requires certain atmospheric turbulence conditions for its successful aerial dissolution. The speed cannot be too high or too low. Calm winds may enable formation of smoke close to the water surface, which can endanger involved personnel. Thus, wind behavior forecasts are crucial for ISB: from the start to the end, especially in the Arctic, where storms are possible. (API 2015; IPIECA 2016)

Wave Turbulence High wave activity makes the method inefficient. With currents more than 0.5 m/s as well as wave height above 90–120 cm for wind-based waves, and 300 cm for swells, ISB is not recommended, as an OSR option. All these parameters are influential on fire boom holding capacity. If it fails to contain and hold oil in one place, OSR operations will not succeed. (ART 2012; API 2015; IPIECA 2016; Michel et al. 2005)

Light Effect During the Polar Night period in the Arctic region, oil spill response using ISB can be still applicable. It can be realized safely and efficiently at night but only in cases when: vessel traffic is far from the accident site and no booms are applied. (IPIECA 2016) As for solar radiation for ISB, it affects the smoke removal from the place of operation. The solar energy enables thermal air mass movement so that warmer air rises in relation to colder one, taking with it part of the smoke. (API 2015) Considering solar radiation influence on the oil, photooxidation is a process, which is dependent on the amount of UV. It plays an important role in breaking down hydrocarbon molecules, in particular (polycyclic aromatic compounds) PACs. Another effect of photooxidation is the generation of emulsion stabilizing components that may enhance the formation of stable emulsion.

Salinity Salinity does not decrease or enhance the burning ability of oil, but only influences its buoyancy. Water of more salty seas is denser and, hence, it makes oil more buoyant. In more sweet seas with brackish water, oil floating ability can be much lower due to difference in densities of both substances. An example of brackish sea water is the southern part of the Kara Sea, a salty one – the Barents Sea. Water, where number of salt parts per thousand (ppt) is less than 35, is considered brackish; whereas water with ppt higher than 35 is saline. Note that water with ppt below 0.5 is freshwater, but this condition is not met at Arctic seas. (NWE 2017)

Precipitation and Visibility Frequent precipitation storms and rapid weather changes is very common for the Arctic. All sorts of precipitations is possible: snow, rain, drizzle, fog, and combinations of those. Rain in any form can decrease the efficiency of ISB. The largest effect is inhibition of hydrocarbon vapor formation during oil burning. (API 2015; IPIECA 2016) For effective oil spill response, it is crucial to observe, where the spill can be contained with booms and vessel operations can be conducted around it for further ISB. Good visibility is not only important for people's health and safety, but also for steering vessels and aircraft that are supporting the oil abatement operations. The horizontal visibility should be at least 4 km for operational flying, and 300 m for site works. (Al-Majed et al. 2012; Potter et al. 2012; Walton and Jason 1999)

11.2.3 Technological Limitations of In Situ Burning

Oil Type and Its Viscosity Only hydrocarbon vapors are ignitable and burnable. Hence, oil types, which become vaporized easier, or have higher vapor pressure, are easier to ignite. They are mostly lighter petroleum products: low to medium viscose with American Petroleum Institute (API) gravity more than 32° or less than 860 kg/m^3 density. (Michel et al. 2005) Too light oils, such as gasoline with API more than 50°, are however unsafe to burn. Same applies to too heavy types, for example crude oil or HFO with API less than 20°, or density more than 930 kg/m^3, which are hard to ignite. (API 2015; API 2017; Federici and Mintz 2014; Michel et al. 2005) Table 11.1 contains several examples of heavy, medium and light oils and their flammability properties (API 2016; IPIECA 2016; POLARIS 2013).

Table 11.1 Flammability and ISB efficiency of different types of petroleum products and crude oils

Oil type	Oil product	API gravity, °	Flammability
Light ($\rho < 870\ kg/m^3$)	Gasoline	65	very high
	Diesel fuel	40	high
	Light crude	30	high
Medium ($\rho = 870 \div 920\ kg/m^3$)	Medium crude	25	moderate
	Heavy crude	20	moderate
Heavy ($\rho > 920\ kg/m^3$)	Heavy fuel oil	19	moderate
	Diluted bitumen	18	moderate
	Bitumen	8	low
	Sinking oils	7	very low

The thicker, denser, and more viscose the oil, or in other words heavier, the less combustible it is. The lighter the oil – less dense, less viscose, more volatile; the more ignitable it is. For example diesel fuel has low density and high flammability, whereas heavy bitumen is hardly ignitable.

Thickness of the Film Thickness of oil slick, as another ISB limiting parameter, is crucial and primary for ignition and combustion maintenance. The recommended thickness is between 1 to 10 mm, depending on oil type – most importantly, its viscosity. Lighter, less viscose oils have much stronger ability to evaporate and create required inflammable hydrocarbon vapors (compared to the thicker heavier oils), and can be ignited already with 1 mm layer. Heavy types and, especially, non-fresh, emulsified slicks are to be fired at thicknesses from 4 up to 10 mm. (ART 2012; API 2015; API 2017) The most common recommended minimal ignitable thickness is from 2 to 3 mm, preferably 5 mm. Herders may be used to increase slick thickness, but they require calm conditions, with wind speeds below 1.5 m/s and negligible wave activity. (IPIECA 2016; Vorobiev et al. 2005)

Water Content in Oil Due to the influence of waves, currents and winds – weathering processes, small drops of water become incorporated in oil. The percentage of one into the other may vary, forming water-in-oil emulsions. As a practical matter, oil slick water content in the form of emulsion should be lower than 20–25% (Al-Majed et al. 2012; Buist et al. 2013; IPIECA 2016). Higher water intakes make oil too foamy and stable. It can create so-called "chocolate mousse", super emulsion. It is hard to ignite, tends to burn slowly and fire is easily extinguished. Another drawback is that burning these emulsions produces more burn residues. All these make ISB process inefficient with highly water saturated emulsions. The exception may take place with some crude oils: e.g. paraffinic ones. They can be treated with in situ burning with a higher percentage of water. (ART 2012; API 2015; Michel et al. 2005) Cases of 60–80% water intake are known, 90% are also possible. These

Table 11.2 Effect of water content in the oil slick to ISB

Water content in oil, %	Effect on ISB efficiency
< 12.5	low
12.5÷25	medium
> 25	high

emulsions are formed fast and stay stable. In order to ignite such substances, emulsion breaking is needed. This can be done via water removal through boiling or using specially designed chemicals. Both options work and help to create a layer of unemulsified oil covering the emulsified oil slick. Only after these measures, fire-starting becomes possible. (Walton and Jason 1999) Table 11.2 above shows the negative effects of water emulsification of the oil towards ISB operations.

The higher the water content is in oil, the harder it is to ignite it. For example, when the water content is above 25% accelerants may be needed. Accelerants include diesel fuel or gelled gasoline. (API 2015; Walton and Jason 1999) Physical transformation of oil in emulsion is also important for fire booms. Experiences from field operations have shown that, if the emulsion has a viscosity below 1000 mPas, there is a risk for larger rate of boom leakage compared to more viscous emulsions or oils. This is important to take into account, when booming low-viscosity oils, especially in relation to towing speeds. Same applies to mechanical OSR – its boom applications. (ART 2015; Vadla and Sörheim 2013)

Specific Parameter – Fire Safety Requirements Special country-dependent approvals are required in order for OSR authorities to be allowed to conduct ISB operations at sea nearby population settlements. Burn plans with specific information about the site, weather forecasts, oil type, fire safety and various other organizational matters have to be presented: logistics, smoke plume behavior forecasts, air monitoring, safety distances downwind for vessels and personnel, OSR vessel orientation in relation to the smoke, communication practices, and others. Besides, ISB training for operators should be in place before actual oil spill response. (API 2015; API 2017) Factors as distance from populated areas, nature reserves and vessel traffic, movement trajectories of the smoke, and proximity of other possible combustibles are noted. As regards safe distances downwind from ISB operations, they vary depending on wind patterns and intensity. Higher winds make safe distances larger, whereas low winds decrease them. (EPA 1999; IPIECA 2016) The ISB site must be at least 1–2 km away from closest settlements (Michel et al. 2005).

Window of Opportunity Enclosure of oil between ice increases the window of opportunity (WOP). On average, WOP is 24–48 h, which applies to all OSR methods (Vorobiev et al. 2005). Due to ice-trap conditions, with at least 70% of ice, and lower evaporation in the Arctic the WOP may be extended up to 72 h and more. For heavy oils WOP is shorter, for light and medium ones – it can be longer. (EPPR 2015; Nordvik 1995; Fritt-Rasmussen and Brandvik 2011; Singsaas and Lewis

2011). The upper tested limit is documented to be more than 30 days under very calm conditions with no emulsification taking place and a high film thickness, after which the successful ignition and combustion were executed. (API 2016; Dickins 2015). Thus, the limit for successful oil spill response is broader in ice-infested waters than in open sea conditions.

There are several physical and chemical processes related to the spilled oil, as a result shortening the WOP: evaporation of volatile hydrocarbon fractions; dispersion of oil into droplets; oxidation of the slick edges resulting tar balls formation; emulsification; biodegradation; and evaporation. When more than 25–30% of all volatile components have evaporated, no ignition is possible (IPIECA 2016). The fresher the oil is, the more successful and efficient combustion process becomes. No fire promoters, no booms to concentrate oil are required in this case. (EPA 1999)

Distance from the Shore Distance from the mainland can be a challenge in terms of delivering equipment and responders at place. However, open water conditions and safe distance from the shore are a must for ISB operations. (API 2015) Since this method requires less arrangements than with mechanical OSR in terms of logistics and infrastructure, especially, when oil is naturally contained in ice, operations can be performed in very remote areas of the Arctic sub-regions (ART 2012). Safe distance from nearby communities and sensitive industrial objects is a crucial factor, since the amount of air pollutants in the smoke, including particulates, carbon monoxide, sulphur dioxide and polycyclic aromatic compounds can be dangerous for human health. (Al-Majed et al. 2012). Closest settlements should be no nearer than 1–2 km away from the ISB site, with common cases for distance restriction of 8 km (Michel et al. 2005). This is, however, a country-dependent decision.

11.3 Mechanical Oil Spill Response

11.3.1 Overview of Mechanical Response

Brief Description The main constituent tasks in mechanical recovery are to localize and concentrate the spill via floating structures (booms) or natural containment (e.g. by ice); and to pump recovered oil from the sea surface onboard of a vessel, where the oil is temporally stored for further disposal. The first task is conducted via boom application, while the other is done via skimmers. (JIP 2014; Wilkinson et al. 2017) Mechanical OSR is a widely accepted maritime option in use for more than 50 years (Al-Majed et al. 2012; Federici and Mintz 2014). Despite the fact that oil recovery usually requires profound planning of personnel work and utilization of multiple pieces of equipment, with operations executed from a special response vessels as well as fishing ones (ART 2015); according to Ramboll Barents (2010) and ART (2012), mechanical clean-up is considered as the first OSR option among local authorities of all eight Arctic states, namely: USA, Canada, Denmark, Iceland,

Norway, Sweden, Finland, and Russia. Due to fast oil spreading and breakage into separate slicks during mechanical oil spill response, the main operational strategy is to work as close as possible to the spill source. Unlike ISB, it does not require any approval on state, federal and local levels. (API 2017; DNV 2015) Most tactics of mechanical OSR are designed to feed oil to the skimmer. However, there are variations of skimming systems: e.g. two vessels with boom, single vessel with outrigger, three vessels of opportunity with boom, single vessel in ice. (EPPR 2017a)

Examples of Response Equipment

Booms, as an OSR support tool, may be inflated by air or designed with an embedded flotation material; closely attached to the vessel, towed by it distantly or set in passive and self-sufficient drift (ART 2015; STAR 2014b). They are made from durable fabrics, which can withstand wind, waves, acids, alkalis, oil and oil products, which they do not adsorb. The symmetric configuration of every piece is usually designed so that the direction of their towing is insignificant and they can be mutually attached to each other by locks forming a long line, up to 500–600 m, creating a swath width of 100 m, with maximum 200 m. The time required for the boom installation is usually counted as 60 min for preparation works and about 30 min for actual works of loading it on the water surface (Ramazanov 2015). There are more than 150 types of boom systems, currently available on the market in application to the Arctic Ocean climate conditions. However, no matter what type it is, the main task of the boom is to contain oil and stop its spreading at sea, prevent oli from reaching the coastline of islands or the mainland and set a certain direction of oil slick movement. As it occurs under external environmental conditions, spilt oil quickly spreads over the marine surface, forms a thin layer and loses its original "fresh" oil spill thickness. Thus, booms installed around the oil slick also serve as an oil containment pool, which concentrates oil to the thickness values required for OSR. This is especially important within first 72 h after the spill incident. Usually, there are three configurations for oil collection on water: U-, V-, and J-types. Apart from common vessel organized installations, booms can be also delivered by helicopters (Al-Majed et al. 2012; ART 2012; ART 2015; Parshentsev 2006; Vorobiev et al. 2005) It is important to consider that booms may be towed by a vessel only at certain speeds: 1 knot, which is equivalent of about 1.85 km per hour (ART 2015; Singsaas and Lewis 2011). This is rather slow and comprises nearly twice as slow as the average pace of human walking. According to ART (2015) and Vorobiev et al. (2005) the fastest solutions allow the upper maximum speed level of 3 knots. Otherwise, with a higher speed, oil spill will leak from under the boom. The leakage not only depends on the towing speed, but also environmental hydrodynamic conditions (wave pattern, wind speed, water currents) and boom parameters (buoyance to weight ratio, heave and roll response, and initial draft). The most common five leakage reasons in booms are related to oil entrainment, drainage, critical accumulation, boom submergence, and splash-over (Li et al. 2016).

Skimmers, the major equipment in mechanical oil recovery at sea, is a hydraulically powered mechanism equipped with rotational flat material structure, designed to adsorb viscose oil substance on its surface without addition of chemical agents. The main principle of work is the difference between physical properties of oil and water, which distinguish with density and molecular coupling with surface of various materials (Vorobiev et al. 2005). Historically, skimmers were engineered for open water surfaces (ART 2012). The latest innovation advances in the design are related to cold and icy environmental conditions, such as heating system introduction against frost formation; and better absorbing surfaces of the rotating oil collectors. Otherwise, no revolutionary advances in the basic skimming technology are expected. (Al-Majed et al. 2012; Federici and Mintz 2014; STAR 2014a; ART 2015) As regards technical parameters of skimmers, there are two crucial aspects to consider: pace of rotation and sorption capacity of the skimming surface. Skimmer designs work with most oil types, however, there is an upper limit, after which the oil does not have enough time to stick to the brush fibers, no matter how strong its oleophilic properties are. (Al-Majed et al. 2012; Federici and Mintz 2014) The following four skimmer types are presently produced and professionally applied in operations: vacuum, weir, mechanical and oleophilic. Vacuum and weir skimmers have a drawback of intaking more water than oil but are still utilized to collect light and medium viscosity oils via vacuum sucking mechanism or gravity forces phenomenon to the recovery tank. The skimmer is applicable for very viscous heavy oil types and therefore equipped with a conveyor belt to transport it in special buckets to the storage tank. All these types – vacuum, weir and mechanical skimmers become clogged with ice particles and fail to operate well in the cold sea conditions: extremely low temperatures and freezing. The most used and efficient skimmers type in Arctic seas is oleophilic. It has either a drum, a belt, a brush, a rope mop or a disc in its structure, which rotates in contact with the upper layer of the water surface. This rotor absorbs oil on its surface and transports it for further treatment. Usually, light and medium oil types may be collected with this skimmer. (ART 2012; Li et al. 2016; Wilkinson et al. 2017) Brush and drum brush designs of this skimmer have the highest perspective in terms of application in the ice-infested waters (Singsaas and Lewis 2011). They have the largest oil sorbing capacity due to the material, organized with millions of brush bristles to provide massive surface contact area; which are capable to collect oil of any viscosity. They can be self-propelled, mounted, portable and towable. Those attached to the vessel have higher capacity and resistance towards the sea ice, with examples of skimmer working surface area of 16 m along the ship stern (Lampela 2011). The following models have been designed for use in the ice environment: "Polar Bear", 60 m^3/h, with custom weir and brush drum; "Polaris", 70 m^3/h, with brush drum; and "LRB 150", 90 m^3/h, with oleophilic brush drum, respectively by Desmi, Framo and Lamor manufactures. Ice management and low temperature preparedness are something distinguishable, which is included in the skimmers design, and support successful mechanical oil abatement operations (ART 2015). "Polar Bear" and "Polaris" can operate in ice cover of 70%; whereas "LRB 150" can withstand 90% of ice in the working zone. (Federici and Mintz 2014; Rytkönen and Sassi 2015; Singsaas et al. 2008, 2010).

Storage tanks is another component of mechanical oil recovery, used for temporary storage of recovered oil (ART 2015). Storage tanks can be of two types: independent on-water storage and onboard vessel storage. The on-water storage (barges, bladders) is only applicable in open water, without any ice present. The onboard storage is usually installed on vessel deck or below it. As a practical rule, it should be twice as large as the effective daily recovery capacity of the skimmer, should have strong pumping system, especially if handling heavy oils, and storage tank heating devices since the cold temperature may influence the oil pumping properties. (STAR 2014a)

Weir as well as vacuum skimmers, as mentioned in this section above, have a drawback of taking excessive amount of water along with collected oil – therefore the storage tank fills its capacity faster, compared to oleophilic skimmer, with declared free water content of 2% (ART 2012; LAMOR 2015). To solve the problem of abundant amount of recovered sea water, a temporary storage system with water separation process, or decanting, is often introduced (ART 2012). This measure maximizes the tank storing capacity by removing recovered water (ART 2015). Depending on the calculations and the state of oil on the water surface, the storage tanks may be of different size. If there is an example case of "chocolate mousse", with 10% oil and 90% water in it, and the incident report says that oil spills is of 1000 t, the size of the storage tank should be at least 10,000 t capacity. In Arctic conditions, these tanks are sometimes heated to make it flow easier. Usually, the heaters maintain 90 °C in the tank. (Vorobiev et al. 2005)

Waste handling and disposal of the recovered oil is required to complete the mechanical oil spill response. It is usually done via a burning facility either onshore or sending to a refinery. Oftentimes, vessel based incineration facilities, mounted on specially equipped places, is the preferred option. Currently, there are three main types of the burners in ascending order according to their capacity: rotary cup burner – hundreds of barrels of oil per day (BOPD), augmented burner – less than 10,000 BOPD, pneumatic flare – 10,000 and more BOPD. Rotary cup burner is rather light and can be used both onboard and inland. The system's main part is the rotating fan, which blows the air in the cup and atomizes oil to be sent to the combustor. Augmented burner is a good option due to its simplicity in design. The major disadvantage is that it has only small-scale application capacity. This burner type refers to chimney-style floating incinerator, which has an aerator to support the combustion process. The last burner type, with commonly used prototypes in oil and gas industry for well testing, pneumatic flare, requires additional equipment with pressurized air to atomize the recovered oil in the system but can be also mounted onboard. This option is recommended for large-scale oil spills. (ART 2015)

Response platforms with built in equipment and capacity to transport trained personnel, as the last component, but surely not least is usually included in mechanical OSR operations. Depending on the water conditions at sea and harshness of the marine environment, the response vessels may be chosen differently: for open water – regular response ships, for thick ice seas – double hull ice-strengthened vessels of ice breaker type. In addition, there should be aerial support for monitoring and guiding the work, small multipurpose vessels and various OSR related equipment: sorbent spreaders, dispersant sprinklers, pumps, pipes, oil squeezing devices for sorbents, temporary storage tanks or sacks, incinerators, chain saws, personal protection equipment and other materials. (ART 2015; Lampela 2011; Slaughter et al. 2017; Vorobiev et al. 2005)

Efficiency of Oil Removal and Required Time Claimed oil recovery efficiencies in ideal conditions reach as high as 80 to 95% (Li et al. 2016). In operational practice, the reported equipment efficiency ranges from 10 to 30% depending on the environmental conditions. In its turn, 30% is often an optimistic forecast; 5–15% is an actual down to earth expectation. (API 2017; EPPR 1998; Vorobiev et al. 2005). Nevertheless, mechanical OSR in icy marine environment is possible. (ART 2015; Li et al. 2016; Singsaas and Lewis 2011) The method is time consuming. The drawback here is the width of the skimmer and its small oil surface catch – hence, low skimmer capacity and low efficiency to collect oil. It is possible to solve this problem, by attaching the boom to it and increasing its grasp width, but at the same time slowing down the speed of work to 1 to 3 knots, or about 2–6 km per hour. (ART 2015; Federici and Mintz 2014; JIP 2014; Slaughter et al. 2017; Vorobiev et al. 2005) Operations can be a matter of days to months to fully complete oil recovery. Oftentimes, oil spreads faster than it is recovered. (API 2015; BP 2018; Slaughter et al. 2017; Wilkinson et al. 2017)

Sustainability of the Method

Environmental Aspect With application of this OSR, oil is mechanically and permanently removed from the surface of the marine environment. In comparison with ISB, mechanical OSR does not add visibly to air pollution on site, since incineration mostly happens onshore. Same is applicable to water contamination. (Al-Majed et al. 2012) However, minor impacts are possible due to application of the response vessels and equipment. Increased noise pollution can be one of characteristic mechanical OSR impacts. Unrecovered oil can also cause substantial marine contamination. (JIP 2014) Recovered oil, often mixed with seawater and ice, is sent further for downstream separation and further treatment. This represents a challenge in the Arctic conditions: large amount of the mixture needs to be temporarily stored and disposed afterwards. Transporting the mixture onboard, storing it in tanks supplied with heating, and disposing it then is energy intensive. It also requires onshore facilities to melt, separate and utilize oil, which are limited in these geographical latitudes. (Al-Majed et al. 2012; Lampela 2011; Slaughter et al. 2017; Wenning et al. 2018; Wilkinson et al. 2017)

Social Aspect Personnel health and safety issues is number one aspect with mechanical oil recovery, since oil spill operations are organized in the harsh Arctic ambient conditions and can last for weeks. (DNV 2015) Among possible hazards are physical – cold stress, snow blindness due to light, safety accidents due to darkness and icing; chemical – exposures to oil via inhalation or dermal contact; biological – contact with polar bears or walruses; and others. (EPPR 2017b)

Economic Aspect In general, mechanical method, as a whole technique, requires a lot of funds, especially, to carry personnel and to ship the equipment to far distant areas. In general, mechanical recovery of oil from the sea surface is usually an event, which requires profound planning of personnel work and utilization of multi-component set of equipment, with operations executed from a special response vessel (ART 2015). The availability of the logistics in the Arctic should be well-thought for this OSR method: food, fuel, accommodation, storage tanks and burning facilities either onboard or onshore for final utilization of recovered oil and produced waste management system. (ART 2015; Federici and Mintz 2014; STAR 2014a; Wenning et al. 2018) In addition to this, there should be response vessels with built-in equipment and capacity to transport trained personnel. Depending on the water conditions at sea and harshness of the marine environment, the vessels may be chosen differently: for open water – regular response ships, for thick ice seas – double hull ice-strengthened vessels of ice breaker type. Besides, there should be aerial support for monitoring and guiding the work, small multipurpose vessels and various OSR related equipment: sorbent spreaders, dispersant sprinklers, pumps, pipes, oil squeezing devices for sorbents, temporary storage tanks or sacks, incinerators, chain saws, personal protection equipment and other materials. (ART 2015; Lampela 2011; Vorobiev et al. 2005) An estimated cost of recovering 1 ton of oil from water is 8500 euros (Li et al. 2016).

11.3.2 Environmental Challenges for Mechanical Response

Ambient Temperature For mechanical OSR method, low temperatures are an inhibiting factor, especially, for oil skimming and pumping. As any technical mechanism, the skimmer has certain limits of its working environment, particularly in icy-cold temperatures. To avoid freezing of its moving parts due to seawater spray, provide allowed operational temperature for collecting and transporting viscous oil, and maintain its declared efficiency; circulative heating system via hot pipe water supply should be constantly in operation. (ART 2015; Singsaas and Lewis 2011) Dealing with heavy oils also becomes challenging. They solidify in freezing conditions because of increased viscosity. Collecting them with skimmers and transporting with pumps become difficult and, in most cases, inapplicable. Instead, nets and other suitable devices are utilized for oil recovery and appropriate solutions are found for storage and transfer. (DNV 2015; Wilkinson et al. 2017) Recommended operational temperature for mechanical oil spill response is within the range of −5 to −18 °C (EPPR 2017a).

Ice in Water For mechanical oil recovery, starting already at 11% ice coverage boom functionality and overall operations are jeopardized. Some of ice is captured in the boom site opposite to the vessel towing direction. Booms can be teared, strained, stretched and damaged to withhold the spilled oil. Nevertheless, booms are applicable with ice conditions up to 30%. (DNV 2015; Wilkinson et al. 2017) For mechanical recovery, ice effects are demonstrated in Fig. 11.2.

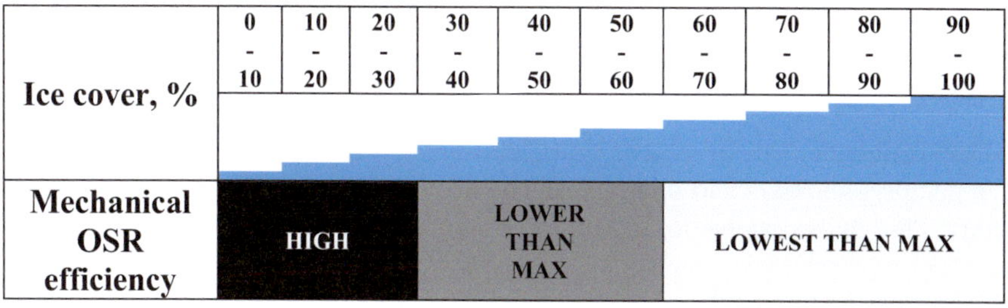

Fig. 11.2 Effects of ice cover to mechanical recovery

Ice coverage with more than 30% of total surface water area in the oil spill site is usually an obstacle for the boom installation. Independent booms lose their usability (Singsaas and Lewis 2011). In 30–60% ice-covered waters, boom systems with short attachment to the response vessel are applied. This construction has high maneuverability and strong mounting arm, which is convenient for OSR operations in icy waters. If the ice-free water surface comprises less than 40%, application of the boom is completely excluded. No artificial oil slick containment is possible. (DNV 2015; Potter et al. 2012; Økland 2000) In this case, ice naturally contains oil, decreasing turbulence influence from the waves, limits its spreading, forms walls for the oil pool and creates separate pockets of collected oil slicks. All these preserve oil for longer periods of time, making it possible to apply mechanical recovery long after oil spill happens. (ART 2012, 2015; JIP 2014; Wilkinson et al. 2017) For obvious reasons, the device effectiveness is lower in ice-covered waters compared to open water surface. Skimmers, as an OSR tool, in its turn also have its limitations in ice-infested waters. The most important aspect for mechanical clean-up is provision of the physical contact between the skimmer and the oil slick. Thereby, skimmer surface should be freed from interfering ice particles or blocks, otherwise the encounter rates are reduced. In this sense, several developments in the skimmer device features have been made to solve this problem. As one of the examples, an improved skimmer body may press ice blocks with a metallic screen-like structure, submerge the ice under the sea water level and, thus, reach the oil-skimmer surface contact. Another solution is built around ice processing, when some particles end up in the skimmer; where separation of oil from ice is automatically arranged (ART 2015). As one more limitation, presence of ice can significantly reduce the efficiency of mechanical OSR facilities and make it often ineffective. In comparison with ideal open water conditions, ice-infested waters decrease applicability of the method essentially. (ART 2015) Starting already at 10% of the ice to free surface water ratio, which stands for a few drifting ice parts, skimmers lose their maximum oil recovery capacity (Dickins 2015). Up to 30% ice coverage regular open-water skimmers may be operated. In the ice range of 30–70%, specialized skimming equipment should be deployed. (DNV 2015) In addition, dealing with oil contaminated ice is a problem, which is an energy challenging task. Pumps may become clogged by ice that causes them to break down. Adapted pumps and heating systems for transport and storage are required, for example, heating coils. When separating oil from ice and melting it, only small amounts of ice are economically acceptable. (ART 2015; DNV 2015; Lampela 2011; STAR 2014a; Vorobiev et al. 2005)

Wind Influence Due to sea water droplets transferred after splashing with wind by air pumps, skimmers and booms undergo icing and accumulate frozen ice on its body. This causes booms to fail in maintaining its buoyancy. Wind can also move the boom and release it from anchor. Same applies to vessel and equipment stability. High winds provide unfavorable conditions for OSR operations. It is difficult to provide personnel safety, keep the vessels in correct position and towing speed, and rely on boom oil containing capacity and skimmers efficiency. (DNV 2015) When wind blows faster than 15 m/s in open water, mechanical recovery is not possible (BP 2018).

Wave Turbulence Mechanical OSR, as ISB, becomes inefficient in high wave activity. The more energy water has, the less efficient boom applicability becomes. Above 2–3 m wave height in open water and currents more than 0.5 m/s booms start to submerge, break and fail with oil spill containment. (API 2015; BP 2018; DNV 2015; Vorobiev et al. 2005)

Light Effect Unlike with ISB, daylight availability is usually a limitation for mechanical OSR. Light can affect working shifts of operating personnel, especially during times of the year with the lack of it – Polar nights. (BP 2018; DNV 2015; Vorobiev et al. 2005) However, in Norway, for example, the contingency plans include OSR operations in darkness with a reduction factor of 50%, provided that aerial remote sensing is available.

Salinity Similarly to ISB, salinity affects the mechanical recovery process only by influencing oil buoyancy. Independent on spilled oil type, the oil is more buoyant in more saline sea water conditions in comparison with brackish waters. (NWE 2017)

Precipitation and Visibility Despite some advances in night vision equipment, visibility of about 1800 m, approximately 1 nautical mile, is a limiting parameter for the operations. (BP 2018; Slaughter et al. 2017) Fog, low clouds, snow storms, darkness limit mechanical OSR operations significantly, including crew safety considerations (DNV 2015).

11.3.3 Technological Limitations of Mechanical Response

The functionality of the method, as also noticed with all other OSR methods, strictly depends on oil physical state and surrounding environmental and oceanographic conditions. It is of technical relevance in response skimmer operations to acquire data about oil slick volume, viscosity, thickness of the oil layer and its sea-conditioned temperature. (Al-Majed et al. 2012; Federici and Mintz 2014)

Oil Type and its Viscosity The skimmer technology is effective with a wide variety of oil types, including strongly emulsified oils (API 2017). The only limitation is related to boom holding capacity, in similar way as in with fire-proof booms. Results from field operations with emulsions of a viscosity below 1000 mPa·s have shown that a degree of boom leakage is higher compared to more viscous emulsions or oils. This is important to take into account: e.g. by reducing the towing speed with low-viscose oils to avoid too significant leakage. (ART 2015; Vadla and Sörheim 2013)

Thickness of the Film In mechanical oil recovery, the oil film thickness should be at least starting from the range of 1–2 mm and up to 10 mm, which is oftentimes reached by application of booms. If it is less than 1 mm for free drifting oils, oleophilic skimmers become inefficient. (API 2017; Vorobiev et al. 2005)

Water Content in Oil With increased emulsification and weathering, oil becomes significantly viscous (DNV 2015). When the state reaches characteristics of the "chocolate mousse" with viscosity more than 15,000–20,000 mPa·s, the only option to remove oil from water is via application of mechanical methods. As efficacy of the mechanically recovered oil in comparison to the amount of recoverable oil on the seas surface may be about 20%, it means only one fifth of the oil is to be collected. The rest four fifth is left in the marine environment to biodegrade naturally, possibly with some of the oil stranded onshore. (Vorobiev et al. 2005)

Specific Parameter – Storage Tank Capacity Most skimmers intake more seawater than oil material: both free and emulsified sea water. Therefore, substantial capacities of storage facilities are needed. Once the storage tanks are full, they must be emptied onshore, unless burning facilities are applied (Slaughter et al. 2017). The round trip to offload the tanks and return them back to the site may be hours. In Arctic conditions, the tanks also require heating systems (API 2016, 2017; DNV 2015). Free water, separated and settled in the storage tanks, can be drained off back into the booms.

Window of Opportunity Depending on ice situation, ambient conditions and oil physicochemical properties, WOP can be from 24 to more than 72 h: sometimes weeks. (EPPR 2015; Federici and Mintz 2014; Singsaas and Lewis 2011; Vorobiev et al. 2005)

Distance from the Shore One of the most crucial considerations during mechanical OSR is its pronounced logistics. The distance from the mainland should not be too long. Delivering and repairing the equipment and transport of the recovered oil is a challenge. Remote areas of the Arctic region require also availability of prepared personnel and equipment nearby an accident area. The operations are labor-intensive, thus, remoteness of the potential oil spill emergency site along the Arctic is a limiting factor. (API 2017; ART 2015; DNV 2015)

11.4 Dispersants Use for Oil Spill Response

11.4.1 Overview of Dispersants Use

Brief Description Dispersants, as the main chemical method agent, are sometimes considered as an alternative OSR option apart from in situ burning and the mechanical oil spill response options in the Arctic. Finland and Sweden avoid using them since they are forbidden due to poor environmental state of the Baltic Sea; whereas Denmark, Iceland, Norway, US and Russia allow application of dispersants after receiving special environmental permission from a local controlling authority. (Ramboll Barents 2010) The method became the most popular in the 1990s. Dispersants are stockpiled with equipment for their application by three out of five OSR centres of global coverage – Clean Caribbean Cooperative, East Asia Response Limited, and Oil Spill Response Limited. Over 70 oil spills in the last 30 years were combatted via dispersant treatment. (Vorobiev et al. 2005; Potter et al. 2012; Prince 2015)

Dispersants are chemical, liquid substances – oily light to dark brown colour, with anionic and neutral surface-active agents and additives in a hydrocarbon solvent. Their purpose is to support oil separation into particles. The colloid system in the given case is oil, spilt as the oil slick, in water; where oil is initially the dispersed phase, and water is the dispersing medium. The surface-active agents, or surfactants, serve to lower the surface tension at the oil-water interface and dissolve the oil part of the system in 20 to 70 μm neutrally buoyant droplets. If their size is more than 100 μm, they resurface back to the water surface and form a slick again (EPPR 2015; Potter et al. 2012; Wilkinson et al. 2017). The small droplets spread through the water column, in the surface mixed layer. This is needed to break the oil film on the water surface in order to speed oil biodegradation and diffusion via turbulence action of marine currents and waves. Initial concentrations of dispersed oil droplets may be as high as 30–50 parts per million (ppm). However, after several hours, the oil concentration is lowered up to 1 ppm and less. This becomes available for dissolution and utilization, which is done by sea biological mechanisms, in particular, indigenous Arctic marine petroleum-fed bacteria (Dickins 2015). This is an effective natural method of oil spill clean-up that has happened for millions of years. (Doshi et al. 2018; BP 2018; Potter et al. 2012; Prince 2015; Slaughter et al. 2017; Vorobiev et al. 2005; Wegeberg et al. 2017) As recent studies proved, it also works in the Arctic environment (ART 2012).

Oftentimes, dispersants are applied when environmentally prioritized areas are endangered. Special priority refers to the coastal areas, shorebirds and animal survival. For instance in 1996 responsible authorities of Wales decided to apply dispersants promptly to avoid a 35 thousand ton oil spill reaching the shoreline. Even though in the short term, dispersants and correspondingly a large number of oil micro-droplets form a highly concentrated "cloud" of oil droplets in the water column, temporary toxic for marine species; immediate bacterial mobilization due to widely available petroleum feedstock, in the long term, functions appropriately to transform oil. Furthermore, since the interfacial area between oil and water has increased dramatically, other degrading processes, including dissolution, contribute to oil impact minimization. Usually, already after several days, first Arctic bacterial communities inhabit the droplets and start the degrading process. The time, which it would take to execute the same and relieve marine wildlife from the long-lasting negative impacts, if leaving the oil slick on the sea surface or if it reaches the shoreline, differs remarkably. As an example, oil-contaminated shore may stay affected for 10–15 years, whereas recovery of the water column at sea happens within weeks (Makhutov et al. 2016; Prince 2015). Some review articles mention half-lives of physical dispersion to take place within one day, from 4 to 24 h (Li et al. 2016). The evaluation about the environmental impacts of the dispersant use in the Wales case showed that this was the best possible OSR solution to be employed in this particular case.

Examples of Response Equipment The dispersants are usually easy to deploy and are sprayed via means of airplanes, helicopters and vessels; with rare cases of being injected subsea. The capability of different dispersant application platforms have different advantages and disadvantages depending on spill situations. For instance, airplanes have high spraying capacity, but are limited in terms of loading space, which causes long transit time between the dispersant loading stations and the place of accident. One part of dispersant volume treats up to 20–30 parts of oil. It can be easily used in a large scale, being a fast and economically feasible option, especially for remote areas. When spraying dispersants from the plane over at sea, documented consumption counts for about 50 liter per hectare (Prince 2015). Recommended dispersant viscosity for aerial application is more than 60 mPa·s (ExxonMobil 2014). One of advantages of the method is in preventing "chocolate mousse" formation – a state of oil which is hardly biodegradable and of high stability. (ART 2012; BP 2018; EPPR 2017b; Lampela 2011; Vorobiev et al. 2005) In subsea applications, dispersant-to-oil ratio (DOR) is 1 part per 100 (BP 2018; Slaughter et al. 2017).

Efficiency of Oil Removal and Required Time The term "removal" is not applicable with dispersants, since oil stays in the marine environment after the OSR operations (API 2015). The post-treatment and final oil spill elimination is directly linked to natural biodegradation processes (Potter et al. 2012). Thus, it makes sense to discuss sea surface oil elimination instead of OSR efficiency. In Doshi et al. (2018), dispersant use with aerial application was put as the fastest in comparison with ISB and mechanical treatment of oil spill. Airborne chemical treatment of oil is done 25 m above the water level and can be 40 times more efficient and several times more rapid than mechanical recovery. (API 2017; ART 2012; Vorobiev et al. 2005) There are two different perspectives on its speed: for airplanes it is 275 km/h, for vessels – 13 km/h. The response vessel speed is much slower and, thus, dispersant spraying is applied only for small scale spills near shore. A chemical carrier plane can arrive to the place of accident in 4–8 h. Dispersant consumption of the aircraft is 50 liters per hectare, with maximum carrying capacity of 19,000 liters. (API 2017) A laboratory documented surface efficiency of dispersants in test facilities equals 90% for fresh and weathered oils under cold conditions (Belore et al. 2009; Wilkinson et al. 2017).

Sustainability of the Method

Environmental Aspect There are three issues to focus on: toxicity of dispersants, toxicity of dispersed oil, and biologically unavailable oil compounds. Recent studies have shown that modern dispersants are low toxic, biodegradable and rapidly diluted. Thus, they do not synergistically increase the negative impact from oil contamination. There are laboratory works conducted to develop new materials based on biological surface-active agents (Li et al. 2016). All in all, many recommendations incline towards dispersant application due to their high long-term benefit to the environment. The last statement stands by the result of comparing two scenarios: one – oil spill is naturally weathered without any OSR method being used, and two – dispersants are utilized to treat oil. In the long run, dispersants make oil biologically available much faster than it happens due to natural degrading processes. Increasing the surface area of oil in water, dispersants enable oil degradation in the time of weeks rather than without them over the period of several months or longer. Moreover, applying dispersants also eliminates a risk of contaminating coastal areas. Thus, dispersant toxicity is put second place after toxicity of oil itself. During the first 2 h after oil slick dispersion, there is a peak concentration of hydrocarbons in water column. This can have the most negative effects and exposure: for example, on fish eggs and other biota in the affected water layer (DNV 2015; JIP 2014; Wilkinson et al. 2017). As for the Arctic conditions, it is believed that there is no difference between ecosystem sensitivity in the Arctic zone and any temperate climate region – biological organisms are as vulnerable. (ART 2012; Dickins 2015; Prince 2015; Wilkinson et al. 2017)

Social Aspect As the harsh Arctic environment may affect responders health and safety, it is important to prepare for potential physical, chemical, biological and other hazards. (DNV 2015; EPPR 2017b) Personal protective equipment should be applied due to possible inhalation and dermal contact (API 2017).

Economic Aspect An estimated cost of application of dispersants per 1 ton of oil in offshore conditions is 5000 euros (Li et al. 2016). This is higher than for ISB but lower than for mechanical recovery. The cost structure eliminates two expensive and work intensive parameters: organization of clean-up operations with related involvement of people and equipment, and arranging waste management of collected oil-polluted material (Prince 2015).

11.4.2 Environmental Challenges for Dispersants Use

Ambient Temperature When lowering ambient and oil temperatures, viscosity of oil increases. The lower the temperature, the higher the viscosity, the lower the chemical efficiency is. (DNV 2015) Dispersants are possible to apply and biodegra-

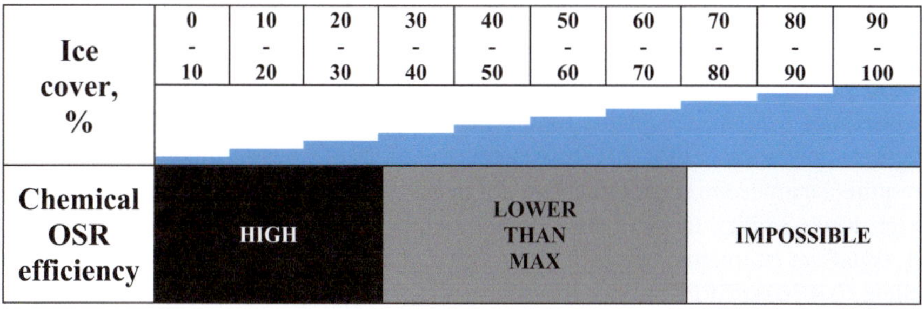

Fig. 11.3 Effects of ice cover to dispersant use

dation may occur effectively in the harsh and freezing conditions of the Arctic Ocean. (ART 2012; EPPR 2015; Vorobiev et al. 2005) In cases when dispersants are sprayed from vessels, freezing of valves and other operational OSR components is possible (DNV 2015).

Ice in Water For dispersant use, ice is an obstacle with dampening effect, which hinders the wave action needed to initiate the mixing processes needed for successful dispersion. (ART 2012; Lampela 2011) There are several scenarios about ice coverage for chemical OSR, as seen in Fig. 11.3.

With ice cover up to 30%, OSR operations are possible, with the fastest dispersant spraying approach applicable – airplane mounted sprayers. From 30 to 70%, ice coverage inhibits efficient airborne spraying, unless it is done by helicopters, which have higher manoeuvrability compared to planes. The method can still be organized but with slower speeds. If applied from vessels within this ice range, artificial mixing energy additional to natural levels have shown to be effective: for instance, from vessel propeller (Daling et al. 2010). Sea waves influence is lower essentially in ice cover 30–50% (Potter et al. 2012). Ice situation with less than 30% free water surface may make chemical OSR inappropriate solution, unless manoeuvrable spray arms are available. (Daling et al. 2010, 2012; DNV 2015; Lewis and Daling 2007)

Wind Influence Wind is a limiting factor for OSR operations, with a favorable value range of 4 to 12 m/s (ITOPF 2011; Potter et al. 2012). Stormy weather is not a limitation for dispersants functionality, unlike it is for skimmers and booms. In contrast, sea storm is favorable for mixing oil with the chemicals. However, still there is an upper limit 12–15 m/s, above which it becomes challenging or unsafe to spray dispersants (DNV 2015; EPPR 2017a). Below 4 m/s, the ambient conditions are also limiting. It is too calm and inadequate amount of mixing energy available to support successful dispersion rates. (Vorobiev et al. 2005)

Wave Turbulence Dispersion requires sufficient mixing energy in order to break oil in droplets and to mix them into the water column. Therefore, the process is highly dependent on natural sources of such energy, being sea current or wind flows, or artificial ones – one of the most common examples of which is an underwater vessel propeller, sometimes applied along with mineral particles to enhance the mixing. (ART 2012; Dickins 2015) Thus, non-calm sea with intensive waving is a good prerequisite for chemical response: wave heights between 2 and 4 m, and sometimes 7 m, are acceptable for airborne dispersant delivery, and up to 3 m – for vessel application (EPPR 2017b; Wegeberg et al. 2017). Vessel-mounted equipment require a calmer sea state, but above 0.5 m wave height (Slaughter et al. 2017). Too high wave activity, in particular above the limit of 3–5 m wave height, may negatively affect interaction between dispersant and oil material and inhibit targeting the slick. (BP 2018; EPPR 2017a; Slaughter et al. 2017)

Light Effect For better visibility and safety during OSR operations, daylight is required. Both vessels and aircrafts require confirmation of slick location before dispersant use and when applying the chemicals. (BP 2018; DNV 2015; EPPR 2017c; Slaughter et al. 2017) Although as experience shows OSR operations are also possible in darkness with a reduction factor of 50%, provided that aerial remote sensing is available for guiding.

Salinity The number of salt parts per million of sea water is important, since dispersants efficiencies are salinity dependent. In brackish waters, efficiency of dispersants decreases. In commercial scale, dispersants are mostly developed for seawater: for salinities 25 to 40 ppt. (ART 2012; DNV 2015; SL Ross 2010; Wegeberg et al. 2017) However, there exist also dispersants formulated for use in low-salinity and fresh water.

Precipitation and Visibility From practical point of view, no precipitations and high visibility is as required for aerial dispersant use as it is for aerial ISB (BP 2018; DNV 2015; Potter et al. 2012). Aerial visibility should be at least 2–5 km, operational in situ – 200–900 m (EPPR 2017a).

11.4.3 Technological Limitations of Dispersants Use

Oil Type and its Viscosity There is a limitation towards highly viscose gel-like oils, above 20,000 mPa·s – it results in inefficiency of dispersants (Canevari et al. 2001; Potter et al. 2012; Prince 2015). In this sense, oil type and its liquid state should be taken into account, when preparing chemical treatment of the slick. (ART 2012; Dickins 2015; Vorobiev et al. 2005) Usually any oil type with the range of viscosity between 2000 and up to 10,000 mPa·s is suitable for the method. The efficiency lies in values from 5 to 70%. Less than 2000 mPa·s is acceptable, and oil disperses well, with the highest efficiency – more than 70%. (Federici and Mintz 2014) Too high viscosity makes dispersion inefficient, since surfactants are washed away by the sea before they penetrate into the oil (EPPR 2017b; Prince 2015; Wedeberg et al. 2017). Table 11.3 shows how oil type affects efficiency of dispersant use. With API gravity greater than 45°, density less than 800 kg/m^3, chemical OSR is not recommended, whereas with API lower than 10° dispersants become inefficient. (API 2017; IPIECA 2015)

Table 11.3 Dispersibility of different types of petroleum products and crude oils

Oil type	Oil product	API gravity, °	Dispersability
Light (ρ < 870 kg/m³)	Gasoline	65	high, but OSR is not recommended
	Diesel fuel /light crudes	40	high
Medium (ρ = 870÷920 kg/m³)	Medium crude	25	high
Heavy (ρ > 920 kg/m³)	Heavy fuel oil	19	moderate-to-low
	Bitumen	8	absent

Thickness of the Film Chemical dispersion occurs well with low oil slick thickness: from 0.1 to 1.0 mm; but work in oil thicknesses up to 10 mm. (API 2017; Vorobiev et al. 2005).

Water Content in Oil After emulsification, oil becomes less available for dispersants. (Prince 2015; Øksenvåg et al. 2018) The emulsion with water content higher than 25% is too stable and difficult to disperse (Al-Majed et al. 2012; Buist et al. 2013).

Specific Parameter – Depth of the Surface Mixed Layer Since the oil droplets spread in the upper layer of water, there is a limitation to the depth of the sea in situ of the oil slick. The deeper the surface mixed layer is, the better uniform spreading will take place. (Prince 2015; Vorobiev et al. 2005) Dispersants are not approved for shallow areas (EPPR 2015). The minimum depth of the layer, at which chemical OSR might be approved is 10–20 m (IPIECA 2015). Most of the time, it is at least more than 60 m. This depth guarantees better uniform spreading in water column. (Prince 2015; Vorobiev et al. 2005)

Window of Opportunity Arctic WOP may be extended longer in comparison with lower latitudes and warmer climatic zones because of reduced weathering rates (Potter et al. 2012). However, the best use of dispersants and the highest efficiency is observed when sprayed on fresh spilt oil within 72 h after the accident. From this point of view, after recognizing an incident, the decision to apply the chemical method should be prompt (API 2017; Federici and Mintz 2014; JIP 2014; Lampela 2011; Slaughter et al. 2017).

Distance from the Shore The method is applicable remotely and does not require special arrangements for logistics and infrastructure, which is advantageous in perspective of existing infrastructure along coastlines of the Northeast and Northwest Passages. (Vorobiev et al. 2005) However, if dispersants are applied from vessels, then resupply of the chemical can become a logistical issue (DNV 2015; Slaughter et al. 2017).

11.5 Physical Oil Spill Response

11.5.1 Overview of Physical Response

Brief Description The main tool of the physical OSR is sorbents. Sorbents are solid or liquid materials, which can intake liquids from the surrounding medium via two physicochemical mechanisms: absorption or adsorption. (Doshi et al. 2018; EPA 1999; ExxonMobil 2014; Federici and Mintz 2014) The liquids, or sorbate, can be both water and oil by its origin. In OSR, sorbents are commonly solid materials and are available in many geometrical forms – powders, rolls, pads, pom-poms, booms, and others. Independent on its form, water-repelling and oleophilic properties are required. Reuse of some material is possible, but rarely organized. The reasons are time consumption and required manpower. Synthetic sorbents can be recycled several times. (EPA 1999; ExxonMobil 2014) Sorbents are frequently used to recover small amounts of oil in proximity to the shore. Operations are usually organized to do the final cleaning of oil traces; clean water surfaces of environmentally sensitive areas, or in cases, when mechanical recovery or other primary OSR methods are not applicable. (EPA 1999; ExxonMobil 2014; ITOPF 2012; Li et al. 2016; Wenning et al. 2018) Physical OSR is considered as a secondary OSR method. It is not often mentioned in major Arctic marine OSR publications. (EPPR 2017a; ITOPF 2012) Therefore, extrapolation of OSR influencing factors are made based on previously reviewed other OSR methods. It regards mainly environmental challenges.

Examples of Response Equipment Sorbents are classified as synthetic and natural. Natural sorbents can be organic and inorganic. Examples of synthetic sorbents are polyurethane, polyethylene and nylon fibres; natural organic – peat moss, straw, feathers, and natural inorganic – vermiculite, wool, clay. (EPA 1999) Depending on given conditions of the spill area, response vessels, fishing nets, storage facilities, sorbent booms, sorbent pads or sorbents of other geometry may be employed. (ITOPF 2012)

Efficiency of Oil Removal and Required Time In comparison with natural analogues, synthetic sorbents usually are more efficient and have better performance in the field: buoyancy, hydrophobicity, sorption capacity, ability to collect sorbents after use (Doshi et al. 2018; Federici and Mintz 2014; ITOPF 2012). Natural sorbents can intake up to 20 g of oil per 1 g of material, whereas synthetic ones – up to 70 g/g. (EPA 1999) With currently developed aerogels, as a subclass of sorbents, which can be both synthetic and natural in its origin – oil sorption rate can be up to 200 g/g for a variety of oils (Doshi et al. 2018). These materials have high surface area, high porosity, and good buoyancy characteristics. All this considered, it makes them ideal OSR sorbents. The only drawbacks are their high production cost, lack of durability after multiple compressions to extract oil from the material, and hydrophilicity. They intake water as well as oil. (Doshi et al. 2018; Liu et al. 2016;

Mahfoudhi and Boufi 2017). The time required for operations depends on oil type and surface area of sorbents. More viscous oils react slower with the sorbent than less viscous types. Smaller size of sorbents creates better exposure opportunities for oil to be sorbed rather than when the material has larger dimensions: e.g. loose strands versus booms. (ExxonMobil 2014; ITOPF 2012) On the other hand, collecting soaked sorbents of small size may be time consuming and labour-intensive, which are both a drawback of OSR operations (Al-Majed et al. 2012).

Sustainability of the Method

Environmental Aspect One of the most vividly expressed environmental drawbacks of sorbents use is generation of excessive amount of waste (ITOPF 2012). Hence, applying this OSR method is a waste management problem: 1 part of sorbent material can generate 70 parts of oil contaminated waste (Al-Majed et al. 2012). After collecting saturated sorbents from the water surface, temporary storage and subsequent logistical organization of sorbents and supportive tools, for instance fishing nets for lifting recovered oil, are required. Disposal can happen two ways: incineration or landfilling. Both require sanitary standards to be met: the first case – air emission (dioxin and PAHs) requirements, the second – sanitary landfill areas inland, usually far from the shoreline and the place of oil spill accident. For successful incineration, water content in the sorbent should be minimal. During sorbent collection stage and lifting the saturated material from water, leakages of incorporated oil are possible due to material sensitivity to mechanical compression. The compression can be also caused by environmental conditions: wind and sea state. (EPA 1999; ITOPF 2012) When dealing with natural sorbent materials, sinking and losing of both the material and incorporated oil is possible. Although natural sorbents are biodegradable, contamination of sea bottom and negative effects to local ecosystems makes them an inappropriate solution (Al-Majed et al. 2012; Federici and Mintz 2014). Synthetic sorbents are environmentally inert, including their unavailability for biodegradation. Some sorbents are enriched with nutrients, which enhance biodegradation rates. (EPA 1999; ExxonMobil 2014)

Social Aspect Practical health and safety considerations for personnel are as relevant as for other OSR options. With loose sorbents, powder form, protection of eyes and respiration channels is required. When dealing with booms saturated with oil, due to oil releases vessel deck floors may become slippery and cause accidents. (ITOPF 2012)

Economic Aspect In terms of economic expenses of materials, natural sorbents are usually less expensive due to their availability in large amounts. On operational level, physical OSR for large oil spills can be rather costly due to number of people, equipment and logistics needed. It is time consuming and labour-intensive process, which requires large storing capacity onboard for excessive amounts of produced hazardous waste. (Al-Majed et al. 2012; ITOPF 2012) Therefore, physical OSR is usually organized for treating small-scale oil spills and in nearshore areas. (EPA 1999; ExxonMobil 2014)

11.5.2 Environmental Challenges for Physical Response

Ambient Temperature and Ice in Water There is insufficient evidence and gap in the Arctic oil spill response literature regarding sorbent applicability in the Arctic conditions: ice-infested waters and low temperatures. The only mentioning is dated of 1975 by Logan et al.

Wind Influence Wind can affect operations in several ways. Some sorbents due to their light weight and small size can be blown away and carried out of the operational area. (ExxonMobil 2014) Some large sorbent structures, in particular booms, can be lifted above the surface, teared and squeezed so that some of oil intakes return back, resulting in secondary contamination (ITOPF 2012).

Wave Turbulence Wave action does not affect sorbents use, unless it is in the form of boom structures. With booms, holding capacity depends on the wave height in the same way, as it does for boom application in ISB and mechanical recovery. (Al-Majed et al. 2012; Federici and Mintz 2014) Sorbent booms are usually light and not durable. High wave activity can break them apart in a scale of several hours but also cause mechanical compression of booms, thus releasing some of the collected oil back to the marine environment. Towing such booms by vessels is also not desirable because the structures are too unreliable and can be easily torn due to created tension of the pull. (ITOPF 2012)

Light Effect Direct UV radiation may degrade synthetic sorbents, if they are left in outdoor temporary storage places of open type (ITOPF 2012). Although this influence may be not so pronounced in the Arctic geographical area (Øksenvåg et al. 2018).

Salinity Both oil and sorbents are dependent on salinity levels of seawater in terms of their capacity to stay on top of the water surface – their buoyancy. If sorbents become too buoyant, their efficiency to intake oil may be reduced. External efforts are needed to force them down in direct contact with oil slick in order to recover the spill. (ExxonMobil 2014) On the other hand, once saturated with oil and some amounts of water, sorbents should stay afloat to be visible and recoverable (ITOPF 2012).

Precipitation and Visibility Synthetic sorbents are usually manufactured with white colour in order to provide better visibility in OSR operations. (ExxonMobil 2014) Snow storms, or other types of weather events are not desirable for sorbent application due to their amphipathic properties (DNV 2015).

11.5.3 Technological Limitations of Physical Response

Oil Type and Its Viscosity Light oils, which have low viscosity, are treated and incorporated better in the sorbent material than heavy and weathered viscous oil types. (Al-Majed et al. 2012; EPA 1999; ExxonMobil 2014; ITOPF 2012)

Thickness of the Film Sorbents are applicable for films, left after ISB or mechanical recovery, which are usually less than 1 mm. Once oil film becomes too thin, no sorption can occur any more (ExxonMobil 2014).

Water Content in Oil Emulsified oils are a limiting factor for high sorbent efficiency. There can be no physical OSR organized around heavily weather oil, or oil in the state of "chocolate mousse". (ITOPF 2012) For this parameter, the requirement is the same as with chemical OSR.

Specific Parameter – Incompatibility with Dispersants After being treated with dispersants, which make interface tension between the oil and water lower, oil loses its initial properties. This affects directly on sorbent application. The oil tends not to stick on the material surface, as it usually does prior the contact with dispersants. (ExxonMobil 2014) There is incompatibility with chemical OSR method. In case of mechanical OSR, sorbent application is only possible after all mechanical recovery operations are completed. With simultaneous use of both response approaches, skimmers and pumping systems may intake sorbents and become blocked by the material. (ITOPF 2012) Incompatibility is also characteristic for combining dispersant use and mechanical recovery, depending on skimming surface material. After being in contact with the chemical, oleophilic surface of skimmers, for instance, cannot recover oil from water. (ITOPF 2011) This, however, cannot be said about disc skimmers, which maintain same degree of efficacy with chemically treated oils (Strøm-Kristiansen et al. 1996).

Window of Opportunity Fresh oil spills are preferred, while oil slicks are still not weathered and emulsified. (ExxonMobil 2014) The WOP is same as with cases of other OSR methods (API 2017).

Distance from the Shore Optimal areas of sorbents application are shoreline or nearshore areas, where distance is between the spill accident and mainland is minimal (Wenning et al. 2018). Logistics is usually a limiting factor for sorbents. Hence, it is usually applied with small scale accidents. For large spills, there is a need to collect the material with recovered oil onboard and retrieve sorbents for the next cycle of application. The collection is done with help of nets and belt skimmers. The retrieval is performed by wringing facilities. These works are labour-intensive, but also generate large amount of solid waste, especially in case sorbents are not to be recycled, but thrown away. Recycling of used sorbents is usually not preferred due to its higher cost in comparison with using fresh sorbent materials. Thus, there is excessive production of solid waste materials to be further disposed. (ExxonMobil 2014; Li et al. 2016)

11.6 Summary

Based on various parameters of primary (in situ burning, dispersant use and mechanical response) Arctic OSR methods reviewed in this chapter, it is possible to make the following conclusions:

1. From an *environmental* perspective:

   There is a number of natural limitations, which can create unfavorable conditions for OSR operations, including extremely low temperatures, excessive ice cover, high wind and wave activity, daylight unavailability, brackish seawater, precipitation storms and low visibility. Traditionally, temperature, ice and darkness are put first in this list. However, all these parameters can cause critical inhibition for successful OSR. Here is an upper boundary, beyond which no OSR operations are organized: air temperature (−18 °C), wind (15 m/s), wave height (4 m), light availability (darkness) and visibility (air: 4,0 km, water: 0,3 km). A variable set of unfavorable scenarios is possible with ice cover at sea and the other parameters.
2. From a *technological* perspective:

   Oil type and its physicochemical properties define the level of difficulty for oil spill response procedures based on oil behavior in the marine environment. Such parameters as oil API gravity, oil/emulsion viscosity and solidification properties, film thickness and water content can force choosing one or another abatement method, including no oil spill response – natural attenuation. For instance, heavy fuel oil (API gravity below 10°) or highly weathered and emulsified (water content of 50%) oil can only be collected by low efficient mechanical OSR. Distance from the shore is also a technological limitation: far offshore spills can be left for natural attenuation due to accident remoteness and lack of OSR resources in nearby region.
3. From a *performance* perspective:

   In ideal conditions, all three methods perform well, with cited efficiencies of more than 90% for in situ burning, mechanical and chemical OSR solutions. However, ideal conditions on both ambient and oil parameters are at least 50% unlikely for the whole Arctic maritime area (EPPR 2017a). As a practical example of mechanical recovery rate drop down in operational conditions, 95% of oil slick during an oil spill accident stays in the marine environment, leading to only 5% encounter rate (JIP 2014). Other OSR methods may experience similar tendencies, keeping however their efficiencies higher than with mechanical OSR (IPIECA 2015). None of reviewed primary methods remove oil completely from the environment and make oil-contaminated waters crystal clear. (Vorobiev et al. 2005)
4. From an *economic* perspective:

   In descending order, from mechanical to chemical and in situ burning OSR, for organizational and logistical reasons, application of skimmers bares the highest economic costs – 8500 euro/ton. Second place is chemical OSR with 5000-euro cost of operations per each treated ton of oil, and third place of 2700 euro/ton – in situ burning.
5. From a *sustainability* perspective:

   All OSR methods should be applied with caution by oil spill responders, including dealing with possible physical, chemical, biological and other hazards. By-effects to the environment consist of air emissions with in situ burning OSR, water column intoxication – chemical OSR, noise pollution and oil surface water contamination – mechanical OSR. All the effects have a temporary but excessively pronounced character. Despite their negative impacts, primary OSR methods eliminate much larger possible consequences of the oil spill, if left to be in the marine environment without any response: this may lead to long-term effects on biota and bioaccumulation on an ecosystem level (Kingston 2002).

In the Arctic marine environment, among existing OSR solutions, in situ burning OSR response is the most applicable for ice infested waters with large coverage and remote oil spill accidents (DNV 2015). Dispersants use can be applied with situations, when the weather is windy, the sea is open from ice, the hydrological state is characterized by high waves (2 m) and large turbulence. No ISB or mechanical recovery are possible in these conditions. (API 2017; Li et al. 2016; Slaughter et al. 2017) Mechanical and physical OSR works the best in proximity to the mainland with relatively calmer sea and weather conditions, including ice-infested waters. (ART 2015)

In all comparative studies of OSR methods, analyzing each method against the other – the general conclusion in relation to the Arctic marine environment is that the best operational time of the year is summer and the best OSR solution is not a singular approach but a suite of all primary abatement methods, with possible application of either of secondary methods (e.g. use of sorbents) in post-cleaning. Each combination is justified depending on the variable seasonal and geographical conditions as well as physicochemical state of the spilled oil. In practice, there exist established contingency plans already in place in advance of the spill accident. The plans are based on response analysis of the different mitigation technologies for relevant spill scenarios, including types of oil, release conditions, environmental conditions, time of the year, biological resources and other parameters. Usually decision making is conducted with help of selection tools, such as Spill Impact Mitigation Assessment – SIMA. It is used to justify choice of one or another OSR method for each particular spill case, taking into account socioeconomic and environmental conditions. (API 2017; DNV 2015; Doshi et al. 2018; IPIECA 2016; Wilkinson et al. 2017) All these is done to ensure an overall least damage to the marine environment, people's health and safety.

References

Allen, A. A. (1988). *In-situ burning: A new technique for oil spill response*. Woodinville: Spiltec.

Al-Majed, A. A., Adebayo, A. R., & Hossain, M. E. (2012, December 30) A sustainable approach to controlling oil spills. *Journal of Environmental Management, 113*213–227. ISSN 0301-4797, https://doi.org/10.1016/j.jenvman.2012.07.034.

American Petroleum Institute (API). (2015). *Field operations guide for in-situ burning of on-water oil spills*. Available: http://www.oilspillprevention.org/~/media/Oil-Spill-Prevention/spillprevention/r-and-d/in-situ-burning/guide-for-isb-of-on-water-spills.pdf. Last accessed 19 Oct 2018.

American Petroleum Institute (API). (2016). *In situ burning: A decision's maker's guide*. Available: http://www.oilspillprevention.org/~/media/Oil-Spill-Prevention/spillprevention/r-and-d/in-situ-burning/api-technical-report-1256-in-situ-burnin.pdf. Last accessed 19 Oct 2018.

American Petroleum Institute (API). (2017). *Factsheets about in-situ burning*. Available: http://www.oilspillprevention.org/oil-spill-research-and-development-cente. Last accessed 19 Oct 2018.

Arctic Oil Spill Response Technology – Joint Industry Programme (JIP). (2014). *Environmental impacts of Arctic oil spills and Arctic spill response technologies*. Available: http://neba.arcticresponsetechnology.org/report/chapter-4/. Last accessed 29 Jan 2019.

Arctic Response Technology (ART). (2012). *Summary of report: Spill response in the Arctic offshore*. Available: http://www.arcticresponsetechnology.org/wp-content/uploads/2012/11/FINAL-printed-brochure-for-ATC.pdf. Last accessed 16 Dec 2016.

Arctic Response Technology (ART). (2015). *Mechanical recovery in ice – Summary report*. Available: http://www.arcticresponsetechnology.org/wp-content/uploads/2015/08/ACS-Mechanical-Recovery-of-Oil-in-Ice-Feasiblity-Report-Final-1208.pdf. Last accessed 16 Dec 2016.

Belore, R. C., Trudel, K., Mullin, J. V., & Guarino, A. (2009). Large-scale cold water dispersant effectiveness experiments with Alaskan crude oils and Corexit 9500 and 9527 dispersants. *Marine Pollution Bulletin, 58*, 118–128.

BP Canada Energy Group ULC (BP). (2018). *Oil spill response plan*. Available: https://www.bp.com/content/dam/bp-country/en_ca/canada/documents/NS_Drilling_Pgm/Oil_Spill%20_Response_Plan_Annexes_A_to_G.PDF. Last accessed 19 Feb 2019.

Buist, I. A., Potter, S. G., Trudel, B. K., Ross, S. L., Shelnutt, S. R., Walker, A. H., & Scholz, D. K. (2013). *In situ burning in ice-affected waters: State of knowledge. Report 7.1.1. Joint industry programme*. Available: http://www.arcticresponsetechnology.org/wp-content/uploads/2013/10/Report-7.1.1-OGP_State_of_Knowledge_ISB_Ice_Oct_14_2013.pdf. Last accessed 6 Oct 2016.

Canevari, G. P., Calcavecchio, P., Becker, K. W., Lessard, R. R., & Fiocco, R. J. (2001). Key parameters affecting the dispersion of viscous oil. *International Oil Spill Conference Proceedings: March 2001, 2001*(1), 479–483.

Daling, P. S., Holumsnes, A., Rasmussen, C., Brandvik, P. J., & Leirvik, F. (2010). Development and field testing of a flexible system for application of dispersants on oil spills in ice. In *Proceedings of the thirty-third AMOP technical seminar on environmental contamination and response* (pp. 787–814). Ottawa: Environment Canada.

Daling, P. S., Brandvik, P. J., Singsaas, I., & Lewis, A. (2012) *Dispersant effectiveness testing of crude oils weathered under various ice condition*. In Proceedings at Interspill Spill Conference, London 2012, 13–15 March.

Det Norske Veritas (DNV). (2015). *Oil spill response in the Barents Sea South East*. Available: https://www.norskoljeoggass.no/globalassets/dokumenter/miljo/barents-sea-exploration-collaboration/basec-rapport-7c%2D%2D%2D%2Dstatusrapport-om-oljevern-i-barentshavet-sorost.pdf. Last accessed 19 Feb 2019.

Dickins, D. (2015). *Overview and background of oil spill response issues covered*. Available: http://www.npcarcticpotentialreport.org/pdf/tp/8-1_Overview_and_Background_of_Oil_Spill_Response_Issues_Covered.pdf. Last accessed 27 Apr 2016.

Doshi, B., Sillanpää, M., & Kalliola, S. (2018). A review of bio-based materials for oil spill treatment, *Water Research, 135*, 262–277. ISSN 0043-1354, https://doi.org/10.1016/j.watres.2018.02.034.

Emergency Prevention, Preparedness and Response (EPPR) Working Group of the Arctic Council. (1998). *Field guide for oil spill response in Arctic waters*. Environment Canada, Yellowknife, NT Canada, 348 pp (Contributing authors E. H. Owens, L. B. Solsberg, M. R. West, & M. McGrath).

Emergency Prevention, Preparedness and Response (EPPR) Working Group of the Arctic Council. (2015). *Guide to oil spill response in snow and ice conditions*. Available: https://oaarchive.arctic-council.org/handle/11374/403. Last accessed 2 Jan 2016.

Emergency Prevention, Preparedness and Response (EPPR) Working Group of the Arctic Council. (2017a). *Circumpolar oil spill response viability analysis*. Available: https://oaarchive.arctic-council.org/bitstream/handle/11374/1928/2017-05-09-EPPR-COSRVA-guts-and-cover-letter-size-digital-complete.pdf?sequence=1&isAllowed=y. Last accessed 19 Oct 2018.

Emergency Prevention, Preparedness and Response (EPPR) Working Group of the Arctic Council. (2017b). *Field guide for oil spill response in Arctic waters.* Available: http://www.pws-osri.org/wp-content/uploads/2018/08/EPPR_Field_Guide_2nd_Edition_20171.pdf. Last accessed 22 Feb 2019.

Emergency Prevention, Preparedness and Response (EPPR) Working Group of the Arctic Council. (2017c). *Overview of measures specifically designed to prevent oil pollution in the Arctic marine environment from offshore petroleum activ*ities. Available: https://oaarchive.arctic-council.org/bitstream/handle/11374/1962/2017-05-05-EPPR-overview-of-measures-to-prevent-oil-pollution-offshore-petroleum-report-complete-A4-size-DIGITAL.pdf?sequence=1&isAllow. Last accessed 19 Feb 2019.

Environmental Protection Agency (EPA). (1999). *Understanding oil spills and oil spill response.* Available: https://www7.nau.edu/itep/main/HazSubMap/docs/OilSpill/EPAUnderstandingOilSpillsAndOilSpillResponse1999.pdf. Last accessed 19 Oct 2018.

ExxonMobil Research and Engineering Company (ExxonMobil). (2014). *Oil spill response – field manual.* Available: https://cdn.exxonmobil.com/~/media/global/files/energy-and-environment/oil-spill-response-field-manual_2014_e.pdf. Last accessed 19 Feb 2019.

Farmahini Farahani, H., Torero, J. L., Jomaas, G., & Rangwala, A. S.. (2019). Scaling analysis of ice melting during burning of oil in ice-infested waters. *International Journal of Heat and Mass Transfer, 130,* 386–392. ISSN 0017-9310, https://doi.org/10.1016/j.ijheatmasstransfer.2018.10.110.

Federici, C., & Mintz, J. (2014). *Oil properties and their impact on spill response options.* Available: https://www.bsee.gov/sites/bsee.gov/files/osrr-oil-spill-response-research/1017aa.pdf. Last accessed 6 Oct 2016.

Fritt-Rasmussen, J., & Brandvik, P. J. (2011). Measuring ignitability for in situ burning of oil spills weathered under Arctic conditions: From laboratory studies to large scale field experiments. *Marine Pollution Bulletin, 62*(8), 1780e1785.

Fritt-Rasmussen, J., Brandvik, P.J., Villumsen, A., & Stenby, E. H. (2012, March). Comparing ignitability for in situ burning of oil spills for an asphaltenic, a waxy and a light crude oil as a function of weathering conditions under arctic conditions. *Cold Regions Science and Technology, 72,* 1–6. ISSN 0165-232X, https://doi.org/10.1016/j.coldregions.2011.12.001.

Gelderen, L., Brogaard, N. L., Sørensen, M. X., Fritt-Rasmussen, J., Rangwala, A. S., & Jomaas, G. (2015, November). Importance of the slick thickness for effective in-situ burning of crude oil. *Fire Safety Journal, 78,* 1–9. ISSN 0379-7112, https://doi.org/10.1016/j.firesaf.2015.07.005.

International Association of Oil & Gas Producers (IPIECA). (2014). *Oil spill waste minimization and management. Good practice guidelines for incident management and emergency response personnel.* Available: http://www.ipieca.org/resources/good-practice/oil-spill-waste-minimization-and-management/. Last accessed 19 Oct 2018.

International Association of Oil & Gas Producers (IPIECA). (2015). *Dispersants: Surface application.* Available: http://www.ipieca.org/resources/good-practice/dispersants-surface-application/. Last accessed 2 Mar 2019.

International Association of Oil & Gas Producers (IPIECA). (2016). *Controlled in-situ burning of spilled oil.* Available: http://www.ipieca.org/resources/good-practice/controlled-in-situ-burning-of-spilled-oil/. Last accessed 19 Oct 2018.

International Tanker Owners Pollution Federation Limited (ITOPF). (2011). *Use of dispersants to treat oil spills.* Available: https://www.itopf.org/knowledge-resources/documents-guides/document/tip-04-use-of-dispersants-to-treat-oil-spills/. Last accessed 2 Mar 2019.

International Tanker Owners Pollution Federation Limited (ITOPF). (2012). *Use of sorbent materials in oil spill response.* Available: https://www.itopf.org/knowledge-resources/documents-guides/document/tip-08-use-of-sorbent-materials-in-oil-spill-response/. Last accessed 29 Jan 2019.

Kingston, P. F. (2002) Long-term environmental impact of oil spills. *Spill Science & Technology Bulletin, 7*(1–2), 53–61. ISSN 1353-2561, https://doi.org/10.1016/S1353-2561(02)00051-8.

LAMOR. (2015). *Arctic solutions.* Available: http://www.lamor.com/wp-content/uploads/2015/10/lamor_arctic_solutions.pdf. Last accessed 30th Oct 2019.

Lampela, K. (2011). *Report on the state of the art oil spill response in ice.* Available: http://www.environment.fi/download/noname/%7BA7D0124E-7054-4F52-AFF5-6BA1B51ACAA3%7D/59270. Last accessed 18 Jan 2017.

Lewis, A., & Daling, P. S. (2007) *Evaluation of dispersant spray systems and platforms for use on spilled oil in seas with ice present* (JIP Project 4, Act. 4.21). SINTEF Materials and Chemistry, Marine Environmental Technology, SINTEF report A 16088, 21 p.

Li, P., Cai, Q., Lin, W., Chen, B., & Zhang, B. (2016, September 15). Offshore oil spill response practices and emerging challenges. *Marine Pollution Bulletin, 110*(1), 6–27. ISSN 0025-326X, https://doi.org/10.1016/j.marpolbul.2016.06.020.

Liu, H., Geng, D., Chen, Y., & Wang, H. (2016). Review on the aerogel-type oil sorbents derived from nanocellulose. *ACS Sustainable Chemistry & Engineering, 5*(1), 49–66. https://doi.org/10.1021/acssuschemeng.6b02301.

Logan, J. W., Thornton, D., & Ross, L. S. (1975). *Oil spill countermeasures for the southern Beaufort Sea: Appendix.* Available: https://www.researchgate.net/publication/280925729_Oil_Spill_Countermeasures_for_the_Southern_Beaufort_Sea_Appendix. Last accessed 1 Mar 2019.

Mahfoudhi, N., & Boufi, S. (2017). Cellulose. *24*, 1171. https://doi.org/10.1007/s10570-017-1194-0.

Makhutov, N. A., Lebedev, M. P., Bolshakov, A. M., Zakharova, M. I., Glyaznetsova, Yu. S., Zueva, I. N., Chalaya, O. N., Lifshits, S. H. (2016). *Forecast of emergencies at oil and gas facilities and elimination of consequences of emergency oil spills in arctic climate.* Available: http://arcticaac.ru/docs/4%2824%29_2016_Arctic/090_099_ARCTICA%204%2824%29_2016.pdf. Last accessed 17th Feb 2017.

McLeod, W., & McLeod, D. (1972). *Measures to combat offshore Artie oil spills.* Offshore Technology Conference

Michel, J., Scholz, D., Warren, S. R. Jr., & Walker, A. H. (2005). *In situ burning: A decision's maker's guide.* Available: https://www.api.org/oil-and-natural-gas/environment/clean-water/oil-spill-prevention-and-response/~/media/4BDBD6AABD534BF1B88EB203C6D8B8F4.ashx. Last accessed 19 Oct 2018.

New World Encyclopedia (NWE). (2017). *Freshwater.* Available: http://www.newworldencyclopedia.org/entry/Freshwater. Last accessed 6 Nov 2018.

Nordvik, A. B. (1995, March). The technology windows-of-opportunity for marine oil spill response as related to oil weathering and operations. *Spill Science & Technology Bulletin, 2*(1), 17–46. https://doi.org/10.1016/1353-2561(95)00013-T.

Økland, J. K. (2000). *Recovery of oil spills in marine Arctic regions.* Available: http://www.fargisinfo.com/referanser/LinkedDocuments/D036.pdf. Last accessed 19 Feb 2019.

Øksenvåg, J. H. C., McFarlin, K., Netzer, R., Brakstad, O. G., Hansen, B. H., & Størseth, T. (2018). *Biodegradation of spilled fuel oil in Norwegian marine environments.* Available: https://www.kystverket.no/globalassets/beredskap/akutt-forurensning/biodegradation-of-spilled-fuel-oils-in-norwegian-marine-environments_final.pdf. Last accessed 7 Feb 2019.

Parshentsev, S. A. (2006). *Analysis of tight force of rope of helicopter external sling while deployment mobile slick bar system on water surface.* Available: http://cyberleninka.ru/article/n/metod-rascheta-sily-natyazheniya-trosa-vneshney-podveski-vertoleta-pri-ustanovke-mobilnoy-sistemy-bonovyh-zagrazhdeniy-na-vodnoy. Last accessed 8 Feb 2017.

POLARIS Applied Sciences, Inc. (2013). *A comparison of the properties of diluted bitumen crudes with other oils.* Available: https://crrc.unh.edu/sites/default/files/media/docs/comparison_bitumen_other_oils_polaris_2014.pdf. Last accessed 23 Oct 2018.

Potter, S., Buist, I., Trudel, K., Dickins, D., & Owens, E. (2012). *Spill response in the Arctic offshore. Prepared for the American petroleum institute and the joint industry Programme on oil spill recovery in ice.* Available: http://www.dfdickins.com/pdf/Spill-Response-in-the-Arctic-Offshore.pdf. Last accessed 17 Feb 2017.

Prince, R. C. (2015). *Oil spill dispersants: Boon or bane?* Available: http://pubs.acs.org/doi/abs/10.1021/acs.est.5b00961. Last accessed 16 Feb 2017.

Ramazanov, D. C. (2015). *Оценка эффективности боновых заграждений для ликвидации аварийных разливов нефти при пересечении магистральным трубопроводом водных преград* (Evaluation of boom effectiveness for oil spill response in water-crossing sections of oil pipelines). Available: http://www.lib.tpu.ru/fulltext/c/2015/C11/V2/327.pdf. Last accessed 9 Feb 2017.

Ramboll Barents. (2010). Improvement of the emergency oil spill response system under the Arctic conditions for protection of sensitive coastal areas (case study: The Barents and the white seas). Pilot project (Vol. I). Murmansk, Russia.

Rytkönen, J., & Sassi, J. (2015). *Novel materials for sustainable oil spill response*. Available: https://tapahtumat.tekes.fi/uploads/f29c2613/Jukka_Sassi-8624.pdf. Last accessed 16th Dec 2016.

Singsaas, I., & Lewis, A. (2011). *Behavior of oil and other hazardous and noxious substances (HNS) spilled in Arctic waters (BoHaSA)*. Available: http://www.arctic-council.org/eppr/wp-content/uploads/2012/07/Final-Report-BoHaSA_23-02-20111.pdf. Last accessed 2 Jan 2016.

Singsaas, I., Leirvik, F., Johansen, B. (2008). *Oil in Ice - JIP*. Available: https://www.sintef.no/globalassets/project/jip_oil_in_ice/dokumenter/publications/jip-rep-no-8-report-lamor-task-3.1-final_2010.pdf. Last accessed 18th Jan 2017.

Singsaas, I., Sörheim K.R., Daae R.L., Johansen, B., Solsberg, L. (2010). *Oil in Ice - JIP*. Available: https://www.sintef.no/globalassets/project/jip_oil_in_ice/dokumenter/publications/jip-rep-no-21-field-report-2009-mechanical-final_2010.pdf. Last accessed 18th Jan 2017.

SL Ross Environmental Research (SL Ross). (2010). *Literature review of chemical oil spill dispersants and herders in fresh and brackish waters*. Prepared for the US Dept. of the Interior, Minerals Management Service, Herndon, VA, USA. 60 pp.

Slaughter, A. G., Coelho, G. M., & Staves, J. (2017). *Spill impact mitigation assessment in support of BP Canada Energy Group ULC*. Available: https://www.bp.com/content/dam/bp-country/en_ca/canada/documents/NS_Drilling_Pgm/Scotian%20Basin%20Exploration%20Project%20SIMA-NEBA%20-%20Final%20(17NOV17).pdf. Last accessed 19 Feb 2019.

Spill Tactics for Alaska Responders (STAR). (2014a). *Mechanical recovery – Containment and recovery*. Marine recovery Available: http://dec.alaska.gov/spar/ppr/star/final/20_B_III_B_MarineRecovery.pdf. Last accessed 16 Dec 2016.

Spill Tactics for Alaska Responders (STAR). (2014b). *Mechanical recovery – Containment and recovery. Basic booming tactics*. Available: http://dec.alaska.gov/spar/ppr/star/final/12_SectionB_III_B_BasicBoomTactics.pdf. Last accessed 16 Dec 2016.

Strøm-Kristiansen, T., Daling, P. S., Brandvik, P. J., & Jensen, H. (1996). *Mechanical recovery of chemically treated oil slicks*. In Proceedings of 19th Arctic and Marine Oilspill Program Technical Seminar, AMOP, Environment Canada, June 12–14.96, Calgary, 15p.

Vadla, R., Sörheim, K. R. (2013). *Oseberg Öst crude oil – properties and behavior at sea*. SINTEF report. Available: https://www.sintef.no/en/publications/publication/?pubid=CRIStin+1268744. Last accessed 21 Oct 2019.

Vergeynst, L., Wegeberg, S., Aamand, J., Lassen, P., Gosewinkel, U., Fritt-Rasmussen, J., Gustavson, K., & Mosbech, A. (2018). Biodegradation of marine oil spills in the Arctic with a Greenland perspective. *Science of the Total Environment*, *626*, 1243–1258. ISSN 0048-9697, https://doi.org/10.1016/j.scitotenv.2018.01.173.

Vorobiev, U. L., Akimov, V. A., Sokolov, U. I. (2005). *Предупреждение и ликвидация аварийных разливов нефти и нефтепродуктов* (Oil and oil products spill prevention and response) (p. 368). Moscow: In-oktavo.

Walton, W. D., & Jason, N. H. (1999). *In-situ burning of oil spills: Workshop proceedings*. Available: https://www.gpo.gov/fdsys/pkg/GOVPUB-C13-7eec9bd700f4bfccb27367ce806094e0/pdf/GOVPUB-C13-7eec9bd700f4bfccb27367ce806094e0.pdf. Last accessed 19 Oct 2018.

Wegeberg, S., Fritt-Rasmussen, J., & Boertmann, D. (2017). *Oil spill response in Greenland: Net Environmental Benefit Analysis, NEBA, and Environmental monitoring*. Available: https://dce2.au.dk/pub/SR221.pdf. Last accessed 19 Feb 2019.

Wenning, R. J., Robinson, H., Bock, M., Rempel-Hester, M. A., & Gardiner, W. (2018). Current practices and knowledge supporting oil spill risk assessment in the Arctic. *Marine Environmental Research*, *141*, 289–304. ISSN 0141-1136, https://doi.org/10.1016/j.marenvres.2018.09.006.

Wilkinson, J., Beegle-Krause, C. J., Evers, K.-U., Hughes, N., Lewis, A., Reed, M., & Wadhams, P. (2017). *Oil spill response capabilities and technologies for ice-covered Arctic marine waters: A review of recent developments and established practices*. Available: http://nora.nerc.ac.uk/id/eprint/518179/1/Wilkinson.pdf. Last accessed 19 Feb 2019.

World Wildlife Fund (WWF). (2011). *Lessons not learned. 20 years after the Exxon Valdez disaster.* Available: https://wwf.fi/mediabank/983.pdf. Last accessed 23 Mar 2019.

Chapter 12
The Role of Supply Vessels in the Development of Offshore Field Projects in Arctic Waters

Antonina Tsvetkova

Abstract This in-depth qualitative comparative case study aims to explore how supply vessels participate in offshore logistics operations and facilitate the development of offshore oil/gas field projects in response to contextual influence. The study presents the development of two offshore field projects located in the Arctic context: the Southwest Barents Sea (Norway) and the Southeast Barents Sea (the Pechora Sea, Russia).

The analysis of two empirical cases reveals that supply vessels play a crucial role in the activities like monitoring the offshore operations and contextual conditions, improving coordination of the logistics process, developing a common understanding of the current situation, ensuring response operations to any emergencies, supporting training exercises, and others. Thus, supply vessels facilitate the anticipation of possible emergencies that might occur and changes in allocating transportation resources. Further, the findings reveal that anticipation is both a strong driver and great challenge for providing the resilience of offshore field development among uncertainties and complexities in Arctic waters.

The study provides an understanding of how multi-functionality of supply vessels enables value-creating activities of offshore logistics and resilience building processes. It is further shown that building resilience can have trade-offs for both the logistical planning and allocation of transportation resources.

Keywords Offshore service · Supply vessels · Arctic shipping · Icy waters · Oil and gas · Maritime logistics

---

A. Tsvetkova (✉)
Faculty of Logistics, Molde University College, Molde, Norway
e-mail: antonina.tsvetkova@himolde.no

E. Pongrácz et al. (eds.), *Arctic Marine Sustainability*,
Springer Polar Sciences, https://doi.org/10.1007/978-3-030-28404-6_12

12.1 Introduction

Maritime logistics operations are essential for the scale and implementation of offshore exploring, developing, and producing oil and gas activities. In order to ensure continuous offshore activities, the installations need to be supplied regularly. Maritime transport plays a bridging role involving suppliers, warehouses, port terminals, platforms, and customers. Appropriate resources and distinct competencies are required to gain a high level of functionality and make offshore operations both economically and technically sustainable. One of the most demanded and multifunctional resources are the offshore supply vessels (OSVs). OSVs are those primarily "workhorses" of offshore field development projects that carry a wide variety of cargo and operate like the "trucks of the ocean" ensuring a tight connection between offshore facilities and offshore field activities (Kaiser 2010).

The literature on offshore operations largely concentrates on vessel routings issues with cargo deliveries and development of a coherent schedule for OSVs between offshore installations and onshore facilities (Fagerholt and Lindstad 2000; Aas et al. 2007; Halvorsen-Weare et al. 2012; Sopot and Gribkovskaya 2014). However, offshore supply operations and vessel routines are often studied in isolation from their adoption in practice through the development of models and scenarios to find an optimal routing policy and fleet composition. Without considering the influence of contextual factors on how offshore supply operations are deployed and managed, their feasibility and resilience may become problematic, particularly about the use of supply vessels (Borch and Batalden 2015). Thus, it seems like there is a lack of understanding of how offshore logistics operations are implemented in practice and developed over time as a result of the interaction between oil and gas field activities and context.

Further, OSVs are usually chartered rather than owned by oil and gas companies. The formal competence of logistics planners within oil and gas companies has traditionally been low. The reason for this could be that logistics is not a core activity for these companies (Aas et al. 2009; Kaiser 2010). At the same time, Aas et al. (2008) argue that outsourcing of vessel routings and daily coordination of supply flows through several logistics providers might weaken operational efficiency (Aas et al. 2008). Despite the contributions of previous research on the routing aspects, there is still a lack of understanding of the role of OSVs in developing offshore field projects. This also includes how the design of OSVs and its ability to execute cargo deliveries can affect logistics operations and address issues associated with the development of offshore field projects (Aas et al. 2009).

Motivated by the above-mentioned shortcomings in the literature, *the present study aims to explore how supply vessels participate in offshore logistics operations and facilitate the development of offshore oil/gas field projects in response to contextual influence.*

In doing so, the study presents two empirical cases of the development of offshore oil installations in different Arctic waters – in the Southwest Barents Sea (Norway) and the Southeast Barents Sea (the Pechora Sea, Russia). Recently oil and gas companies move their offshore operations further north and into the remote Arctic area. The motivation for this is that the Arctic contains approximately 13% of the world's undiscovered oil resources and about 30% of undiscovered natural gas resources. Roughly, 87% of Arctic's oil and natural gas resources (360 billion barrels oil equivalent) are located offshore (Budzik 2009; USGS 2008). However, the Arctic Seas are characterized by numerous challenging conditions, which oil and gas companies have to face when managing their maritime operations, including severe natural conditions (low temperatures, icing, polar lows, darkness), remoteness from harbours and land infrastructure, and vulnerability of the ecosystem (Milaković et al. 2015). In these conditions, the deployment of high-risk projects significantly intensifies uncertainty and complexity; increases transport costs and risks of disruptions in supplying and emergencies, mainly related to oil spills, platforms with many people on board and vessels. Both cases illustrate the role of supply vessels in response to contextual challenges both oil companies face when developing their offshore oil field projects in the Arctic Seas. Further, this study reflects on the logistics systems of both cases and specifically focuses on the functional demands of OSVs providing supply logistics, safety and emergency preparedness to search-and-rescue (SAR) operations and oil spill response.

The study is organized as follows: the next section outlines the field of offshore logistics in more details. This is followed by the research method. Then the context and two empirical cases are presented. The findings are discussed in the following section. The chapter concludes with implications for practice and future research opportunities.

12.2 Offshore Operations Management in Oil/Gas Industry: Existing Literature

The offshore oil and gas industry presents a logistics system whose purpose is to supply offshore operations with all necessary cargoes and services (Milaković et al. 2015) regularly as planned and in a cost-efficient way (Fagerholt and Lindstad 2000; Aas et al. 2009). These operations include delivery of various suppliers (e.g. equipment, fuel, water, and food); personnel transportation to/from and between offshore installations; rig towing; rig anchors placing and retrieving; as well as delivery of waste, empty load carriers from the platform to the land, and so on. There is a number of transportation modes, which participate in offshore oil and gas activities, including helicopters, crew boats and offshore supply vessels (OSVs). In the oil and gas industry, these kinds of operations constitute so-called "upstream" logistics, wherein the end of customers for OSVs are offshore drilling and production installations. Unlike "downstream" logistics when bringing oil and gas to onshore customers, the main challenge of offshore supply operations is to design an upstream chain that is highly responsive and flexible to avoid a shortage of suppliers or a buildup of return cargo (Aas et al. 2009).

Figure 12.1 illustrates the bridging role of supply vessels operating between the onshore supply base and the offshore field.

Overall, the value of the offshore logistics system is perceived as maritime transportation of goods and services, which plays a bridging role in the connections involving suppliers, warehouses, port terminals, and customers, with the focus on optimization and efficiency (Borch and Batalden 2015).

The optimization focus particularly addresses vessel routing issues with cargo deliveries and resource configuration aiming at finding an optimal routing policy and coherent schedule of supply vessels for servicing offshore installations (Fagerholt and Lindstad 2000; Aas et al. 2007; Halvorsen-Weare et al. 2012; Sopot and Gribkovskaya 2014). The efficiency focus is primarily found in concepts like leanness, just-in-time and logistics system design, which aim to achieve a low cost of cargo deliveries and to make the relationships between offshore activities and onshore facilities more efficient and integrated (Medda and Trujillo 2010; Borch and Batalden 2015). However, a common understanding of context provided by these research streams is limited to suppliers, demands of offshore installations, oil/gas companies-operators and/or customers. The external and institutional environment does not matter.

At the same time, knowledge of the maritime geography and contextual conditions of operations is important for the strategy setting and implementation of the company (Panayides 2006; Tsvetkova and Gammelgaard 2018). The feasibility of offshore field project development may become challenged without taking into account the impact of contextual factors. In particular, the possible volatility of conditions increases risks and makes it difficult to map cause-effect relations for decision-making. Within a maritime context, the environment description should include broad maritime-specific factors especially to map cause-effect relations for decision-making (Kristiansen 2005; Borch and Batalden 2015).

Another maritime logistics value has been recognized in providing extra support regarding safety at sea, environmental considerations and security standards according to international rules and regulations (Kristiansen 2005). Most offshore emergencies and accidents occur as a result of human activities and organizational factors (Antonsen 2009; Skogdalen et al. 2011; Bergh et al. 2013; Mendes et al. 2014).

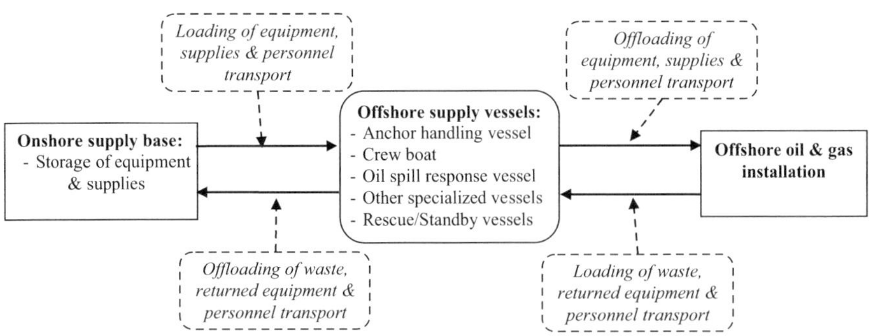

Fig. 12.1 An offshore upstream logistics supply system. (Adopted from Milaković et al. 2015)

Further, the majority of accidents involve vessels attendant the offshore operations due to some form of control failure and vessel damages occurred in the bad weather conditions (Sii et al. 2003). It is common that supply vessels have multi-task functions with extra duties. In practice, oil and gas companies decide on their own how to use the OSVs and what properties the vessels should have due to contextual issues like weather conditions, the amount of equipment needed and the distance from the shore and onshore facilities. Due to the complexity of offshore operational issues, the offshore logistics system should be constituted through the discovery and deployment of specific value-creating activities (Borch and Batalden 2015).

The recent research points out that the supply vessels' ability to execute cargo deliveries and their potential capability to take extra functions without increasing logistics resources involved and hence costs may allow meeting unforeseen challenges, mitigating the possibility of accidents and increasing operational effectiveness (Borch and Kjerstad 2018; Tsvetkova and Borch 2019). This capability of the OSVs determines a perspective to make the development of offshore operations resilient. McManus et al. (2008) claim that resilience is a function of an organization's situation awareness, management of keystone vulnerabilities and adaptive capacity in a complex, dynamic and interconnected environment. This study views resilience as the ability within an entire organization to respond, monitor and anticipate threats to normal operations (Hollnagel et al. 2006). This capability makes it possible to foresee any changes that are likely to occur, plan the logistics resources in an optimal way, and produce a wanted outcome during the offshore operations. According to this perspective, preparedness to the variability of resource allocation within the offshore logistics system is regarded as an advantage, and, then, resilience is achieved by controlling variability rather than by constraining it (Hollnagel et al. 2006; Hollnagel 2011).

However, the knowledge of how the OSVs can contribute to the resilient development of the offshore field projects and facilitate value-creating activities is still very limited in the literature (Aas et al. 2009). Further, Borch and Kjerstad (2018) assert that previous research often disregards both specialization of supply vessels and their ability of multi-functionality that are able to create more maritime logistics values like a need for redundant resources and an increase of flexibility in offshore operations to meet unforeseeable situations, respond to contextual factors and provide safety.

12.3 Operating in the Arctic Environment: Context Description

The present study addresses the role of supply vessels in facilitating the development of offshore field projects in the Arctic Seas. The Arctic harsh natural conditions make offshore supply operations challenging, impose limitations of logistics resources and demand on a special design of OSVs. Table 12.1 represents Arctic-specific challenges, considered as being most significant for ensuring regular offshore logistics operations in this study.

Table 12.1 Arctic challenges for offshore operations, including possible effects and implications for OSV capacities

Challenges	Possible Effects and Factors	Implications for OSV capacities
Remoteness and infrastructure limitations	Long distances from supply base to the drilling field; Lack of self-sufficiency; Lack of communication that is a challenge for commercial operations and emergency response	Loading capacity vessel (deck, tank, passenger capacity); Speed; Advanced navigation tools, incl. ice radar; Additional communication through satellites; Helicopter platforms and fueling arrangements; Depot storage capacities; Hospital facilities; SAR & oil spill capacities
Cold climate and weather conditions: *low temperatures, polar lows, high density fog (especially in summers), snow, ice*	Low speed of OSVs; Higher safety risk; Low predictability of polar lows; Icing – challenges to deck machinery, equipment, work on deck, icing of antennas; Snowdrift; Reduced visibility	Ice class hull; Winterization of vessels; Cover of superstructure; Low temperature design properties for equipment and materials; Equipment for removing ice; Electrical heating of windows, decks, valves, etc.
Winter darkness	Challenges for navigation and operation; Lack of daylight; Combination of darkness and coldness; Noise and vibration that increase fatigue and reduce sleep quality	Night navigation tools; Strong search lights; Illumination of vessel; Reduced noise and vibration in vessel; Comfort in cabins and social zones; Welfare facilities; Added crew for reduced time of duty
Vulnerability of eco-systems	High risk of pollution; Emission to sea and air from drilling units and service vessels; Accidents ending in pollution; Noise in compartments; Propeller noise	Hull integrity; Zero emission to sea through waste treatment facilities; Engines using fuel causing lowest possible emission to air; Storage capacity for garbage; Oil spill response capacity on board; Reduction of noise through better propeller system and electricity as fuel

(continued)

Table 12.1 (continued)

Challenges	Possible Effects and Factors	Implications for OSV capacities
Stakeholder complexity	Involvement of many interest groups like shareholders; Insurance companies; Fishing and other industries with conflicting interests; Environmental organizations; Endogenous people; Disputes between local and national governments; Possible delays in decision-making	Safe navigation tools; Lowest possible emission to air; Zero emission to sea; Environment friendly fuel; Hull integrity against leakage of oil products and sinking; Oil spill recovery capacity; Security measures against destructive actions; Functions that nay serve the local communities of the region

Adopted from Borch (2018)

Existing literature on offshore logistics operations in Arctic waters has emphasized that the field location in different Arctic regions can be characterized by various challenges, determines distinct features of the logistics system and needs specific requirements to OSVs (Borch et al. 2012; Borch and Batalden 2015; Milaković et al. 2015; Borch and Kjerstad 2018; Tsvetkova and Borch 2019). Through emphasizing Arctic-specific challenges to offshore operations, the present study provides deeper insights into what specific capacities of OSVs are demanded to create more values of maritime logistics in the Arctic Seas.

12.4 Method

12.4.1 Research Design

An in-depth qualitative case study approach was chosen to explore the role of OSVs in developing offshore oil/gas field projects in response to the Arctic contextual influence. Two cases of Norwegian and Russian offshore field projects were selected in different environmental settings: (1) the Southwest Barents Sea (Norway) and (2) the Southeast Barents Sea (the Pechora Sea, Russia).

The case study research method was applied as the most appropriate for looking deeply into contextual challenges in different Arctic environments and highlighting the complexity of offshore operations in Arctic waters. It also allowed providing a detailed portrayal and rich description (Halinen and Törnroos 2005) of how OSVs are engaged in the development of both offshore field projects and facilitate emergency preparedness to SAR and possible oil spills. This method helped reveal the potential influence of the contextual settings on OSVs' services and functions.

Further, the benefits of case studies have been illustrated in prior research, stemming primarily from their information richness and the ability to answer how and why questions (Eisenhardt 1989; Yin 2009).

12.4.2 Data Collection

Qualitative data was collected from multiple sources as recommended in previous SCM research (Voss et al. 2002). Semi-structured and in-depth face-to-face interviews were conducted with senior managers within the case companies, who played an influential role in carrying out the offshore supply operations and outsourcing decision-making processes. The interviews took place in Bergen, Kristiansand, Oslo and Tromsø during two periods: June 2017, September 2017. The use of the semi-structured interview guide enabled asking additional questions by providing clarifications of certain issues or helping to keep the interview focus on the intended topics, especially when the respondents had difficulty in expressing their opinions and views. Most of the interview answers were recorded with the consent of the respondents, while making hand-written notes, transcribed, validated with the respondents, and consequently analyzed. Further, the interview data was supported by personal observations within the case companies and onboard the vessels.

In addition, secondary data were collected from internal archival materials, the case companies' annual reports, vessel-shore communication, press releases and official websites. The documentary sources helped in the preparation for interviewing and, later on, analyzing the empirical data to complement the findings. Using several different types of data sources allowed increasing the internal consistency and validity of data through data-triangulation (Voss et al. 2002).

12.4.3 Data Analysis

The cases were analyzed based on the identification of key differences and similarities of the development of two offshore projects located in different environmental settings to reveal and compare the role of OSVs. Then, the cross-case analysis was conducted in order to move from the two specific cases to the identification of a general perception of the role of OSVs in developing offshore field projects in Arctic waters.

Further, the empirical analysis was conducted by an iterative process of reading, coding, and interpretation of the transcribed personal interviews and observation notes (Hall and Nordqvist 2008) of two case studies. This iterative approach helped identify patterns across both case studies, as well as ensure consistency and improve inter-rater reliability (Pagell and Krause 2005).

12.5 Case Presentation

12.5.1 Case 1. Southwest Barents Sea (Norway)

12.5.1.1 Location

The offshore operation took place in the South-Western Norwegian sector of the Barents Sea, 85 km (53 nm) northwest of the town Hammerfest, in water depths of 341 m. The discovery was made with the first exploration well in 2000. The oil production with offshore loading started in March 2016. The field produced about 100,000 barrels of oil per day in 2017.

12.5.1.2 Natural Conditions

The weather conditions in this region cause challenges for offshore operations mostly in winter. Average temperatures vary between 0 °C and 2.7 °C. The extreme minimum air temperatures can be between −30 °C and −34 °C. The minimum average temperature in January/February is about −15 °C. Average wave height is about 2 m.

There is a low probability of sea ice and icebergs in the region (NORSOK 2007). Polar lows are relatively common in the South-Western Barents Sea to affect somehow offshore operations.

12.5.1.3 Resource Configuration and Infrastructure

Offshore Operations

The operator employed a floating, production, storage, and off-loading platform to develop the field. This platform was designed to both withstand icing and ensure that rain and snow drain naturally from the walls and roofs. Its geostationary capability allowed oil tankers to move relative to the platform in response to prevailing weather conditions. The hose was longer than what is normally used, almost 400 m. The platform was housed in an enclosure to prevent icing.

The offshore operation was served by a fleet consisted of two supply vessels, one standby vessel and three shuttle tankers (see Table 12.2).

Onshore Infrastructure

The main onshore supply facility was a supply base in the town Hammerfest, which was responsible for inbound deliveries of all suppliers to the base and waste transportation from the base by using trucks. Two extra supply bases for emergency

Table 12.2 OSVs and roles in the development of offshore operations in the Southwest Barents Sea

Vessel	Ice class	Characteristics of vessels	Additional equipment
"Esvagt Aurora"	+1A1, ICE-1C	Contingency vessel: towing, anchor handling, oil spill response, cargo transportation	Oil-detecting radar and infrared cameras, winterized by oil boom systems and skimmers, Fire Fighter I; Rescue capacity up to 320 survivors
"Stril Barents"	+1A1	Supply/standby vessel: towing, anchor handling, oil spill response, cargo supply	Oil-detecting radar and infrared cameras, oil recovery; Rescue capacity up to 240 survivors + hospital on board
"Skandi Iceman"	+1A1, ICE-1B	Anchor handling tug supply vessel: anchor handling	Oil-detecting radar and infrared cameras, winterized by oil boom systems and skimmers, Firc Fighter I and II; ROV hangar lifted with launch & recovery system, external lighting covered with LED design; Rescue capacity up to 300 survivors

equipment were located farther in other municipalities: Hasvik and Måsøy. As noted by a senior manager of the operator:

> We faced some challenges for ensuring the effectiveness of offshore operations like the lack of infrastructure for support, as well as long distances between the harbor and onshore facilities. We had to establish our own regional office and helicopter base in Hammerfest to be closer to the offshore activity.

The availability of onshore resources like helicopters and hospitals was extremely limited: 88 km to Hammerfest, 219 km to Tromsø, and in Longyearbyen. That increased potential risk factors.

12.5.1.4 Operational Process Management

Logistics Process

The distance from the main supply base was 88 km (53 nm) with a normal transit time of 4 h that might increase in case of bad weather. The supply vessels were constantly on the move between the main supply base and the offshore platform. The logistics process was coordinated in meetings between the platform's offshore operation manager, logistics supervisor and the store master, and linked up to the regional office and helicopter base. As emphasized by a senior manager of the operator:

> Due to weather conditions and unforeseen situations, there were frequent changes of plans and delays in cargo transport that caused frequent changes of orders and many stand by hours for new orders from the logistics coordinators.

At the same time, the logistics process was quite predictable, although long distances increased risks to ensure the resilience of offshore operations.

Emergency Preparedness Resources

Further, there were a number of challenges to SAR and oil spill preparedness such as close location to the coast because of short drifting time to the shore at the field location; long distances and limited infrastructure; long response time for resources from the Norwegian Sea and the North Sea; limited light conditions and icing during winter season. In order to address these challenges, the operator used a set of integrated systems to detect and monitor immediate discharges from the field platform like satellites, planes and helicopters (equipped with synthetic aperture radars, side-looking airborne radars, infra-red detectors), installation (with radars, infrared inspections), sensors on the subsea templates, land-based installations (with high frequency radar systems). As emphasized by a senior manager of the operator:

> The development of offshore operations is a high-risk project. In order to ensure the resilience of offshore operations, the logistics system includes not only cargo transportation but also emergency preparedness. Our logistics department has to monitor, anticipate and adequately respond to any possible emergencies and learn consequences. Anticipation is the most challenging area for logistical planning. The task is to allocate the transportation resources like vessels and helicopters in an optimal way that makes it possible to foresee possible changes in case of an emergency. But it is very difficult to overview all the risks and decrease the number of unplanned ad-hoc solutions. As a result, the logistics system always has to be on the alert for emergencies.

All supply vessels were equipped with infrared cameras and oil-detecting radars. Two supply vessels had a spreading capacity. As noted by a senior manager of the operator:

> Nothing is so urgent or important that we cannot find the time to do it safely.

To increase coherence and test the field contingency equipment, the operator conducted an exercise in SAR and oil spill response in September 2012. As told by a senior manager of the operator:

> The exercise of 2012 in the Barents Sea near the offshore operations allowed us to verify the effectiveness and functionality of contingency plans. We could integrate 30–40 fishing boats into the coastal contingency plan that had to tow equipment for collecting, recording and storing oil. That made it possible to increase emergency response actions along the coastlines.

One emergency response and rescue vessel "Esvagt Aurora" was hired for stand by duties near the field platform and was equipped with oil spill protection, optimized for cold conditions and winter operations. Advanced monitoring equipment on board made it possible to discover and follow up any oil spills, under any light conditions.

Two other supply vessels were equipped with extra systems of oil spill protection to be able to detect and monitor immediate oil spills at the field platform.

12.5.2 *Case 2. Southeast Barents Sea (the Pechora Sea, Russia)*

12.5.2.1 Location

The offshore operation took place in the Southeastern part of the Barents Sea (the Pechora Sea), at a distance of 60 km (32 nm) from the shore. The water depth in this area is 19 m to 20 m. The field was discovered in 1989. The oil production with offshore loading started since spring 2013. The total volume of oil shipped from the field in 2014–2015 amounted to more than 1.1 million tons.

12.5.2.2 Natural Conditions

The annual average temperature is −4 °C and the temperature minimum is −50 °C. Wind strengths reach up to 40 m/s. Average wave height is about 2 m, but an extreme wave height can be up to 12–14 m. Some individual waves can achieve approximately 26 m, while an extreme wave crest was estimated at 15.5 m above calm water. Polar lows are relatively uncommon in the Southeast Barents Sea to affect somehow offshore operations.

This area is characterized by extremely low temperatures and strong ice loads. It is ice-free for 110 days a year and the cold period lasts 230 days from the end of October/middle November until the end of July/early August. The first year ice is one of the characteristics of the Pechora Sea. Ice thickness is up to 1.7 m. The most extensive ice cover period is observed in March–April when the whole sea surface is covered with ice (see Table 12.3). Some characteristics of fast and drift ices are presented in Table 12.4 (Bauch et al. 2005).

12.5.2.3 Resource Configuration and Infrastructure

Offshore Operations

The operator employed an ice-resistant platform that was specially adapted to operate in harsh natural conditions and to withstand maximum ice loads. The platform was equipped to cover all technological operations, including well drilling,

Table 12.3 Ice parameters in the Pechora Sea (Bauch et al. 2005)

Ice Parameters	Early dates	Average dates	Late dates
1. Beginning of ice freeze-up	25.10	18.11	23.12
2. Fast freeze-up	23.12	22.02	11.04
3. Beginning of fast ice break-up	05.04	23.05	07.07
4. Total disappearance of ice cover	10.04	19.05	30.08
5. Duration of ice-covered season	131	213	272

Table 12.4 Parameters of fast and drift ices in the Pechora Sea (Bauch et al. 2005)

Fast ice:	
Extent, km	3–15
Average thickness, cm	110
Drift ice thickness, cm:	
Average	80
Maximum	145
Size of ice fields, km:	
Average	1,4
Maximum	17,5
Continuity, units	10
Hummocks, %	60–90

production, processing, storage, the offloading of oil into tankers, and heat and power generation. The platform's residential module was designed to accommodate up to 200 people, all year round. Due to the depth of the sea (19.2 m), the facility—at 126 m^2 the size of two football fields, and weighing 500,000 tons—was installed directly onto the seabed and reinforced with a protective 45,000 cubic-meter-plus stone berm (weighing 120,000 tons) ensuring the well cluster had no direct contact with the water.

As emphasized by a senior manager of the operator:

> It is estimated on average 4.2 thousand tons of materials and equipment to be delivered to the offshore installation for the development of one well. This is the equivalent of a large railway train loaded with pipes, chemicals, bulk materials, and so on required for the well needs. In order to estimate the scale of offshore operations in this field, it needs to take into account that 32 wells are expected to be drilled by 2023. Up to 2018, 22 wells were in operation. In 2015, the total cargo volume was about 100 thousand tons. That amount included about 30 thousand tons of drinking water and approximately 25 thousand tons of diesel fuel that were delivered from the supply base. Further, approximately 12 thousand tons of cuttings were transported from the platform to the shore in 2015 as the platform operation principle envisaged zero discharges.

The platform was served by the fleet consisted of two supply vessels, own rescue (stand-by) vessel and two shuttle tankers. All vessels are capable of operating in the ice conditions (see Table 12.5).

As noted by a senior manager of the operator:

> To estimate the need in supply vessels, logistics planning focuses primarily on volumes of cargo traffic. At the same time, we are monitoring the current situation based on available real-time, statistical and historical data to be able to change the fleet composition.

All the OSVs involved were specifically designed to operate without the icebreaker escort in 1.7-m thick drifting ice in temperatures as cold as −35 °C. Further, they combined high maneuverability, a powerful propulsion system, environmental friendliness, and large passenger capacity (during an evacuation, the vessel can hold up to 150 people). All the OSVs were multifunctional and could not only carrying a various type of cargo and providing a number of specialized services as a supplement like anchor handling, rig moves, towing, ROV and others but also work as stand by vessels to support in SAR and oil spill recovery operations.

Table 12.5 OSVs and roles in the development of offshore operations in the Southeast Barents Sea

Vessel	Ice class	Characteristics of vessels	Additional equipment
"Yury Topchev"	Icebreaker 6, +1A1, ICE-15	Platform service vessel, supply operations	Oil spill recovery, fire fighter 1, Rescue capacity – up to 85 survivors
"Vladislav Strizhov"	Icebreaker 6, +1A1, ICE-15	Platform service vessel, supply operations	Oil spill recovery, fire fighter 1, Rescue recovery – up to 85 survivors
"Aleut"	Icebreaker 6, +1A1, ICE-10	Offshore standby vessel, anchor handling, tugging	Dynamic positioning system; oil recovery system, fire fighter 1, winterized cold – 30 °C, breaking of 1-meter thick ice; deck glaciation avoiding Rescue capacity – up to 34 survivors
"Balder Viking"[a]	+1A1, ICE-10	Supply vessel, anchor handling, tugging	No oil spill response system
"Murman"	Icebreaker 6	Rescue vessel	OSR equipment; Additional equipment for oil spill recovery

[a]The vessel "Balder Viking" was engaged in the project during the several winter periods

During several winter periods with more severe conditions, the vessel "Balder Viking" was temporarily engaged in the project to provide support to the supply vessels that were constantly on duty.

Shift personnel was transported during 2.5 h from the project air support base in the town Arkhangelsk by using AN-24 turbo-prop aircrafts towards the settlement Varandey. Then, 4 MI-8 AMT helicopters were used for personnel transfer from the settlement Varandey (60 km away) to the platform with a flight time of 35 min. The operator outsourced Varandey airport, owned by another oil company. Personnel traffic was about 10,000 people.

Onshore Infrastructure

The closest supply base was located in the town Murmansk at a distance of 980 km from the offshore platform. The supply base consisted of several smaller bases being leased by the operator from the third parties. The operator executed a specialized subsidiary at the supply base to be responsible for warehousing of materials, equipment and foodstuff to be transported to the platform. As noted by a senior manager of the operator:

> Despite the availability of its own logistics operator, contractual relationships and interaction with several third parties, as well as working from several distant bases made supply support of the platform very complicate. It was a challenge for our subsidiary, and we decided to construct our own onshore supply base in Murmansk.

An alternative base could be located in the settlement Varandey at the distance of 60 km from the platform. However, another oil company owned all the facilities there.

12.5.2.4 Operational Process Management

Logistics Process

The supply vessels were constantly on the move between the platform and the supply base. The distance from the supply base in Murmansk was 980 km that took a minimum 4–6 days roundtrip time for the supply vessels. Long distance was one of the main challenges for ensuring regular support of the platform. As noted by a senior manager of the operator:

> The supply vessels' capacity was a significant factor for us due to lack of infrastructure and long distance between the offshore platform and the onshore supply base. The supply vessels were the only available and reliable resources compared to helicopters that could operate for many different purposes, e.g. to carry necessary supplies and to take extra equipment onboard like fire-extinguishing, oil spill response and hospital facility. It was essential to ensure the resilience of the field project development in harsh Arctic conditions.

Ice Handling Management

The logistics process was challenged primarily by two issues for developing the offshore operations due to the presence of ice for more than 230 days per year:

1. Loading/offloading operations:

The access to the offshore platform by OSVs and oil tankers was often challengeable due to ice drifting or high waves, especially in the winter period. A wide ice-free zone was formed behind the platform along the direction of the current. The vessels used this pass to avoid large ice loads while loading/offloading operations. As emphasized by a senior manager of the operator:

> The platform had only two offloading centers at opposite corners. They were along the direction of the current to make loading/offloading operations safe. However, the wind force could change ice drifting in direction of the wake behind the platform. This often created some difficulties and sometimes could cause even possible emergencies. We had to wait for so-called weather windows to load/offload cargoes and to pump the oil. These windows lasted for a maximum of six hours; then the current direction could change again, and ice squeezing against the vessels prevented them from holding the loading/off-loading position until the conditions were favourable again.

Thus, the loading/offloading operations could be done only when the wake was at the side of loading/offloading centers.

2. Ice rubbling:

The ice was often accumulated at a side of the offshore platform hull consolidating and forming ice rubble that impeded the supply vessels and oil tankers to get close to the platform. As noted by a senior manager of the operator:

> It was quite hard to remove the ice rubble because it formed again and again. The OSVs could not get close to the platform. The length of the cargo cranes was about 50m that sometimes could be not enough, as the width of the ice rubble could be the same. Lifting cargo from the vessels was often difficult and risky.

Further, the ice rubbles increased the load on the platform's structure and could reduce the horizontal stability of the platform.

Icebreaker "Murman" was constantly on standby and rescue duty near the offshore platform. This vessel was also hired for ice management like ice breaking, ice monitoring and providing data about ice conditions and ice floe movements. That allowed foreseeing possible changes in the utilization of the vessels and other logistics resources that were likely to occur. As emphasized by a senior manager of the operator:

> The main task of cooperation between the platform's offshore manager and the logistics coordinators focused on operational risk assessments and monitoring of the current situation. Both ensuring real-time data and anticipating some challenges that could cause accidents or delays in offshore operations were essential for optimal utilization of the logistics resources. We knew where the resources were at any given time to enable them to respond quickly. It also allowed us to be prepared for possible changes in resource allocation, e.g. the vessel schedule, due to the weather and ice conditions.

Emergency Preparedness Resources

The maximum size of possible oil spills from the offshore platform was officially calculated to be 1500 tons of oil for wells and 10,000 tons for oil tankers. The impacted area could reach 140,000 km^2 and more than 3000 km of shoreline.

There were two vessels "Aleut" and "Murman" serving as stand by guard and emergency preparedness vessels. Both vessels were equipped with additional equipment for oil spill recovery in ice conditions. Further, all the OSVs involved in the project development were interchangeable in executing supply functions and/or standby duties.

According to the oil spill response plan, the following resources would involved: two platform service vessels "Yury Topchev" and "Vladislav Strizhov" with first line oil recovery facilities and sea boom fixing; icebreaker "Murman" for fixing sea booms; a number of boom-laying boats, a tanker-barge for oil collection; oil skimmer boats, bank protection booms, depot at the settlement Varandey in 60 km away and some other small boats. As emphasized by a senior manager of the operator:

> Our logistics department had to constantly keep abreast of the offshore activities – monitoring, anticipating, learning logistical demands and regularly training for different scenarios – to be prepared to respond strategically in case of any possible emergency in the most rapid and optimal way.

The OSVs were equipped with a special navigation system onboard. These systems allowed everyone in the project – personnel on the platform, the staff of the oil company, crews on board, suppliers, vessel operators and logistics coordinators – be aware of the current situation: real-time data, weather and ice conditions, vessel position and so on. It was essential to be better prepared for any possible emergencies. As noted by a senior manager of the operator:

> We used an integrated navigation system and geographical information system onboard the supply vessels. The combination of both systems provided us with an ability for finding more effective logistics solutions. It was essential when all offshore operations could be carried out during a very short operating window before the direction of drifting ice changed again. When all key persons monitored the same picture, e.g., how the ice moved, decision-making on the offshore operations was much more efficient. It was possible for us to foresee possible consequences and changes that might occur in case of accidents in order to re-allocate the transportation resources, to re-route the vessels and make sure the emergency preparedness is kept for the platform.

All the service and supply vessels had a capacity to accommodate up to 90 persons on board in case of an emergency evacuation from the offshore platform. As emphasized by a senior manager of the operator:

> It would be a real challenge to perform the evacuation operation in so high waves and drifting ice as we had at the site of the offshore operations. Through anticipation of operational risks, communication and sharing of real-time data, our logistics coordinators increased the situational awareness to understand, interact and predict.

For SAR operations, the helicopters at the settlement Varandey were available. They were also the local emergency services like the fire service provide assistance. Further, they assisted in rescue missions and the search for missing persons. As noted a senior manager of the operator:

> We faced a number of issues with the use of helicopters, e.g. due to the abundance of fogs. Sometimes we had to use the supply vessels for personnel transportation. These issues with helicopters increased operational risks and negatively influenced the cargo capacity of our supply vessels. And our experience of operating in Arctic waters showed that the supply vessels were the only resources, which could ensure a special capacity for ensuring emergency preparedness far away from the onshore facilities in Arctic waters.

12.5.3 Summary

The two cases illustrated how supply vessels participated in the development of both offshore oil filed projects and created values of maritime logistics operations in different Arctic contexts.

Table 12.6 represents briefly both cases and compares the natural conditions, challenges to cargo transportation, SAR operations and oil spill response, roles of the OSVs involved, as well as logistics processes and tasks.

Table 12.6 Overview of two empirical cases

	Case 1: Southwest Barents Sea (Norway)	Case 2: Southeast Barents Sea (Russia)
Start of operations	2000	2013
Production start	Since 2016	Since 2014
Water depth	341 m	19–20 m
Ice conditions	No ice, icebergs available during winter; Icing; Heavy snow falls	Ice period during 230 days between November–July; Average ice thickness 1.7 m; drifting ice thickness 0.8 m (max.1.45 m); Hummocks; icing
Weather conditions	Annual average temperature 0 °C/+2.7 °C; Average January air temp.−15 °C (extreme minimum −34 °C); Polar lows; Average wind speed 27 m/s; Average wave height 2 m	Annual average temperature + 4 °C; Average February temperature − 17.4 °C; Minimum −50 °C; Average July temperature + 6.5 °C (max. + 26 °C); Wave height: 3.2–4.7 m (extreme up to 15.5 m); Wind strengths up to 40 m/s
Distance from the shore	88 km	60 km
Distance from the supply base	88 km from the town Hammerfest (about 4 hours of roundtrip time)	980 km from the town Murmansk (about 4–6 days roundtrip time)
Number of OSVs and other vessels	2 OSVs with low ice class; 1 rescue vessel	3 OSVs with high ice class; 1 stand-by vessel; 1 icebreaker
Number of helicopters used	1SAR + 1transport helicopters	4 helicopters
Transport challenges	Long distance; lack of infrastructure; Wind, storms; icing; polar nights	Long distance; lack of infrastructure Ice drifting; ice rubbling; high waves; polar nights
Challenges to SAR and oil spill response	Close to the coast due to very short drifting time in this area; Long response time for resources from the Norwegian Sea and North Sea; Light conditions (polar nights); Icing	Close to the national parks; Long response time for resources from Murmansk; Rapidly changing ice and weather conditions: Ice drifting, high waves; Ice rubble formation; Polar nights
Characteristics of OSVs	Narrow specificity with a focus on concrete functions like towing, anchor handling; Equipped with special facility onboard for working in harsh conditions	Multifunctional and interchangeable; Designed specifically to operate in harsh ice conditions; High maneuverability and ability to operate without icebreaker assistance in 1.7-m thick ice Increased capacity of the OSVs due to lack of infrastructure and long distance

(continued)

Table 12.6 (continued)

	Case 1: Southwest Barents Sea (Norway)	Case 2: Southeast Barents Sea (Russia)
Logistics process	OSVs are constantly on the move; Coordination and collaboration between operating units; Frequent changes of plans and delays in cargo deliveries due to bad weather conditions and unforeseen situations	OSVs are constantly on the move; Changes in the vessels' schedule due to bad weather conditions, ice rubbles; Necessity to wait for weather windows for loading/offloading operations; Collaboration between operating units; Development of a common understanding of the current situation between operating units
Role of OSVs	Cargo delivery; Facilitation in coordinating the logistics process; Emergency preparedness to oil spills and SAR operations; Ensuring of the resilience of the oil field project development	Cargo delivery; Facilitation in coordinating the logistics process; Ice handling management; Emergency preparedness to oil spills and SAR operations; Ensuring of the resilience of the oil field project development
Logistics tasks	Monitoring, anticipating and adequately responding to any possible emergencies; Readiness to re-allocate the transportation resources; Being on the alert for emergencies	Focus on changes in needs of cargo volumes due to growing offshore activity; Operational risk assessments and monitoring of the current situation; Providing real-time data and anticipating challenges that could cause accidents or delays in offshore operations; Readiness to re-allocate the transportation resources and re-route the vessels in an optimal way in case of possible accidents

12.6 Discussion

The primary function of OSVs is to ensure a regular connection between the supply base and the installations. At the same time, the analysis of both cases shows that the role of the supply vessels encompasses more aspects that are essential: addressing issues to operate in Arctic waters; facilitating the coordination of the logistics process; managing ice handling; supporting emergency preparedness to oil spills and SAR operations; ensuring the resilience of the offshore field project development.

12.6.1 OSVs' Services in Addressing Issues to Operate in Arctic Waters

The two cases illuminate logistics challenges related to the different environmental settings of Arctic waters. While Case 1 shows offshore operations in a more favourable Arctic zone, in Case 2 the development of the oil field project is challenged by extremely harsh climatic conditions with the ice period of about 230 days per year. At the same time, both cases reveal the logistics process and development of oil field projects have to cope primarily with the challenges such as long distances from the onshore infrastructure, lack of infrastructure and ice formations. This increases the risks of possible delays in supplying the offshore installations and, thereby, the resilience of the offshore operations as a whole.

The findings reveal that the supply vessels are most available and reliable resources involved in the development of offshore field projects in comparison with the transport resources, e.g. helicopters. Further, the findings identify that the demands for the OSVs' characteristics and capacities increase as the Arctic climatic conditions become more complex and the risks of possible accidents increase. In Case 2, the OSVs were specifically designed for the concrete oil field project to operate in extreme environmental conditions with a high icebreaking capability and could fulfill multiple purposes. In Case 1, the supply vessels had, in contrast, a rather narrow specificity with a focus on concrete functions and could be replaced by the operator with other vessels for other specific functions depending on the needs of the offshore activities.

Further, the case analysis shows that the maneuvering-abilities of the supply vessels are in the high priority during static operations like loading/unloading operations, which are very sensitive to bad weather and the ever-changing direction of drifting ice. Thus, the availability of special vessel capabilities like multi-functionality, high maneuverability, ice strengthening, winterization, and quite high icebreaking ability ensures and expands oil and gas companies' participation in Arctic environments of higher complexity and uncertainty.

12.6.2 Logistics Process Management

The OSVs facilitate in coordinating the logistics process among all the operating units involved – the supply base, the offshore installation, the supply vessels, different sub-contractors, helicopters, and others. They have a capability for installing integrated navigation systems and geographical information systems onboard that allows developing a common understanding of the current situation and, thereby, improving collaboration between all the operating units. Therefore, the findings reveal that monitoring the same picture by all the key persons makes decision-making on the offshore operations more efficient. Therefore, the supply vessels provide support in foreseeing possible consequences and changes that might occur in case of accidents or delays due to bad weather conditions.

12.6.3 Ice Handling Management

The supply vessels can serve a number of tasks related to ice management like ice breaking, ice monitoring, iceberg towing, deflecting sea ice and ice floes, surveillance, operational ice charts, and ice advisory. Further, they are engaged in depicting the current situation by providing real-data about the ice conditions and the ice movements. Therefore, the supply vessels assist in performing operational risk assessments and in anticipating challenges that might cause accidents or delays in the offshore operations.

12.6.4 Emergency Preparedness to Oil Spills and SAR Operations

The findings reveal that challenges for the emergency preparedness and oil spill response capacity relate primarily to long distances and long response time for extra resources, contextual factors like rapidly changing ice and weather conditions, ice rubble formation and transport issues due to very low temperatures and poor daylight conditions. The supply vessels' capacity is a significant factor to address these challenges. In addition to transporting the required volume of cargo, the supply vessels are equipped with extra means for the activities like fire-extinguishing, oil spill response and hospital functions. This extra equipment may adversely affect the carrying capacity of the supply vessels.

In case 2, we can see that the demands for safety and environment precautions are higher due to the presence of drifting ice and high waves. Ice management can play an active helpful role in ensuring emergency preparedness. By hiring the OSVs with a high icebreaking capability and towing equipment, the operators can tow away the ice, e.g. with oil, from the offshore platform and then to an area where it is possible to collect oil. Further, the iceberg-towing task performed by the OSVs when changing the trajectory of iceberg drifts makes it possible to protect the conventional offshore installations from the collision threat.

Through sharing real-time data and providing communication between all the operating units, the supply vessels contribute to the anticipation of operational risks and possible accidents. Their services and capabilities increase the awareness of what is going on and facilitate the decision-making process to make it faster and more efficient. Therefore, the supply vessels help the logistics systems be on the alert for emergencies.

12.6.5 Ensuring Resilience of the Oil Field Project Development

The cases studied here reveal that the supply vessels due to their multi-functionality facilitate in creating a number of maritime logistics values in Arctic waters. Their involvement is crucial for the activities like monitoring the offshore operations and contextual conditions, facilitating the coordination of the logistics process and the development of a common understanding of the current situation, ensuring response operations to any emergencies, supporting training exercises to improve preparation for different scenarios and some others. All these activities are an essential foundation for the anticipation of consequences and changes that might occur.

The findings reveal that the anticipation is one of the most challenging tasks for the logistics department because it means an overview of all the risks to respond quickly to all kinds of possible emergencies and accidents during offshore operations. That is not realistic in practice. Therefore, the logistics department has to be constantly on the alert. At the same time, the anticipation includes monitoring of where the transportation resources are at any given time to enable them to respond quickly when it is needed. Hence, the anticipation facilitates the decision-making process on re-allocation and re-routing the transportation resources in an optimal way in case of accidents not to increase a need for redundant resources. These activities enable building resilience into the development of offshore field projects in Arctic Seas, primarily through ensuring emergency preparedness and emergency handling. This finding is consistent with the theoretical assumptions of the recent research (Hollnagel et al. 2006; Hollnagel 2011) by illustrating this effect in the offshore practices. Further, both cases emphasize the interconnectedness of the context and operational process for a better understanding of how the participation of the supply vessels in facilitating value-creating activities is affected by the contextual settings of different Arctic environments. This is coherent with assumptions by Borch and Kjerstad (2018) and Milaković et al. (2015).

12.7 Conclusion and Implications for Theory and Practice

This study presents the investigation of how supply vessels participate in offshore logistics operations and facilitate the development of offshore oil/gas field projects in response to the contextual challenges of different Arctic environments. The analysis of two empirical cases reveals that the supply vessels play a crucial role in the activities like monitoring the offshore operations and contextual conditions, improving coordination of the logistics process, developing a common understanding of the current situation, ensuring response operations to any emergencies, supporting training exercises, and others. Therefore, the supply vessels facilitate the anticipation of possible emergencies that might occur and changes in allocating transportation resources. Further, the findings reveal that anticipation is both a strong driver and great challenge for providing resilience of offshore field project development among uncertainties and complexities in Arctic waters.

The study responds to calls for conducting more empirical and case study-based research within offshore logistics (Borch and Batalden 2015; Milaković et al. 2015) and provides an in-depth understanding of the supply vessels' role in developing offshore field projects in particular empirical settings. Two cases illustrate how Arctic natural conditions impose limitations on logistics resources and offshore operations, as well as demand on specific services of OSVs, e.g. ice management, icebreaking capabilities, special design to meet cold-climate operating needs, emergency preparedness and, simultaneously, keeping cargo capacity due to long distances and lack of infrastructure. Further, the analysis of both cases identifies that contextual influence demands on a better logistical planning through increasing competencies in the uncertainties and possible risks to enable re-allocating transportation resources and re-routing supply vessels in an optimal way in case of accidents.

In contrast to the previous research with a focus on vessel routing issues and cost level issues of supplier delivery (Fagerholt and Lindstad 2000; Aas et al. 2007; Sopot and Gribkovskaya 2014; Borch and Batalden 2015), this study emphasizes a special role of the supply vessels and their multi-functionality in value-creating activities of offshore logistics and resilience building processes. The findings reveal that building resilience can have trade-offs for both the logistical planning and, as a result, allocation of the transportation resources.

Reflection on the contextual challenges and the role of the supply vessels in offshore operations is crucial for decision-making within logistical planning and strategic actions, including an optimal allocation of the transportation resources and anticipation of all possible risks and changes that might occur. By considering all maritime-specific factors and OSVs' specialization and capabilities, managers will gain a better understanding of how to manage offshore operations in particular settings. This is particularly relevant when developing offshore field projects in environments full of uncertainties and complexities to be able to respond quickly to any possible emergencies. Good collaboration between the operating units involved is crucial for making effective decisions and developing a common understanding of the current situation. This requires high competence of logisticians to use and allocate all transportation resources in an optimal way, as well as coordinate activities and resources between different operational units in integrated planning and operation context.

12.8 Limitations and Future Research

This study shows a number of demanded capabilities of the OSVs in two environmental settings of Arctic Seas. Future research should expand the geographical scope of Arctic Seas and include more case studies on how supply vessels contribute to the development of offshore oil/gas field projects in other Arctic environments to learn more about the offshore logistics processes.

There is also a need for more empirical research on vessel technology and equipment needed as to specialization of OSVs and multi-functionality. Therefore, future research may extend knowledge about how these capabilities of OSVs contribute to value-creating activities of offshore projects like the need for redundant capacities and ensure the resilience of offshore operations to respond to contextual challenges and mitigate the possibility of unforeseen situations and emergencies.

References

Aas, B., Buvik, A., & Cakic, D. J. (2008). Outsourcing of logistics activities in a complex supply chain: A case study from the Norwegian oil and gas industry. *International journal of procurement management, 1*(3), 280–296.

Aas, B., Gribkovskaia, I., & Halskau, Ø. (2007). Routing of supply vessels to petroleum installations. *International Journal of Physical Distribution and Logistics Management, 37*(2), 164–179.

Aas, B., Halskau, Ø., Sr., & Wallace, S. W. (2009). The role of supply vessels in offshore logistics. *Maritime Economics and Logistics, 11*(3), 302–325.

Antonsen, S. (2009). The relationship between culture and safety on offshore supply vessels. *Safety Science, 47*, 1118–1128.

Bauch, H. A., Pavlidis, Y. A., Polyakova, Y. I., Matishov, G. G., & Koç, N. (Eds.). (2005). *Pechora Sea environments: Past, present, and future*. Meeresforsch: Ber. Polarforsch.

Bergh, L. I. V., Ringstad, A. J., Leka, S., & Zwetsloot, G. I. J. M. (2013). Psychosocial risks and hydrocarbons leaks: An exploration of their relationship in the Norwegian oil and gas industry. *Journal of Cleaner Production, 84*, 824–830.

Borch, O. J. (2018). *Offshore service vessels in high Arctic oil and gas field logistics operations – Fleet configuration and the functional demands of cargo supply and emergency response vessels*. R&D-Report, vol 22. Nord University, Bodø.

Borch, O. J., & Batalden, B.-M. (2015). Business-process management in high-turbulence environments: The case of the offshore service vessel industry. *Maritime Policy & Management, 42*(5), 481–498.

Borch, O. J., & Kjerstad, N. (2018). The offshore oil and gas operations in ice infested water: Resource configuration and operational process management. In L. P. Hildebrand & L. W. Brigham (Eds.), *Sustainable shipping in a changing Arctic* (pp. 401–425). Springer.

Borch, O. J., Westvik, M. H., Ehlers, S., & Berg, T. E. (2012, December 3–5). Sustainable Arctic field and maritime operation. In *Proceedings of the offshore technology conference*, Houston

Budzik, P. (2009). Arctic oil and natural gas potential. In *U.S. Energy Information Administration, Office of integrated analysis and forecasting, oil and gas division.*

Eisenhardt, K. M. (1989). Building theories from case study research. *The Academy of Management Review, 14*(4), 532–550.

Fagerholt, K., & Lindstad, H. (2000). Optimal policies for maintaining a supply service in the Norwegian Sea. *Omega, 28*(3), 269–275.

Hall, A., & Nordqvist, M. (2008). Professional management in family businesses: Toward an extended understanding. *Family Business Review, 21*(1), 51–65.

Halvorsen-Weare, E. E., Fagerholt, K., Nonås, L. M., & Asbjørnslett, B. E. (2012). Optimal fleet composition and periodic routing of offshore supply vessels. *European Journal of Operational Research, 223*, 508–517.

Halinen, A., & Törnroos, J.-Å. (2005). Using case methods in the supply of contemporary business networks. *Journal of Business Research, 58*(9), 1285–1297.

Hollnagel, E. (2011). Epilogue: RAG – The resilience analysis grid. In E. Hollnagel, J. Pariès, D. Woods, & J. Wreathall (Eds.), *Resilience engineering in practice – A guidebook*. Farnham: Ashgate Publishing Limited.

Hollnagel, E., Woods, D., & Leveson, N. (2006). *Resilience engineering. Concepts and precepts*. Farnham: Ashgate Publishing Limited.

Kaiser, M. J. (2010). An integrated systems framework for service vessel forecasting in the Gulf of Mexico. *Energy, 35*, 2777–2795.

Kristiansen, S. (2005). *Maritime transport, safety management and risk analysis*. Burlington: Elsevier Butterworth-Heinemann.

McManus, S., Seville, E., Vargo, J., & Brunsdon, D. (2008). Facilitated process for improving organizational resilience. *Natural Hazards Review, 9*(2), 81–90.

Medda, F., & Trujillo, L. (2010). Short-sea shipping: An analysis of its determinants. *Maritime Policy & Management, 37*(3), 285–303.

Mendes, P. A. S., Hall, J., Matos, S., & Silvestre, B. (2014). Reforming Brazil's offshore oil and gas safety regulatory framework: Lessons from Norway, the United Kingdom and the United States. *Energy Policy, 74*, 443–453.

Milaković, A.-S., Ehlers, S., Westvik, M. H., & Schütz, P. (2015). Chp 20: Offshore upstream logistics for operations in Arctic environment. In S. Ehlers, B. E. Asbjornslett, O. J. Rodseth, & T. E. Berg (Eds.), *Maritime-port technology and development* (Vol. 2014, pp. 163–170). Leiden: CRC Press.

NORSOK (2007) NORSOK N-003: Actions and action effects. http://www.standard.no/PageFiles/1149/N-003e2.pdf

Pagell, M., & Krause, D. R. (2005). Determining when multiple respondents are needed in supply chain management research: The case of purchasing and operations. In K. A. Weaver (Ed.), *A new vision of management in the 21st century. Proceedings of the Academy of Management Conference* (pp. 5–10).

Panayides, P. M. (2006). Maritime policy, management and research: Role and potential. *Maritime Policy & Management, 33*(2), 95–105.

Sii, H. S., Wang, J., & Ruxton, T. (2003). A statistical review of the risk associated with offshore support vessel/platform encounters in UK waters. *Journal of Risk Research, 6*(2), 163–177.

Skogdalen, J. E., Utne, I. B., & Vinnem, J. E. (2011). Developing safety indicators for preventing offshore oil and gas Deepwater drilling blowouts. *Safety Sciences, 49*, 1187–1199.

Sopot, E., & Gribkovskaya, I. (2014). Routing of supply vessels to with deliveries and pickups of multiple commodities. *Procedia Computer Science, 31*, 910–917.

Tsvetkova, A., & Borch, O. J. (2019). *Arctic oil and gas field logistics and offshore service vessel capacities: The case of the Norwegian and Russian High Arctic* (R&D-Report) (Vol. 35). Bodø: Nord University.

Tsvetkova, A., & Gammelgaard, B. (2018). The idea of transport independence in the Russian Arctic: A Scandinavian institutional approach to understanding supply chain strategy. *International Journal of Physical Distribution and Logistics Management, 48*(9), 913–930.

Voss, C., Tsikriktsis, N., & Frohlich, M. (2002). Case research in operations management. *International Journal of Operations & Production Management, 22*(2), 195–219.

USGS. (2008). Circum-Arctic resource appraisal: Estimates of undiscovered oil and gas North of the Arctic Circle. In *United States Geological Survey* (Fact Sheet, 2008–3049).

Yin, R. K. (2009). *Case study research: Design and methods*. Thousand Oaks: Sage.

Chapter 13
Special Rules for the Arctic? The Analysis of Arctic-Specific Safety and Environmental Regulation of Offshore Petroleum Development in the Arctic Ocean States

Daria Shapovalova

Abstract Following the announcement of vast petroleum resources in the Arctic waters, politicians and commentators called for the adoption of an Arctic treaty establishing a harmonised approach to developing petroleum resources in the fragile and harsh circumpolar environment. Five Arctic Ocean coastal States (Canada, Greenland/Denmark, Norway, Russian Federation, and the United States) have all either expressed interest in developing or are already producing Arctic offshore resources. While some of these States have an established history of offshore petroleum development, the development in the Arctic waters presents a unique set of challenges requiring additional regulation. In addition to the general petroleum legal regime, each of these four States has developed some Arctic-specific regulations to establish more stringent safety and environmental rules compared to more conventional locations. The chapter identifies such Arctic-specific rules and provides a comparative analysis of safety and environmental rules developed specifically for the Arctic.

Keywords Arctic oil and gas · Arctic governance · Oil and gas law · Environmental law

13.1 Introduction

Following the announcement of vast petroleum resources in the Arctic waters (USGS 2009), politicians and commentators called for the adoption of an Arctic treaty establishing a harmonised approach to developing petroleum resources in the fragile and harsh circumpolar environment (Verhaag 2002; Koivurova and Molenaar

D. Shapovalova (✉)
University of Aberdeen, Aberdeen, Scotland, UK
e-mail: dshapovalova@abdn.ac.uk

E. Pongrácz et al. (eds.), *Arctic Marine Sustainability*,
Springer Polar Sciences, https://doi.org/10.1007/978-3-030-28404-6_13

2010; EU Commission 2008). The Arctic Ocean States were quick to pronounce, through the 2008 Illulisat Declaration, their acceptance of the United Nations Convention on the Law of the Sea as the framework treaty for the regulation of marine Arctic. While there is no comprehensive treaty regulating oil and gas development in the Arctic, national laws of the coastal States establish a detailed legal framework. The current level of production in the region remains relatively low, and it is appropriate to examine the national regulations to identify best practices in addressing Arctic challenges and room for harmonisation (Baker 2012, 2017).

Five Arctic Ocean coastal States (Canada, Greenland/Denmark, Norway, Russian Federation, and the United States (US)) have all either expressed interest in developing or are already producing Arctic offshore resources. While some of these States have an established history of offshore petroleum development, operating in the Arctic waters presents a unique set of challenges requiring additional regulation. In Greenland, all offshore areas are located in the Arctic, and the four remaining States are engaged in offshore production in the areas well south of the Arctic Circle. In addition to the general petroleum legal regime, each of these four States has developed some Arctic-specific regulations to establish more stringent safety and environmental rules compared to the conventional locations.

The legal regime for offshore petroleum development in the Arctic has been examined from national (Pelaudeix and Basse 2017; Henderson and Loe 2014; Mikkelsen and Langhelle 2008) and international perspectives (Shapovalova 2019; Johnstone 2015; Baker 2013). Baker further examined the possibility of harmonization of petroleum regulation in the Arctic focusing on the US and Canada (2012), and the Arctic Council compiled the best practices in the prevention of oil spills and the use of standards in the Arctic (EPPR 2017a, b). This chapter examines the national legislation and builds on the existing literature by focusing on the safety and environmental regulations adopted by the Arctic States specifically for petroleum development in the Arctic.

To this end, the chapter first examines the Arctic as a future resource base and its vulnerability to pollution from offshore drilling. It then identifies and analyses the Arctic-specific petroleum regulations for the prevention of routine and accidental pollution across the Arctic Ocean States.

13.2 The Arctic as the Future Petroleum Province

Despite the discovery of the potentially vast petroleum resources in the Arctic waters and the reports of the 'race for resources', the current level of production in the region remains relatively low (Shapovalova and Stephen 2019). However, the economic development of the region, with the primary focus on oil and gas, is featured in all the Arctic Ocean States' policy documents. To elucidate on the importance of the Arctic as the potential future petroleum province, this section examines the past and current efforts to develop offshore petroleum recourses in the region in the five Arctic Ocean States in light of their Arctic policies. For Greenland, which being wholly located in the Arctic and understandably does not have an Arctic policy, oil and mineral strategy is examined instead.

Russia is estimated to have the largest share of the Arctic resources, around 58%. The US holds the share of approximately 18%, Greenland and Norway – 12% each (Henderson and Loe 2014; USGS 2009). All Arctic Ocean States prioritise economic development of the Arctic areas, in particular through offshore petroleum development (CDIAND 2009; NMFA 2014; Russian Government 2014; POTUS 2013; Government of Greenland 2014).

The new Canadian Arctic Policy Framework is currently being developed to replace the dated Northern Strategy (CDIAND 2009) and the Statement on Canada's Arctic Foreign Policy (Government of Canada 2010). The Northern Strategy acknowledged the 'renewed interest in the offshore, including a new era of oil and gas exploration in the deeper waters of Beaufort Sea'. In the 1970s and '80s, the Canadian government invested heavily in exploring offshore petroleum resources through Panarctic Oils, a partnership between the government and private companies (Masterson 2013). While many discoveries were made in the '80s, the costs of developing the oil and getting it to the market made the large-scale production unfeasible (Natural Resources Canada 2007). In 2014–2016, a number of companies withdrew from exploration and relinquished their licenses in the Canadian Arctic waters (CBC News 2014, 2015; Shell 2016). As of November 2018, there is no offshore production of petroleum in the Canadian Arctic. The main physical challenges to large-scale offshore hydrocarbon production in the Canadian Arctic are harsh weather conditions, the lack of infrastructure, and distance to markets. Additionally, it is unlikely that any new licenses will be issued at least until 2021 due to the moratorium announced in 2016 (US-Canada Joint Arctic Leaders' Statement).

In the neighbouring US Arctic, the current offshore production is limited to the installations on artificial islands, but more might follow from the two exploration developments currently in place (BOEM 2018a). Exploration drilling in the Alaska Outer Continental Shelf (OCS) started in 1976, peaking in 1984–1985, and has been in stagnation ever since (BOEM 2018b). With the oil price crash, high operating costs in the American Arctic conditions and scarce infrastructure, no substantial production followed (LeVine et al. 2014). In 2008, the Department of Interior sold close to 500 offshore leases in the Alaska OCS, including the infamous Lease 193 in the Chukchi Sea, acquired by Shell (Secretary of Interior 2013). Exploratory drilling on the lease was postponed for a few years due to multiple court challenges (NVPH v Salazar 2010; NVPH v Jewell 2014), failed oil spill containment equipment testing (Secretary of Interior 2013), and the fiasco with the drilling rig *Kulluk*, which ran aground in a storm (The New York Times 2014). Just over a month after Shell finally started drilling in 2015, it issued a statement on ceasing any 'further exploration activity in offshore Alaska for the foreseeable future' citing insufficient resources found, high costs, and 'challenging and unpredictable federal regulatory environment' (Shell, 2015). More companies followed in exiting the province following Shell's decision (The Local Norway 2015). The 2017–2022 Leasing Programme includes Cook Inlet closer to existing infrastructure in Alaska but excludes Chukchi and Beaufort Seas due to the 'opportunities for exploration and development on existing leases', lack of industry interest, 'current market conditions, and sufficient existing domestic energy sources already online or newly accessible' (BOEM 2016a). The 2019–2024 draft proposal programme, in contrast, includes vast areas of the Alaska OCS (BOEM 2018c).

Over in the European Arctic, Norway is the leader in offshore petroleum development. From around 1980, the petroleum activities on the Norwegian continental shelf 'gradually moved north' (Nordtveit 2015) to the Norwegian and Barents Seas. The Norwegian parliament opened the Barents Sea for oil and gas activities in 1979 (Norwegian Ministry of Environment 2014). Two years later, the first field was discovered, which later formed the Snøhvit natural gas development, the first Norwegian Arctic field to be developed. Snøhvit, operated by Equinor (formerly – Statoil) is reported to contain enough gas to sustain production 'through 2050, and beyond' (Offshore Energy Today 2018). Equinor is also planning to produce oil from the Johan Castberg development 100 km north of Snøhvit, projected to be one of the biggest oil producing fields in Norway (Equinor 2018). Current oil production in the Norwegian Arctic is limited to the *Gøliat* platform in the Barents Sea, operated by Eni. It started producing oil in March 2016, using a floating, production, storage and off-loading unit, specifically designed for Barents conditions (Eni Norge 2018). Production on the *Gøliat* had to be halted several times due to technical issues and faults discovered by the Norwegian Petroleum Safety Authority (Nilsen 2016; PSA 2017). High hopes have been placed on Norway to provide 'a template for responsible oil and gas development in the Arctic' (Emmerson 2011). The 2014 'Norway's Arctic Policy' emphasises the need for a knowledge-based approach to further petroleum development and sets the targets of improved oil spill prevention preparedness and response. The Norwegian petroleum sector is a major source of State revenue and with maturing fields in the South, it is expected that the petroleum activities in the North will increase in the next few decades. Developing reserves in the Barents Sea will undoubtedly require effective cooperation with neighbouring Russia.

Although Russia was the first country to start producing oil in the Arctic offshore, the development has proven to be more costly and time-consuming than expected. The Arctic is important to Russia for resources, shipping, but also for its national identity and assertion as a leader in Northern development. The depleting resources onshore push the Russian industry north and offshore to maintain its leading position as a petroleum exporter. Yet, the technology and infrastructure of the Russian companies are limited. Despite postponing a few large petroleum projects in the Russian Arctic, the long-term outlook for offshore production is ambitious (Sidortsov 2017). The first platform to produce oil in the Arctic offshore was the Russian *Prirazlomnaya* in the Pechora Sea. With the first shipment to the European market in April 2014, the production in 2016 totalled 2.2 mln tons (Gazprom 2018). The production has been delayed several times 'due to lack of environmental approval and deficiencies in drilling equipment' (Lunden and Fjaertoft 2014). Further, the economic feasibility of *Prirazlomnaya* is reportedly dependent on national tax breaks and subsidies (Lunden and Fjaertoft 2014). Despite these setbacks, *Prirazlomnaya* serves as a proud example for the Russian State industry and could become a hub for further developments in the Pechora Sea. Although no offshore gas fields have been developed yet, Russia is establishing itself a major player in the liquefied natural gas (LNG) market, with Yamal LNG project delivering gas to the Asian markets via the Northern Sea Route. Currently, only State companies with over 5 years of experience of operating on the Russian shelf can hold licenses on the Russian Arctic shelf (Subsoil Law art 10 1992). Western companies are further limited in their involvement by the EU and US sanctions. Thus, for example, ExxonMobil confirmed that it is pulling out of its Arctic cooperation with Rosneft' in the Kara Sea because of the sanctions (Staalesen 2018). Statoil's partnership with Rosneft' for offshore development in the Sea of Okhotsk and the Barents Sea is also on hold until the sanctions are lifted (Bloomberg 2015). The sanctions make Russian authorities look for partnership in the East – Rosneft' reportedly signed preliminary agreements for Arctic development with Chinese CNPS, Japanese INPEX, and Vietnamese PetroVietnam (IEA 2014). Overall, Russian Arctic policy acknowledges the need to extend hydrocarbon development operations north and offshore. The earlier documents prioritise geological prospecting with a view to finalising the outer limits of the Russian continental shelf in the Arctic Ocean (Medvedev 2008; Russian Government 2009). The policy direction thereafter shifted to the exploitation of offshore resources, with a focus on LNG (Putin 2013; Russian Government 2014). At the same time, the policy acknowledges the ageing infrastructure and the limitations of Russian technology for developing such resources (Russian Government 2014).

Following the adoption of the 2009 Act on Greenland Self-Government, the powers to administer rights over offshore petroleum resources in Greenland were devolved from Copenhagen to Nuuk. Pelaudeix notes that tapping into offshore petroleum resources is vital to building a self-sustaining economy and gaining financial independence from Denmark (2017). Indeed, Greenland's Oil and Mineral Strategy indicates that the goal of the country's licensing strategy is to 'cultivate and maintain industry interest in oil exploration activities' (Government of Greenland 2014). To attract companies, Greenland has designed a rather lenient legal framework for both licensing and environmental protection (Pelaudeix 2017; Shapovalova and Stephen 2019). However, at the latest licensing round, held in December 2017, no bids from companies were received. The Government commented that 'the absence of interest was expected due to the global recession within the exploration industry' (Government of Greenland 2017). Additional hurdles are the location and scarce infrastructure in the region as well as past failed attempts at discovering oil in Greenlandic waters. Cairn Energy, the only company to drill in Greenlandic waters, found no commercial discoveries after exploring eight wells in 2010–2011 (The Telegraph 2013).

To date, the production of petroleum in the Arctic offshore is limited. However, at least, in some parts of the Arctic, the activities are accelerating with Russia and Norway leading the way. Each Arctic State is motivated by a different set of priorities, which inevitably dictate the way offshore petroleum activities are regulated. Although developing vast resources is a desirable enterprise, the risks associated with such development are very high. The next section makes the case for special safety and environmental rules for Arctic drilling.

13.3 Regulating Petroleum Development: Challenges in the Arctic Waters and Environmental Harm

In line with the international legal obligations[1] and in order to balance the development with environmental protection, each State exploiting petroleum resources have developed a set of rules to ensure the safety of personnel and operations, and environmental protection. It is the level and the rigour of such rules that differ from State to State based on the numerous factors, such as political and economic priorities, the maturity of the legal regime, and external obligations (e.g., EU membership). States adopt different models of granting access to companies to exploit their natural resources: from limiting access to national oil companies only to offering licenses via bids and auctions (Shapovalova and Stephen 2019). After a company gets a license (or another legal right) to access the resources and before it starts exploratory works, it needs to conduct an environmental impact assessment (EIA) and obtain further permits. Generally, the safety and environmental regulations of offshore petroleum development can be divided into prescriptive and performance-based. Prescriptive systems 'set specific technical or procedural requirements', while performance-based (or goal-oriented) systems 'identify functions or outcomes for regulated entities but allow them considerable flexibility to determine how they will undertake the functions and achieve the outcomes' (Dagg et al. 2011). In the Arctic, Russian legal framework is characterised as highly prescriptive (Sidortsov 2017), Norwegian and Greenlandic – as performance-based, US and Canada – as hybrid systems with the elements of both (Dagg et al. 2011; Baker 2012).[2] Experts agree that performance-based systems are more appropriate for the Arctic as they 'more flexible, allowing new (and more effective) technology and practices to be adopted as they emerge' (PAME 2014).

Regardless of the system adopted the main aims of safety and environmental regulation is to prevent and respond to accidents, and to minimise the routine emissions and discharges from the petroleum development operations. To achieve these aims, States may set various requirements on companies, including: to develop exploration and production plans for approval by the regulator; to limit the discharges into air and water; to have access to certain equipment intended to e.g., stop oil flow after a blowout or to rescue personnel from a rig after an accident. The State acting through a special agency would normally evaluate any documents submitted by a

[1] Under the United Nations Convention on the Law of the Sea (1982) States have a duty to protect and preserve the marine environment (art. 193); and to take measures to minimize pollution, including that from offshore petroleum installations (art. 194(3)).

[2] The US historically used the prescriptive system, only recently adopting more performance-based regulation. See Sect. 13.4.2.

company before issuing a permit to begin exploration or production drilling. After the operations begin such State agency has a system in place to monitor the compliance and enforce the rules if there are any non-conformities.

States involved in offshore petroleum development usually have a nationwide legal regime for safety and environmental protection, rather than having separate laws for each development area. However, as the Arctic petroleum riches rose to the attention of experts and environmental groups (WWF 2009; AMAP 2010), the calls for special rules for the Arctic emerged (Pew Charitable Trusts 2013). Experts argue that although Arctic is not a homogenous region in terms of geology, climatic conditions, and infrastructure, applying the usual rules for safety and environmental protection might not be enough to protect the fragile northern ecosystems. Regional organisations, such as the Arctic Council, issued a number of documents recommending such special rules – e.g., the Arctic Offshore Oil and Gas Guidelines (PAME 2009, 2014). Industry associations started developing Arctic-specific standards – e.g., International Organisation of Standardisation standards for Arctic Offshore Structures (ISO 2010) and the American Petroleum Institute standards on Arctic structures and pipelines design (API 1995). Eventually, all Arctic coastal States have either adopted these regulations and standards to a varying degree or developed their own Arctic-specific environmental and safety petroleum regulations.

Drilling in the Arctic requires additional regulation for a number of reasons related to the resilience of the northern ecosystems to pollution and the difficulty of conducting clean-up and rescue operations in the region. The Arctic ecosystems are often described as unique and fragile (AEPS 1991). They are unique in that many of their landscapes and seascapes are pristine, their flora and fauna are very distinctive and highly adapted to the cold and dryness (CAFF 2001). Moreover, the Arctic is home to many endemic species and is, in the global context, a 'significant component of the diversity of life on Earth' (CAFF 2001). Many Arctic indigenous communities depend on clean land and waters for subsistence. Petroleum extraction activities have impacts on the Arctic ecosystem at all of their stages (PAME 2009).

The biggest risk to the marine environment, however, is posed by a possible large-scale oil spill in the Arctic Ocean (AMAP 2010). The US Bureau of Ocean Energy Management estimates that in the course of the proposed production in the Chukchi Sea there is a 75% chance of one or more spills of more than 1000 barrels of oil (BOEM 2015). Potential oil spill recovery in the Arctic presents an additional challenge due to the lack of natural light, low temperatures, and strong winds, as well as the lack of infrastructure capabilities (DNV GL 2014; Kokorin et al. 2008). Oil from a spill in the cold Arctic waters at the end of the drilling season could get trapped in or under the ice and thus be impossible to clean up or even detect (WWF 2014; Payne et al. 1990). Effective Arctic oil spill recovery operations would require advanced planning, international cooperation, trained personnel, and sufficient infrastructure. Additionally, any clean-up operations would have to depend on the weather and aerial observation availability (NEB 2011). Further, the lack of infrastructure in the High North means that any response operation would have to be delayed by the time needed to gather assets and personnel.

Environmental destruction caused by an oil spill would be more severe and last a longer period of time in the Arctic than the impact in more temperate climates due to the lower rate of oil biodegradation in cold temperatures (EPPR 2017c). If spilled, oil would 'persist in the Arctic environment for decades' (Steiner 2010). Some characteristics of the Arctic environment and wildlife species 'exacerbate the potentially negative consequence of an oil spill to Arctic waters' (WWF 2007). The vulnerability of seabirds and some marine mammals is stressed because of the danger of oil fouling the feathers or fur, which such species depend on for insulation. During the 1989 *Exxon-Valdez* oil spill in Alaska, this was the main cause of the death of over 250,000 seabirds and thousands of marine mammals, subsequently leading to a nesting failure, which resulted in a loss of thousands of additional birds (Steiner 2010). Generally, species population recovery after an incident may be slowed because many species in the Arctic have longer lifespans and slower generational turnover (AMAP 1998).

Experimental studies and case histories of spills 'leave no doubt that oil is toxic to aquatic organisms' (AMAP 2010). Phytoplankton, zooplankton, and their consumers can be destroyed immediately on contact by oil and are especially vulnerable during extended periods of exposure (Peterson et al. 2003). The Arctic marine food chain is rather compressed and simple, and 'long-term disruptions in the productivity of primary species (…) may seriously affect ecosystem function and sustainability of these northern systems' (Nelleman et al. 2001).

The exploration activities, involving rig emplacement and drilling, are associated with fuel discharges and risk of blowouts (PAME 2009). The development and production phase further extends the risks of the blowouts and causes short and long-term negative effects on the seabed, reproduction of fish, climate, and acidification (PAME 2009). The resilience of the Arctic marine ecosystem to withstand risks associated with oil and gas development is described as 'weak' (Lloyds 2011). It remains unclear how the Arctic ecosystem would respond to these disturbances due to the 'fragmented nature of much of the existing knowledge' about the Arctic marine regions (CAFF 2013). In the face of these challenges, it is imperative to prevent accidents and account for the Arctic conditions, such as the presence of ice, shorter drilling seasons, and the lack of infrastructure in some areas. The following sections outline the petroleum regulatory systems in the Arctic States, and identify and analyse Arctic-specific regulatory requirements.

13.4 Special Rules for Arctic Petroleum Development

As discussed in Sect. 13.2, all five Arctic Ocean States have either expressed interest in developing or are already producing Arctic offshore resources. States with an already established history of offshore petroleum development, such as the US, Canada and Norway, already have established legal regimes for petroleum development. In addition to the general petroleum legal regime, each of these four States has developed some Arctic-specific regulations to establish more stringent safety and environmental rules compared to more conventional locations. Russia also has a well-established legal regime for petroleum regulation, yet large-scale offshore production is a relatively new development. Finally, Greenland has only recently developed its legislation for offshore petroleum development following the Self Rule Act in 2009.

13.4.1 Canada

The management of Canadian Arctic offshore oil and gas resources is mainly exercised under federal statutes and regulations. The Canada Petroleum Resources Act (CPRA 1985) regulates the allocation of rights for resources and collection of royalties and the Canada Oil and Gas Operations Act (COGOA 1985) regulates safety and environmental protection. The Canada Environmental Protection Act regulates, *inter alia*, pollution that might violate international treaties, including air and water pollution (CEPA 1999). A number of regulations were adopted under the Oil and Gas Operations Act to further administer various aspects of development.[3] Although some progress has been made in awarding greater authority to the territorial and indigenous governments (Campbell and Cameron 2016), offshore resource management largely remained under the regulation of the federal government (CDIAND 2009).

The National Energy Board (NEB) has federal authority over safety and environmental aspects of petroleum activities in most of the Canadian waters, including the Canadian Arctic (NEB 2018). The Board's responsibilities include overseeing activities throughout their cycle starting with the seismic surveying (NEB 2015), issuing authorisations for exploration and production works, and abandonment (NEB 2016). After a company acquires an exploration license and before it commences any drilling activity, it must apply to the NEB for the Operating License and Authorisation for Work (COGOA sec 10). Before granting these, the NEB conducts a technical review. The documents that must accompany any application are outlined by the Canada Oil and Gas Drilling and Production Regulations (2009).

In April 2010, days after the *Deepwater Horizon* blowout, the NEB initiated a review of safety and environmental regulations for offshore petroleum activities in the Canadian Arctic. In the process of knowledge collection, the Board held a number of meetings with over 200 participants representing local residents, government, and industry (NEB 2011). The review covered the prevention, preparedness, response to accidents, and lessons learned from a number of major accidents associated with offshore petroleum activities.

[3] E.g., Canada Oil and Gas Drilling and Production Regulations SOR/2009–315; Canada Oil and Gas Certificate of Fitness Regulations SOR/96–114; Canada Oil and Gas Geophysical Operations Regulations SOR/96–117; Canada Oil and Gas Installations Regulations SOR/96–118; Canada Oil and Gas Diving Regulations SOR/99–60; Canada Oil and Gas Spills and Debris Liability Regulations SOR/87–331; Canada Oil and Gas Operations Regulations SOR/83–149.

Following the Review, the NEB adopted Filing Requirements for Arctic Drilling, which are consistent with the general legal framework but include additional documents to address unique Arctic environment (NEB 2014). Thus, the oil spill contingency plan has to account for unique Arctic conditions and consider inter alia spill trajectory under ice cover; operational limitations caused by unique Arctic conditions such as waves, ice, visibility; personnel training under Arctic conditions; potential role in response operations for Arctic communities. The blowout prevention and well control systems have to be tested under conditions similar to those in the Arctic. An important additional requirement for Arctic drilling is the Same Season Relief Well (SSRW) policy of the NEB. Although it is non-statutory, the NEB requires the applicant to demonstrate the 'capability to drill a relief well to kill an out-of-control well during the same drilling season' (NEB 2014). Adopted in 1976 (Pew Environmental Group 2011), it is perceived as a significant regulatory obstacle for offshore drilling in the Canadian Arctic (Henderson and Loe 2014). It is largely opposed by the industry, who argued their position during the Review (NEB 2011). It increases the costs and reduces the already short ice-free drilling window due to the operators having to schedule enough time at the end of the season to safely drill a relief well in case there is a blowout.

Gas flaring and venting are included in the definition of 'waste' under the COGOA, which generally prohibits the disposal of waste during oil and gas operations (sec 18). Flaring and venting are prohibited unless expressly permitted by the NEB or are necessary due to an emergency situation, in which case the Board must be notified (Canada Oil and Gas Drilling and Production Regulations sec 67). There are no additional regulations as to air pollution in the Arctic.

Not directly applicable to the upstream operations, but still relevant is the Arctic Waters Pollution Prevention Act (CAWPPA 1985). The Act applies to shipping in the Canadian Arctic waters and establishes safety control zones and allows further establishment of requirements for ships entering such zones in the Arctic waters (secs 11–18).

In 2016, Prime Minister Trudeau and then-US President Obama announced the moratorium on new licenses in the Arctic waters (US-Canada Joint Arctic Leaders' Statement). Following a consultation process, the Government decided to cooperate with the Northern partners in developing a new governance framework 'for a science-based, life-cycle impact assessment review every 5 years that takes into account marine and climate change science.' It further decided to negotiate a Beaufort Sea oil and gas 'co-management and revenue-sharing agreement with the governments of the Northwest Territories and Yukon, and the Inuvialuit Regional Corporation' (Government of Canada 2018a). The Government also launched strategic environmental assessments in the Beaufort Sea region, Baffin Bay and Davis Strait in order to assess cumulative effects of any potential petroleum activities using science and traditional knowledge (Government of Canada 2018b).

As demonstrated, the regulation of offshore petroleum activities in the Canadian Arctic is more stringent than elsewhere in Canadian waters due to additional regulations adopted by the NEB and the CAWPPA. These additional requirements mostly refer to companies accounting for unique Arctic conditions when planning their operations. They further impose an onerous SSRW requirement that many companies find economically prohibitive in the current oil market. In the nearest future, following the consultation processes, additional regulatory requirements for Canadian Arctic drilling are expected to emerge to fulfil the promise of science-based regulation.

13.4.2 United States

For the purposes of 'Arctic' regulation, the US legislation includes all territories north of the Arctic Circle as well as the Bering Sea and the Aleutian Island chain to the south (US Arctic Research and Policy Act 1984). According to the division of powers between federal and Alaska state authorities, any development more than three miles offshore falls under federal jurisdiction (US Submerged Lands Act 1993). This chapter focuses on federal laws as the vast amount of undiscovered oil and gas reserves are reportedly situated in the Alaska OCS, beyond the three-mile State limit (BOEM 2016b).

The offshore oil and gas regime in the US comprises of a number of regulations, including those dealing with resource ownership, exploration and development, protected species, environmental standards and permits. The federal exploration and development regime has changed significantly after the 2010 *Deepwater Horizon* blowout in the Gulf of Mexico. The special investigation concluded that the regulatory oversight of oil and gas industry 'required reforms' (National Commission on the BP Deepwater Horizon Oil Spill and Offshore Drilling 2011), a view supported by many scholars and stakeholders (Peterson et al. 2012; Houck 2010; DNV 2010a). As a result, what used to be the Minerals Management Service has been restructured into the Bureau of Ocean Energy Management (BOEM), the Bureau of Safety and Environment Enforcement (BSEE), and the Office of Natural Resource Revenue in an effort to separate revenue generation and independent safety and environmental oversight (US Secretarial Order 3299 2010).

The primary federal law regulating offshore petroleum development is the US Outer Continental Shelf Lands Act (1953). It outlines the main stages of decision-making starting with the development of a nation-wide 5-year Leasing Programme by the Secretary of Interior.

The development of the Programme is a federal activity that requires an Environmental Impact Statement under the US National Environmental Policy Act (1969). The 2017–2022 Programme does not include the Chukchi and Beaufort Seas while acknowledging the 'uniqueness of the Arctic' and the need for continued scientific research on risk minimisation and identification of areas that warrant additional protection. The need for further research along with 'opportunities for exploration and development on existing leases', a lack of industry interest, 'current market conditions, and sufficient existing domestic energy sources already online or newly accessible' were identified as the main reasons for the delay in offering Arctic OCS leases for sale in 2017–2022 (US DOI 2015).

Before any exploration activities commence, the leaseholder must submit an Exploration Plan for approval by the Secretary of Interior (43 USC §1340(c)(1)). Any drilling associated with exploration may require a separate permit (43 USC §1344 (a)(1)). During these stages, companies must obtain a number of permits: air, water, noise and other pollution that the activities may cause.

US offshore oil and gas regulation is characterised by extensive incorporation of standards developed by the private sector. Over 100 American Petroleum Institute (API) and other industry standards have been incorporated by reference by the BSEE (30 CFR 250.198). API represents the industry and develops non-binding standards and recommended practices. Once these are incorporated in federal regulation they become a requirement of law (30 CFR 250.198 (a) (3)).

The pre-*Deepwater Horizon* regulatory framework for offshore oil and gas safety was characterised largely by prescriptive regulations (Dagg et al. 2011). Since April 2010 there has been a shift to a performance-based system. In July 2016, the DOI adopted a new Rule for Arctic offshore oil and gas development (US Arctic OCS Final Rule 2016). It was described as the 'biggest step towards performance-based system' (Gentile 2016). This chapter focuses on this 2016 Arctic Drilling Rule, which has not been as widely considered in the literature before. The overall regulatory framework applicable to petroleum development in the US Arctic waters is examined in more detail in Canuel (2017) and Levine et al. (2014).

The 2016 Arctic Drilling Rule is the first Arctic-specific national legislation adopted for offshore petroleum operations in the US. It establishes additional requirements to acknowledge the unique climatic and logistical conditions of the US Arctic OCS. The Rule is only applicable to exploratory drilling from mobile offshore drilling units (MODUs) in the OCS within the Beaufort Sea and Chukchi Sea Planning Areas, where most of the undiscovered oil and gas resources are reportedly located. This means that most of the current petroleum developments, conducted off artificial islands, are exempt from the Rule. The new regulations are designed to help 'ensure the safe, effective, and responsible exploration of Arctic OCS oil and gas resources while protecting the marine, coastal, and human environments, and Alaska Natives' cultural traditions and access to subsistence resources' (US Arctic OCS Final Rule 2016). The Rule was drafted with active engagement from all stakeholders, in particular, the Alaskan local community (US Arctic OCS Final Rule 2016).

The main requirements additional to the existing under the general offshore legal framework are:

1. to develop an Integrated Operations Plan (30 CFR 550.204). The Plan must explain how the proposed exploratory drilling is 'fully integrated from start to finish in a manner that accounts for Arctic OCS conditions'. This includes information on vessel equipment and design; overall schedule of operations; weather and ice forecasting and management; operational safety principles; contractor management and oversight; preparation and staging of oil spill response assets; and impact on local community. The Plan must be submitted 90 days prior to the filing of the Exploration Plan and does not need to be

approved. Rather, it is intended as a 'conceptual, informational document' that ensures that the operator planned to address risks at an early stage;
2. to 'have access to and ability to promptly deploy Source Control and Containment Equipment' (30 CFR 250.471). The Rule explains that while such equipment is readily available in the Gulf of Mexico 'due to the level of activity in the area', it is not the case in the Arctic. Such equipment includes but is not restricted to a 'capping stack, cap and flow system, and containment dome'. To avoid potentially discouraging innovative technologies, this requirement is performance-based and allows the operator to propose an 'alternate technology' if the operator can demonstrate that its 'response capabilities able to stop or capture the flow of an out-of-control well';
3. to 'have access to a separate relief rig' (30 CFR 250.472). The rig must be located in a way that it could 'arrive on site, drill a relief well, kill and abandon the original well, and abandon the relief well prior to expected seasonal ice encroachment at the drill site and in no event later than 45 days after the loss of well control'. This requirement is performance-based, whereby the operator might use alternate technology if he can show that it 'will meet or exceed the level of safety and environmental protection' provided by a relief rig;
4. to have the capability to 'predict, track, report and respond to ice conditions and adverse weather events'. This includes notifying BSEE of any sea ice movement that might affect the operation (30 CFR 250.188(c));
5. to 'effectively manage and oversee contractors'. This requirement is implemented by including relevant information into the Plan' (30 CFR 550.204(b), (f), (g)).

The Rule is not intended to prevent operators from exploring the Arctic for resources. Instead, the DOI states that 'additional clarity and specificity (…) should assist the oil and gas industry to plan better and to more effectively conduct exploratory drilling on the Arctic OCS with lower risk' (US Arctic OCS Final Rule 2016). Nevertheless, the new regulations have been criticised by the industry. Head of the API called it an 'unfortunate turn' that would 'continue to stifle offshore oil and natural gas production' (McGwin 2016). As expected, the biggest opposition from the industry is to the stand-by rig for drilling a relief well in an event of an accident, which reportedly adds $1 million a day to the operation costs (McGwin 2016). The Rule was met more warmly by environmental NGOs. Oceana welcomed the new regulations, albeit suggesting they 'do not ensure safe and responsible operations' in the Arctic (LeVine 2016). The Pew Charitable Trusts commented on the draft Rule (2015), and although not all of their suggestions have been adopted, positively commented on the Final Rule (2016).

Although the US joined Canada in announcing a moratorium on new licenses in 2016, President Trump has since overturned the decision (The White House 2017). The current administration has also expressed no interest in adopting new regulations for offshore petroleum industry. Instead, there is a possibility some regulations are going to be rolled back (BSEE 2018). The 2016 Arctic Drilling Rule adds significant challenges for the industry with its requirements of more advanced planning, access to relief rig, and accounting for ice presence. It is in line with the Arctic Council's Arctic Offshore Oil and Gas Guidelines (PAME 2009) and represents one of the highest standards among the Arctic jurisdictions.

13.4.3 Norway

Norway is the world's third largest exporter of oil and gas (IEA 2012), with all of its petroleum production located offshore. The Norwegian petroleum sector is active in the North, Norwegian, and Barents Seas. Petroleum development is the largest sector of the Norwegian economy, responsible for nearly 40% of the country's exports (Statistics Norway 2016). This is supplemented by the export of specialised services to the global oil industry, which became an important industry 'in its own right' (Alvik 2013).

The Norwegian petroleum regulation system is based on the Ten Oil Commandments formulated in the wake of oil and gas development on the country's continental shelf (Norwegian Ministry of Environment 2010). These principles include environmental concerns: consideration for the 'protection of nature and the environment', and the general prohibition of gas flaring. Relevant for the Arctic, the Commandment 9 requires that 'an activity plan must be adopted for the area north of the 62nd parallel which satisfies the unique socio-political factors associated with that part of the country'. The Commandments are now considered to have been largely fulfilled (Norwegian Ministry of Environment 2010).

The Ministry of Petroleum and Energy (MPE) is responsible for the petroleum sector as a whole. Subordinate to the Ministry, the Norwegian Petroleum Directorate *inter alia* advises the Ministry; exercises authority over petroleum development; issues relevant regulations; collects fees from the petroleum industry; and manages the State's ownership interests in State companies, Equinor (formerly Statoil) and Petoro.

The main legal basis for oil and gas development in Norway is the Petroleum Act (1996). It outlines the conditions of granting licenses, production of petroleum, liability for pollution damage, and safety requirements. Under the authority of the Act, the Petroleum Regulations were adopted (1997). They largely follow the structure of the Petroleum Act, providing details on the procedure and specific content or requirements. The Ministry of Labour and Social Affairs is responsible for the working environment, safety, and emergency preparedness of the Norwegian petroleum industry. The Petroleum Safety Authority (PSA), subordinate to the Ministry, regulates: drilling and well technology; process and structural integrity; logistics and emergency preparedness; occupational health and safety; health, safety, and environmental management. The Petroleum Act requires that all the activities 'be conducted in such a manner as to enable a high level of safety to be maintained and further developed in accordance with the technological development' (sec 9–1). There are a number of regulations adopted under the PSA framework.[4] Most of them are accompanied by non-binding guidelines that promote the use of standards. Standards set a performance level and alternative solutions can be utilised as long as the same performance level is demonstrated (DNV 2010b) The Norwegian regime is thus largely performance-based or goal-oriented (Nordtveit 2015; Hanson 2010).

In addition to an exploration or production license, a number of additional approvals from the government are required before and after the field is set in production. The most important is the Plan for Development and Operation (Petroleum Act sec 4–2, 1997). The Plan must be approved by the Ministry of Petroleum and Energy and must contain 'an account of economic aspects, resource aspects, technical, safety-related, commercial and environmental aspects, as well as information on how the facility may be decommissioned and disposed of' (Petroleum Act sec 4-2 1997) at the end of the production. The detailed content of the Plan is outlined in the Petroleum Regulations (sec 19–21). In advance of the Plan submission, the licensee must present to the Ministry of Petroleum and Energy its plan for environmental impact assessment (including transboundary impact) (Petroleum Regulations sec 22 and 22a-c 1997).

Many regard the so-called 'Norwegian system' as the leader in environmental sustainability and safety (Hanson 2010). The Norwegian Environmental Agency, working under the Ministry of Climate and Environment, exercises inspection and enforcement powers under the Pollution Control Act (1981). The Act prohibits all pollution except that, which is explicitly allowed by obtaining a permit from the Environmental Agency (sec 7, 11).

The Norwegian offshore petroleum legislation does not generally distinguish between Arctic and non-Arctic waters, but the weather conditions in the Barents Sea are milder than in the Canadian, US, and the rest of the Russian Arctic waters. However, in addition to the general Norwegian legal framework, some Norwegian Arctic areas are subject to the Integrated Management Plans (IMPs). This approach allows for integrated, holistic management of various aspects of oceans-use, such as petroleum development, fisheries, and shipping. Such plans have been adopted for the North Sea and Skagerrak (Norwegian Ministry of Environment 2012), the Norwegian Sea (Norwegian Ministry of Environment, 2008), and the Barents Sea and Lofoten (Norwegian Ministry of Environment 2005, 2014).

The IMP for the Barents Sea and Lofoten was updated in 2014. The Plan is inte gral for petroleum regulation in the Norwegian Arctic. First, it identifies 'particularly valuable and vulnerable areas' which have been identified as 'being of great importance for biodiversity and for biological production in the entire Barents Sea–Lofoten area, and where adverse impacts might persist for many years' (Hoel 2010).

[4]Regulations Relating to Health, Safety and the Environment in the Petroleum Activities and at Certain Onshore Facilities (12 February 2010); Regulations Relating to Management and the Duty to Provide Information in the Petroleum Activities and at Certain Onshore Facilities (29 April 2010); Regulations Relating to Design And Outfitting of Facilities, etc. in the Petroleum Activities (29 April 2010); Regulation Relating to Conducting Petroleum Activities (29 April 2010).

As a result, vast areas including around Lofoten and Northern Barents Sea remain closed to petroleum activities (Hansen and Midtgard 2008; Norwegian Ministry of Environment 2014). The IMP introduces a restriction on drilling locations, prohibiting operations in the 'marginal ice zone' It is a dynamic zone, the boundaries of which are defined based on the 'ice persistence' criterion – where ice covers 'more than 15% of the sea surface' (Norwegian Ministry of Environment 2014). It has been noted that the plan does not cover the entire Barents Sea ecosystem as it is limited by the Norwegian-Russian maritime border (Olsen et al. 2007).

The Norwegian industry is working under the 'Zero Discharge Policy' in an attempt to minimise emissions and discharges. The policy, formulated in the 1990s (Norwegian Ministry of Environment 1996) with the participation of government and industry, 'allows the discharge of some chemicals, while discharges of other more harmful chemicals are prohibited' (Harlaug Olsen et al. 2011). For the areas of the Barents Sea and Lofoten, the rules are more stringent than on the rest of the NCS (Knol 2011). The 2006 IMP states that 'during normal operations, no discharges of any substances with a negative impact on the environment are permitted from petroleum installations'. This position was elaborated further in the updated IMP in 2014.

While the IMPs primarily focus on environmental regulation, the standards for safety are defined in the petroleum regulations are largely developed by the industry. To ensure the highest standards for petroleum production in the Arctic in both Russian and Norwegian waters, the Barents 2020 project was established (Norwegian Government 2006). It provides non-binding standards, guidance and recommendations on all aspects of petroleum development in the Barents Sea. The Barents 2020 project did not create new standards *per se*, but it had identified the existing appropriate standards for use in the area (DNV GL n.d.).

Thus, while generally the petroleum legislation in Norway applies to all areas on the Norwegian Continental Shelf, some areas around Lofoten and Northern Barents Sea are subject to additional regulation in accordance with the relevant IMP. Such additional regulation includes leaving certain areas off-limits for the industry due to the presence of ice and stricter limits on routine pollution. That said, Norwegian petroleum regulation already established strict standards, comparable to e.g., 2016 US Arctic Drilling Rule, including mandatory access to a relief rig (Activities Regulations sec 86). Norway further promotes the use of appropriate safety standards in the broader Barents region by cooperating with industry and Russia *inter alia* through the Barents 2020 project.

13.4.4 Greenland

Since the majority of maritime areas in Greenland are located in the Arctic, it would be unreasonable to expect separate rules for Arctic petroleum development. The regulation of petroleum is performance-based and instead of prescriptive rules, sets out a general legal framework in the Greenland Mineral Resources Act (2009).[5] Greenland's Mineral Authority is responsible for all aspects of offshore drilling including licensing, permitting, and safety and environmental regulation and enforcement.

After a company obtains a license, it needs to obtain further governmental approvals before commencing any seismic testing or drilling. A company must first prepare a location-specific environmental impact assessment report and reports on the socio-economic impact of the proposed activity (Government of Greenland 2009). The Environmental Protection Plan specifies how the waste will be managed during the activities, including sewage, chemicals, and minor oil spills (Government of Greenland 2009). The Greenlandic guidelines require that such documents are consistent with 'good international practice' and further refer to international documents such as AOOGG, Safety Management Guidelines (PAME 2014), OSPAR Guidelines for Monitoring Environmental Impact (Government of Greenland 2018). Finally, an Emergency Response Plan is required to outline procedures for dealing with a major oil spill. After the relevant permits are granted, any drilling is subject to the Mineral Resources Act, which sets out the general legal framework. This general framework is supplemented by more specific provisions in the Petroleum Drilling Guidelines (Greenland BMP 2011), which are considered minimum requirements (Dagg et al. 2011). Most of the contents in the Guidelines mirror the Norwegian NORSOK regulations and standards with regards to risk assessment, well integrity, and drilling facilities (Greenland BMP 2011). The Guidelines take into account the Arctic conditions in that they require ice studies and ice management plan to be submitted along with the EIA. The risk assessment needs to include the hazards of icebergs and packed ice. The Guidelines establish a Dual Rig policy, essentially mirroring the SSRW requirement in Canada, Norway, and the US (Greenland BMP 2011).

Overall, environmental and safety petroleum regulation in Greenland is comparable to Norway in that it is performance-based and requirements vary on the case-by-case basis. However, this type of regulation requires that the regulator is informed, trained and experienced in risk assessment (PAME 2014), which might not be the case in Greenland as much as it is in Norway.

13.4.5 Russia

While Russian legislation establishes stringent requirements for the companies with regards to gaining access to petroleum licenses in the Arctic waters, it does not provide substantial additional requirements in the environmental or safety regulation of petroleum operations (Shapovalova and Stephen 2019; Sidortsov 2017).

The legal framework for offshore petroleum development is constantly evolving and consists of federal laws, codes, Presidential and Governmental decrees, and

[5] For detailed examination of Greenland's environmental petroleum regulation see Mosbech et al. 2017.

other normative documents issued by the relevant agencies. The AMAP estimated that there are over 800 documents regulating environmental protection and natural resources use in Russia (AMAP 2010). The main legislative basis for offshore hydrocarbons development is the Subsoil Law (1992), the Continental Shelf Law (1995), the Exclusive Economic Zone Law (1998) and the Environmental Protection Law (2002). The Subsoil Law requires that license holders on the continental shelf blocks be legal entities with at least five years of experience in operating on the Russian continental shelf, and with a share owned by the Russian State of no less than 50% (Subsoil Law art 10 1992). The petroleum regulation system in Russia is highly prescriptive and State-centric. The State governs offshore petroleum production through the Ministry of Energy (Minenergo) and the Ministry of Natural Resources and Environment (Minprirody). Whilst generally the resource development is under the joint competence of federal and regional authorities (The Constitution of the Russian Federation art 72 1993), the Arctic offshore fields are subject to exclusive federal regulation (Subsoil Law art 2.1 1992).

Environmental regulation of offshore petroleum activities in Russia is primarily based on the Environmental Protection Law (2002) and the relevant provisions of the Subsoil Law and the Continental Shelf Law. The Environmental Protection Law does not contain any specific provisions for northern development (Ignatyeva 2013). It is largely based on the polluter-pays principle and includes economic incentives for companies to limit their pollution (art 16).

The Subsoil Law established State supervision of the operation to prevent safety and environmental violations (art 38). The Federal Service on Supervision in the Sphere of Natural Resources Use (Rosprirodnadzor) monitors compliance with environmental regulations and licences. Rosprirodnadzor issues safety rules for equipment and processes for offshore petroleum development (Rosprirodnadzor 2013, 2014). Sidortsov highlights the lack of Arctic-specific norms Russian petroleum regulation with the exceptions of some provisions on operating in ice and cold weather conditions (2017). Indeed, the Rosprirodnadzor rules prescribe that drilling must be ceased in the presence of floating ice or a storm (Rosprirodnadzor, items 69, 200 2014); decision on the construction and exploitation of ice-resistant platforms must be made considering their performance in low temperatures and the presence of ice (Rosprirodnadzor items 6–10 2014). The rescue operations must consider evacuation equipment and procedures in ice conditions (Rosprirodnadzor item 90 2014). This is in line with the requirement of Continental Shelf Law for a license to include information on methods for oil spill response in ice conditions (if the drilling is conducted in such conditions) (art 8).

There have been reports of developing Arctic-specific standards for materials used in offshore petroleum exploration and production (Kichanov 2016), but none have been adopted so far.

Regulation of petroleum development in the Russian Arctic is, in principle, different from that in the other Arctic States due to its highly prescriptive nature. The difficulty with such an approach is that the legislation might not keep up with the rapid development of technology in the offshore industry (Sutton 2014). This means that prescriptive standards may not be sufficiently up-to-date to address current issues. However, with the growing interest in developing the Arctic offshore resources and the ever-developing legal framework, there is room for the adoption of higher standards and more specialised rules.

13.5 Analysis and Conclusions

Petroleum activities in Arctic waters, while potentially economically attractive, pose significant risks to the environment. These risks are exacerbated by the unconventional locations, climatic conditions, lack of infrastructure, and the unique features of the Arctic ecosystems. In addition, the existing knowledge of the Arctic ecosystems is fragmented, and the full effects of large-scale industrial developments are not known. The offshore petroleum production in the Arctic waters is currently occurring in Russia and Norway. In Greenland and the American Arctic, the production is currently on hold but is set to increase in the long-term outlook based on the States' policies. Pelaudeix compared Canada, Norway, and Greenland, found a strong correlation between the energy strategies and the offshore petroleum legal framework (2017). Having considered Russia and the US as well, this chapter confirms Pelaudeix's findings. All Arctic coastal States have adopted regulations specifically addressing the peculiarities of operating in the Arctic, but these regulations vary in extent and priorities.

The common issues in the Arctic-specific regulatory requirements in the coastal States are the SSRW, prevention of and response to oil spills. The SSRW, or access to an additional drilling rig, allowing to regain control of a well blowout within the same drilling season (or 12 days – in Norway), is featured in all the Arctic coastal legislation, except for Russia. The Arctic Offshore Oil and Gas Guidelines includes relief well arrangements as a necessary element of oil spills planning (PAME 2009) as in case of a blowout it is imperative to stop the uncontrolled flow of oil as soon as possible, and definitely before the drilling season ends due to incoming ice.

The statutory requirements and the use of specialised industry standards regarding drilling and oil spill prevention are also a running theme in the Arctic-specific regulations. The use of standards is adopted to a varying degree in all Arctic jurisdictions and there may be room for harmonization (DNV GL n.d.; Baker 2012). Russia remains an outlier with its distinct highly prescriptive regulatory system. While Russian law places strict restrictions on the access to Arctic shelf licensing, there does not appear to be a corresponding concern for addressing safety and environmental risks in the legislation.

With regards to oil spill response, the performance-based system in Norway and Greenland may allow the regulator to place additional requirements on the case by case basis. The US Arctic Drilling Rule enhances the requirement for spill response equipment in the Arctic compared to the general petroleum legal framework. There is room for knowledge exchange and further international cooperation on response operations, techniques, and infrastructure on a bilateral basis or through the Arctic Council.

Finally, as is the case in Norway, some areas of the Arctic waters might need to remain closed for petroleum activities or have stricter routine pollution requirements. Science-based approach and integrated management are required to identify the areas that are particularly vulnerable and justify additional regulation.

The regulation of upstream petroleum in the offshore areas range from a highly prescriptive system in Russia, hybrid systems in the US and Canada, and mostly performance-based system in Norway and Greenland. While the complete harmonization of the Arctic States' petroleum regulation is not feasible or desirable, it is important to identify and acknowledge best practices to promote their wider adoption (Baker 2012). While the performance-based system is identified as the preferred one for Arctic operations (PAME 2014), the limitations of the knowledge, capacity, and experience of the regulator must be recognised appropriately.

It is important to acknowledge, there is no need for Arctic-specific legislation in all the coastal States due to a variety of regulatory systems and the geophysical conditions. Thus, in Norway, the conditions across the whole continental shelf are more homogenous that on Canada or the US. Nevertheless, the challenges of operating in the Arctic, such as the presence of ice, shorter drilling seasons, and the lack of infrastructure in some areas, must be addressed. All five States have implemented special requirements for operating in the Arctic waters.

References

AEPS (Arctic Environmental Protection Strategy). (1991). http://library.arcticportal.org/1542/1/artic_environment.pdf. Accessed 8 Dec 2018.

Alvik, I. (2013). Norway. In K. Talus & E. G. Pereira (Eds.), *Upstream law and regulation: A global guide* (pp. 367–381). London: Globe Business Publishing.

AMAP (Arctic Monitoring and Assessment Programme). (1998). *AMAP assessment report: Arctic pollution issues*. https://www.amap.no/about/the-amap-programme/amap-assessment-reports. Accessed 8 Dec 2018.

AMAP (Arctic Monitoring and Assessment Programme). (2010). *Assessment 2007: Oil and gas activities in the Arctic: Effects and potential effects* (Vol. 1). https://www.amap.no/about/the-amap-programme/amap-assessment-reports. Accessed 8 Dec 2018.

API (American Petroleum Institute). (1995). *RP 2N planning, designing and constructing structures and pipelines for Arctic conditions*. http://www.techstreet.com/api/standards/api-rp-2n?product_id=1894499#document. Accessed 8 Dec 2018.

Baker, B. (2012). Offshore oil and gas regulation in the Arctic: Room for harmonization? *The Yearbook of Polar Law, IV*, 475–504.

Baker, B. (2013). Offshore oil and gas development in the Arctic: What the Arctic council and international law can—And cannot—Do. *ASIL, 107*, 275–279.

Baker, B. (2017). The Arctic offshore hydrocarbon hiatus of 2015: An opportunity to revisit regulation around the pole. In C. Pelaudeix & E. M. Basse (Eds.), *Governance of Arctic offshore oil and gas* (pp. 148–166). Abingdon: Routledge.

Bloomberg. (2015). *Statoil seeks broader Rosneft ties despite Russia sanctions*. https://www.bloomberg.com/news/articles/2015-08-04/statoil-seeks-broader-rosneft-ties-in-russia-as-sanctions-stifle. Accessed 8 Dec 2018.

BOEM. (2015). *Oil spill risk in the Chukchi Sea outer continental shelf the 75-percent figure: What does it mean?* https://www.boem.gov/BOEM-Newsroom/Risk-and-Benefits-in-the-Chukchi-Sea.aspx. Accessed 8 Dec 2018.

BOEM. (2016a). *2017–2022 outer continental shelf oil and gas leasing proposed final program.* https://www.boem.gov/2017-2022-OCS-Oil-and-Gas-Leasing-PFP/. Accessed 8 Dec 2018.

BOEM. (2016b). *Assessment of undiscovered oil and gas resources of the nation's outer continental shelf.* https://www.boem.gov/2016-National-Assessment-Fact-Sheet. Accessed 8 Dec 2018.

BOEM. (2018a). *Alaska OCS region.* https://www.boem.gov/Alaska-Region/. Accessed 8 Dec 2018.

BOEM. (2018b). *Alaska historical data.* http://www.boem.gov/AKOCS_Wells_Drilled_per_Year/. Accessed 8 Dec 2018.

BOEM. (2018c). *2019–2024 OCS oil and gas leasing draft proposed program and potential exclusion areas: Arctic.* https://www.boem.gov/NP-DPP-Map-Alaska/. Accessed 8 Dec 2018.

BSEE (US Bureau of Safety and Environmental Enforcement). (2018). *Oil and gas and sulfur operations in the outer continental shelf-blowout preventer systems and well control revisions.* https://www.federalregister.gov/documents/2018/05/11/2018-09305/oil-and-gas-and-sulfur-operations-in-the-outer-continental-shelf-blowout-preventer-systems-and-well. Accessed 8 Dec 2018.

CAFF (Conservation of Arctic Flora and Fauna). (2001). *Arctic Flora and Fauna: Status and conservation.* https://www.caff.is/assessment-series. Accessed 8 Dec 2018.

CAFF (Conservation of Arctic Flora and Fauna). (2013). *Arctic biodiversity assessment: Status and trends in Arctic biodiversity.* https://www.caff.is/assessment-series. Accessed 8 Dec 2018.

Campbell, A., & Cameron, K. (2016). Constitutional development and natural resources in the north. In D. A. Berry, N. Bowles, & H. Jones (Eds.), *Governing the North American Arctic: Sovereignty, security, and institutions.* Basingstoke/New York: Palgrave Macmillan.

Canuel, E. T. (2017). Alaska and offshore hydrocarbon extraction. In C. Pelaudeix & E. M. Basse (Eds.), *Governance of Arctic offshore oil and gas.* Abingdon: Routledge.

CAWPPA (Canada Arctic Waters Pollution Prevention Act). RSC 1985 c A-12.

CBC News. (2014). *Chevron puts Arctic drilling plans on hold indefinitely.* http://www.cbc.ca/news/canada/north/chevron-puts-arctic-drilling-plans-on-hold-indefinitely-1.2877713. Accessed 8 Dec 2018.

CBC News. (2015). *Imperial oil, BP delay Arctic oil drilling plans indefinitely.* http://www.cbc.ca/news/canada/north/imperial-oil-bp-delay-beaufort-sea-drilling-plans-indefinitely-1.3129505. Accessed 8 Dec 2018.

CDIAND (Canada Department of Indian Affairs and Northern Development). (2009). *Canada's northern strategy: Our north, our heritage, our future.* http://www.northernstrategy.gc.ca/cns/cns-eng.asp. Accessed 8 Dec 2018.

CEPA (Canada Environmental Protection Act). SC 1999 c 33.

COGOA (Canada Oil and Gas Operations Act). RSC 1985 c O-7.

Continental Shelf Law Russia – Federal Law no 187-FZ 'On the continental shelf of Russian federation' (30 November 1995).

CPRA (Canada Petroleum Resources Act). RSC 1985 c 36 2nd Supp.

Dagg, J., et al. (2011). *Comparing the offshore drilling regulatory regimes of the Canadian Arctic, the US, the UK, Greenland and Norway.* www.pembina.org/reports/comparing-offshore-oil-and-gas-regulations.pdf. Accessed 8 Dec 2018.

DNV. (2010a). *Key aspects of an effective US offshore safety regime.* https://www.scribd.com/document/246972281/DNV-Position-Paper-on-Key-Aspects-of-an-Effective-U-S-Offshore-Safety-Regime-22-July-2010. Accessed 8 Dec 2018.

DNV. (2010b). *OLF/NOFO – Summary of differences between offshore drilling regulations in Norway and US Gulf of Mexico.* https://www.norskoljeoggass.no/contentassets/4ed0dc5bf143402ea7ef4c6fb9953eb2/report-no-2010-1220-summary-of-differences-rev-02-2010-08-27-signed.pdf. Accessed 8 Dec 2018.

DNV GL. (2014). *Challenges and best practice of oil spill response in the Arctic.*

DNV GL. (n.d.). *Barents 2020 project*. https://www.dnvgl.com/oilgas/arctic/barents-2020-reports.html. Accessed 8 Dec 2018.
Emmerson, C. (2011). *The future history of the Arctic*. London: Vintage.
Eni Norge. (2018). *Gøliat*. http://www.eninorge.com/en/Field-development/Goliat/Development-solution/The-platform%2D%2D-FPSO/. Accessed 8 Dec 2018.
Environmental Protection Law Russia – Federal law no 7-FZ 'On environmental protection' (10 January 2002).
EPPR (Emergency Prevention, Preparedness, and Response). (2017a). *Overview of measures specifically designed to prevent oil pollution in the Arctic marine environment from offshore petroleum activities*. https://oaarchive.arctic-council.org/handle/11374/1962. Accessed 8 Dec 2018.
EPPR (Emergency Prevention, Preparedness, and Response). (2017b). *Standardization as a tool for prevention of oil spills in the Arctic*. https://oaarchive.arctic-council.org/handle/11374/1951. Accessed 8 Dec 2018.
EPPR (Emergency Prevention, Preparedness, and Response). (2017c). *Field guide for oil spill response in Arctic waters* (2nd ed.). https://oaarchive.arctic-council.org/bitstream/handle/11374/2100/EPPR_Field_Guide_2nd_Edition_2017.pdf?sequence=12&isAllowed=y. Accessed 8 Dec 2018.
Equinor. (2018). *Johan Castberg*. https://www.equinor.com/en/what-we-do/new-field-developments/johan-castberg.html. Accessed 8 Dec 2018.
EU Commission. (2008) Communication: The European Union and the Arctic region COM (2008) 763 final 7.
Exclusive Economic Zone Law Russia – Federal law no 191-FZ 'On the exclusive economic zone of the Russian federation' (17 December 1998).
Gazprom. (2018). *Prirazlomnoe*. http://www.gazprom.com/projects/prirazlomnoye/. Accessed 8 Dec 2018.
Gentile, G. (2016). Performance versus prescription in new US Arctic rules: Fuel for thought. *The Barell Blog*. http://blogs.platts.com/2016/08/15/arctic-oil-gas-new-rules-performance-prescription/. Accessed 8 Dec 2018.
Government of Canada. (2010). *Statement on Canada's Arctic foreign policy*. http://international.gc.ca/world-monde/assets/pdfs/canada_arctic_foreign_policy-eng.pdf. Accessed 8 Dec 2018.
Government of Canada. (2018a). *Arctic offshore oil and gas*. https://www.rcaanc-cirnac.gc.ca/eng/1535571547022/1538586415269. Accessed 8 Dec 2018.
Government of Canada. (2018b). *Northern oil and gas report 2017*. https://www.aadnc-aandc.gc.ca/DAM/DAM-INTER-HQ-NOG/STAGING/texte-text/oil_gas_2017_1528132190313_eng.pdf. Accessed 8 Dec 2018.
Government of Greenland. (2009). *Exploration and exploitation of hydrocarbons in Greenland strategy for licence policy*. https://web.law.columbia.edu/sites/default/files/microsites/climate-change/files/Arctic-Resources/Oil-and-Gas/greenland%20strategy%20for%20license%20policy.pdf. Accessed 8 Dec 2018.
Government of Greenland. (2014). *Greenland's oil and mineral strategy 2014–2018*. https://naalakkersuisut.gl/~/media/Nanoq/Files/Publications/Raastof/ENG/Greenland%20oil%20and%20mineral%20strategy%202014-2018_ENG.pdf. Accessed 8 Dec 2018.
Government of Greenland. (2017). *The Baffin Bay oil and gas licensing round 2017*. http://naalakkersuisut.gl/en/Naalakkersuisut/News/2017/12/1512_baffinbugten. Accessed 8 Dec 2018.
Government of Greenland. (2018). *International reference documents*. https://www.govmin.gl/environment/international-reference-documents. Accessed 8 Dec 2018.
Greenland BMP (Bureau of Minerals and Petroleum). (2011). *Drilling guidelines* https://govmin.gl/images/stories/petroleum/110502_Drilling_Guidelines.pdf. Accessed 8 Dec 2018.
Greenland Mineral Resources Act. (2009).
Hansen, H. O., & Midtgard, R. M. (2008). Going North: The new petroleum province of Norway. In A. Mikkelsen & O. Langhelle (Eds.), *Arctic oil and gas: Sustainability at risk?* (pp. 200–239). New York/London: Routledge.

Hanson, A. L. (2010). Offshore drilling in the United States and Norway: A comparison of prescriptive and performance approaches to safety and environmental regulation. *The Georgetown Environmental Law Review, 23*, 555–575.

Harlaug Olsen, G., et al. (2011). Challenges performing risk assessment in the Arctic. In K. Kenneth Lee & J. Neff (Eds.), *Produced water: Environmental risks and advances in mitigation technologies* (pp. 521–536). New York: Springer.

Henderson, J., & Loe, J. (2014). *The prospects and challenges for Arctic oil development*. The Oxford Institute for Energy Studies. https://www.oxfordenergy.org/wpcms/wp-content/uploads/2014/11/WPM-56.pdf. Accessed 8 Dec 2018.

Hoel, A. H. (2010). Integrated oceans management in the Arctic: Norway and beyond. *Arctic Review of Law and Politics, 1*(2), 186–206.

Houck, O. A. (2010). Worst case and the deepwater horizon blowout: There ought to be a law. *The Environmental Law Reporter, 40*, 11034–11040.

IEA (International Energy Agency). (2012). *Energy policies of IEA countries: Norway 2011 review*.

International Energy Agency, 'Russia 2014: Energy Policies beyond IEA Countries' (2014).

Ignatyeva, I. A. (2013). Environmental assessment in Russian law: Can it promote sustainable development in the Russian Arctic? *The Yearbook of Polar Law, 5*(1), 321–336. https://doi.org/10.1163/22116427-91000128.

ISO (International Organization for Standardization). (2010). *ISO 19906:2010 petroleum and natural gas industries – Arctic offshore structures*. https://www.iso.org/standard/33690.html. Accessed 8 Dec 2018.

Johnstone, R. L. (2015). *Offshore oil and gas development in the Arctic under international law: Risk and responsibility*. Leiden: Brill Nijhoff.

Kichanov, M. (2016). Siberian innovations will be tested in the Arctic. *Kommersant*. https://www.kommersant.ru/private/pdoc?docsid=3023750. Accessed 8 Dec 2018.

Knol, M. (2011). The uncertainties of precaution: Zero discharges in the Barents Sea. *Marine Policy, 35*(3), 399–404. https://doi.org/10.1016/j.marpol.2010.10.018.

Koivurova, T. & Molenaar, E. J. (2010). *International governance and regulation of the Marine Arctic (WWF)*. https://www.worldwildlife.org/publications/international-governance-and-regulation-of-the-marine-arctic-three-reports-prepared-for-the-wwf-international-arctic-program. Accessed 8 Dec 2018.

Kokorin, A. O., Karelin, D. V., & Stetsenko, A. V. (eds.). (2008). The impact of climate change on the Russian Arctic and paths to solving the problem (WWF). http://assets.panda.org/downloads/wwf_arctica_eng_1.pdf; Accessed 8 Dec 2018.

LeVine, M. (2016). *Government finalizes safety and prevention rules for Arctic ocean exploration drilling* (Oceana USA). http://usa.oceana.org/press-releases/government-finalizes-safety-and-prevention-rules-arctic-ocean-exploration-drilling. Accessed 8 Dec 2018.

LeVine, M., Van Tuyn, P., & Hughes, L. (2014). Oil and gas in America's Arctic Ocean: Past problems counsel precaution. *The Seattle University Law Review, 37*(4), 1271–1370.

Lloyds. (2011). *Drilling in extreme environments: Challenges and implications for the energy insurance industry*. https://www.lloyds.com/~/media/lloyds/reports/emerging-risk-reports/lloyds-drilling-in-extreme-environments-final3.pdf. Accessed 8 Dec 2018.

Lunden, L. P., & Fjaertoft, D. (2014). *Government support to Upstream oil & gas in Russia: How subsidies influence the Yamal LNG and Prirazlomnoe projects*. Geneva/Oslo/Moscow: The International Institute for Sustainable Development/WWF.

Masterson, D. M. (2013). The Arctic Islands Adventure and Panarctic Oils Ltd. *Cold Regions Science and Technology, 85*, 1–14. https://doi.org/10.1016/j.coldregions.2012.06.008.

McGwin, K. (2016). *Obama's offshore rules a rig too far for oil industry*. The Arctic Journal. http://arcticjournal.com/oil-minerals/2438/obamas-offshore-rules-rig-too-far-oil-industry. Accessed 8 Dec 2018.

Medvedev, D. (2008). Fundamentals of the state Arctic policy until 2020 and further no PR-1969.

Mikkelsen, A., & Langhelle, O. (Eds.). (2008). *Arctic oil and gas: Sustainability at risk?* New York/ London: Routledge.
Model JOA. (n.d.). *Norwegian Model JOA agreement concerning petroleum activities*. https://www.regjeringen.no/globalassets/upload/OED/Vedlegg/Konsesjonsverk/k-verk-vedlegg-1-2-eng.pdf. Accessed 8 Dec 2018.
Mosbech, A., Boertmann, D., Wegeberg, S., et al. (2017). The interplay between environmental research and environmental regulation of offshore oil activities in Greenland. In C. Pelaudeix & E. M. Basse (Eds.), *Governance of Arctic offshore oil and gas*. Abingdon: Routledge.
National Commission on the BP Deepwater Horizon Oil Spill and Offshore Drilling. (2011). *Deep water: The Gulf oil disaster and the future of offshore drilling: Report to the president*. Washington, DC.
Natural Resources Canada. (2007). *Arctic oil and gas* [factsheet]. http://www.gac.ca/PopularGeoscience/factsheets/ArcticOilandGas_e.pdf. Accessed 8 Dec 2018.
NEB (National Energy Board Canada). (2011). *Spill response gap study for the Canadian Beaufort Sea and the Canadian Davis Strait*. https://apps.neb-one.gc.ca/REGDOCS/Item/Filing/A30372. Accessed 8 Dec 2018.
NEB (National Energy Board Canada). (2014). *Filing requirements for offshore drilling in the Canadian Arctic*. https://www.neb-one.gc.ca/bts/ctrg/gnthr/rctcrvwflngrqrmnt/cncrdnc/2015-01cncrdnc-eng.html. Accessed 8 Dec 2018.
NEB (National Energy Board Canada). (2015). NEB's role in Arctic offshore seismic exploration. https://www.neb-one.gc.ca/nrth/pblctn/nbrlrctcssmcxprtnfs-eng.html. Accessed 8 Dec 2018.
NEB (National Energy Board Canada). (2016). *The NEB's lifecycle approach to protecting the environment*. https://www.neb-one.gc.ca/sftnvrnmnt/nvrnmnt/lfcclpprch/index-eng.html. Accessed 8 Dec 2018.
NEB (National Energy Board Canada). (2018). *COGOA map*. https://www.neb-one.gc.ca/nrth/dvltn/mg/mp01xl-eng.jpg. Accessed 8 Dec 2018.
Nelleman, C., Kullerud, L., & Vistnes, I., et al. (2001). *GLOBIO: Global methodology for mapping human impacts on the biosphere*. https://www.globio.info/downloads/218/globioreportlowres.pdf. Accessed 8 Dec 2018.
Nilsen, T. (2016). Groundbreaking lawsuit filed against Norway over Arctic oil drilling. *The Independent Barents Observer*. http://thebarentsobserver.com/en/ecology/2016/10/groundbreaking-lawsuit-field-against-norway-over-arctic-oil-drilling. Accessed 8 Dec 2018.
Nilsen, T. (2017). Goliat Oil: On-off-on-off-on-off. *The Independent Barents Observer*. https://thebarentsobserver.com/en/industry-and-energy/2017/02/goliat-oil. Accessed 8 Dec 2018.
NMFA (Norwegian Ministry of Foreign Affairs). (2014). *Norway's Arctic policy: Creating value, managing resources, confronting climate change and fostering knowledge. Developments in the Arctic concern us all*. https://www.regjeringen.no/globalassets/departementene/ud/vedlegg/nord/nordkloden_en.pdf. Accessed 8 Dec 2018.
Nordtveit, E. (2015). Regulation of the Norwegian upstream petroleum sector. In T. Hunter (Ed.), *Regulation of the upstream petroleum sector: A comparative study of licensing and concession systems* (pp. 132–158). Cheltenham/Northampton: Edward Elgar.
Norwegian Government. (2006). *Barents 2020: A tool for a forward-looking high north policy*. https://www.regjeringen.no/globalassets/upload/ud/vedlegg/barents2020e.pdf. Accessed 8 Dec 2018.
Norwegian Ministry of Environment. (1996). *Environmental policy for a sustainable development – Joint efforts for the future* [Storting White Paper 58].
Norwegian Ministry of Environment. (2003). *On the petroleum industry* [Storting White Paper 38].
Norwegian Ministry of Environment. (2005). *Integrated management of the marine environment of the Barents Sea and the sea areas off the Lofoten Islands* [Storting Report 8].
Norwegian Ministry of Environment. (2008). *Integrated management of the marine environment of the Norwegian Sea* [Report 37 to the Storting].
Norwegian Ministry of Environment. (2010). *An industry for the future: Norway's petroleum activities* [Storting White Paper 28].

Norwegian Ministry of Environment. (2012). *Integrated management of the marine environment of the North Sea and Skagerrak* [Storting White Paper 37].

Norwegian Ministry of Environment. (2014). *Update of the integrated management plan for the Barents Sea–Lofoten area including an update of the delimitation of the marginal ice zone* [Storting White Paper 20].

NVPH (Native Village of Point Hope) v Jewell. (2014). 44 ELR 20016 9th Circ.

NVPH (Native Village of Point Hope) v Salazar. (2010). WL 2943120 D Alaska.

Offshore Energy Today. (2018). *Statoil takes next step in Barents Sea project with Investment of over $642M*. https://www.offshoreenergytoday.com/statoil-takes-next-step-in-barents-sea-project-with-investment-of-over-642m/. Accessed 8 Dec 2018.

Olsen, E., Gjøsæter, H., Røttingen, I., et al. (2007). The Norwegian ecosystem-based management plan for the Barents Sea. *ICES Journal of Marine Science, 64*, 599–602.

PAME (Protection of the Arctic Marine Environment). (2009). *Arctic offshore oil and gas guidelines*. https://oaarchive.arctic-council.org/handle/11374/62. Accessed 8 Dec 2018.

PAME (Protection of the Arctic Marine Environment). (2014). Systems safety management and safety culture: Avoiding major disasters in Arctic offshore oil and gas operations. https://oaarchive.arctic-council.org/handle/11374/418. Accessed 8 Dec 2018.

Payne, J. R., McNabb, G. D., Jr., & Clayton, J. R., Jr. (1990). Oil weathering behavior in Arctic environments. In E. Sakshaug, C. C. E. Hopkins, & N. A. Øritsland (Eds.), *Proceedings of the pro Mare symposium on polar marine ecology, Trondheim, 12–16 May 1990*.

Pelaudeix, C. (2017). Governance of offshore hydrocarbon activities in the Arctic and energy policies: A comparative approach between Norway, Canada and Greenland/Denmark. In C. Pelaudeix & E. M. Basse (Eds.), *Governance of Arctic offshore oil and gas*. Abingdon: Routledge.

Pelaudeix, C., & Basse, E. M. (Eds.). (2017). *Governance of Arctic offshore oil and gas*. Abingdon: Routledge.

Peterson, C. H., Rice, S. D., Short, J. W., et al. (2003). Long-term ecosystem response to the Exxon Valdez oil spill. *Science, 302*(5653), 2082–2086.

Peterson, C. H., Anderson, S. S., Cherr, G. N., et al. (2012). A tale of two spills: Novel Science and policy implications of an emerging new oil spill model. *Bioscience, 62*(5), 461–469. https://doi.org/10.1525/bio.2012.62.5.7.

Petroleum Act Norway. (1996). Act no 72 relating to petroleum activities.

Petroleum Regulations Norway. (1997). Regulations to act relating to petroleum activities.

Pew Charitable Trusts. (2013). *Arctic standards: Recommendations on oil spill prevention, response, and safety*. https://www.pewtrusts.org/en/research-and-analysis/reports/2013/09/23/arctic-standards-recommendations-on-oil-spill-prevention-response-and-safety. Accessed 8 Dec 2018.

Pew Charitable Trusts. (2015). *A letter to Janice Schneider*. https://www.pewtrusts.org/-/media/assets/2015/05/pew_comments_doi_arcticstandards_27may2015.pdf?la=en&hash=94E16769B86AACA442967BFE8F67CCA40DA8136D. Accessed 8 Dec 2018.

Pew Charitable Trusts. (2016). *Pew: New offshore drilling standards help protect U.S. Arctic Ocean*. https://www.pewtrusts.org/en/about/news-room/press-releases-and-statements/2016/07/07/pew-new-offshore-drilling-standards-help-protect-us-arctic-ocean. Accessed 8 Dec 2018.

Pew Environmental Group. (2011). *Becoming arctic ready: Policy recommendations for reforming Canada's approach to licensing and regulating offshore oil and gas in the Arctic*. https://www.pewtrusts.org/en/projects/protecting-life-in-the-arctic/arctic-science/arctic-science-initiatives/becoming-arctic-ready. Accessed 8 Dec 2018.

Pollution Control Act Norway. Act no 6 concerning protection against pollution and concerning waste.

POTUS (President of the United States). (2013). *National strategy for the Arctic region*. https://obamawhitehouse.archives.gov/sites/default/files/docs/nat_arctic_strategy.pdf. Accessed 8 Dec 2018.

PSA (Petroleum Safety Authority Norway). (2017). *Audit of Gøliat*. http://www.ptil.no/audit-reports/audit-of-goliat-article12764-889.html. Accessed 8 Dec 2018.
Putin, V. (2013). Strategy for the Russian Arctic zone development and ensuring state security until 2020.
Rosprirodnadzor. (2013). *No 34077 safety rules in oil and gas industry*.
Rosprirodnadzor. (2014). *No 101 safety rules for offshore oil and gas facilities*.
Russian Government. (2009). Order No 1715-r 'On the Russian energy strategy for the period up to 2020.
Russian Government. (2014). Decree No 366 'On the approval of the state programme "On the socio-economic development of the Russian arctic zone until 2020"'.
Secretary of Interior. (2013). *Review of Shell's 2012 Alaska offshore oil and gas exploration program*. https://www.doi.gov/news/pressreleases/department-of-the-interior-releases-assessment-of-shells-2012-arctic-operations. Accessed 8 Dec 2018.
Shapovalova, D. (2019). International governance of oil spills from upstream petroleum activities in the Arctic: Response over prevention? *International Journal of Marine and Coastal Law, 34*(2), 1–30.
Shapovalova, D., & Stephen, K. (2019). No race for the Arctic? Examination of interconnections between legal regimes for offshore petroleum licensing and level of industry activity. *Energy Policy, 129*, 907–917.
Shell. (2015). Shell updates on Alaska exploration. http://www.shell.com/media/news-and-media-releases/2015/shell-updates-on-alaska-exploration.html. Accessed 8 Dec 2018.
Shell. (2016). *Shell contributes offshore rights near Lancaster Sound to the nature conservancy of Canada*. https://www.shell.ca/en_ca/media/news-and-media-releases/news-releases-2016/nature-conservancy-of-canada.html. Accessed 8 Dec 2018.
Sidortsov, R. (2017). The Russian offshore oil and gas regime: When tight control means less order. In C. Pelaudeix & E. M. Basse (Eds.), *Governance of Arctic offshore oil and gas*. Abingdon: Routledge.
Staalesen, A. (2018). They found one of Russia's biggest Arctic oil fields, but now abandon it. *The Independent Barents Observer*. https://thebarentsobserver.com/en/industry-and-energy/2018/03/they-found-one-russias-biggest-offshore-arctic-oil-field-now-abandon-it. Accessed 8 Dec 2018.
Statistics Norway. (2016). *This is Norway 2016: What the figures say*. https://www.ssb.no/en/befolkning/artikler-og-publikasjoner/this-is-norway-2016. Accessed 8 Dec 2018.
Steiner, R. (2010). Risks to Arctic ecosystems. *The Circle, 3*, 13–16.
Subsoil Law Russia – Federal Law 'On Subsoil' No 2395-1 (21 February 1992).
Sutton, I. (2014). *Offshore safety management* (2nd ed.). Oxford: Elsevier.
The Constitution of the Russian Federation (12 December 1993).
The Local Norway. (2015). *Norway's Statoil pulls out of Alaskan Arctic*. https://www.thelocal.no/20151117/norways-statoil-pulls-out-of-alaskan-arctic. Accessed 8 Dec 2018.
The New York Times. (2014). The Wreck of the Kulluk. http://www.nytimes.com/2015/01/04/magazine/the-wreck-of-the-kulluk.html. Accessed 8 Dec 2018.
The Telegraph. (2013). *Cairn prepares to resume oil search off Greenland*. https://www.telegraph.co.uk/finance/newsbysector/energy/oilandgas/10255743/Cairn-prepares-to-resume-oil-search-off-Greenland.html. Accessed 8 Dec 2018.
The White House. (2017). *Remarks by president trump at signing of executive order on an America-first offshore energy strategy*. https://www.whitehouse.gov/briefings-statements/remarks-president-trump-signing-executive-order-america-first-offshore-energy-strategy/. Accessed 8 Dec 2018.
US Arctic OCS Final Rule. (2016). *Oil and gas and sulfur operations on the outer continental shelf*—Requirements for exploratory drilling on the Arctic outer continental shelf 81 FR 46478.
US Arctic Research and Policy Act. (1984). Pub L 98–373, title I, §112, 31 1984, 98 Stat 1248.
US Arctic Research and Policy Act 1984. 15 USC 67 §4101—11.

US DOI. (2015). *Interior department cancels Arctic offshore lease sales*. https://www.doi.gov/pressreleases/interior-department-cancels-arctic-offshore-lease-sales. Accessed 8 Dec 2018.
US National Environmental Policy Act 42. USC §4332(2)(c).
US Outer Continental Shelf Lands Act 43. USC §1331—56.
US Secretarial Order 3299 (19 May 2010).
US Submerged Lands Act (1993) 43 USC § 1301.
US-Canada Joint Arctic Leaders' Statement. (2016). http://pm.gc.ca/eng/news/2016/12/20/united-states-canada-joint-arctic-leaders-statement. Accessed 8 Dec 2018.
USGS (United States Geological Survey). (2009). *Circum-Arctic resource appraisal: Estimates of undiscovered oil and gas north of the Arctic circle*. http://library.arcticportal.org/1554/. Accessed 8 Dec 2018.
Verhaag, M. A. (2002). It is not too late: The need for a comprehensive international treaty to protect the Arctic environment. *The Georgetown Environmental Law Review, 15*, 555–579.
WWF. (2007). *Oil spill response challenges in Arctic waters*. http://assets.panda.org/downloads/nuka_oil_spill_response_report_final_jan_08.pdf. Accessed 8 Dec 2018.
WWF. (2009). *Not so fast: Some progress in spill response, but US still ill-prepared for Arctic offshore development*. https://c402277.ssl.cf1.rackcdn.com/publications/401/files/original/Not_So_Fast_Some_Progress_in_Spill_Response__but_US_Still_Unprepared_for_Arctic_Offshore_Development.pdf?1345754373. Accessed 8 Dec 2018.
WWF. (2014). *Modeling oil spills in the Beaufort Sea exploring the risk: What would happen if oil spills in the Beaufort Sea?* http://awsassets.wwf.ca/downloads/wwf_beaufort_sea_oil_spill_modelling_summary_report.pdf. Accessed 8 Dec 2018.

Part IV
Local Communities

Chapter 14
Increasing Shipping in the Arctic and Local Communities' Engagement: A Case from Longyearbyen on Svalbard

Julia Olsen, Grete K. Hovelsrud, and Bjørn P. Kaltenborn

Abstract Increasing ship traffic in the Arctic has a broad range of impacts on coastal communities' wellbeing and the natural environment. Despite a number of existing national and international efforts to mitigate the risks and secure the benefits of this development, the role of local initiatives and arrangements is still understudied. Focusing on the town of Longyearbyen, situated on the Svalbard Archipelago, this chapter examines the impacts of and responses to the considerable growth in shipping activities comprising marine tourism, cargo (supply), fishing, research and Search and Rescue vessels. Since the settlement's establishment in 1906, Longyearbyen has seen shipping play an important role in the community's development by serving as a vital transport link between the Archipelago and the mainland. The impacts of recent growth in ship traffic, coupled with environmental changes and an ongoing transition from a coal dominated economy toward tourism, research and education, challenge the local capacity to accommodate such growth. The analysis of empirical data indicates that local, bottom-up engagement serves as a support mechanism for institutional response strategies and enables local adaptive capacity. At the same time, community engagement is sensitive to demographic trends that influence the scope and efficiency of actions.

Keywords Shipping · Arctic · Longyearbyen · Local community · Local engagement · Adaptive capacity

J. Olsen (✉) · G. K. Hovelsrud
Nord University, Bodø, Norway
e-mail: julia.olsen@nord.no; grete.hovelsrud@nord.no

B. P. Kaltenborn
Norwegian Institute for Nature Research, Lillehammer, Norway
e-mail: bjorn.kaltenborn@nina.no

E. Pongrácz et al. (eds.), *Arctic Marine Sustainability*,
Springer Polar Sciences, https://doi.org/10.1007/978-3-030-28404-6_14

14.1 Introduction

The Barents area and its adjacent terrestrial areas (including Svalbard and Franz Josef Land) (AMAP 2017), are experiencing multiple changes, including a considerable growth in ship traffic. Europeans and *Pomors*[1] have navigated the Barents Sea for centuries (e.g. Arlov 2003). Recently, reduced sea ice extent and a decrease in the number of days with sea ice cover (Overland et al. 2017; Borch et al. 2016), coupled with an increasing interest in Arctic marine resources and tourism attractions, have increased shipping activities. Currently, the Norwegian and Barents Seas have the highest concentration of Arctic shipping activities (Eguíluz et al. 2016), including all types of vessels operating in Arctic waters (PAME 2009, p. 3). In fact, about 80% of all Arctic shipping crosses Norwegian waters (St. Meld. 31 2015–2016).

With reduced sea ice, new areas of the Northern Barents Sea have become accessible to marine tourism, fishing and research activities. A recent evaluation of future Arctic development shows that with the opening of previously icebound areas, activity levels will continue to increase in those parts of the Arctic (Borch et al. 2016). Further growth may be possible with the emergence of a trans-Arctic shipping route across the North Pole, which, according to Smith and Stephenson (2013), may occur by mid-century (see also Farré Buixadé et al. 2014).

At the same time, these waters challenge maritime safety efforts due to a lack of supportive infrastructure, long travel distances and severe weather conditions (Marchenko et al. 2016). Increasing shipping activities require new safety and environmental guidelines and a strengthening of Search and Rescue (SAR) and emergency preparedness services, which are necessary to reduce the risk of shipping operations and to avoid loss of life, health and environmental damage. Several important steps have already been taken to address these issues, including a sectorial agreement on SAR within the Arctic Council. The Arctic Search and Rescue Agreement delimits the Arctic region between all the circumpolar states (Arctic Council 2011; ratified in 2013). As a result, significant improvements were made to the SAR-system within the Barents area, including the Svalbard Archipelago, which plays a key role in SAR operations for the Western Sector of the Arctic (Marchenko et al. 2016).

Moreover, the impacts of shipping development will be felt in the Arctic port towns and local coastal communities that provide supportive infrastructure and host increasing numbers of visitors (e.g. Davydov and Mikhailova 2011; Olsen and Nenasheva 2018; Stewart et al. 2015). However, despite the general growth in ship traffic across the Arctic (e.g. Dawson et al. 2018; Borch et al. 2016) and the attention given to such activities, knowledge about the local implications of, and responses to this growth remain scarce. Little is known about whether Arctic communities in the Barents area, which was historically navigable, can benefit from

[1] Russian settlers living by the White Sea.

these changes while limiting the threats to their wellbeing, local environment and natural resources.

To increase the available knowledge on this topic, this study investigates whether and how shipping activities influence the adaptive capacity of one Arctic community, Longyearbyen, a populace that also represents the administrative center on Svalbard. Based on 36 qualitative interviews with local residents, who are engaged with shipping development and exist within a framework of adaptation and adaptive capacity, we identify (1) the impacts of different types of shipping, including marine tourism; and (2) the aspects of a community's adaptive capacity that emerge in response to such impacts.

14.2 Background and Context

14.2.1 Shipping Perspectives for the Svalbard Archipelago

Svalbard marks the northernmost part of Norway, located between 74°N and 81°N in the Arctic Ocean (Fig. 14.1). However, compared to other areas at the same latitude, Svalbard's climate is surprisingly mild due to the presence of the Gulf Stream, a warm Atlantic Ocean current. Moreover, climate change has increased ocean and air temperatures in the Barents Sea and in adjacent areas, impacting hydrological regimes (e.g. Vikhamar-Schuler et al. 2016). Sea ice in the Barents Sea has undergone dramatic changes (MOSJ 2018), noticeably decreasing in both thickness and extent since 1979 (Vikhamar-Schuler et al. 2016). This reduction will likely affect the distribution of ship traffic in the Barents area.

The density of ship traffic near Svalbard is much lower than in the Norwegian Sea and the southern part of the Barents Sea (St. Meld. 32 2015–2016). The traffic has seasonal variations and is dominated by fishing, marine tourism, research and cargo activities (Borch et al. 2016; The Governor of Svalbard 2016). Despite the intensive fishing activities near Svalbard, coupled with the increasing biomass of boreal fish species (Misund et al. 2016), there are no landing or processing facilities for fish or seafood on Svalbard. This is due to the lack of specific regulations for the Svalbard Archipelago, which differs from mainland Norway (e.g. Marine Resources Act; Food Act) (St. Meld. 32 2015–2016). As a result, seafood products are primarily delivered from the mainland. Given the growing possibility of an interest in harvesting sea food, the Norwegian government has considered facilitating the development of seafood on the Archipelago to meet local food and tourism needs (St. Meld. 32 2015–2016, p. 92).

The growth in marine tourism is noticeable in both the number of vessels and in the volume of passengers. Despite the 150-year-long marine tourism history on Svalbard (Nyseth and Viken 2015), the development trends show that Svalbard (and the port of Longyearbyen) is approached by ever-larger cruise ships with a capacity for more than 5000 passengers (Fig. 14.2), but also by a fast-growing

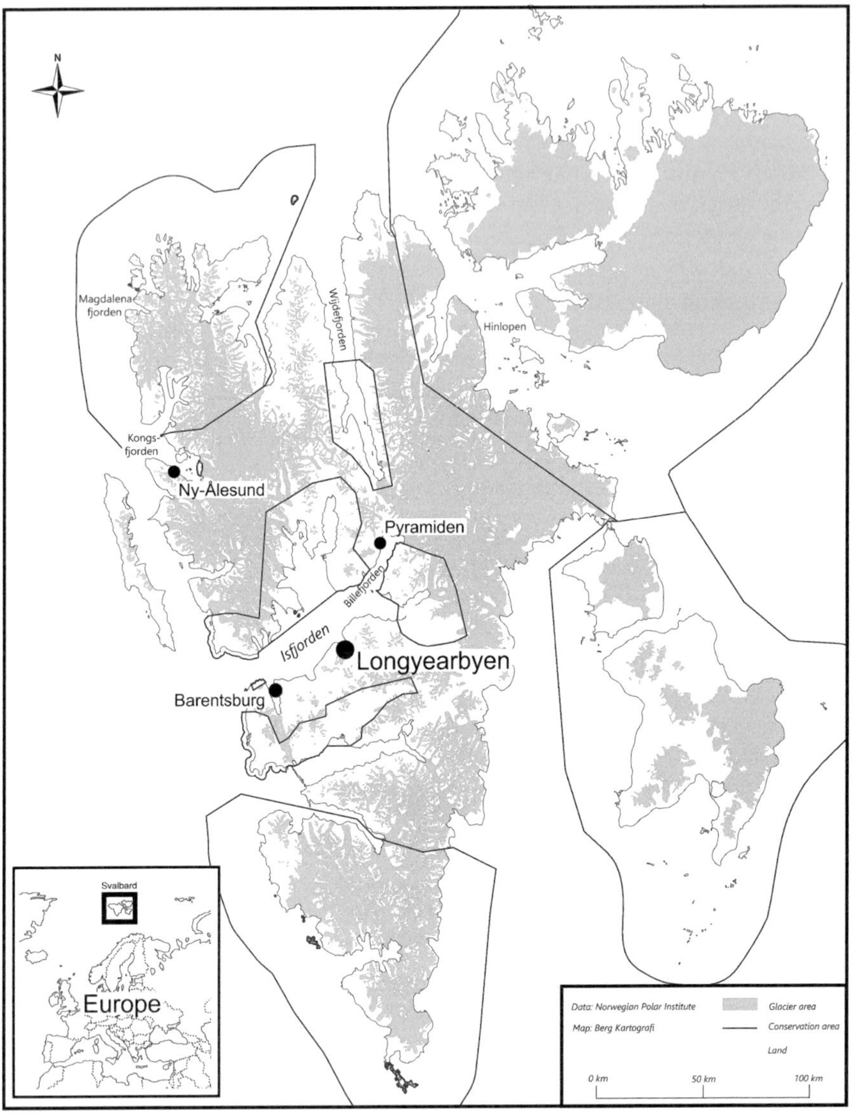

Fig. 14.1 A map of Svalbard

pleasure craft sector (Table 14.1). Moreover, the extension of the navigation season has affected the distribution of vessels in space and time, including increases in fishing vessels and cruise ships sailing northward towards the ice's edge.

A recent estimate of shipping development around Svalbard indicates that the level of activity will continue to increase into 2025 (Borch et al. 2016) and beyond (DNV-GL 2014). Due to its geographical location, Svalbard has no logistical

Fig. 14.2 MSC Preziosa, with over 5000 passengers, arrives at the port of Longyearbyen, *Bykaia* (Town Pier). August 2017. (Photo credit: Julia Olsen)

Table 14.1 Population and shipping trends in Longyearbyen

Year	2000	2002	2004	2006	2008	2010	2012	2014	2016
Population in Longyearbyen and Ny-Ålesund[a]	N/A	1570	1581	1721	1821	2052	2115	2100	2152
Number of passengers	15,899	18,757	21,837	37,085	38,569	40,123	55,091	54,808	75,201
Number of ship calls, including	**166**	**505**	**490**	**799**	**771**	**814**	**812**	**1178**	**1542**
Tourism (passenger) vessels[b]	78	345	374	550	550	566	558	806	1099
Fishing vessels	50	43	20	27	21	8	15	30	32
Cargo vessels (incl. community supply)	5	29	20	78	54	60	52	67	51
Research	28	47	23	64	41	92	108	70	84
Coast Guard and the Governor's vessel	5	41	45	68	89	74	72	74	110
Pilot[c]	N/A	N/A	N/A	N/A	N/A	N/A	N/A	96	142

[a]Statistics Norway report the population for both Longyeabyen and Ny-Ålesund. Ny-Ålesund is an international research community with 43 residents, as of 2015. The statistics are not available (N/A) for the year 2000

[b]The number of passenger vessels includes overseas cruise vessels, expedition cruises, day-trip cruises and pleasure crafts. The last two groups stand for the major portion of number of ship calls (approx. 50–80%)

[c]The pilot boat started operating in 2014

Sources: Port of Longyearbyen (2018) and SSB (2016)

function for shipping operations along the Northeast Passage (NEP). According to Smith and Stephenson (2013), the prognosis for an ice-free Arctic Ocean by the mid-century will place the archipelago on the Trans-Polar Route, the new Arctic route between East and West (see also Farré Buixadé et al. 2014). This area is characterized by a lack of supportive infrastructure and services, long travel distances, severe and unpredictable weather conditions (Marchenko et al. 2016) and the long polar night in winter. In the event of accidents, response times may vary from a few hours to a few days (The Governor of Svalbard 2016).

The Norwegian government has applied several local measures to reduce the risk of unwanted events and to avoid loss of life and environmental damage. They entail strengthening emergency preparedness, developing maritime services around the archipelago (e.g. Marine Automatic Identification System (AIS)-stations), and issuing regulations. For example, since 2012, shipping has been locally regulated by restrictions on vessel type and fuel use (particularly directed towards vessels sailing in East Svalbard), as well as by compulsory pilotage services for certain types of crafts (Borch et al. 2016).

Given current shipping trends and future perspectives, Longyearbyen represents a potentially crucial port for shipping infrastructure and a SAR base. Hence, for this study, we have examined current local perspectives and impacts of shipping growth to understand whether and how the community responds and adapts to them.

14.2.2 Case: Longyearbyen, Svalbard

Longyearbyen is the world's northernmost town at 78°N (SSB 2016) and is the hub of administration, transportation, and business on the Svalbard Archipelago. It comprises the Governor's office, the University Centre on Svalbard (UNIS), diverse services and industries (Viken 2008, p. 139) and, as noted above, has a major deep-sea port with supportive infrastructure and SAR facilities. Longyearbyen is usually described as a rotation community of 2200 (Table 14.1) inhabitants from 46 nations with a 7-year average residence time (SSB 2016). This has major implications for local demographics and the community viability of Longyearbyen.

The settlement was established in 1906 as a "company town" (Fig. 14.3) where the Norwegian coal mining company, "*Store Norske Spitsbergen Kulkompani*," historically controlled most aspects of community life. With the onset of uncertainty about the future of coal production in the late 1980s (e.g. Arlov 2003), Longyearbyen began a period of transition toward tourism, education and research. A major reduction in coal mining activities occurred in 2017 due to the closure of the Svea Mine (e.g. Pedersen 2017). This politically-guided transition is evident in the port of Longyearbyen, as mining-related shipping is steadily decreasing while research and tourism-related shipping activities consistently increase (See Table 14.1).

Longyearbyen's geographical location, remoteness and logistic complexity amplify its dependence on ship traffic for socio-economic development. Since the establishment of the settlement until the opening of the airport in 1975 (Fig. 14.3.

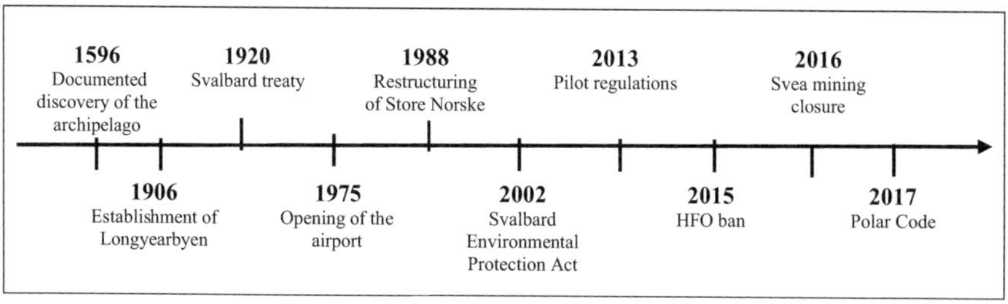

Fig. 14.3 Timeline of historical events related to community and shipping development

Timeline), marine vessels have been the primary transportation link with the mainland, as well as the area's main source of supplies and mobility. Today, shipping services remain crucial for local activities and development, though regular and stable year-round airline connections have substituted for some of these services. Until the previous decade, the Longyearbyen community was accustomed to marking a period between "the last and the first boat," a span of time in which the community remained isolated through winter once sea ice created a natural barrier to shipping.

Despite yearly variations, gradual reductions in the sea ice of Isfjorden outside Longyearbyen (e.g. Muckenhuber et al. 2016, Teigen et al. 2011) has resulted in recent year-round town accessibility for supply vessels (see Bring 2016, for example). The extension of the navigation season has also become more noticeable in the port of Longyearbyen, where some expeditions and day-long cruises begin their seasons earlier and earlier in the spring (in 2017, the navigation season for these vessels commenced as early as March), though most traffic still occurs in the summer months. The increase in shipping associated with tourism is also apparent in the number of tourists and crew members, which has increased fivefold since the beginning of the century (Table 14.1). In addition to tourism vessels, the port of Longyearbyen is used for community supply, fishing and research vessels. These new trends in shipping distributions present a potential challenge for SAR and have resulted in the extended presence of the Governor's SAR vessel from 6 to 9 months (St. Meld. 32 2015–2016) in addition to the constant presence of the Coast Guard. With the new pilotage regulations of 2012, growth in the number of pilot boat calls has been linked to increased shipping in the port of Longyearbyen (Table 14.1).

Longyearbyen has four main docking facilities: Old Pier (*Gamlekaia*), Coal Pier (*Kullkaia*), Town Pier (*Bykaia*) and Tourist Pier (*Turistkaia*), the last of which is a floating dock for small passenger vessels. *Bykaia* is currently used for marine tourism, fishing, research vessels, cargo ships and the Coast Guard (Multiconsult 2014). Despite several docking options, the increasing number of vessels challenges port capacity because the relatively high volume of vessels arrives during the short summer season (St. Meld. 32 2015–2016). Further development of infrastructure and facilities is one major task for the Norwegian government on the archipelago (St. Meld. 32 2015–2016).

14.3 Theoretical and Conceptual Framework

14.3.1 Conceptualizing Local Community

The concept of "local community" is described, identified, approached and defined in multiple ways in the literature. In this chapter, we align ourselves with Haugen and Villa (2016) and Aarsæther's (2014) definition of community as geographically bounded, where physical proximity facilitates interactions. "Community" includes shared perceptions of challenges and duties, experiences, and tasks, which all contribute to shaping local institutions (administrative or voluntary organizations), and the interactions between people who feel attached to a place or an area (Haugen and Villa 2016, p. 18).

Place attachment has been found to be a driving force in addressing community concerns, which in turn may enable adaptive responses (Akama et al. 2014; Hovelsrud et al. 2018). Place attachment may also be a strong motivator for living with risk of infrastructural disruptions, such as avalanches and other weather-related risks (Hovelsrud et al. 2018). Earlier research has shown that the level of place attachment in Svalbard is a predictor of how serious Longyearbyen residents consider environmental impacts (Kaltenborn 1998).

Moreover, local social relations will be affected by a number of multi-scalar processes and changes in political, economic, cultural and other systems (Haugen and Villa 2016, p. 21). Those changes are particularly noticeable in 'company-town' communities that are more dependent on international markets and external labour (Haugen and Villa 2016, p. 28; Valestrand 2016). Following this section's discussion, we will address community settings and whether they shape the determinants of local adaptive capacity and responses to increasing ship traffic.

14.3.2 Adaptation and Adaptive Capacity

To understand how Longyearbyen responds to the current impacts of increased shipping, we align ourselves with the literature on human dimensions of Arctic change and employ the concepts of adaptation and adaptive capacity to describe the strategies and activities used to address current change and/or plan for changes (e.g. Hovelsrud and Smit 2010; Smit and Wandel 2006). A bottom-up approach is usually applied to study communities' adaptive capacity in order to access community perspectives and to empirically identify how a particular community experiences changing conditions (Smit and Wandel 2006; see also Hovelsrud and Smit 2010; Keskitalo et al. 2011). Focusing on the local level, this study investigates community characteristics in order to understand which dimensions of adaptive capacity manifest in the context of increasing shipping activities in Longyearbyen. Adaptation is considered within the context of multiple stressors or factors acknowledging that climate change is rarely the only factor to which communities adapt (e.g. Leichenko and O'Brien 2008).

It is increasingly recognized in the climate adaptation literature that adaptation is a process taking place along multiple dimensions to address cumulative and interacting consequences of changing environmental, political and socio-economic conditions in a community (e.g. Hovelsrud and Smit 2010). Such processes include barriers, limits and options that emerge cross-scale and involve multiple sectors (e.g. municipalities, tourism, energy), and actors (e.g. businesses, individuals and policy makers) (AMAP 2017, pp. 219–252). These complex adaptation processes are context-dependent and vary within and between communities. The potential for conflict is clear; adaptation for one individual, business or sector may create challenges for others. In our case area, this manifests in differing interests and responses to increased shipping; a local business owner might celebrate higher rates of activity while a local individual might find it challenging to contend with hordes of cruise ship tourists in the town. Their respective adaptive responses and strategies will also vary. It is, therefore, imperative to understand the local context in which adaptation processes take place, including the local residents' perceptions and responses.

Adaptation, as an act, response or strategy, is closely connected to the concept of adaptive capacity, a dynamic, case-specific attribute that characterizes a community's ability to adapt to multiple changes (e.g. Smit and Wandel 2006, Smit et al. 2010). Brown and Westaway (2011) emphasize adaptive capacity's link to adaptation, describing the concept as "*the precondition necessary to enable adaptation to take place.* [I]t *is a latent characteristic that must be activated to effect adaptation*," (Brown and Westaway 2011). Adaptive capacity comprises several determinants, usually grouped as subjective (e.g. values, perception of risk, place attachment) and objective (resources, governance, income) dimensions (e.g. Bay-Larsen and Hovelsrud 2017; Wolf et al. 2013), or as endogenous (local, individual) and exogenous (governance, decisions-beyond-individual-control) factors (Wesche and Armitage 2010).

The determinants are specific to culture and place (Hovelsrud and Smit 2010) and to scale (Brown and Westaway 2011). Adaptive capacity determinants are shaped by processes and interactions across scales and dimensions (cf. Wesche and Armitage 2010, p. 186) and will differ between communities (Smit and Wandel 2006, p. 287). Each individual determinant and their interconnections shape local adaptive capacity (e.g. Olsen and Nenasheva 2018). When combined and activated, these dimensions enable adaptive capacity (see also Bay-Larsen and Hovelsrud 2017).

14.4 Methods

This study is guided by a case study research design that investigates a contemporary social phenomenon in depth (Yin 2014). The primary source data for this study was generated through interviews. We began data collection by reviewing secondary sources on shipping trends in the Svalbard area in order to get an overview of

Table 14.2 Description of the types of the interviews and participants (interviewees) in Longyearbyen

Type of interview	Residents involvement in the study
19 personal semi structured interviews with pre-defined topics and questions	Six residents involved with marine cruise development
	Five residents involved with the development of harbor facilities and other types of shipping-related activities
	Four residents involved in decision-making
	Two residents engaged with NGOs
	One seasonal worker
17 personal unstructured interviews with pre-defined topics only	Two residents partly employed in the summer tourism industry
	Six residents involved in local services that serve tourism needs (stores, museums and cafés)
	Five residents employed in the shipping industry
	Four residents with practical and/or historical knowledge on shipping development

the range of such activities. This information was then used as a basis for a research protocol and an interview guide, as well as to identify potential interviewees in Longyearbyen.

The secondary data was generated from a scientific literature review, document analysis (e.g. White papers, statistical data on Svalbard), media review (local newspapers, web pages of involved organizations) and a review of relevant maps. Review of relevant maps provided a useful source of data to gain an overview of shipping routes, historical sea ice extension trends and geographical locations and sites on the Archipelago (see for example TopoSvalbard, Marinetrafic, Polarview). Finally, social media was included in order to understand inhabitants' perceptions and attitudes toward the growth in shipping traffic, particularly in marine tourism.

Primary data was generated during fieldwork from interviews with local residents (Table 14.2). In total, the first author interviewed 36 residents through 19 personal semi-structured interviews, and 17 unstructured interviews. As suggested by van Bets et al. (2017), a marine community model guided our selection of the interviewees. According to this model, a marine community comprises a user community (industrial stakeholders, researchers, port authorities and local inhabitants) and a policy community (cross-scale institutional stakeholders). A diverse range of stakeholders were interviewed, but our approach differs from van Bets et al. (2017) in that our study was designed to interview the local stakeholders, the residents of Longyearbyen, involved and engaged in shipping and its related operations on Svalbard.

The majority of interviewees were selected during the secondary data collection process (during media and social media review). The interviewees were contacted several weeks prior the fieldwork in order to schedule the personal interviews and to provide background information about the project. In addition, a snowball technique was applied during the fieldwork (Blaikie 2010, p. 179), i.e. we asked our interviewees to identify other potential stakeholders who could participate in the study.

To ensure access to a broad range of residents, the fieldwork took place on two occasions: prior to the summer navigation season (in April 2017), and during the summer navigation season (in July–August 2017) when port turnover was at its highest. During the springtime, the research topic was discussed with residents that are most often absent or busy in the summer but are directly involved in shipping operations, including representatives from the marine tourism industry, cargo services, pilot services, Search and Rescue, local decision-making organizations, NGOs, unions and other relevant representatives within the community. In April 2017, the port of Longyearbyen was characterized by low turn-over while it prepared for the summer shipping season of May to September. However, the navigation season for the day-long cruises and a few expedition vessels had already started in March/April. During the summer season, seasonal workers, local guides, and port employees were interviewed.

Two interview guides were used during the fieldwork. The first was semi-structured with a set of open-ended questions. This interview guide was revised during and after the fieldwork in April to include more case-specific questions, which in turn were asked during the summer season. The interview guide contained open-ended questions under the following categories: changes in social and ecological systems, changes in shipping patterns (season, boat types, number of visitors, supporting infrastructure), main impacts of shipping activities, organization of decision-making systems, opportunities for future development. At the end of each interview, we invited the interviewees to provide additional comments or feedback on the project. The second guide was used to cover related local aspects of shipping development and contained topics such as perceptions and attitudes toward the growing number of vessels in the Svalbard area and the features of navigation in Arctic waters.

Almost all interviews were recorded, and detailed notes were taken during unstructured interviews when the option to record was not available. The data was collected in Norwegian, English and Russian. The interview data was thematically analyzed in NVivo, a software program (Bazeley and Jackson 2013). A set of predefined and emerging themes (codes) that correspond with the interview guide, collected data and theoretical basis were used for data analysis. To follow anonymity conventions, we employ a coding system (L1–L36) when citing our interviewees in this chapter.

14.5 Findings: Community Engagement and Adaptive Capacity

14.5.1 Increased Shipping: Diversity, Impacts and Responses

Our empirical data show that the locally identified impacts of shipping activities vary depending on the type of activity and its seasonality. The increasing number of port calls is challenging for the harbor infrastructure, SAR and emergency preparedness, town facilities, local services and community livelihoods. Local value-creation is a critical component of evaluating positive impacts of such growth. For those involved in local infrastructure and port development, it is "*important to meet the demands of the shipping industries we have today; mainly tourism, but also Search and Rescue, such as the Coast Guard and the Governor's vessel and Norway's new research vessel, the Kronprins Haakon. The large ships require a lot of space and capacity,*" (L22, also L25). At the same time, local decision-makers are more concerned with the impacts on environment and navigation safety: "*We get less ice, meaning tourism, fishing and transportation shipping will increase. For us, the concerns are twofold: environment and safety,*" (L12, also L4).

The main impacts associated with increasing marine tourism in Longyearbyen and Svalbard waters are over-crowding, pollution, and visitors' inappropriate behavior on sites. Despite the extension of the navigation season for marine tourism activities, the local impact of such growth is felt mostly during the summer navigation period, when the community simultaneously hosts tourists and crewmembers from overseas expeditions, day-long cruises and pleasure crafts (L10). Several residents described the increasing number of marine tourism visitors as follows: "*The community of Longyearbyen has little capacity; few facilities for passengers. They are not suited for such a large number of tourists,*" (L33, also L19). Another interviewee suggested that local infrastructural needs should be taken into account when allocating resources for tourism development, such as signs, sidewalks and other harbor facilities: "*...there is not a lot of infrastructure for tourism...but should we use money to build infrastructure for the tourist industry or for local needs, for example, a school?*" (L18).

While the majority of marine tourism vessels operating in Svalbard waters visit the port of Longyearbyen, only a few fishing vessels approach the town (Table 14.1). This is due to the fact that there are no fish landing facilities on the Archipelago. Those who use the port are usually trying to avoid bad weather conditions in the open ocean and/or need medical assistance and services for their vessels (L22). However, despite the small volume of vessels, there are a number of potential impacts of fishing in Svalbard waters that are felt and identified locally. Unlike marine tourism, fishing occurs year-around in areas with little or no connection to the community of Longyearbyen. These activities cause concern among the local population because they provide little-to-no value creation in the community itself while also polluting the environment with marine litter (L7, L11, L34) (Table 14.3).

The number of calls by supply vessels is directly related to local economic development (including construction work and/or supply for a particular industry) and varies from year to year. From a local development perspective, the extension of the navigation season toward year-round accessibility is a positive change, as it covers community needs for food and goods deliveries, as well as asphalt, construction materials, and fuel. No impacts have been identified locally from research-related and SAR-vessels, with the exception of increasing call volume, which challenges the port's capacity. The presence of a Coast Guard vessel is usually described as a response to the increasing shipping activities in Svalbard waters but is not usually correlated with any specific impacts. For those

Table 14.3 Locally identified effects and impacts to increasing shipping (L1–L36)

Type of shipping	Effects	Positive impacts	Negative impacts
General shipping	Need for development of harbor infrastructure and town facilities	Co-beneficial for local needs Extends the ability to accommodate several vessels	Conflict with cultural and natural heritage Challenges current infrastructure capacity
	Constant improvement of local preparedness and search and rescue	Cooperation between local population and voluntary organizations Implementation of proactive institutional measures Development of navigation services to avoid accidents	Expensive High reliance on SAR facilities, not all of which are well-informed on response time and difficulties of SAR operations Lack of SAR facilities for bigger boats
	Marine pollution and emission/ disturbance of marine species	New regulations decrease negative impacts, but also limit visiting opportunities Shift toward new types of fuel New types of vessels; constant improvement to reduce environmental impacts	Marine litter, emission and water pollution threaten vulnerable Arctic nature and wildlife Ballast water may precipitate the introduction of new species
	Increase in number of community visitors (crewmembers and tourists)	Increasing demand for more seasonal workers, especially in the tourism industry New economic and employment opportunities Focus on sustainable development Local value creation: contribution to the "Svalbard environmental protection fund" (environmental tax)	Overcrowding Threatens local environment Affects community's lifestyle Engenders fear that the area will become a mass-tourism destination Inappropriate behavior of some community visitors

(continued)

Table 14.3 (continued)

Type of shipping	Effects	Positive impacts	Negative impacts
Marine tourism, including			
Overseas cruises	Increase in number of community visitors (crew members and tourists) Direct connection between time spent in the town and local value creation (more time in the harbor = higher value creation locally and less people pollution)	Increasing number of tourism-related facilities/activities/product variety at stores that are used by locals Local engagement in hosting activities before (cleaning the town) and during the season (guides, bus drivers, helping in the stores) Established network between local actors who provide services for cruise vessels	Challenges existing infrastructure, town facilities and available human resources Concern for losing a local sense of wilderness and becoming a new destination of "mass tourism" Inappropriate behavior among some visitors (e.g. entering private houses, taking pictures of residents, blocking driveways) Low value-creation compared to other types of tourism For some stores, overseas cruise tourists are unprofitable; they spend less locally than other types of tourists
Expedition cruises	Increase in number of community visitors (crewmembers and tourists) Increase in focus on SAR in the Arctic Increase in focus on environmental impacts	Greater contributions to local value creation (visitors stay in hotels/spend more on clothing) Actively limits the impact on the natural environment (increased awareness about Arctic nature; tourists are informed on visited sites and participate in beach cleaning) Actively involved and part of SAR, preparedness	Emission and pollution due to longer sailings patterns May disturb wildlife in fjords Social wear and tear (*Sosialslitasje*), i.e. marine tourists that experience wild nature and isolation can encounter other vessels and tourist groups
Day-trip cruises	Became one of the main sources of local mobility in Isfjorden Offers trips and meets tourism demands outside the high tourism season (early spring-late autumn)	Increases awareness about Arctic nature Offers cheaper trips to locals and to students Increases local mobility, especially when it is impossible to drive scooters	May disturb wildlife in fjords, especially early in the season when sea ice is still present

Pleasure crafts (excluding day-trip vessels for under 12 passengers)	Fast-growing sector with a lack of regulations	Increases awareness about Arctic nature Participate in beach cleaning activities and in research projects	Not all vessels are equipped for severe Arctic conditions Difficult to monitor vessel activities due to absence of tracking (not all have AIS) Cases of marine species disturbance
Other types of shipping			
Community supply	Year-round supply services Supplements tourism industry High dependence on weather conditions	Improves food security Cheaper delivery services compared to air transportation Major supplier for marine tourism industry (day-trip cruises) Useful for sending garbage from the archipelago to the mainland	Monopolized service leads to higher prices Goods can be damaged or lost Can be delayed in delivering crucial goods
Fishing	Possible economic opportunities (landing, production, distribution and tourist fishing trips) Possible increase in local food availability Possibility for Longyearbyen to be an Arctic hub for fish/other species distributions Accidents and pollution	Increases community access to marine resources/possible improvement of local food security Possibility to establish local economic and employment options (including fish landing facilities and logistical organizations) Improvement of local preparedness and search and rescue	Limited value creation locally Increases marine litter Increases need for year-around preparedness and SAR services Possible conflicts between nations over marine resources Immigration concerns

reasons, these two types of vessels (research and Cost Guard) are not presented in the table.

Table 14.3 illustrates the broad range of effects and impacts of ship traffic in the port of Longyearbyen and in Svalbard waters as identified by interviewees (L1–L36). The table is structured to capture the effects and impacts that are specific to shipping in general and to a particular type of shipping activity.

14.5.2 Local Residents' Engagement in Adaptive Responses in Longyearbyen

Given these identifiable impacts, Longyearbyen faces a dilemma in balancing the growth of shipping with protecting the natural environment and improving the harbor and town infrastructure and facilities. All of these tasks must be accomplished while also providing well-functioning preparedness and Search and Rescue (SAR) services. Moreover, several concerns derive from residents who would like to see benefits from increased shipping (e.g. local value creation), especially from marine tourism. These individuals are of the opinion that a cruise vessel arriving in Longyearbyen is worth more than the vessels that just pass by (L10, L7). They acknowledge, "*This is our source of living here. Many experienced people are involved*" (L35), referring to key stakeholders and representatives from Longyearbyen who are involved in the development.

Further analysis identifies a number of adaptive responses that have been developed locally (as bottom-up responses) in order to mitigate negative impacts while securing the benefits of ship traffic growth in the port of Longyearbyen. These responses primarily comprise anticipatory measures that directly address the increase in the number of vessels and community visitors. These measures are divided into the following categories: preventing environmental harm, strengthening preparedness and SAR, improving visitor management systems, improving infrastructure and information dissemination, mapping and evaluating the socio-economic opportunities of fishing activities.

Preventing Environmental Harm To prevent environmental harm, several residents who are involved in shipping and tourism industry, as well as decision-makers, cooperate and map the possible threats from vessels operating in Svalbard waters and the impacts of increasing numbers of visitors on local natural environment sites (L8, L10, L12). Still, major accidents and/or oil spills in remote areas present major environmental threats. As was stressed by one interviewee, "*If we get a bigger oil spill on Svalbard…it will be extremely challenging. Thus, both regulations and practices work to prevent such situations*," (L12). Moreover, the fast-growing marine tourism industry adds a new dilemma to what and how Svalbard can be experienced by the tourist; "*It is difficult to find balance between experiencing and protecting*," (L22).

Marine litter is partially compounded by increasing marine activities, especially fishing activities in the Barents Sea and near Svalbard, but it is also carried with ocean currents from elsewhere. Numerous littered beaches have been observed by both community members and tourists. To address this environmental concern, public bodies, local residents and tourist industries have engaged in beach-cleaning initiatives. Locals are highly aware of this challenge and are eager to contribute to its resolution. Cruise visitors from some expedition cruises and pleasure crafts have also been proactive, using information about environmental damage to orchestrate participation in beach-cleaning activities as a part of the cruises' itinerary (L36).

Strengthening Preparedness and SAR Changing patterns of vessel distribution in remote areas (i.e. places that are difficult to access in the event of an accident) (L4), but also of cruise visitors' mobility on land (sometimes on landing sites due to polar bear danger) require better preparedness systems and SAR (L6). Improvement of maritime safety is a continuous process that involves a number of international and national stakeholders, but also local residents.

Locally, over 60 community members are involved in the Red Cross, which plays an important role in SAR (L4, L6). Voluntary members are trained for different types of rescue operations and can aid in the field when the assistance is needed. A previous head of the organization designed the "*dropkit:* Arctic Survival Kit," which contains necessary equipment, water and blankets that can be used before rescue services arrive. However, the Red Cross' capacity is limited during the summer navigation period by the absence of some of members that usually take a vacation during summertime.

Improving the Visitor Management System to Limit Societal Impacts Although under constant improvement, the visitor management system facilitates and welcomes diverse cruise vessels with capacities of over 5000 passengers. As mentioned by one of the interviewees, residents involved in the tourism industry are usually concerned about "*the amount of time the cruise vessel spends in a port, the facilities it uses in the town and the excursions' capacity*," (L7). This management system is supported by a well-established cooperation network of over 70 local companies that aim to develop Longyearbyen and Isfjorden as tourist destinations. Much of the work targets the improvement of visitor information and services, as well as the development of supportive infrastructure.

Information distribution to ship-owners, community visitors and the local population presents another important component of this system. Recently developed "community guidelines" for Longyearbyen are characterized by local residents' involvement. In addition to community guidelines, the local population actively participates in a number of organized workshops, initiatives, public hearings and conferences. As representatives from the local tourism office noted, "*The majority* [of community members] *should be on the development of the visitor management system. We need this joint discussion about tourism growth,*" (L7). Social media presents another source of local information distribution that informs and receives feedback and questions from residents and key stakeholders. Prior to the arrival of an overseas cruise ship and after its departure, information is sent to residents, especially those involved in the cruise network (via e-mail and Facebook) about *inter alia*, the size of the boat, how long it will stay in the town and how the visitors are distributed to avoid "overcrowding" (L8, also L10, L30, L35).

Local host services have developed in order to limit the impacts of overcrowding (i.e. a large number of people in a particular place at a particular time). The primary aim is to support an even distribution of people in time and space while offering community services. Examples of such responses include welcoming facilities in the harbor area, where visitors receive information about the place, sightseeing options and open hours of museums and shops; tourist information in the town center, where guests can access the Internet, order excursions and learn about the city; town service facilities, which correspond their opening hours with cruise schedules. Moreover, one of the interviewees mentioned, "*When we have 'massive visits,' we do not have enough guides to cover the demand. Then locals are recruited*," (L10). This is also common for bus drivers (L2) and for extra assistance in the stores (L35).

Improving Infrastructure Improvement of infrastructure in the port and town area has emerged in response to the growth in the number of vessels and community visitors. Several interviewees stress that there has been almost no development in infrastructure despite the rapid increase in ships using the port. "*Already, in 1996, there was a need to expand the harbor. In 2006, the port capacity reached its limit. Since then, the activities have increased by 165%*," (L22, also L25). After national acknowledgement of a much-needed improvement in port infrastructure and capacity, a number of local residents, who are involved in local shipping and infrastructure development, began drafting a strategic plan for the Longyearbyen port. In addition, they address a need to improve infrastructure and facilities along the designated route from the harbor to the town, including sidewalks, signage and information boards (L7, L10, L22). The absence of facilities and information irritates both visitors and locals. As several participants of this study noted, mapping needs and developing solutions to better welcome community visitors is ongoing. At the same time, infrastructure development is a complex task for land management; "*…there are many processes going on* [within infrastructure projects] *because there are many changes in the city*," (L10, also L22).

Mapping and Evaluating the Socio-Economic Opportunities of Fishing Activities The question of potential local benefits from the northward movement of fish and other marine species is critical for several local stakeholders. One of the emerging responses to the increasing fishing activities in the Svalbard area is local stakeholder discussions of scenarios around fish-landing facilities and logistical options for marine product export to global markets (L7, L9). "*I believe that the fishing industry is perhaps the only mature segment that has the power to set a new industry here, assuming that the legislation falls into place. Should we succeed, we have to make strategies around what kind of marine industry we are going to have up here*," (L11). Even though it is ultimately a national government decision, the possibility of Longyearbyen fish landing facilitates have sparked business ideas from a number of stakeholders. The possibilities include local use of marine resources, the development of operational cycles, "*branding and developing niche products*" (L11), and distribution to global markets.

14.5.3 Motivating Factors for Community Engagement

The adaptive responses in Longyearbyen identified above are characterized by the engagement of community residents and local stakeholders. This phenomenon was described by one interviewee as such: "*Longyearbyen is known to have many people with high engagement and strong meanings and who have a clear vision of how things should be done,*" (L12).

Our further analysis of the empirical data identifies the mechanisms behind the strong engagement of the residents in this unusual remote, international and highly fluctuating community. Those mechanisms can be divided into four main motivating factors for community engagement in local responses. These are (1) a shared place connection, (2) the perception of the changing natural environment, (3) established cooperation practices (networks, voluntary initiatives) across a wide group of local stakeholders and the local population, and (4) the ability to influence decision-making. In this part, we present a summary of how these community factors manifest as motivation for response engagement.

Connection to Place One of the motivations for the residents' engagement in adaptive responses is their connection to place. Interviewees say that many who live on Svalbard tend to stay there longer than they planned. "*I planned to be here only one year and then return to the mainland. But it did not happen,*" (L8, similar for L11). While others explain this emotional tie to a place as getting "*Svalbardbasillen,*" "*the Svalbard virus.*" It is an expression that describes people who visit Svalbard and tend to come back. "*I come here each summer, I got Svalbardbasillen,*" (L36). Moreover, given the unusual configuration of the settlement, people who live in Longyearbyen for more than 30 days receive local status (L7, L15). One of the residents who had lived in the community for a couple of decades was joking about this fact in the following way: "*Back in 1997, I was asked by a mining worker whether I was a tourist. I told them that I had been living in Longyearbyen for 5 years. He replied that I still was a tourist,*" (L15).

Perception of the Changing Natural Environment Increasing environmental consciousness has been identified as another motivating factor in responding to growing ship traffic. The local population has experienced a rapid change in the local environment (e.g. sea ice reduction and disappearance, new types of fish in the fjords) and has witnessed marine litter. One of the interviewees told us, "*Before we could drive snowmobiles to the other side of Advent fjord...We have not seen sea ice in many years here,*" (L8). While another was surprised at the fact that, "[they were] *fishing for new fish species that were not here 6 years ago,*" (L2). The residents who have been experiencing these changes in the local environment are concerned that some types of cruise vessels, driven by demand, will operate in newly opened, remote and vulnerable areas. Another interviewee told us, "*It is important that the tourists take care of their trash. We have another attitude toward nature here,*" (L22).

Cooperation Practices The next factor, cooperation practices, refers to the community's setting. Being a remote, isolated community increases the need to help one another. As one of the interviewees mentioned: "*Those who live in the North are used to rough nature; people know that they are vulnerable, know that they need to help each other, and I think it develops a special culture,*" (L4). This finding also reflects established local social and institutional networks, as well as voluntary initiatives: "*I believe that we have a culture within the environment so that we get strong no matter what appears. Even though there is a new manager in a big business, the person will not be able to 'rock' the fundament,*" (L7).

Influence Decision-Making This last factor is described by interviewees as an ability to influence decision-making (L12, L2). Some residents state that the influence of local and national decision-making systems is due to the community's size and the absence of regional political levels on Svalbard: "*It's fun with local politics in small towns. You get to have a say and you will be heard and get more attention…We have a shorter route to the national level,*" (L2).

14.6 Discussion

The findings illustrate the connection between local motivation factors and community engagement in local adaptive responses for the case of Longyearbyen. To elaborate on these findings, the following discussion illustrates the way in which the empirically identified determinant of 'community engagement' shapes local adaptive capacity in the context of increasing shipping activities.

The concept of engagement, when applied to human responses, can take place across several dimensions, from the personal to the collective, and may differ in the way it is activated (bottom-up vs. top-down) (e.g. Udofia et al. 2015; Moser and Berzonsky 2015; Leonard et al. 2016). On the one hand, top-down engagement in adaptation frameworks (e.g. Moser and Pike 2015, p. 112) is described as an overarching process that involves the public in matters of public concern. By presenting a typology of engagement with climate change, Moser and Berzonsky (2015) argue that there are different types of engagement, ranging from personal awareness and support (cognitive) toward more concrete public actions (civic and political). This process also refers to community involvement in processes such as decision-making via consultation and public meetings (e.g. Udofia et al. 2015).

On the other hand, the conceptualization of engagement at the community level refers to bottom-up processes of community engagement, which is described in the environmental change literature as community agency (e.g. Leonard et al. 2016). According to Brown and Westaway (2011), this agency refers to a community's ability to act collectively in addressing a particular concern, also known as collective action (e.g. Karlsson and Hovelsrud 2015). This type of engagement is characterized by "*strategic thinking and action, negotiating the social landscape, and collective efficacy*," (Leonard et al. 2016, p. 18).

The discussion in this study addresses bottom-up community engagement in relation to strategies undertaken by local actors and community members contributing to effective responses (e.g. Karlsson and Hovelsrud 2015). Our analysis shows that, in addition to previously established institutional responses, local adaptive responses have been taken by stakeholders and community members in order to address the diversity of impacts from increased shipping in the port of Longyearbyen and in Svalbard's waters (Table 14.3). We have illustrated that these local adaptive responses are characterized by community members' engagement (regardless of their residence time in the community and/or their nationality and professional backgrounds) and present a supportive mechanism for institutional (top-down) responses.

14.6.1 Community Engagement and Adaptive Capacity

Earlier studies (e.g. Brown and Westaway 2011; Karlsson and Hovelsrud 2015) argue that there is a connection between community engagement (community agency) and local adaptive capacity, as the ability to engage in collective strategies determines and shapes local adaptive capacity (Karlsson and Hovelsrud 2015, p. 95). Brown and Westaway (2011) argue that agency (in our study this is community engagement), access to resources and structural aspects (contextual attributes) are three main dimensions of adaptive capacity.

Our empirical analysis indicates that the community's engagement in adaptive responses is activated by four motivating factors that derive from community settings: place connection, perception of the changing natural environment, established cooperation practices across a wide group of stakeholders and the ability to influence decision-making. In the adaptation literature, such case-specific, motivating factors are often referred to as social capital, which comprises social processes and relationships and enables community engagement (e.g. Hovelsrud et al. 2018). Because of strong engagement deriving from social capital, Longyearbyen exhibits community characteristics, despite its unusual constellation of transient labour and its international profile. In addition to the defined motivation factors, this conformity can be partially explained by the area's geographical location and remoteness; people in Longyearbyen share the notion of isolation, finding themselves "in the same boat." Although the Longyearbyen community comprises individuals from over 40 different countries, community connection is facilitated by the citizens' love for the nature and wilderness of the Arctic and by the attractive job opportunities Svalbard offers without the need for a work visa from the Norwegian State (SSB 2016).

Longyearbyen also includes people with long-term residence, the so-called "Svalbardianere," who have been described as the "community glue" and the keepers of local, in-depth experience and knowledge. This "glue" is expressed through place attachment, a concept supported by other studies arguing that the uniqueness of place persists despite globalization, high mobility and interconnectedness (e.g. Escobar 2001; Amundsen 2015).

Place attachment is often described as a psychological bond to a particular place that can be ranked from weak to strong (Kaltenborn 1998, p. 173). It is mostly emotional but can also contain functional dimensions such as resource dependency. Place attachment is not an expression of how people perceive and respond to changes per se, but place attachment can influence how people experience change. The role of place connection in shaping adaptive responses is discussed by several scholars (e.g. Hovelsrud et al. 2018, Amundsen 2015) and is also applicable to the community of Longyearbyen, where community members develop adaptive responses despite a short residence period. Place attachment is expressed through a shared Svalbard identity and a sense of pride in belonging to Longyearbyen (Low and Altman 1992), which contributes to quality of life and well-being (see also Adger et al. 2013). At the same time, in a contemporary, globalized world—where people are more mobile and are often part of several communities—the phenomenon of "multiple belonging" (Haugen and Villa 2016) influences interactions between people and place.

Place connection affects peoples' perceptions of the natural environment (Kaltenborn 1998) and presents another motivation for engagement in adaptive responses. The observed changes in the natural environment and the negative impacts that derive from increasing shipping activities have influenced this perception. Although an earlier study showed that increasing shipping elicited fewer concerns than other types of human activities (Kaltenborn 1998, p. 181), the growth has resulted in a focus on keeping the shipping footprint as small as possible by supporting strict environmental legislation, industry guidelines and recent community participation in developing "community guidelines" (AECO 2018). It is also noteworthy that the community's participation in beach cleaning initiatives is not a new phenomenon (Kaltenborn 1998), however the practice's development within the last year is a product of the marine cruise industry's contribution to environmental conservation, as well as the environmental awareness of community visitors and tourists.

Established cooperation practices across a range of stakeholders are closely connected to the ability to influence decision-making. These two motivating factors—cooperation and decision-making influence—represent important aspects of social capital (e.g. Hovelsrud et al. 2018) that enable adaptive responses to increasing ship traffic. Here, established cooperation practices are applicable to industrial networks (e.g. Cruise Network, which, in uniting over 30 local stakeholders, becomes an actor with the ability to participate in and influence a decision-making process), but also to voluntary initiatives (e.g. the Red Cross).

Finally, though we describe Longyearbyen as a unique Arctic community due to its transient labor force and unique political situation, we are still able to identify local community characteristics, i.e. motivating factors that also define a social group as a local community (see Haugen and Villa 2016). Moreover, the empirical results have produced evidence that those motivations activate community engagement in adaptive responses, which, in turn, strengthen local adaptive capacity. Hence, given the integration of numerous components, we argue that the community engagement found in Longyearbyen is a dimension of communities' adaptive capacities. This dimension, according to Brown and Westaway (2011, p. 325) (described as one's agency), is "*one's independent capability or ability to act on one's will.*" Our study shows that this ability is shaped by contextual variables, such as social capital.

14.7 Conclusion

According to recent projections (e.g. Borch et al. 2016), shipping development in the Barents area will continue to increase and expand in space and time due to a number of changing conditions, including sea ice reduction. The same development is documented to have a broad range of impacts on coastal communities' wellbeing and the local natural environment. Both positive and negative impacts have been identified for the community of Longyearbyen (See Table 14.3).

The application of a community-based approach allows us to assess perspectives on Arctic shipping development by assessing adaptive capacity at the local level. For, the effects from increasing shipping are first and foremost felt at the local level, and it is also at this level that adaptive responses emerge to mitigate the most salient negative impacts of change, while enhancing the positive ones.

We have derived three main conclusions from our analysis:

1. Given the current scenarios for shipping development in the Arctic, it is of particular importance that plans develop proactively. The strategic role of Longyearbyen as a hub for projected activities in Arctic Trans-Polar routes, and as a hub for SAR and emergency preparedness in the Barents Sea, supports this emphasis. The expansion of marine tourism activities in the Barents area will most likely be felt on Svalbard.
2. There is a growing need to understand the complexity of possible impacts of increased shipping and its local adaptive responses. Although the current engagement in adaptive responses of Longyearbyen's local population presents a supportive mechanism for locally established institutional and industrial response, we argue that such engagement is sensitive to community fluctuation and other dynamic community settings, e.g. demographic trends.
3. Using the framework of adaptation and adaptive capacity, the analysis of empirical data reveals that local engagement in local adaptive responses

strengthens the adaptive capacity. This high engagement of such transitory community is activated by a number of motivating factors: place attachment, perception of the changing natural environment, established cooperation practices across a wide group of stakeholders and the ability to influence decision-making.

The results of this study can be used for current and future recommendations in managing ship traffic in the port of Longyearbyen and in Svalbard's territorial waters. The study may also be useful as a guideline for methodological and theoretical approaches to assessing local perspectives of shipping development in other Arctic regions.

Acknowledgement The authors gratefully acknowledge the participants involved in this study for sharing their meanings and experiences with us. We are particular thankful to Norwegian Pilot Services (Norwegian Coastal administration), Longyearbyen Port Authorities and Cruise Network Svalbard for providing access to and sharing practical knowledge. We thank Håvard Berg for creating a map of the case area, Thina M. B. Mohus for checking the references and Malinda Labriola for reviewing the earlier versions of this chapter. We are particularly appreciative of the comments by an anonymous reviewer that helped significantly to improve the paper.

This work was supported by The Research Council of Norway under Svalbard Science Forum Arctic Field Grant 269947 and under Large-scale Programme on Climate Research Grant 255783 'Sustainable coasts. Public and scientific perceptions of drivers affecting coastal zone management and ecosystem services'.

References

Aarsæther, N. (2014). *Viable communities in the North?* Gargia conferences 2004–2014. https://doi.org/10.7557/5.3201. Accessed 25 Apr 2018.

Adger, W. N., Barnett, J., Brown, K., et al. (2013). Cultural dimensions of climate change impacts and adaptation. *Nature Climate Change, 3*(2), 112–117.

AECO. (2018). *Guidelines*. https://www.aeco.no/guidelines/. Accessed 15 June 2018.

Akama, Y., Chaplin, S., & Fairbrother, P. (2014). Role of social networks in community preparedness for bushfire. *International Journal of Disaster Resilience in the Built Environment, 5*(3), 277–291.

AMAP. (2017). *Adaptation actions for a changing Arctic: Perspectives from the Barents Area*. Arctic Monitoring and Assessment Programme (AMAP), Oslo, Norway, pp. xiv + 267.

Amundsen, H. (2015). Place attachment as a driver of adaptation in coastal communities in Northern Norway. *Local Environment, 20*(3), 257–276.

Arctic Council. (2011). *Agreement on cooperation on aeronautical and maritime search and rescue in the Arctic*. https://oaarchive.arctic-council.org/handle/11374/531. Accessed 25 Apr 2018.

Arlov, T. (2003). *Svalbards historie*. Trondheim: Tapir Akademisk Forlag.

Bay-Larsen, I., & Hovelsrud, G. (2017). Activating adaptive capacities: Fishing communities in northern Norway. In G. Fondahl & G. Wilson (Eds.), *Northern sustainabilities: Understanding and addressing change in the circumpolar world* (pp. 123–134). Cham: Springer.

Bazeley, P., & Jackson, K. (2013). In J. Seaman (Ed.), *Qualitative data analysis with NVIVO* (2nd ed.). London: SAGE.

Blaikie, N. (2010). *Designing social research*. Cambridge: Polity Press.

Borch, O. J., Andreassen, N., Marchenko, N., et al. (2016). *Maritime activity in the high north – Current and estimated level up to 2025*. Utredning nr. 7. Bodø: Nord University.

Bring. (2016). *Seilingsplan Tromsø – Svalbard 2016*. http://www.msupply.no/Userfiles/Upload/files/KK-828-10_2015_Seilingsplan_Tromsø%20-%20Svalbard%202016_PRINTres.pdf Accessed 30 Nov 2018.

Brown, K., & Westaway, E. (2011). Agency, capacity, and resilience to environmental change: Lessons from human development, Well-being, and disasters. The. *Annual Review of Environment and Resources, 36*, 321–342.

Davydov, A., & Mikhailova, G. V. (2011). Climate change and consequences in the Arctic: Perception of climate change by the Nenets people of Vaigach Island. *Global Health Action, 4*(10). https://doi.org/10.3402/gha.v4i0.8436.

Dawson, J., Pizzolato, L., Howell, S. E. L., et al. (2018). Temporal and spatial patterns of ship traffic in the Canadian Arctic from 1990 to 2015. *Arctic, 71*(1), 15–26.

DNV-GL. (2014). *Prognoser for skipstrafikken mot 2040*. Report number: 2014–1271. DNV Gl Maritime. http://www.kystverket.no/globalassets/nyheter/2015/november/prognoser_2040-rev.e-2018-02-14-002.pdf Accessed 17 Oct 2016.

Eguíluz, V. M., Fernández-Graciaet, J., Irigoien, X., et al. (2016). A quantitative assessment of Arctic shipping in 2010–2014. *Scientific Reports, 6*, 30682. https://doi.org/10.1038/srep30682.

Escobar, A. (2001). Culture sits in places: Reflections on globalism and subaltern strategies of localization. *Political Geography, 20*(2), 139–174.

Farré Buixadé, A., Stephenson, S., Chen, L., et al. (2014). Commercial Arctic shipping through the northeast passage: Routes, resources, governance, technology, and infrastructure. *Polar Geography, 37*(4), 298–324.

Haugen, M. S., & Villa, M. (2016). Lokalsamfunn I Perspektiv. In M. Villa & M. S. Haugen (Eds.), *Lokalsamfunn*. Oslo: Cappelen Damm.

Hovelsrud, G., & Smit, B. (2010). *Community adaptation and vulnerability in the Arctic regions*. Dordrecht: Springer.

Hovelsrud, G. K., Karlsson, M., & Olsen, J. (2018). Prepared and flexible: Local adaptation strategies for avalanche risk. *Cogent Social Sciences, 4*, 1460899. https://doi.org/10.1080/23311886.2018.1460899.

Kaltenborn, B. P. (1998). Effects of sense of place on responses to environmental impacts: A study among residents in Svalbard in the Norwegian high Arctic. *Applied Geography, 18*(2), 169–189.

Karlsson, M., & Hovelsrud, G. K. (2015). Local collective action: Adaptation to coastal erosion in the Monkey River village, Belize. *Global Environmental Change, 32*, 96–107. https://doi.org/10.1016/j.gloenvcha.2015.03.002.

Keskitalo, C., Dannevig, H., Hovelsrud, G., et al. (2011). Adaptive capacity determinants in developed states: Examples from the Nordic countries and Russia. *Regional Environmental Change, 11*, 579–592.

Leichenko, R., & O'Brien, K. (2008). *Environmental change and globalization: Double exposures*. Oxford: Oxford University Press.

Leonard, R., McCrea, R., & Walton, A. (2016). Perceptions of community responses to the unconventional gas industry: The importance of community agency. *Journal of Rural Studies, 48*, 11–21.

Low, S. M., & Altman, I. (1992). Place attachment. In I. Altman & S. M. Low (Eds.), *Place attachment* (Human behavior and environment (Advances in theory and research)) (Vol. 12, pp. 1–12). Boston: Springer.

Marchenko, N. A., Borch, O. J., & Markov, S. V., et al. (2016). *Maritime safety in the high north – Risk and preparedness*. Paper presented at the ISOPE-2016. The twenty-sixth (2016) international offshore and polar engineering conference, Rhodes (Rodos), Greece. 26 June–2 July 2016.

Misund, O. A., Heggland, K., Skogseth, R., et al. (2016). Norwegian fisheries in the Svalbard zone since 1980. Regulations, profitability and warming waters affect landings. *Polar Science, 10*(3), 312–322.

Moser, S. C., & Berzonsky, C. (2015). There must be more: Communication to close the cultural divide. In K. L. O'Brien & E. Silboe (Eds.), *The adaptive challenge of climate change*. New York: Cambridge University Press.

Moser, S. C., & Pike, C. (2015). Community engagement on adaptation: Meeting a growing capacity need. *Urban Climate, 14*(1), 111–115.

MOSJ. (2018). *Sea ice extent in the Barents Sea and Fram Strait Environmental monitoring of Svalbard and Jan Mayen*. Retrieved from http://www.mosj.no/en/climate/ocean/sea-ice-extent-barents-sea-fram-strait.html

Muckenhuber, S., Nilsen, F., Korosov, A., et al. (2016). Sea ice cover in Isfjorden and Hornsund, Svalbard (2000–2014) from remote sensing data. *The Cryosphere, 10*(1), 149–158.

Multiconsult. (2014). *Strategisk havneplan for Longyearbyen*. Vedtatt i Longyearbyen lokalstyre sak 3/14 11.02.14. http://portlongyear.no/wp-content/uploads/2017/02/Strategisk-Havneplan.pdf. Accessed 1 Oct 2016.

Nyseth, T., & Viken, A. (2015). Communities of practice in the management of an Arctic environment: Monitoring knowledge as complementary to scientific knowledge and the precautionary principle? *Polar Record, 52*(1), 66–75.

Olsen, J., & Nenasheva, M. (2018). Adaptive capacity in the context of increasing shipping activities: A case from Solovetsky, northern Russia. *Polar Geography, 41*(4), 241–261.

Overland, J., Walsh, J., & Kattsov, V. (2017). Trends and feedbacks. In *Snow, Water, Ice and Permafrost in the Arctic (SWIPA) 2017* (pp. 9–24). Oslo: Arctic Monitoring and Assessment Programme (AMAP).

PAME. (2009). *Arctic Marine Shipping Assessment 2009 Report (AMSA)*. https://oaarchive.arctic-council.org/handle/11374/54. Accessed 25 May 2013.

Pedersen, T. (2017). The politics of presence: The Longyearbyen dilemma. *Arctic Review on Law and Politics, 8*, 95–108.

Port of Longyearbyen. (2018). *Statistics of Port Longyear*. http://portlongyear.no/statistics-port-longyear/. Accessed 25 Mar 2018.

Smit, B., & Wandel, J. (2006). Adaptation, adaptive capacity and vulnerability. *Global Environmental Change, 16*(3), 282–292.

Smit, B., Hovelsrud, G., Wandel, J., et al. (2010). Introduction to the CAVIAR project and framework. In G. Hovelsrud & B. Smit (Eds.), *Community adaptation and vulnerability in the Arctic regions* (pp. 1–22). Cham: Springer.

Smith, L. C., & Stephenson, S. R. (2013). New trans-Arctic shipping routes navigable by midcentury. *Proceedings of the National Academy of Sciences, 110*(13), 1191–1195.

SSB. (2016). *Dette er Svalbard. Hva tallene forteller. Statistisk sentralbyrå*. https://www.ssb.no/befolkning/artikler-og-publikasjoner/dette-er-svalbard-2016. Accessed 20 Feb 2017.

St. Meld. 32. (2015–2016). *Svalbard. Det kongelige Justis- og beredskapsdepartementet*. https://www.regjeringen.no/no/dokumenter/meld.-st.-32-20152016/id2499962/. Accessed 25 Nov 2017.

Stewart, E., Dawson, J., & Johnston, M. (2015). Risk and opportunities associated with change in the cruise tourism sector: Community perspectives from Arctic Canada. *The Polar Journal, 5*(2), 403–427.

Teigen, S. H., Nilsen, F., & Skogseth, R. (2011). *Heat exchange in the sea west of Svalbard. Manuscript, paper IV in Water mass exchange in the sea west of Svalbard*. Ph.D. thesis, Department of Geophysics, University of Bergen, Bergen, 172 pp.

The Governor of Svalbard. (2016). *Svalbard. Ros-analyse*. https://www.sysselmannen.no/globalassets/sysselmannen-dokument/skjemaer/ros-analyse-svalbard-2016.pdf. Accessed 19 Apr 2018.

Udofia, A., Noble, B., & Poelzer, G. (2015). Community engagement in environmental assessment for resource development: Benefits, emerging concerns, opportunities for improvement. *Northern Review, 39*, 98–110.

Valestrand, H. (2016). Gruvebyen Bjørnevatn: Industrisamfunn, omstilling og lokalsamfunn. In M. Villa & M. S. Haugen (Eds.), *Lokalsamfunn* (pp. 322–343). Oslo: Cappelen Damm.

Van Bets, L. K. J., Lamers, M. A. J., & van Tatenhove, J. P. M. (2017). Collective self-governance in a marine community: Expedition cruise tourism at Svalbard. *Journal of Sustainable Tourism, 25*(11), 1583–1599.

Viken, A. (2008). The Svalbard transit scene. In J. O. Barenholdt & G. Granas (Eds.), *Mobility and place: Enacting northern European peripheries* (p. 139). London: Ashgate Publishing.

Vikhamar-Schuler, D., Førland, E., & Hisdal, H. (2016). *Kort oversikt over klimaendringer og konsekvenser på Svalbard.* https://cms.met.no/site/2/klimaservicesenteret/rapporter-og-publikasjoner/_attachment/9559?_ts=1559b5c5534S. Accessed 16 Apr 2018.

Wesche, S., & Armitage, D. R. (2010). As long as the sun shines, the rivers flow and the grass grows: Vulnerability, adaptation and environmental change in Deninu Kue traditional territory, Northwest Territories. In G. K. Hovelsrud & B. Smit (Eds.), *CAVIAR—Community adaptation and vulnerability in the Arctic regions* (pp. 163–189). Dordrecht: Springer.

Wolf, J., Allice, I., & Bell, T. (2013). Values, climate change, and implications for adaptation: Evidence from two communities in Labrador, Canada. *Global Environmental Change, 23*(2), 548–562.

Yin, R. K. (2014). *Case study research. Design and methods* (5th ed.). Thousand Oaks: SAGE Publications.

Chapter 15
Arctic Search and Rescue: A Case Study for Understanding Issues Related to Training and Human Factors When Working in the North

Derek D. Rogers, Michael King, and Heather Carnahan

Abstract With increased development in Arctic regions (e.g., oil and gas, tourism, fisheries, shipping) the risk to the people in this region in case of emergency needs to be mitigated. Search and rescue in the Arctic is a critical but often ignored aspect of safe development in the Arctic. However, the logistics and training of Arctic SAR are very unique when compared to other regions. In this chapter we will review how the harsh environment of working in Arctic regions affects the ability of rescue technicians to perform the technical skills required for search and rescue. The chapter will be built around a 2013 investigative report of the death of a Canadian Forces Search and Rescue Technician that occurred on a mission near Nunavut. We will review the main points made by the investigative report and extrapolate what was learned about human factors issues related to working in Arctic regions that has application to all human activities in this region. The need for emergency procedures that are designed for Arctic operations will also be addressed. In summary, literature will be reviewed and based on this, recommendations for training, equipment and operational procedures will be made.

Keywords Search and rescue · Human factors · Emergency response · Hazard identification · Risk assessment

D. D. Rogers
Total Response Solutions, Halifax, NS, Canada

M. King · H. Carnahan (✉)
School of Maritime Studies, Marine Institute, Memorial University of Newfoundland, St. John's, NL, Canada
e-mail: hcarnahan@mun.ca

E. Pongrácz et al. (eds.), *Arctic Marine Sustainability*,
Springer Polar Sciences, https://doi.org/10.1007/978-3-030-28404-6_15

15.1 Part I: Introduction and Case Study

There is an increasing level of commercial and recreational development in Arctic regions of the world. The Arctic is defined as the region above the 66°33′ latitude line and spans approximately 14.5 million square kilometers. A critical but often less recognized aspect of safe development in the Arctic is Arctic search and rescue (SAR). The risk to the people in Arctic regions (e.g., oil and gas, tourism, fisheries, shipping) needs to be mitigated and emergency response for these persons is especially underdeveloped. There are obvious characteristics of the Arctic, such as size and isolation that pose challenges for SAR. There is only one highway in the Arctic (Dalton Highway, Alaska Route 11) and aircraft are generally not based in the Arctic or far north. As a result, SAR by helicopter or fixed wing aircraft requires significant travel in the order of thousands of kilometres and this poses unique challenges to SAR that can be life threatening.

In this chapter, we will review how the harsh environment of working in Arctic regions affects the ability of rescue technicians to perform the skills required for SAR. The issues discussed in this chapter will also have implications for development in the Arctic that involves humans working in or travelling through this harsh environment. In Part 1, we will first describe a case study of an Arctic search and rescue mission that ended in the tragic death of a rescue technician. In this section we will propose modifications to how SAR is performed in the Arctic. Then in Part 2, we will discuss some of the human factors that should be considered for both SAR and other persons travelling to the Arctic. This chapter will clearly show that Arctic-SAR requires an updated approach to emergency response planning, equipment and its use. Solutions to these problems must be found if there is to be successful development in the Arctic.

15.1.1 Case Study

This case study is based on the Minister of National Defence, Canadian Forces Flight Safety Investigation Report (2013). On a calm October morning in 2011, a father and son from Ogloolik (Nunavut, Canada) began their seasonal walrus hunting trip in the Hecla Strait. As experienced hunters, they routinely prepared their equipment and loaded their 4-meter aluminum outboard boat. Included in their equipment was a mix of old and new; traditional seal skin clothing along with an advanced satellite beacon and modern rifles. Their equipment and preparations would serve them well in the challenges that were to come.

The hunt proved to be a success; the men harvested a walrus and went about field dressing their kill for the return trip home, which was only 90 min travel. In the fading light, the weather started to turn, temperatures dropped, and the sea conditions worsened with increasingly larger waves. These factors alone would have been challenging enough, but the extreme cold led to a simple but debilitating mechanical failure of the boat. A frozen outboard motor stranded them amongst thick pans of ice, which undulated with the waves and threatened to crush their boat. They soon realized that their situation was life threatening. They had no choice but to call for help and they activated their satellite beacon. Although relatively close to home, the isolation of the region meant that help was thousands of kilometers and several hours away.

Their call for help was received via satellite by the regional Joint Rescue Coordination Centre, which put into motion a massive rescue effort to save them. This included local search assets and the mobilization of fixed wing aircraft from the Rescue Squadrons in Trenton (Ontario 2800+ km) and Winnipeg (Manitoba 2300+ km), along with a Rotary Wing Asset in Gander (Newfoundland 4500+ km). The initial plan was to parachute in a rescue team to provide immediate support for the hunters, while the helicopter leapfrogged (i.e. regular fuel stops) along the North East coast of Canada towards the rescue team and hunters in Igoolik.

The first aircraft on scene was Call Sign Rescue 340, a 435 Squadron CC-130 Hercules from Winnipeg. Communications with the hunters determined that immediate assistance was required. The Winnipeg crew were at the end of their shift and their CC-130 had limited fuel. Given the deteriorating weather, the rescue was handed over to Rescue 323, the 424 Squadron CC-130 Hercules out of Trenton. The Trenton team was led by Sargent Janik Gilbert with support from Master Corporals Lahaye-Lemany and Journeyman. During this time, a rescue Helicopter from 103 Squadron Gander was already working its way up the east coast but was several hours away. The only way to get to the stranded father and son was a parachute jump and a rescue plan was quickly assembled. The team planned to parachute to the disabled vessel with survival and medical equipment to sustain the father and son and themselves until the recovery helicopter could arrive. However, as the rescue team deployed into the fading day light, no one was properly prepared or equipped for the severe conditions and challenges the SAR Technicians (SAR Techs) would face from the ice and water. A high wind, low light parachute jump into the Arctic Ocean is the epitome of a high risk, low frequency event. The jump did not go as planned.

Only Master Corporals Layn-Lemany and Journeyman made it to the stranded hunters while Sargent Gilbert was blown off course and did not land with the team. The SAR Techs and hunters spent 5 h in an open raft exposed to temperatures well below freezing. At this time they did not know the fate of Sargent Gilbert. Over 5 h after entering the water the 103 Squadron Gander helicopter was on scene. In 10 m seas and high wind, the helicopter rescue was equally challenging but the 103 Squadron managed to recover the 2 SAR Techs and hunters. At this time, they then began to search for Sargent Gilbert and found in him close by, in the water – unresponsive in a flooded survival suit. He was pronounced dead on the aircraft. Although the hunters were saved, this was a crushing blow to the rescue team. The mission exposed numerous flaws in the Canadian National SAR system Arctic response.

Despite the massive mobilization of resources and the years of experience of the SAR personnel the rescue ended in tragedy. A full investigation of the SAR response was conducted and several issues were raised by the board of inquiry including: (i) planning for hazard identification and risk assessment, (ii) operating procedures and emergency protocols, and (iii) training and equipment. The inappropriate SAR response is a microcosm for the misunderstood challenges facing northern commercial operations and regional development. The next sections will use this tragedy as a case study to elaborate the corrective actions of these issues in order to improve SAR response in Arctic conditions. Further, this discussion will inform how to improve the safety of commercial operations and regional development in Arctic conditions.

15.1.2 Findings and Corrective Actions

The inability to perceive and identify the unique risks of Arctic offshore operations placed the crew at a significant disadvantage and was an overarching contributing factor in the mission fatality. The Arctic is inherently one of the most extreme environments on earth and acknowledging the uniqueness of the environment is a critical first step to understanding hazard identification and risk assessment. Worst-case scenarios, such as what occurred in this case study, can be prepared for by taking this perspective.

15.1.3 Hazard Identification and Risk Assessment

The rescue personnel on this mission were well trained and experienced but they did not have a custom hazard identification and risk assessment for the operation. Successful risk identification requires experience, education, and an in-depth knowledge of the region combined with a detailed understanding of the operational tasks. The generic approach taken in this case was the all-threats approach, which is utilized by the Military. An all-threats approach addresses an emergency according to the specific training protocol that is required to mitigate the specific emergency. While this seems effective, one cannot train for every possible emergency scenario and this approach discourages skill transfer (i.e. applying learned skills to novel emergency situations). The all threats approach failed to accurately assess the level of risk for the crews involved and to identify the specific hazards, such as the flight duration (causing SAR-tech overheating) and landing on mixed water/ice pans, which define the typical terrain in the Arctic.

An alternative and potentially more effective approach, which the safety industry has embraced, is a specific hazard identification and safety case development approach. This provides personnel and organizations with a structured process of analysis and preparation, where multiple skill sets and equipment are employed to prepare for crises during SAR rescue. While this process takes on many different forms, its key elements involve three major steps (i) identifying the hazards (ii) assigning a value based on frequency and possible impact (see ISO 2013; and (iii) designing custom emergency response plan to mitigate the hazards by applying a diverse set of widely applicable skills. Using this approach would have mitigated the response plan to include overheating as a potential threat to survival and the preparedness for landing off course into water rather than onto ice. Indeed, the Arctic is different from Antarctica as the majority of the surface is covered with and ice/water mixture instead of land.

Collectively, this approach has the capacity to create an emergency protocol that includes the procedures and resources for specific crisis reaction during SAR. In the current case study, SAR personnel did not have a full understanding of the risks involved with the parachute deployment onto Arctic terrain (i.e. mixed water and ice pans), as such they were not completely prepared. Effective hazard identification and assessment would have allowed for the SAR personnel to create a more effective and safe emergency protocol equipped with crises reaction plans.

15.1.4 Emergency Response Planning, Operating Procedures & Emergency Protocols

An effective Emergency Response Plan (ERP) for Arctic operations has to establish standardised operating procedures and a detailed communications plan to ensure safe coordination and the most effective utilization of assets in an emergency. ERPs have many designs and formats but the main components include: Planning, equipment and technology, coordination and communication (International Organization for Standardization 2013) and training, drills and exercises, and education. Most commercial and government organizations have generic ERPs that do not account for regional challenges, such as isolation, extra Arctic-specific SAR equipment, and Arctic terrain. In the Arctic, these challenges can be deadly and warrant special considerations and treatments for the ERP. Below we discuss the planning, equipment, and technology that should be taken into consideration when compiling an Arctic ERP.

15.1.5 Arctic SAR Planning for Isolated Rescues

ERPs for Arctic operations should have unique considerations during planning. The major factors affecting SAR in the Arctic are the prolonged travel durations, additional equipment and parachute deployment onto Arctic terrain. Air support is generally not based in the far north and rescue requires significant travel to reach northern regions. For example, rescue times can be as long as 9 days (Power et al. 2016) and a significant portion of that involves travel to the emergency site. Travelling for several consecutive hours poses a difficulty to the proper use of a survival suit. If the survival suit is not vented for heat, then excessive sweating occurs, which leaves the SAR potentially dehydrated and without proper thermal protection. If the survival suit is vented, then there is the risk of improper donning prior to deployment. While it is protocol to confirm and amend potential water entry points, this process can be complicated by additional equipment blocking visibility of points of entry. Further complicating this issue is the fact that on board flight engineers or loadmasters are not trained to recognize all seal points in a fully equipped SAR Tech. This is particularly important for the last SAR Tech that exits the aircraft. In the case study presented above, Sargent Gilbert was the last person to leave the aircraft and did not recognize that his suit was not sealed. Thus, Arctic SAR ERPs should be updated to modify existing protocols to include a formal venting and sealing of survival suits with special attention to an Arctic-equipped SAR Tech.

15.1.6 Arctic SAR Equipment and Technology

Reliability of equipment is poor in cold conditions the Arctic and there is limited custom design for durability and functionality under extreme weather cold conditions, such as those in the Arctic. It is important to test any equipment in realistic Arctic conditions before it is put into use (Rahimi et al. 2011). The Arctic terrain is a mixture of water and ice pans. As a result there are many additional pieces of equipment required for Arctic SAR. For example, Arctic SAR Techs must not only carry a standard jump kit but 16 additional items to satisfy the requirements for a water and arctic jump kits. These additional items decrease mobility, increase drag, and increase weight. An increase in items causes a cumbersome survival suit that is difficult to evaluate for water entry points. Further, these factors challenge the ability of the SAR Tech to properly hit their landing target, which is often onto moving ice pans. While we cannot confirm whether these factors directly led to the SAR Tech to miss the landing target, these factors certainly compromise Arctic jump performance. Additional jump training with increased weight and drag may facilitate Arctic SAR jumps or other missions requiring extra equipment. It may be the case that these extra considerations require additional jump training with increased weight and drag. Further, a user may lose dexterity due to cold hands (Ray et al. 2017) or because of heavy mittens. Designs that require the dextrous bare hand will not be effective in a survival situation and may contribute to poor survival outcomes. While the physiological effects of cold temperatures are not specific to the arctic, they are relevant considerations when considering arctic SAR. An increase in items makes for a cumbersome survival suit that is difficult to evaluate for water entry points. Potential solutions to this is to better train flight engineers and loadmasters to recognize water entry points on a survival suit in this context.

15.1.7 Training, Drills and Exercises, and Education

Training personnel to work in this region poses challenges because of the dangers of simulating the Arctic environment. That is, short of engaging in training in Arctic environments and exposing people to the associated risks, it is difficult for people to really understand how the cold, wind, high sea states, and terrain found in this region will affect their ability to do their jobs safely. For example, landing on *moving* ice pans in high winds is unique to the Arctic and cannot be prepared for by practicing landing on stationary ice. Based on principles of specificity of learning (Proteau 1992) and concepts of simulation fidelity (Norman 2014; Grierson 2014) we know that the closer training is to the real life application the more effective it is.

One caveat to this is that specific training is not easy to implement and may require the use of expensive simulators that can model high equipment jumps onto moving ice pans. However, there are several issues complicating this training. For example, SAR crews are often rotating (particularly military crews) training must be ongoing for each new cohort, which can be expensive. Additionally, SAR skills are perishable (Sanli and Ergonomics 2018), which means routine retraining is necessary and that is complicated by the lack of centralized training facilities, centralized command, and general lack of resources. If the Arctic is to be safely developed, commercial and recreational operations need to make a commitment to safety and contribute towards the safety and survival training of the SAR teams.

15.1.8 Coordination and Communication

As well, there are challenges for radio and satellite communications with entire regions being unserved; communications satellites operating in geostationary Earth orbit do not cover some regions in the Arctic and as a result communication between SAR personal was lost which disabled the potential rescue of Sargent Gilbert. Further, communications can be unreliable in regions that do have service because of weather disruptions, such as icing on antennas or heavy sea states (Ho and Fjørtoft 2017). For optimal coordination of SAR missions, these issues should be mitigated with advances in technology as well as increased satellite coverage.

15.2 Part 2: The Arctic Environment and Human Factors

The human factors associated with working in the Arctic is an often less discussed but critical factor for safety. The effect of the harsh environment on factors, such as human performance, is a major limitation to successful development in the Arctic. While there is little research on how the Arctic affects human performance in situ, there are three well-developed research fields that can inform the potential effects of the Arctic on human performance (i) the effect of cold exposure on human performance and physiology (Stocks et al. 2004; Vincent and Tipton 1988; Ray et al. 2017), (ii) the effect of stress on human performance, and (iii) the effect of darkness on human performance and health.

15.2.1 The Effect of Cold on Physiology and Performance

Individuals who work in cold ocean environments often have to perform manual skills with hands that have been exposed to the cold. For example, people in fisheries, aquaculture, off shore oil platforms, diving, and search and rescue perform some of their occupational skills in cold water. In a survival situation, individuals exposed to cold water might have to secure protective gear, use an underwater breathing apparatus, hold onto life preservers, or manipulate rope or tools with cold hands.

Cold water has drastic effects on human physiology. Cold water exposure (e.g., 2 °C) to the hand alone causes cognitive dysfunction (Værnes et al. 1988), intense sensations of pain (Mitchell et al. 2004), thermal discomfort (Geurts et al. 2006), and increased cardiovascular response, such as blood pressure and increased heart rate (LeBlanc et al. 1979). Further, cold exposure increases joint stiffness (Hunter et al. 1952), and slows nervous system and muscle function (Rutkove 2001). As a result cold exposures causes a reduction in grip strength (Chi et al. 2012; Giesbrecht et al. 1995), finger mobility (Hunter et al. 1952), tactile sensitivity (Cheung et al. 2008) and dexterity (Daanen 2009; Muller et al. 2011). This loss of tactile sensitivity will further impair manual function because tactile sensitivity is critical for grasp maintenance (Westling and Johansson 1984). It is unclear whether these factors influenced the death of the SAR Tech but it is likely that after exposure to the cold water via leaking of his survival suit that all of these factors exacerbated the short survival time associated with extreme cold exposure.

Cold exposure duration and the subsequent changes in hand skin temperature are key variables that will determine the effect that cold has on manual performance. If an individual's hands are exposed to cold water for a sufficient amount of time, and critical temperature thresholds are reached, then manual skills will be difficult to perform due to impairment in hand function. Cheung et al. (2003) showed that immersing the hands in 10 °C water will start to show signs of impairment between 120 s and 300 s. Power et al. (2018) showed that exposure to 2 °C can lead to impairment in as little as 60 s. Cold exposure can greatly impair the performance of manual skills and can lead to accidents and even death. Indeed, 94% of injured electricians and telecommunication workers who work in cold environments attribute cold conditions as a primary cause of their injury (Päivinen 2006). It is unclear however, how training and other solutions can ameliorate the performance decrements seen after exposure to cold water.

Research on methods to improve cold-related performance decrements indicates that it is possible to improve one's response to the cold and enhance performance, which will in turn increase safety during emergency situations. Repeated cold exposure habituates cardiovascular factors such as blood pressure (Glaser et al. 1959), heart rate (Tipton et al. 2000), and respiratory frequency (Tipton et al. 2000; Barwood et al. 2013). Similarly, anxiety (Barwood et al. 2014), pain (Cheung and Daanen 2012), thermal discomfort (Geurts et al. 2006) habituate after repeated exposure to the cold. Importantly, repeated exposure to cold water is effective in increasing breath-holding performance in cold water (Barwood et al. 2007). Additional research on cold exposure and human performance is paramount if we are to improve training practices for persons working and living in the Arctic. Currently, recommendations for safe exposure to cold environments involve the use of outer-wear, but this recommendation has limitations.

15.2.2 *Effective Use of Outerwear*

Using effective outerwear is an important way to maintain thermal comfort. However, it is difficult to design outerwear that is waterproof while allowing water vapor to escape from inside the clothing. Condensation of water vapor inside the clothing reduces the thermal efficiency of the protective clothing and leads to cooling the body (Meinander and Hellsten 2004). Anecdotally, this also leads to improper use of survival clothing whereby zippers are sometimes undone, and if not secured during an emergency or sudden immersion in water, can compromise equipment efficacy. Indeed, this was the major issue leading to the death of the SAR Tech discussed in Part 1. Lastly, the functionality of protective outerwear highly depends on appropriate sizing and accurately donning procedures and this may not always be effectively performed under stress. Although this does not directly apply to the case study in Part 1, it remains important for other emergency situations.

15.2.3 *The Effect of Working in the Dark*

In addition to exposure to cold temperatures, another human factors issue is the months of continual darkness in the Arctic; the North Pole is dark for 6 consecutive months starting in late September. Occupational darkness has both immediate and long-term effects for health and safety. One immediate effect of darkness is reduced or inhibited visual communication. Verbal communication may be reduced in a windy environment or by helicopter rotors; this situation could be ameliorated by the addition of non-verbal communication. Indeed, verbal communication is enhanced by gesturing (Driskell and Radtke 2016) and persons communicating under stress use more hand gestures than during non-stressful conditions (Kelly et al. 2011). However, during helicopter rescue at night and in poor weather engenders a situation where both verbal and auditory communication are jeopardized. Further, motor performance is influenced by the removal of visual feedback (Keele and Posner 1968) and perception of objects and spaces is reduced in low light conditions (Boyce 2014). If low light conditions persist, such as during the winter months, circadian rhythm patterns are disrupted (Arendt 2012) and reduced exposure to light can increase the risk of depression, which in turn affects work performance (Lerner et al. 2010). Station latitude has been correlated with symptoms of seasonal affective disorder (Palinkas et al. 1996) and it is estimated that 6.5% of men and 13.1% of women in the military working in the Arctic are affected with seasonal affective disorder (Rosen et al. 2002). Given that mental health is typically under reported in similar occupations, such as the military (Hoge et al. 2004), then attention to mental health by those working in the Arctic is especially important. It is clear that acute and chronic low light conditions would present challenges to the success of search and rescue missions and occupational health in the Arctic.

15.3 Summary

The above considerations have to be addressed and planned for due to the very specific nature of the challenges that require a customized SAR Arctic response. Any commercial operation expanding into the region will need to account for these factors with the consideration that commercial operations will increase the incidence of SAR response. Effective ERPs, rescue protocols, and equipment are currently lacking for Arctic operations; as such it is difficult to provide effective emergency response. The development of Arctic ERPs for response agencies will not only better equip the response teams but will help drive the development of training programs and equipment requirements.

The only component of the case study in Part 1 that was not singled out for improvement was the bravery and dedication of the SAR Tech team. The death of Sargent Janik Gilbert was not the first in the Arctic and unfortunately it probably will not be the last. The Arctic is the last great frontier, as development expands into this new territory it is vital that this growth is given the proper support and oversight to ensure the safety and success of new ventures for people in the north.

References

Arendt, J. (2012). Biological rhythms during residence in polar regions. *Chronobiology International, 29*, 379–394.

Barwood, M. J., Datta, A. K., Thelwell, R. C., & Tipton, M. J. (2007). Breath-hold time during cold water immersion: Effects of habituation with psychological training. *Aviation, Space, and Environmental Medicine, 78*, 1029–1034.

Barwood, M. J., Corbett, J., Green, R., Smith, T., Tomlin, P., Weir-Blankenstein, L., & Tipton, M. J. (2013). Acute anxiety increases the magnitude of the cold shock response before and after habituation. *European Journal of Applied Physiology, 113*, 681–689.

Barwood, M. J., Corbett, J., & Wagstaff, C. R. D. (2014). Habituation of the cold shock response may include a significant perceptual component. *Aviation, Space, and Environmental Medicine, 85*, 167–171.

Boyce, P. R. (2014). *Human factors in lighting* (3rd ed.). Boca Raton: CRC Press.

Cheung, S. S., & Daanen, H. A. M. (2012). Dynamic adaptation of the peripheral circulation to cold exposure. *Microcirculation, 19*, 65–77.

Cheung, S. S., Montie, D. L., White, M. D., & Behm, D. (2003). Changes in manual dexterity following short-term hand and forearm immersion in 10 °C water. *Aviation, Space, and Environmental Medicine, 74*, 990–993.

Cheung, S. S., Reynolds, L. F., Macdonald, M. A. B., Tweedie, C. L., Urquhart, R. L., & Westwood, D. A. (2008). Effects of local and core body temperature on grip force modulation during movement-induced load force fluctuations. *European Journal of Applied Physiology, 103*, 59–69.

Chi, C. F., Shih, Y. C., & Ergonomics, W. C. I. J. O. I. (2012). Effect of cold immersion on grip force, EMG, and thermal discomfort. *International Journal of Industrial Ergonomics, 42*, 113–121.

Daanen, H. A. M. (2009). Manual performance deterioration in the cold estimated using the wind chill equivalent temperature. *Industrial Health, 47*, 262–270.

Driskell, J. E., & Radtke, P. H. (2016). The effect of gesture on speech production and comprehension. *Human Factors, 45*, 445–454.

Geurts, C. L. M., Sleivert, G. G., & Cheung, S. S. (2006). Central and peripheral factors in thermal, neuromuscular, and perceptual adaptation of the hand to repeated cold exposures. *Applied Physiology, Nutrition, and Metabolism, 31*, 110–117.

Giesbrecht, G. G., Wu, M. P., White, M. D., Johnston, C. E., & Bristow, G. K. (1995). Isolated effects of peripheral arm and central body cooling on arm performance. *Aviation, Space, and Environmental Medicine, 66*, 968–975.

Glaser, E. M., HALL, M. S., & Whittow, G. C. (1959). Habituation to heating and cooling of the same hand. *The Journal of Physiology, 146*, 152–164.

Grierson, L. E. M. (2014). Information processing, specificity of practice, and the transfer of learning: Considerations for reconsidering fidelity. *Advances in Health Sciences Education, 19*, 281–289.

Ho, T. D., & Fjørtoft, K. (2017). *Communications challenges in the Arctic: Oil and gas operations perspective* (pp. V008T07A001–V008T07A001). ASME.

Hoge, C. W., Castro, C. A., Messer, S. C., McGurk, D., Cotting, D. I., & Koffman, R. L. (2004). Combat duty in Iraq and Afghanistan, mental health problems, and barriers to care. *The New England Journal of Medicine, 351*, 13–22.

Hunter, J., Kerr, E. H., & Whillans, M. G. (1952). The relation between joint stiffness upon exposure to cold and the characteristics of synovial fluid. *Canadian Journal of Medical Sciences, 30*, 367–377.

International Organization for Standardization. (2013). *ISO 3100 risk management*. Geneva: Vernier.

Keele, S. W., & Posner, M. I. (1968). Processing of visual feedback in rapid movements. *Journal of Experimental Psychology, 77*, 155–158.

Kelly, S., Byrne, K., & Holler, J. (2011). Raising the ante of communication: Evidence for enhanced gesture use in high stakes situations. *Information, 2*, 579–593.

LeBlanc, J., Côté, J., Jobin, M., & Labrie, A. (1979). Plasma catecholamines and cardiovascular responses to cold and mental activity. *Journal of Applied Physiology: Respiratory, Environmental and Exercise Physiology, 47*, 1207–1211.

Lerner, D., Adler, D. A., Rogers, W. H., Chang, H., Lapitsky, L., McLaughlin, T., & Reed, J. (2010). Work performance of employees with depression: The impact of work stressors. *American Journal of Health Promotion, 24*, 205–213.

Meinander, H., & Hellsten, M. (2004). The influence of sweating on the heat transmission properties of cold protective clothing studied with a sweating thermal manikin. *International Journal of Occupational Safety and Ergonomics, 10*, 263–269.

Mitchell, L. A., MacDonald, R. A. R., & Brodie, E. E. (2004). Temperature and the cold pressor test. *The Journal of Pain, 5*, 233–237.

Muller, M. D., Ryan, E. J., Kim, C.-H., Muller, S. M., & Glickman, E. L. (2011). Test–retest reliability of Purdue Pegboard performance in thermoneutral and cold ambient conditions. *Ergonomics, 54*, 1081–1087.

Norman, G. (2014). Simulation comes of age. *Advances in Health Sciences Education, 19*, 143–146.

Päivinen, M. (2006). Electricians' perception of work-related risks in cold climate when working on high places. *International Journal of Industrial Ergonomics, 36*, 661–670.

Palinkas, L. A., Houseal, M., & Rosenthal, N. E. (1996). Subsyndromal seasonal affective disorder in Antarctica. *The Journal of Nervous and Mental Disease, 184*, 530–534.

Power, J.T, Kennedy, A. M., Monk, J.F.. (2016). *Survival in the Canadian Arctic recommended clothing and equipment to survive exposure* (pp. 1–15). Offshore Technology Conference, Arctic Technology Conference, St. John's.

Power, C., Ray, M., Luscombe, T., Jones, A., & Carnahan, H. (2018). *A timeline for hand function following exposure to 2 degree Celsius water*. Halifax.

Proteau, L. (1992). On the specificity of learning and the role of visual information for movement control. *Advances in Psychology, 85*, 67–103.

Rahimi, M., Rausand, M., & Wu, S. (2011). *Reliability prediction of offshore oil and gas equipment for use in an arctic environment* (pp. 81–86). IEEE.

Ray, M., Sanli, E., Brown, R., Ennis, K. A., & Carnahan, H. (2017). The influence of hand immersion duration on manual performance. *Human Factors, 59*, 811–820.

Rosen, L., Knudson, K. H., & Fancher, P. (2002). Prevalence of seasonal affective disorder among U.S. Army soldiers in Alaska. *Military Medicine, 167*, 581–584.

Rutkove, S. B. (2001). Effects of temperature on neuromuscular electrophysiology. *Muscle & Nerve, 24*, 867–882.

Sanli, E. A., Carnahan, H., 2018. Long-term retention of skills in multi-day training contexts: A review of the literature. *International Journal of Industrial Ergonomics 66*, 10–17. https://doi.org/10.1016/j.ergon.2018.02.001.

Stocks, J. M., Taylor, N. A. S., Tipton, M. J., & Greenleaf, J. E. (2004). Human physiological responses to cold exposure. *Aviation, Space, and Environmental Medicine, 75*, 444–457.

Tipton, M. J., Mekjavic, I. B., & Eglin, C. M. (2000). Permanence of the habituation of the initial responses to cold-water immersion in humans. *European Journal of Applied Physiology, 83*, 17–21.

Værnes, R. J., Knudsen, G., Påsche, A., Eide, I., & Aakvaag, A. (1988). Performance under simulated offshore climate conditions. *Scandinavian Journal of Psychology, 29*, 111–122.

Vincent, M. J., & Tipton, M. J. (1988). The effects of cold immersion and hand protection on grip strength. *Aviation, Space, and Environmental Medicine, 59*, 738–741.

Westling, G., & Johansson, R. S. (1984). Factors influencing the force control during precision grip. *Experimental Brain Research, 53*, 277–284.

Chapter 16
The Possibilities and Limitations of Tourism Development in Greenland Impacting Self-efficiency and Socio-economic Wellbeing of Coastal Communities

Vishakha Tay

Abstract Almost all of Greenland's population live in coastal areas. Coastal tourism in Greenland has the potential to diversify the local economy from fishing and the fish-processing industry to tourism-based entrepreneurship that possibly could reduce the gender and income inequality in Greenland and the outmigration of the younger generation from small settlements in search of opportunities in large cities, and abroad, and the revitalizaton of traditional culture. Tourism in coastal Greenland exposes the reality of climate change to visitors from the outside world, the consequences of melting ice sheets on people, nature, and the traditional way of life and could promote an alliance between the locals and visitors through honest dialogues and data collection through citizen science-based excursions. Tourism also creates an opportunity to improve and modernize local infrastructure that benefits the local population with a higher standard of living.

The negative impacts of tourism rely heavily on the lack of negotiation power of the indigenous people in Greenland against the pre-established tour operators (from Denmark) and other non-Greenlandic carriers who would receive the largest economic benefits unless strong government policy protects local interests and resources. Responsible tourism and tourists are the building block of sustainable tourism. The limitations of a short tourist season and a bottleneck situation during the high season creates investment that is risky and costly. Thus, a thorough and ongoing social-economic-environmental impact assessment would be mandatory steps in harmonizing tourist expectations and local understanding of sustainable Arctic coastal tourism in Greenland.

Keywords Arctic tourism · Arctic cruise tourism · Skill matching job creation · Social sustainability · Coastal tourism · Nature-culture tourism · Responsible tourism

---

V. Tay (✉)
University of Paris-Saclay, Paris, France

E. Pongrácz et al. (eds.), *Arctic Marine Sustainability*,
Springer Polar Sciences, https://doi.org/10.1007/978-3-030-28404-6_16

16.1 Introduction

16.1.1 Tourism, More than an Industry

According to data published by World Tourism Organization (UNWTO) in 2017, one in ten jobs was tourism related, which generated U.S. $1.6 trillion in export, and accounted for 10% of global GDP from "direct and indirect activities combined". Tourism is more than an income generating industry, it is also one of the pillars of cultural preservation, environmental protection and a medium of peace and security (Source: UNWTO Tourism Highlights 2018 Edition). The 1985 Tourism Bill of Rights and Tourist Code reinforces the "human dimension of tourism" and reiterates the claims that tourism contributes to social, economic, cultural and educational sectors of national societies and improves the international community (World Tourism Organization (WTO) 1985). Tourism has made positive contributions to the preservation of cultures, when globalization has been a force for cultural homogenization (Cohen and Kennedy 2000, p. 226).

Despite all these benefits, economists, economic developers, and governments have either failed to take tourism seriously because of the widespread notion that tourism is only "fun, and games, recreation, leisure, unproductive" (Davidson 1994, pp. 20–21) or, used only as an economic activity, a vehicle of consumerism, when the government focuses on "growth fetishism" (Growth Fetish, Clive Hamilton 2004, p. ix). The precaution from WTO: tourism is an irreplaceable factor of solidarity in the development and dynamic growth of international exchanges, multinational enterprises of the tourism industry should not exploit the dominant positions they sometimes occupy; they should avoid becoming the vehicles of cultural and social models artificially imposed on the host communities; in exchange for their freedom to invest and trade which should be fully recognized, they should involve themselves in local development, avoiding, by excessive repatriation of their profits, or their induced imports, a reduction of their contribution to the economies in which they are established. Partnership and the establishment of balanced relations between enterprises of generating and receiving countries contribute to the sustainable development of tourism and an equitable distribution of benefits of its growth (WTO 1999).

16.1.2 Tourism in Greenland

In Greenland, the average growth for the estimated tourism revenue was 5% in the period from 2006–2015 (source: stat.gl). During the Icelandic tourism boom, Greenland also enjoyed significant growth: 23.8% in 2015 and 9.9% in 2016, respectively, in the number of tourists through the partnership with Icelandair with arranged short trips/day trip from Iceland. Even with a conservative projection of an annual growth of 5% in the estimated tourist revenue it will look as predicted in Fig. 16.1 in the period through to 2030 (Source: stat.gl).

An annual growth of 5% requires an ongoing investment in the infrastructure of the country (number of beds and flight seats over the year) in order to be able to receive the growing number of tourists. This scenario is based on the current airport infrastructure. UNWTO (UN's tourism body) has forecast an annual global tourism growth of 3.3% between 2010 and 2030.

On a practical note, the lack of infrastructure hinders Greenland's efforts to welcome more tourists than its present capacity. However, the expansion of three airports in Nuuk, Illulissat, and Qaqortoq and the port in Nuuk, call for a massive infrastructure overhaul countrywide. These steps are necessary to ensure the supply lines can cope with continued development. Following the footsteps of Iceland's tourism board, a strong year-around tourism sector could strengthen Greenland's economy and thus making the communities more socio-economically viable in the long run. Greenland's geostrategic location between Europe and North America can become the new "tourism hub" growing in importance in the coming years. With the growing interest for the Arctic, Nuuk is in a good position for when it becomes possible to use the North West Passage and be the first choice when politicians, businesses and tourists look to the North. As part of Nuuk City Development project, the mayor of Kommuneqarfik Sermersooq, Asii Narup, has accounced the construction of 4200 new homes over 10 years to welcome new visitors from all over the world (Arctic Circle Conf. 2018).

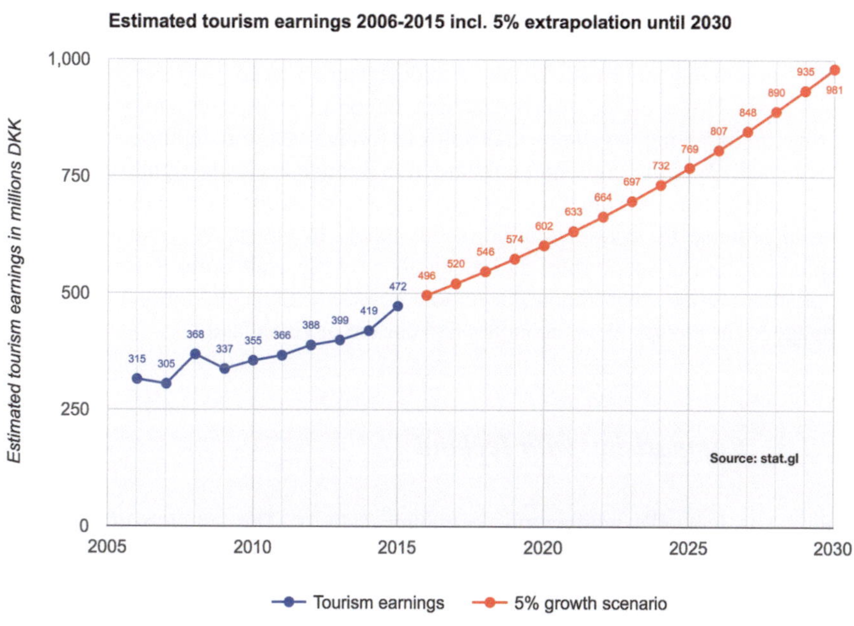

Fig. 16.1 Projected Tourism Growth in Greenland. (Source: Greenland in Figures 2018; www.stat.gl)

As Nuuk continues to grow and expands its capacity for inhabitants, so does the possibility of tourism. Following Iceland's successful marketing strategy combined with a growing global interest in the Arctic as the last frontier, Greenland has a coinciding rise in political, scientific, and societal interest in tourism. Today, tourism is considered as one of Greenland's three economic pillars, next to fishing and mining, and a promising lever for the Arctic nation's future economic development (Source: Stat Greenland). Its successful development could potentially help pave the way for Greenland's financial dependency from the Danish Commonwealth to become an independent Arctic country.

The government of Greenland has been involved with periodic research related to tourism conducted by researchers across towns and settlements to understand the local ideas about tourism, to draw an authentic picture based on local needs and available resources or the lack of them (INUSSUK, Arctic Research Journal 2, 2011). Equipped with research data, and local knowledge, Greenlandic politicians and businesses are hoping and planning for substantial growth in tourism to diversify its economy through sustainable development. With the construction of three transatlantic airports by Air Greenland, the country's national carrier, and a vibrant national tourism campaign (Visit Greenland) constructed by local and international expertise, and the increasing carrying capacity of the capital city Nuuk, a broader societal discussion of how (much) tourism should be developed, in what ways, and by whom, becomes an urgent necessity. The potential tourism practitioners in Greenland are facing the challenges and tremendous potential posed by tourism. Its development could be linked to other spheres of society social-cultural and environmental dimensions, turning tourism from an industry into a potential catalyst for social change. In Greenland, tourism may increase the social capital of local residents through engaging with visitors and developing new skills (George 1999 and Reid 2003). However, the government can compromise the community ownership of tourism by giving developmental rights to foreign investors in hopes of creating tourism activities, facilities, and infrastructure through exclusive rights of activities in delimited areas thus creating foreign competition. Arctic cruise tourism holds a special promise for coastal tourism development in Greenland, and without the implementation of sound rules and regulations for big scale cruise lines to operate in pristine marine environment, in remote coastal settlements, tourism can cause damage to the environment, economy and culture of Greenland.

16.1.3 Comparision with Iceland

Tourism is Iceland's number one economic activity. The strength of Icelandic human capital has been the key component in tourism development. The interplay between local and global informally acquired skills originates from tourism (Nordal and Kristinsson 1996). The informally acquired skills of international exchange in Icelandic society are gathered from the interaction with visiting tourists and the guest workers supporting the tourism related activities. Since the early 1950s,

Iceland has managed to transform into a transatlantic air travel hub utilizing the old US airfield at Keflavík. The role as hub was established through a network that connected the USA (and later Canada) and Europe, as a low cost flying option based on the business of Loftleiðir and in 1953, flights took off between USA and Europe via Keflavík. This business model of using Keflavík as a hub between North American and European cities, helped to build up an international air traffic network for Iceland far beyond what the Icelandic market itself could sustain. As Greenland prepares for three new international airports, this story could become relevant for Greenland's near future as direct flight connections to both North America and Europe; for purposes of developing research, education, government, business and society, could have transforming effect. One can only imagine how internationalized Greenlandic society and its economy could become, and how many international tourists could be in Greenland en route Europe as a stop over in Nuuk, Illulissat, and Qaqortoq, following Iceland's model of Rekykavik as stop over hub for European travel.

In order to create a "rational tourism growth" research, collaboration, and planning are the key factors. With the help of NATA (North Atlantic Tourism Association) and Air Greenland, Visit Greenland (VG 2016) carried out market surveys in 2016–17 in four core markets in Germany, Great Britain, France and USA with the help of the research agency NIT Kiel. Four thousand respondents in each country filled out online questionnaires, which has given valuable insight into the potential of the markets in terms of potential travelers to Greenland. The surveys included question about volcanoes/hot springs/geysers and specific nature and culture based tourism to understand who would rather choose other destinations such as Iceland before Greenland or to show that tourists are exactly sure of what Greenland has to offer (Source: Greenland Tourism Report 2018).

16.2 Possibilities of Tourism

16.2.1 Refocusing on Human Capital Through Tourism (Now) in (Future) Capacity Building in Remote Settlements

Gender and income inequality are widespread in Greenland. Income inequality measured by the so-called Gini-coefficient is about 33.9, 2015 estimate (www.indexmundi.com), while in the Nordic countries it is about 25, and a Sub-Saharan country with an index of 50. Approximately 15% of the population lives in relatively poor families, 10% in poor families and 5% in very poor families (Statistics Greenland 2014). There is a significant difference in the educational achievements between youths growing up in settlements and in towns. Though increasing, the education level in Greenland remains the lowest in the Nordic. More than half of all 25–64 years old have no education beyond lower-secondary education, compared to about 25% in other Nordic countries; about a quarter pursue a vocational education and only 16%, the majority women, pursue higher education. (Source: http://bank.stat.gl/UDEISCPROB). In Greenland, approximately 84% of males have no education apart from primary school. The unemployed part of the workforce has a high proportion of unskilled workers. For the higher education, the unemployment rate is very low. Another distinct feature in the Greenlandic labor market is a large seasonal variation in employment, due to the climate and the geographic dispersion, which limits mobility (Source: Greenland Statistics).

In Greenland, and in many other Arctic countries, there is a rather large group of people who do not have a formal education and, thus, have fewer opportunities in life. But they still have skills: some are specialists in hunting or fishing, some are skilled handicraft men/women, and some are artists, without a formal educational paper to prove it. In contrast to westernized standards of success, some of these people might be perfectly content with their situation and get by on a subsistence economy, navigating in an informal economy, while living a traditional life of a hunter and fisherman. However, they are limited in the labor market because of their lack of documentation for their skills (Kleist and Knudsen 2016). They might be further challenged because they do not wish – or are able to – continue in a formal educational system. Therefore, skill mapping to use the informally acquired skills as well as traditional activities (hunting, fishing, dog sledging, country food cooking) is a crucial factor in order to obtain a fuller picture of how these skills are beneficial in building a socio-economically healthy society in Greenland utilizing its potential workforce (Knudsen 2016).

Although not clearly explicit, arctic nature-culture tourism is based on traditional knowledge, and in current Greenlandic context, it is practical whether and how traditional knowledge can be used – not necessarily according to a western model of society but rather in a modern Inuit society where activities may be combined without compromising the opportunity to continue the traditional activities of Greenlandic society. Greenlandic people have managed to live and thrive in the harsh arctic conditions for centuries and have acquired invaluable skills that passed down from generation to generation through practicing the daily life. However, in present day formal education based employment opportunities, their skills are invisible and unrecognized by the job market. The lack of appreciation of the qualification of this group and failing to map the informally acquired skill sets with potential employment opportunities are a great loss of value to society and to the individual. It is also the root cause of domestic violence in Greenland, where women are most adaptable with their higher education and, thus, more mobile, while men are stuck in the traditional male dominant culture. Any careful observation would make the correlation of nature and culture tour based on the skills and life experience of the Greenlandic people, especially the hunters and fishermen as tour guides and operators, thus lifting the value of self. This perspective attaches different values to local knowledge and counters the westernized economic reasoning that has dominated policies of centralization. Sustainable development should be regarded as the 'third way' in that it transforms the experiences and relations from traditional society and the understanding related to global, industrial developments, which includes a new understanding of the interplay between man and nature (Holm and Rasmussen 2000). It also emphasizes the need for the population to become involved in articulating and developing a sustainable policy for institutions and infrastructures through participatory processes (Hersoug 1999). The 'cultural dimension' is important within educational transition: from home to school (perhaps in another city) to further education (perhaps in another city) and, finally, to the job market (perhaps in another city). Tourism has tremendous potential in bridging these gaps.

16.2.2 Link to Education, Formal Training, and Language Development

In Greenland, the tour guide and tourism education might represent more than just improving the service level of tourism and tourism development, but, more essentially, could be an interesting starting point to get young people back to education and job market perspectives, especially in smaller settlements and remote areas (Ren and Chimirri (2017). The adventure guide education in Campus Kujalleq is based on the Icelandic model and is created in dialogue with different Greenlandic tourism stakeholders. The new program that started in 2013 is an alternative to the Arctic Guide program or the previous government sponsored "Outfitter" program. It is for people who want to help tourists on trips in nature lasting over 8 hours. The aim of providing a qualification for academic studies is realized within the areas of business economics and socio-economics combined with foreign languages and other general subjects. The education program's goal is to develop the students' capacity for in-depth studies and their understanding of theoretical knowledge as tools for analyzing realistic issues (knowledge of safety protocols, operating modern equipment, professionalism, certification and license process) while applying their cultural and traditional knowledge (Source: Campus Kujalleq http://cak.gl/in-english/). Tourism Education Program Campus Kujalleq offers three programs within tourism education: A 2½-year academy profession program, and two guide program of one semester. The arctic guide program focuses on guiding techniques and background facts necessary to a guide working in arctic communities. Through tourism, informally acquired skills under the conditions of the Arctic are increasingly touched by globalization where the foreign tourists and local mingle exchanging their worldviews. Thus, tourism allows Arctic citizens to participate in global networks and transactions.

16.2.3 Gender Equality and Harmony by Balancing the Modern and Traditional Skills

The process of urbanization has led more Greenlanders living in the cities, with a significant rate of out migration from small settlements the capital or even abroad. Women in rural Greenland are generally more educated than men and are more adaptive socially and geographically to seek employment in service sector jobs outgrowing from traditional gender expectations of the society. However, men have been stuck in the old tradition of being a provider through hunting and fishing skills. In the past, these skills were considered as the key activities of the society. The survival of a family is now becoming obsolete as Greenland transitions towards a modern society and imposed quota and restrictions on hunting animals. There has been a shift in family dynamics in Greenland. Greenlandic men now depend on the steady income of educated women to be able to continue to practice the traditional masculine roles in the society (Rasmussen 2014). Since men are less adaptive than women, the lack of mechanized (mines) jobs, left Greenlandic men with fewer options.

The tourism industry in remote coastal Greenland could become a catalyst to encourage participation of women in regional development. By helping women with business grants and entrepreneurship training (bed and breakfast, restaurant/ café, souvenir shop), the government cannot only diversify the economy of remote settlements, but this inclusion of women also helps with the mainstreaming of gender equality based policy and policy measures. In Southern Greenland, more women participate in municipal government, as they are more financially independent and globally connected through tourism. While managing sheep farms, the locals in the South have managed to have an extra income by hosting tourists in their farm house, without substantial investment in a business venture. Additionally, hospitable by traditional custom, sheep farmers have reported a sense of personal wellbeing by connecting with the tourists-the outside world. Therefore, this type of tourism continues to be a favorable medium for personal-socio-economical outlet in south Greenland amongst the sheep farmers.

However, in Northern Greenland, women are not involved in this level of governance due to lack of financial opportunities and lack of global exposure (Rasmussen 2014). As Greenlandic men are more inclined to participate in traditional hunting and fishing, together, both men and women can combine their traditional knowledge/skill and modern education/formal education (English language, communication, service training) as a hybrid form in tourism related entrepreneurships and serve as an intersection for job creation, traditional lifestyle, and integration of local voices, especially voices of women in politics and regional development.

16.2.4 The Concept and Status of 'Knowledge' in the Arctic in Regards to Climate Change

> *If people come to Greenland and see how much the glaciers have been retreating and realize it's for real, and change the way they use energy, then maybe the net benefit will be for the globe, for the climate.* (Malik Milfeldt, the Greenland Tourism and Business Council (smithsonian.com, October, 2011))

People, who share a history of living off the land and the sea as hunters, fishermen, farmers etc., are highly dependent on interpreting the signs of the surrounding environment through the in-situ everyday observations of flora and fauna in the natural world. To ensure a continuation of this way of life, it is essential for these people, and for the communities in which they reside, to learn and to pass on their experiences as these relate to the co-habitation with their environment (Ingold 2000, p. 195). These people are known for possessing in-depth ecological knowledge, an understanding of, and an appreciation for environmental variations and seasonal changes and how these influences, for example, the harvest of local marine mammals upon which local subsistence continues to be sustained (Berkes 2000; Huntington 1998). Traditional knowledge also draws on the experience made across the changing arctic seasons which also reflects the passage of time, through both contemporary and intergenerational transmission of experiences made with the immediate environment (Duerden 2004; Ingold 2000). Others worry about the effects on traditional ways of life. With less summer ice cover, hunters who use dog sleds are limited, says Hanne Nielsen, who teaches Greenlandic and Danish languages in Nuuk: "Climate change has had a really harmful influence on people's lives, not only professional hunters and fishers, because ordinary people also fish and hunt." Climate change is one of the motivations for travel to see peripheric locations Stromberg (2011). Most people do not travel to understand the impact of global warming on their destination but rather out of curiosity to "see them before they disappear." Nature-culture based tourism puts the community as the "human face" during a challenging time and thus invoking sentiments and sense of responsibility into the tourists through honest dialogues and portraits of people and place interconnected via climate change.

In August 2018, an iceberg melting to the north of Greenland made the world news as the oldest and thickest ice in the arctic crumbling, thus opening previously inaccessible sea routes opened up for the first time in the history (The Guardian, August 21st). Although data reliability could not be on per with scientists, part of the US Study of Environmental Arctic Change project, scholars identified Arctic tourism as an important area of human activity that should be monitored as a social component of AON (Fay and Karlsdottir 2011; Kruse et al. 2011). Citizen science based polar expedition is one way to utilize tourism as a force to address the challenges of climate change in coastal Greenland where tourists can participate actively in data gathering. It is significant that the tourism sector globally is already engaged in citizen science research; it is active also in similarly oriented activities through both "conservation tourism" and "participatory environmental research" (Scheepens 2014). Having an "experience" is at the heart of tourism and therefore visitors to the polar regions desire the inclusion of citizen science, an aspect desirable for both the destinations and other tourism stakeholders, which should be important for strategists reliant on observational monitoring. If a critical hurdle to engaging citizen science is the absence of alignment between community and research priorities (Pandya 2012), then tourism may offer a particular opportunity to engage given the potential that exists to align the environmental priorities of visitors, communities and researchers alike (Mason et al. 2000)

Women are more acute than men in internalizing the effects of climate change. Their specific ways of engaging with the changing environment: fishing, skinning the animals (assessing the fat content of seals or marine animals), berry picking, fish drying, collecting eggs, are more engaging on a personal level as women focus more on relationships and social fabric of community explaining the impacts of climate change. However, women are not present to discuss their extensive knowledge in a formal level while tourism may bring an informal yet powerful personal narrative by women to understand the effects of climate change in Greenland. Focusing on male dominated hunting activities and material resources only presents an incomplete understanding of the impacts of climate change and a great deal of women's expertise is lost. Through women-led hunting/fishing/gathering tourism or even creating a social gathering over Kaffemik (social gathering with coffee, and sweets, a recipe for a cozy time in Greenland) to listen to women's stories, the outside world gets the full impacts of the climate change. Women in Greenland are nature adapted, sociable, and focus more on non-material relations, they project the indirect effects of climate change, such as limited ice leading to less connectivity between settlement thus weakening sense of community, food security, education, language vitality, and good housing.

16.3 Limitations of Tourism in Greenland

Daniela Tommasini, a researcher at North Atlantic Regional Studies (NORS) Roskilde University Denmark was appointed by the Ministry of Education and Research of Government of Greenland to conduct almost a decade long research project related to tourism in Greenland (1995–2004). She traveled throughout Greenland to interview tourists, entrepreneurs in the tourism industry and local residents and assessed the socio-economic impact of modern-day tourism, its past development and potential for the future. This chapter is based on her first-hand experience with locals about their dreams and realities involving tourism in Greenland. Her case studies have contributed "Visit Greenland (VG)" marketing company an honest overview of local's perspective in nature-culture based tourism and as a possible tool for the revitalization of marginal communities.

Her fieldwork from various towns and settlements with tourism potential has analyzed the limitations of tourism; extremely short bottleneck tourist season, poor infrastructure, lack of trained personnel, lack of English language skills, competition with foreign tour operators, and little awareness of tourism among local population. The report suggested aiming towards inexpensive investments and infrastructure: "The key words are control, guidance, and limitations, if tourism is not to create problems" (Master Plan 1.3.4.). Nonetheless, due to all the above factors, traveling in Greenland is still an expensive activity and the tourism industry still has not generated any surplus to Greenland, and remains heavily subsidized by the government.

16.3.1 Foreign Competition and Historical Prejudice

During high season competition is very high between local Greenlandic guides and international tour guides who basically work for free for major cruise companies just to gain a unique experience of working in Greenland. Anna Burdenski, a master's student from Aalborg University, Denmark, conducted interviews with the students of Campus Kujalleq in Qaqortoq as part of tourism, her master's thesis. Student # 9 from Campus Kujelleq in Qaqortoq shared both hope and realistic despair, "We have a lot of guides who travel around the world and work on an international basis but they do not have any knowledge about the particularities of Greenland. We see it, as you said; the experience is enough for themselves, so they do not need to get paid and how can you compete with practical free labour? And that is a huge problem. We see this especially within the cruise business. The cruise companies they are very particular type of business, because they smell money, let us put it like that, and it is hard core about the money than it is about the experience. As soon as they would have local tour guide on board who wants to get paid on Greenlandic salary basis, they would be like 'oh no honey, no', and at the same time they are crying to us about 'why are there no Greenlandic guides? Why cannot we have Greenlandic guides?'. As soon as we ask them about how much they offer to pay, they are very surprised about paying them. They are looking at it as it should be a privilege from those guides to work and be with the cruise company. So I think international companies are having a very hard time adapting to that (Burdenski 2018)."

"Even though the Greenlandic society is a very heterogeneous mass, intergeneration negative attributes are stuck with them over long periods of times. While this is setting them back in international recognition, job opportunities with international companies & reputation, it is an immense hurdle to work against this wall of negativities. Fighting against this, takes much more effort, engagement & courage to stand up to. It is an unnecessary setback, which international people have put onto the Greenlandic society. While becoming aware of it & trying to find ways to overcome the burden, it might be undetermined but desired, that tourism as a tool of understanding other cultures, can counteract to prejudices many people, communities & societies are facing" (Bjørst 2008).

At present, all big scale tourism operations in Greenland is foreign owned. One of the major foreign tour company is "Greenland Adventure by Icelandic Mountain Guides." On its website it states; "For many on the Greenland Adventures team, our love for this country has been strong since childhood. In fact, the majority of the team are native Greenlanders that have migrated to Iceland." With the knowledge of English language, and an established track record of guiding business, many foreign tour guide companies are successful businesses in Greenland in comparison to locally owned businesses which lack professional certification, adverting, and language skills.

16.3.2 Language Barrier

The common language of global tourism is English which creates another serious limitation in building confidence among the Greenlandic people in becoming a tour operator. This rudimentary demand is paired with the expressed need for raising service levels and thus calls for improving the relatively low educational level in Greenland. Indeed, Greenlandic people do not feel comfortable in communicating in English and strong initiatives and training courses need to be in place by the government. (Finne 2018) A journalist for "High North News" newspaper, reported that understanding the incompetency it presents in global context, the Government of Greenland declared that it will work to have English replace Danish as primary foreign language in Greenland. An expert committee, convened by the previous government, found that Greenlandic, Danish and English fulfill separate functions in Greenland and are likely to continue to do so for some time to come.

Vivian Motzfeldt, Greenland's Naalakkersuisoq (Minister) for Education, Culture, Research and Church, in a press release from Motzfeldt's department said that the Naalakkersuisut has been working to establish agreements with English-speaking countries like the United States and Canada regarding an exchange of English-speaking teachers to Greenland. "They are to help grow young people's competence so that they in a long-term perspective may become so confident in English that more future teacher students will choose English as one of their specializations."

The future course of Greenland is determined by a political negotiation of 'who we are' and how to realize the ideal. Current debates about language policy are, at the core, about 'whom" we should be.' The growth in tourism industry in Greenland paves the road to language competency, global connectivity, and self realization as an independent nation.

16.3.3 Arctic Cruise Ship Tourism

The presence of the cruise ships in the Arctic water, creates a struggle between environmental and economic concerns in the era of climate change. While cruise tourism does not burden a small coastal settlement with accommodation needs, their presence nonetheless can be overwhelming and unwelcomed to the local community especially if there is lack of communication between the community and the cruise-ship about the arrival, program, and attitude of the tourists visiting the settlements. Coastal Greenlandic communities are not very fond of cruise tourism as it does not generate any substantial income opportunity albeit receiving an influx of people for a short amount of time. Cruise ships usually show up unannounced and it could be during the seasons when villagers travel to hunt or fish in different locations.

Usually, tourists do not buy anything from local community except a few postcards. On the other hand they purchase imported fresh produce that the local community members await for months. Due to lack of infrastructure (sitting area, restaurant, public toilets, etc.) there is no chance to engage in any Kaffemik, traditional dance performance, etc. (local idea about cultural tourism). Although local people make an effort by wearing their festive traditional costume, and display their handicrafts, tourists do not buy them and research investigations revealed that tourists are either ignorant about the significance of the animal bone carvings, due to lack of cultural understanding, imposed ban on animal fur and bones in their home country, or they find them not aesthetically pleasing (Tommasini 2011). Lack of respect for local customs, and treating the inuits as an "object of curiosity" by taking pictures of them and their homes without permission and asking private questions make the encounter between a local and a tourist not so welcoming. Additionally, locals also worry about their community problems getting exposed to foreign travelers, alcoholism and scattering of garbage throughout the settlement. And, most importantly, big cruises are dangerous to marine animals, especially narwhal, when they enter the breeding ground in the fjords.

This is why Arctic Council's one of the working groups, Protection of the Arctic Marine Environment (PAME) has created Arctic Marine Tourism Projects (AMTP). They produced the Best Practice Guidelines in 2015 to analyze and encourage sustainable tourism in the polar water. Among other goals, two topics related to Greenlandic cruise ship tourism:

- Takes into account regional variations, types of vessels and tourism operations and multiple stakeholder perspectives
- Considers the intended audience(s) for best practice guidelines

Due to the challenges and disconnection between cultural understanding, cruise ship tourism has not been successful in remote, and small settlement and the data related to Greenlandic cruise tourism reflect that finding in Fig. 16.2.

Some operators with a longstanding history in the Arctic – in particular members of the Association of Arctic Expedition Cruise Operators (AECO) – tend to be familiar with the risks, issues and considerations related to Arctic marine tourism. Generally, these operators set a positive example, including for those less experienced with the challenges of Arctic marine tourism like pleasure craft. Indeed, many of these experienced operators have good relations and communications with local communities and coastal administrations and conduct their operations in a responsible, safe, cultural and environmentally sustainable way. On their website, AECO declared, "The Association of Arctic Expedition Cruise Operators" (AECO 2018) is an international organization for cruise operators. Our organization's main objective is to ensure that expedition cruises and tourism in the Arctic are carried out with the utmost consideration for the fragile, natural environment, local cultures and cultural remains, while ensuring safe tour operations at sea and on land." To inform the visiting tourists about the local environment, social, and economic situation, AECO has

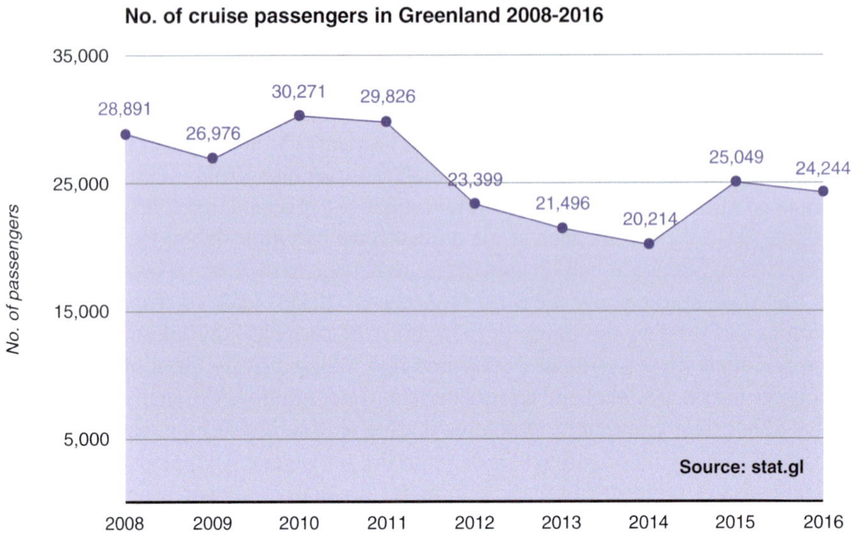

Fig. 16.2 Projected Tourism Growth in Greenland. (Source: www.stat.gl)

created several informative guidelines: visitor guidelines, wildlife guidelines, community guidelines, biosecurity guidelines, etc. In addition, on their website, there is a special section dedicated to "Cultural and Social Interaction" it reads:

> *For a small and sometimes isolated town or settlement in the Arctic, the call of a cruiseship is often a welcome and happy event. Locals may find both ships and their passengers interesting. But tourism in Arctic regions is growing rapidly. Anyone involved needs to be cognizant to the economic, social and cultural impact the growing tourism may have on local communities. Responsibility for respectable interaction and local benefits also rests with the tour operators and visiting guests being respectful and understanding of the local culture.*
>
> ***Respect local cultures:***
>
> - *Work against prejudiced attitudes*
> - *Respect privacy; keep a good distance from private houses and never glance or photograph through private windows*
> - *Talk to and not about people you meet*
> - *Do not visit graveyards or other areas of religious or cultural significance without permission*
> - *Ask before you photograph – a hesitation means NO*
> - *Cairns may be signposts – do not alter them*
> - *Never barter or import banned substances to a community*
> - *You are encouraged to buy local souvenirs and products, but be aware of the legalities of importing/transporting purchases into other countries e.g. CITES – Convention of 3 March 1973 on International Trade in Endangered Species of Wild Fauna and Flora/ The Washington Convention, www.cites.org*
>
> ***Cultural understanding:***
>
> *Tourism is a great way of learning about, promoting and creating tolerance between people of different backgrounds and cultures. When visiting foreign countries and cultures,*

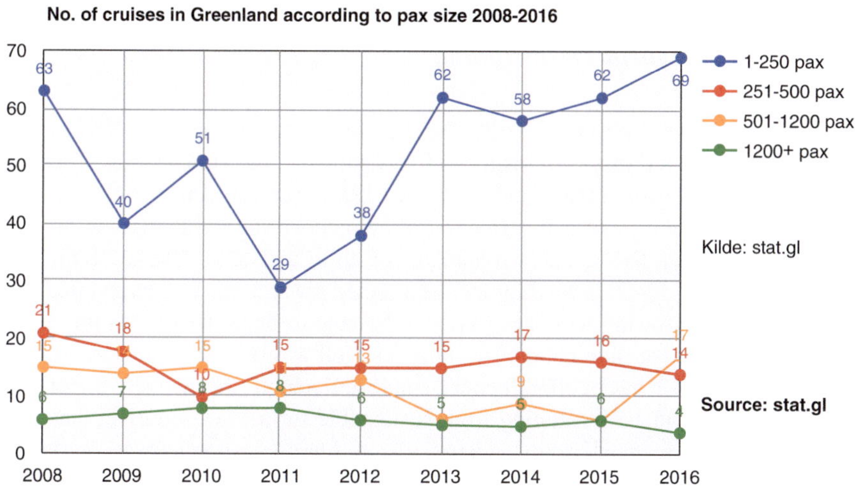

Fig. 16.3 Projected Tourism Growth in Greenland. (Source: www.stat.gl)

guests may find things very different from home. It is important not to judge other cultures based on one's own sense of reality, norms and values, but try to understand that cultures are qualitatively different.

To the contrary, the expedition ships (passengers up to 250 people) have been steadily increasing and if cruise tourism in Greenland is to fulfill an overall strategy for economic and environmental sustainability, expedition tourists fit into that strategy (Fig. 16.3). This type of tourism fits the targeted tourists criteria; nature lovers, culture lovers, pioneering mindset to match an adventure destination such as Greenland. In this type of adventure travel, usually the local fishermen and hunters are involved with their small boats to take the tourists in the inner fjord areas to enjoy the whales and narwhal without causing any disturbance to the wildlife as well as participate in kaffemik, traditional drum dancing in the settlements.

16.4 Conclusion

"We're in a time of reconciliation, and tourism is one of the best ways to get people face to face and getting to know each other. For us in the NWT we've got 33 communities, 11 official languages and each of those communities has their own vibrant and rich culture and unique story that helps define them and everyone of them can benefit from this area of tourism because it's something anyone can participate in if that's what they want. Tourism not only creates and sustains jobs it helps preserve language and culture," says Keith Henry, the president & CEO Indigenous Tourism Association of Canada (interviewed by Ellis Quinn, journalist "Eye on the Arctic" circumpolar news project, Radio Canada International).

16.4.1 *Indigenous Cultural Tourism: A Success Story from North West Canada*

Ellis Quinn reported on Oct 5th, 2018 that government of Northwest Territories' Department of Industry, Tourism and Investment, the Indigenous Tourism Association of Canada (ITAC), and Northwest Territories Tourism (NWTT), signed a memorandum of understanding to develop Indigenous cultural tourism in the territory and included new annual investments of up to $257,000 (Quinn 2018). The majority of the 2018–2019 funding will be directed towards workshops and marketing, to help the growing local businesses to be able to better handle visitors from southern Canada and international visitors, as well as creating greater awareness about what is NWT has to offer. In some small settlements, the unemployment rate is between 5% and 10% resulting from sustainable form of eco-tourism that was thoroughly well planned from land use to monitoring visitor impacts by some community-owned tourism development. The governmental agreement was based on ongoing mentorship in the community, and a strong support system. Understanding the reality of unreliable mining business and to create a balanced, sustainable economy, the government has started to take tourism seriously.

This success story provides hope for the development of tourism in the remote coastal towns and settlements of Greenland. "Tourism is a social/economic phenomenon that acts both as an engine of economic progress and a social force. Tourism is much more than an industry. Tourism is more like a "sector" that impacts a wide range of industries. Tourism is not just businesses or governments – it is people. Supporting rational tourism growth and development needs to be viewed in this broader context (Davidson 1994, p. 26)."

Arctic societies and economies have historically been and continue to be natural resource-based, which have integrated these societies in global networks. Tourism relies increasingly on natural and cultural elements. And this sector would transmit the formally and informally acquired skills of Greenlandic citizens to participate in global networks and transactions.

According to research by Freya Higgins-Desbiolles, it is important to qualify the emphasis on tourism's economic contributions by highlighting its other positive impacts (McLaren 1998; Wearing 2001; Wearing 2002; Scheyvens 2002; Reid 2003) which include improving individual wellbeing, fostering cross-cultural understanding, facilitating learning, contributing to cultural protection, supplementing development, fostering environmental protection, promoting peace and fomenting global consciousness which contributes to the formation of global society (Cohen and Kennedy 2000, p. 212 for the last point; WTO 1999 for the former point). In the 1990s, many analysts acknowledged the power of tourism as a social force. Barnard and Spencer argue that "to ignore tourism in our accounts of culture contact in the 20th century is probably as great an omission as to ignore slavery in the 18th century or colonialism in the nineteenth".

Tourism indeed is an agent of positive change when the pace and scale remain appropriate to the capacity of the community, at current and future times (through careful socio-environmental, assessments). Instead of focusing on mass tourism in Greenland, the focus should be have a complete assessment of the possible negative impact that large numbers of tourists can have on communities. "Factors found to contribute to the success of tourism development in peripheral regions include the presence of a leader, effective private and/or public sector partnerships, the identification and development of specialist attractions, government control and support, good market research, and community involvement" (Blackman et al. 2004, p. 59). Tourism development is often considered as viable alternative for peripheral regions. The realities of tourism are not always clearly understood. The long-term success of the tourism industry depends upon the acceptance and support of the host community (Murphy 1985; Wearing 2001). Successful tourism development does seem possible for peripheral regions but is not a rapid or simple solution: it requires substantial long-term governmental support and extensive training, research and planning processes. The potential exists, however, and a growing number of tourists are seeking the sort of specialized experiences available in peripheral regions (Blackman et al. 2004). How to build and enhance the community capacity for tourism? It is only possible when the big industries (airlines, brand name hotels) work with the local organizations (i.e., tour guide companies) by providing tools for actual practice, including suggested steps (housing deveopment in Nuuk) for building local capacity for development, models and framework and lessons for participants in real tourism development situations by continously sharing the success story of participants and building the confidence in the community, and industry, collectively.

Long before the era of commercialized Arctic tourism, Greenland has always been a magnet for the well known explorers who came to this land to reach the north pole. Their legendary expeditions are still inspirations that lure the modern day travelers to follow their path, or experience the off-the-beaten path and become a "Pioneer" (the current marketing slogan of Visit Greenland). Due to the remoteness of the place and harsh arctic condition, organized tourism has been the practical solution for decades. The current Visit Greenland campaign is based on investigative research to understand local perspective and expectations of people in the settlements perceive tourism, and what they see as the best way to develop and offer a high-quality tourist product, which has the ability to attract tourists to their remote and beautiful areas. Offering a unique experience to visitors may well develop a "niche tourism" for very special tourists in a very special place while helping build a strong local economy and social and cultural sustainability in Greenland.

References

Association of Arctic Expedition Tour Operators (AECO). (2018). https://www.aeco.no/guidelines/visitor-guidelines/. Retrieved on October 5, 2018.

Berkes, F. (2000). Indigenous knowledge and resource management systems in the Canadian subarctic. In F. Berkes & C. Folke (Eds.), *Linking social and ecological systems* (pp. 98–129). Cambridge: Cambridge University Press.

Bjørst, L. R. (2008). En anden verden. *Fordomme og stereotyper om Grønland og Arktis* [Another World. Prejudices and Stereotypes against Greenland and Arctic Regions]. Denmark: Forlaget BIOS.

Blackman, A., Foster, F., Hyvonen, T., Jewell, B., Kuilboer, A., & Moscardo, G. (2004). Factors contributing to successful tourism development in peripheral regions. *The Journal of Tourism Studies, 15*(1).

Burdenski, A. (2018). *An emerging Arctic destination in the worldmaking – The journey of a tour guide in Greenland (tourism master thesis with a specialization in global tourism development)*. Denmark: Aalborg University. http://www.tourismstat.gl/resources/reports/da/r22/Master%20Thesis_Anna%20Burdenski.pdf. Accessed 10 Sept 2018.

Cohen, R., & Kennedy, P. (2000). *Global sociology*. Houndsmills: Macmillan Press.

Davidson, T. L. (1994). What are travel and tourism: Are they really an industry? In W. Theobold (Ed.), *Global tourism: The next decade* (pp. 20–39). Oxford: Butterworth-Heinemann.

Duerden, F. (2004). Translating climate change impacts at the community level. *Arctic, 57*, 204–212.

Fay, G., & Karlsdottir, A. (2011). Social indicators for Arctic tourism: Observing trends and assessing data. *Polar Geography, 34*, 63–86.

Greenland adventures by Icelandic mountain guides (Tour Company). (2018). Retrieved on October 10, 2018, https://www.greenland.is/about-us/

Greenland in Figures. (2018). *Statistics Greenland*. Retrieved from: Greenland in figures 2018. pdf. Accessed 5 Oct 2018.

Finne, A. (2018, June). English won't replace Danish as Greenland's second language. *High North News*. Retrieved from www.arctictoday.com. Retrieved on October 5, 2018.

George, S. (1999). *A short history of neo-liberalism*. Conference on Economic Sovereignty in a Globalising World, Bangkok, 24–26 March. Retrieved on August 10, 2018 from https://www.tni.org/en/article/short-historyneoliberalism

Hamiltom, C. (2004). *Growth fetish*. USA: Pluto Press

Hersoug, B. (1999). Conditions for a sustainable development – Some experiences from the Norwegian Fishing Industry. In H. Petersen & B. Poppel (Eds.), *Dependency, autonomy, sustainability in the Arctic* (pp. 335–350). Aldershot: Ashgate.

Holm, M. & Rasmussen, O. R. (2000). *Knowledge and sustainability as factors in the Greenland process of development* (in Danish: Viden og bæredygtighed som faktorer i den grønlandske udviklingsproces). Working paper no. 157. Geography, Roskilde University.

http://www.europarl.europa.eu/RegData/etudes/briefing_note/join/2014/522332/EXPO-AFET_SP(2014)522332_EN.pdf. Accessed 18 Oct 2018.

Huntington, H. P. (1998). Observations on the utility of the semi-directive interview for documenting traditional ecological knowledge. *Arctic, 51*(3), 237–242.

Ingold, T. (2000). *The perception of the environment*. London: Routledge.

Kleist, K. V., & Knudsen, R. J. (2016). *Sitting on gold: A report on the use of informally acquired skills in Greenland*. Retrieved from bit.ly/2uzHwPg. Accessed 17 Sept 2018.

Knudsen, R. (2016). *Perspectives on skills. Greenland perspective*. University of Copenhagen.

Kruse, J., Lowe, M., Haley, S., Fay, G., Hamilton, L., & Berman, M. (2011). Arctic observing network social indicators project: Overview. *Polar Geography, 34*, 1–8.

Mason, P., Johnston, M., & Twynam, D. (2000). The world wide Fund for Nature Arctic Tourism Project. *Journal of Sustainable Tourism, 8*(4), 305–323.

McLaren, D. (1998). *Rethinking tourism and ecotravel*. West Hartford, Conn: Kumarian Press.

Murphy, P. E. (1985). *Tourism: A community approach*. New York: Methuen.

Naatsorsueqqissaartarfik/Greenland Statistics (stat.gl). (2015). *Befolknings uddannelsesprofil 2015 (Educational profile of the population)*. Retrieved from bit.ly/2Gz8Jqn. Accessed 17 Sep 2018.

Nordal, J., & Kristinsson, V. (Eds.). (1996). *Iceland, the republic: Handbook*. Reykjavik: Central Bank of Iceland.

Pandya, R. E. (2012). A framework for engaging diverse communities in citizen science in the US. *Frontiers in Ecology and the Environment, 10*, 314–317.

Quinn, E. (2018). *cbc, ca. Indigenous cultural tourism: How the North is learning from community success in Southern Canada*. Retrieved on October 10, 2018 from: http://www.rcinet.ca/en/2018/10/06/indigenous-cultural-tourism-how-the-north-is-learning-from-community-success-in-southern-canada/

Rasmussen, Rasmus. (2014). *Gender perspectives on path dependency*. Conference Report on Gender Equality in the Arctic, Akureyri, Iceland, October 2014. Retrieved on October 20, 2018 from https://www.stjornarradid.is/media/utanrikisraduneyti-media/media/nordurslodir/Gender-Equality-in-the-Arctic.pdf

Reid, D. G. (2003). *Tourism, globalization and development: Responsible tourism planning*. London: Pluto Press.

Ren, C., & Chimirri, D. (2017). *Turismeudvikling i Grønland: Afdækning og inspiration (Tourism development in Greenland – Identification and inspiration)*. Denmark: Aalborg University. Retrieved from www.aau.dk/digitalAssets/282/282589_turismeudvikling-i-groenland-rapport_endelig.pdf. Accessed 17 Sept 2018.

Scheepens, S. (2014). *Exploring the potential participation in citizen science, conservation tourism, and participatory environmental research tourism to lead an environmental change in practices*. Master's thesis, Wageningen University: Environmental Policy Group.

Scheyvens, R. (2002). *Tourism for development: Empowering communities*. Harlow, England: Prentice-Hall.

Stromberg, J. (2011). *Climate change tourism in Greenland*. Retrieved from https://www.smithsonianmag.com/travel/climate-change-tourism-in-greenland-74303453/. Accessed 20 Sept 2018.

Tommasini, D. (2011). Tourism experience in the peripheral north-case studies from Greenland (Naalakkersuisut government of Greenland). *Inussuk: Arctic Research Journal, 2*. Retrieved from: https://naalakkersuisut.gl/~/media/Nanoq/Files/Attached%20Files/Forskning/Inussuk/DK%20og%20ENG/inussuk%202%202011.pdf. Accessed 12 Sept 2018.

Visit Greenland (VG). (2016). *Tourism strategy 2016–2019 visit Greenland*. Retrieved April 10, 2018 from, http://corporate.greenland.com/en/about-visit-greenland/strategi-2016-2019

Wearing, S. (2001). *Volunteer tourism: Experiences that make a difference*. Oxon: CABI.

Wearing, S. (2002). Re-centering the self in volunteer tourism. In G. S. Dann (Ed.), *The tourist as a metaphor of the social world* (pp. 237–262). Oxon: CABI.

World Tourism Organization (WTO). (1985). *Tourism bill of rights and tourist code*. Retrieved 08/20/2018 from https://www.e-unwto.org/doi/abs/10.18111/unwtogad.1985.1.hp408706117j8366

World Tourism Organization. (1999). *Resolutions adapted by the general assembly at it's thirteenth session*. Santiago, Chile, 27 September – 1 October, 1999.

World Tourism Organization. (2016). *Compilation of UNWTO recommendations, 1975–2015*. Madrid: UNWTO.

World Tourism Organization (WTO). (2017). *UNWTO annual report 2017*. Retrieved October 10, 2018. https://www.e-unwto.org/doi/pdf/10.18111/9789284419807

Chapter 17
Marine Tourism Development in the Arkhangelsk Region, Russian Arctic: Stakeholder's Perspectives

Julia Olsen, Marina Nenasheva, Karin Andrea Wigger, Albina Pashkevich, Sonja H. Bickford, and Tatiana Maksimova

Abstract The Arkhangelsk region is a strategic area for cruise tourism development in the Russian European Arctic. The region offers its domestic and foreign visitors a large number of unique natural, cultural, and historical sites and provides an opportunity to explore coastal settlements and the region's remote areas. However, it can be said that despite the variety of existing national and regional institutional arrangements, as well as the industry's managerial practices, the sustainable development of marine tourism in the region is highly reliant on local stakeholders, such as local authorities, travel companies, and local providers of hosting/tourism activities. In order to examine the sustainability of the current development practices, this chapter uses the findings from qualitative interviews to understand how cruise tourism in the Solovetsky archipelago is managed locally and regionally. Our study emphasizes the need to implement a communication model based on the cooperation and engagement of all relevant stakeholders as a platform to address sustainability issues inherent in the growth of cruise tourism. The study thus helps to address the problems associated with cruise tourism development in the Arctic and to deepen the discussion related to the peculiarities of tourism destination development in the Russian European Arctic.

J. Olsen (✉) · K. A. Wigger · T. Maksimova
Nord University, Bodø, Norway
e-mail: julia.olsen@nord.no; karin.a.wigger@nord.no

M. Nenasheva
Northern (Arctic) Federal University, Arkhangelsk, Russia
e-mail: m.nenasheva@narfu.ru

A. Pashkevich
Dalarna University, Falun, Sweden
e-mail: alp@du.se

S. H. Bickford
University of Nebraska Kearney, Kearney, NE, USA
e-mail: bickfordsh@unk.edu

E. Pongrácz et al. (eds.), *Arctic Marine Sustainability*,
Springer Polar Sciences, https://doi.org/10.1007/978-3-030-28404-6_17

Keywords Cruise tourism · Russian Arctic · Solovetsky · Stakeholders · Sustainability · Communication

17.1 Introduction

In recent years, cruise tourism has been gradually moving northward, offering passengers a chance to visit and experience the Arctic wildness and coastal settlements (Lück 2010; Dawson et al. 2014; Johnston et al. 2012). The Arctic has fascinated humankind for centuries because of its unique cultural, historical, and natural attractions (Howard 2009; Bystrowska et al. 2017). Military restrictions and climatic and geographical conditions, such as long distances between settlements and the prevalence of sea ice, are just some of the challenges for foreign cruise operators in the Russian High Arctic (Ho 2010; Pashkevich and Stjernström 2014). The ongoing changes in the region's climate have resulted in the decline of sea ice, which in turn has led to an extension of the seasons for different tourism activities, such as overseas and explorer cruises (e.g., Dawson et al. 2018). Simultaneously, due to the political changes in Russia, the Russian Arctic has become a strategic region for the country's tourism development (Pashkevich and Stjernström 2014; Grushenko 2009, 2012). As a result of these climatic and political changes, several Arctic destinations, such as local coastal communities, are increasingly being visited by overseas cruise vessels (Dawson et al. 2018; Pashkevich et al. 2015; Pashkevich and Stjernström 2014).

The potential economic impact of tourism could improve the standard of living of the local population by providing employment opportunities and increased income in the retail and service sectors (Huse et al. 1998; Viken and Aarsaether 2013). However, this is not always the case for all cruise tourism destinations. Examples from the Canadian Arctic illustrate that visitors stay in the communities only for a short time, and consequently the financial benefits for the local retail and service providers are minimal (Stewart et al. 2015). In addition, tourism activities, if not managed and monitored properly, can have a negative impact on the surrounding nature and can change the living conditions of the local population significantly (e.g., Hall et al. 2010). In this regard, it is necessary to find a balance between the expected economic benefits and the possible social and environmental consequences of tourism—more commonly known as the problem of sustainable tourism development (Butler 1999; Viken 2004). With the growing number of visitors to the Arctic, the anthropogenic impact on the ecosystem of the northern territories also increases, potentially leading to the loss of both natural and cultural values.

Climate change is another factor that must be taken into consideration with regard to the sustainability of the cruise industry, in addition to environmental changes, social issues, and cultural loss (Mason 1997; Scott 2011). Sustainable tourism is defined as "tourism that takes full account of its current and future economic, social, and environmental impacts, addressing the needs of visitors, the industry, the environment, and the host communities" (UNEP-WTO 2005, p. 12). Prior studies have analyzed sustainable practices in the development of several Arctic tourism destinations, such as Svalbard (Viken 2011; Van Bets et al. 2017), Alaska (Klein 2011), Canada (Stewart and Draper 2006), and the Russian Arctic (Pashkevich et al. 2015; Pashkevich and Lamers 2015). The study of tourism development in the European Russian Arctic by Pashkevich and Lamers (2015) suggested that further study of this region is required in order to increase our understanding of the effects caused by the changing environment and the opportunities hidden in the development of cruise tourism there.

Expanding on the debate on sustainable cruise tourism in the Arctic, this chapter attempts to improve our understanding of the sustainability of current and ongoing development practices by examining how cruise tourism on the Solovetsky archipelago, Arkhangelsk region, is managed locally and regionally. It adopts a stakeholder approach (Byrd 2007) to identify regional and local stakeholders' actions and opinions on sustainable development in the region, and in order to assess the local stakeholders' perspectives, a case study of the Solovetsky archipelago is used. The study's uniqueness and novelty are to be found in the inclusion of local viewpoints. We contribute to the literature on destination development by gathering and analyzing the insights on sustainable cruise practices from local and regional stakeholders. This enables us to gain a more holistic understanding of the possibilities of sustainable tourism development in the Arkhangelsk region. An additional contribution comes from the proposition of the communication model based on the cooperation and engagement of multiple stakeholders as a platform for addressing sustainability issues inherent in the growth of cruise tourism.

17.2 Study Approach

17.2.1 Sustainability of Cruise Tourism

Tourism is one of the largest industries in the world (McElroy and Potter 2006), of which cruise tourism is the fastest-growing market segment (Van Bets et al. 2017). Cruise tourism is defined as a journey, via water transport, along a specified route with stops at local landing sites, such as port cities or a nature-based tourism attraction (Gibson 2012; Baikina and Valkova 2011). Various types of cruise can be distinguished, such as overseas, explorer, river cruises, etc., depending on the direction and area of navigation, as well as the type of vessel and its purpose. The main characterizing feature of cruise tourism is that the ship acts simultaneously as a means of transport, a place of residence, and a recreation site (Logunova 2013). Studies have highlighted critical factors for the successful development of cruise tourism, including the geographical location of the port, availability of port infrastructure, level of port charges, availability and accessibility of cultural and historical attractions and unique natural landscapes, presence of tourist infrastructure (transport, hotels, guides, etc.), the recognition of the region among cruise tourism providers, and political stability at the destination (Smirennikova 2009; Yakovenko and Lazitskaya 2014). Several of these factors can be influenced directly by the destination, which makes support from both the local authorities and the port administration an important prerequisite for successful cruise tourism development.

Experiences have shown that an increase in the arrival of tourists from larger cruise ships to a port community may pose challenges for sustainable tourism development (Marsh 2012). Because an increase in cruise tourism activities in the Arctic could lead to a decline in the attractiveness of the destinations (Stewart et al. 2007), it could also negatively impact the local population (Stewart et al. 2011) and result in biodiversity loss (Hall et al. 2010). Sustainable tourism development implies a balance between economic benefits, social development, and preservation of the natural environment (Butler 1999; Viken 2004). In addition, important factors for sustainable tourism development are competent and effective management of the tourism sector and the inclusion and involvement of all stakeholders in its planning (Byrd 2007).

Despite the current national strategic focus in Russia on increasing cruise activities and on the known sustainability issues accompanying increased tourism, the sustainability of tourism development has not yet received much attention in the scientific literature (Gairabekov et al. 2017). The existing literature exploring cruise practices devotes much of its focus to the economic aspects of cruise tourism development (e.g., Grushenko 2009), while socio-ecological aspects remain neglected. We find the same approach, with more of an economic focus, in the state's strategies on tourism development (e.g., Russian Federation Government 2014). The strategy's main emphasis is on the economic efficiency of tourism development (here, tourism is an essential component in the country's innovative, financial, and employment base), while the social and environmental aspects of tourism development are practically ignored (Kiyakbaeva 2014; Rassokhina and Seselkin 2015; Seselkin 2014). Moreover, the coastal zone of the White Sea has not been sufficiently studied. As a result, there are no recommendations on the sustainable use of its potential.

17.2.2 Stakeholders' Perspectives

A stakeholder approach is a normative tool used in sustainable tourism development and planning (Sautter and Leisen 1999; Simpson 2001). The fundamental idea of stakeholder theory, in the context of tourism development, is that in order to succeed, the various stakeholders need to agree with the strategic orientation of tourism development. These stakeholders, however, can have complementary and/or conflicting interests and goals, making such an agreement a challenging endeavor (Sautter and Leisen 1999).

Stakeholders' engagement and collaboration in cruise tourism are crucial for the development of the industry locally but also for ensuring the sustainability of the region and industry (Simpson 2001; Byrd 2007). Stakeholders in the case of cruise tourism include operators, travel agencies, ship owners, and their passengers, as well as local tourism providers, retail businesses, and tourism destinations, not to mention public decision makers and the general public of the community. Inspired by London and Lohmann's (2014) categories of cruise stakeholders, for this study we separate them into cruise industry stakeholders (e.g., operators, shareholders, ship owners, and passengers), regional stakeholders (e.g., regulatory officials, authorities, and regional tourism organizations), and local shore-site stakeholders (e.g., attractions, tour operators, local transportation providers, and local business). In this study, we focus on a stakeholder approach in order to assess the role of regional and local shore-site stakeholders and their perspectives on cruise practices.

For example, the local population is a key stakeholder in tourism development and has traditionally been the bearer of cultural capital (Castro and Nielsen 2001; Moller et al. 2004). Factors affecting stakeholders in a tourist destination are financial and resource constraints, political situation and support, policies affecting the management of tourism operations, and the level of collaboration among stakeholders in the region (Jovicic 2014). Thus, participation by local residents in managerial decision making can be ensured by engaging those in the local communities who are interested into the discussions about the strategies and plans for tourism development, as well as securing their active involvement in the tourism activities themselves (see Chap. 14). The effectiveness of using the stakeholders' skills and potential largely depends on the chairperson's ability to build effective dialogue and a system to allow for the participation and involvement of all the actors in the management process (Melnik 2015; Jovicic 2014). Jovicic (2014) states that a fundamental change among stakeholders is needed with regard to ethical interest alignment so that the norms and principles of sustainable tourism can be a common focus for discussions, strategies, and operations. Huxley and Yiftachel (2000) found that the planner, meaning the one facilitating the communication between stakeholders over matters of common concern, fostered communication and engagement through the recognition of diversity, difference, and the common understanding of how communication takes place.

The concept of communicative management has gained weight as a practical application for solving specific problems of planning (Healey 1996; Huxley and Yiftachel 2000). The development of communicative management involves establishing a dialogue and actively integrating local communities into managerial processes (Krasilnikov et al. 2014; Bulkeley 2005; Karkkainen 2002).

With regard to communication, one of the already implemented practices that secure information exchange and strategic planning for destination development is a collaborative cruise network that provides a platform for cooperation between several stakeholders and enables community participation (e.g., Cruise Network Svalbard, see Chap. 14). In Russia, the regional and local governments provide tools to secure the participation of local businesses in tourism development by providing special tax incentives, creating tourist clusters, and stimulating public-private partnerships. As for the Arkhangelsk region, the "Belomorsky tourist cluster" was established a result of cooperation between several regional districts (Lamers and Pashkevich 2015).

However, even today, the local businesses and local inhabitants are not broadly involved in the formal decision-making process concerning tourism development (Melnik 2015; Nekrasova 2016), even though these actors possess competencies that would allow them to participate actively in the interactive management of the tourism industry in their respective regions. More specifically, companies (including tourism companies) that have intellectual and economic capital can and should participate in the conceptual development of strategies and plans for tourism development of territories, in addition to investing in new business projects and opportunities. The Russian Constitution ensures that the appropriate citizens participate in the decision-making process for territorial development programs, as do a number of federal regulatory and legal acts. In addition, specific tools have been developed to ensure citizens' access to reliable information, as well as their participation in discussions about planned projects. These tools include public hearings, public discussions, and public examinations (Nenasheva et al. 2015).

17.2.3 Case Study Settings

This qualitative study has been designed to examine regional and local stakeholders' perspectives on cruise tourism sustainability in the Solovetsky archipelago, Arkhangelsk region, to assess local stakeholders' attitudes, perspectives, and roles in the ongoing development of the industry, as well as to explore future opportunities and potential.

The increasing interest among cruise tourists to explore Arctic Russia is becoming noticeable in the Arkhangelsk region (e.g. Grushenko 2014; Lamers and Pashkevich 2015). Along with its historical, cultural, and natural significance, the geographical location of the Arkhangelsk region (Fig. 17.1), including its coastline on the White Sea and the Barents Sea, offers a wide range of cruise tourism destinations. These include everything from experiencing the wildlife in the High Arctic (e.g., at the Russian Arctic National Park on the northernmost island of Novaya Zemlya and the Franz Josef Land archipelago) to cultural heritage excursions and rural settlements (e.g., the coastal communities of Pomors[1]), including a UNESCO World Heritage Site: Solovetsky archipelago.

Cruise tourism in the Arkhangelsk region is not a new phenomenon. Due to the dense river network and its connection to both the White and the Barents Seas, extensive use of internal and external water communications and transportation have helped to develop a strong connection between the inhabitants of the region and have allowed for the development of tourism, mainly attracting domestic tourists (Nenasheva and Olsen 2018). In contrast to domestic sea tourism options, foreign cruise vessels constitute a new trend, which began at the start of the twenty-first

[1] Russian settlers living by the White Sea.

Fig. 17.1 Map of the study area with adjustment territories

Table 17.1 The main characteristics of passenger vessels (including overseas cruise vessels) to and from Solovetsky (2008–2016)

Year	2008	2009	2010	2011	2012	2013	2014	2015	2016
Total amount of port calls by passenger vessels	**466**	**410**	**516**	**490**	**563**	**545**	**540**	**595**	**596**
Including overseas cruises	1	2	0	3	6	3	4	6	4
Total amount of passengers (thousands)	**22.9**	**27.8**	**31.0**	**33.6**	**30.3**	**30.1**	**62.8**	**78.5**	**74.4**
Including the amount of passengers arriving onboard overseas cruise vessels	102	804	0	1306	2004	1970	1232	3524	2116

century (Lamers and Pashkevich 2015). Despite its location in the European part of the Russian Arctic, the Arkhangelsk region still occupies a modest place in cruise tourism (Toskunina and Smirennikova 2011; Gomilevskaya and Petrova 2017).

Nowadays, the port of Arkhangelsk is visited by three to seven overseas cruise vessels during the summer months (see Table 17.1 for more details on oversea cruise arrivals in Solovetsky). Like in other Arctic destinations, there are certain challenges for cruise development in the Arkhangelsk region. For example, due to weather and sea ice conditions cruise tourism activities on the basin of the White Sea take place only during the summer months, from June to early September and while recent changes in the ice cover have resulted in the expansion of this navigation season (Dumanskaya 2014), only cargo transportation is currently taking advantage of this, with overseas vessels continuing to operate only in the summer months (Olsen and Nenasheva 2018). Other challenges are governance complexity (Pashkevich et al. 2015), a lack of sufficient on-land and offshore infrastructure to accommodate bigger vessels (Grushenko 2014) and insufficient host activities, visitor management and logistical services (Lamers and Pashkevich 2015).

17.2.4 The Solovetsky Archipelago: A Pearl in the Arkhangelsk Region

The Solovetsky archipelago (*Solovki* in Russian), one of the most popular cruise destinations in the Arkhangelsk region, was chosen for this study to illustrate the local stakeholders' perspectives on cruise tourism development (Fig. 17.1). The archipelago's unique natural, cultural, and historical heritage, developed on the foundation of the Solovetsky monastery (Fig. 17.2), led to the inclusion of the archipelago on UNESCO's World Heritage list in 1992 (UNESCO 1992). As with Arkhangelsk, interest in this destination has been growing since the early 2000s, with a rise in visits from Russian and international tourists, pilgrims, and other visitors. This influx of tourists became especially noticeable due to the increased number of people present and visible in the community. Because the key attraction is on an island, access for passenger traffic to and from the archipelago is predominantly

Fig. 17.2 The Solovetsky Monastery, the main attraction of the Solovetsky archipelago. (Photo credit: Julia Olsen)

Fig. 17.3 The cruise vessel *Discovery* anchors near the Solovetsky archipelago. (Photo credit: Maksim Iliin)

by sea. Nowadays, Arkhangelsk is connected with Solovetsky via a few seasonal voyages by the passenger vessel *Belomorie.*

Marine passenger transportation, via vessels from Arkhangelsk to the Solovetsky archipelago, was initially established in the eighteenth century (Popov and Davydov 2003). Since then, vessels have taken pilgrims—and subsequently domestic tourists—to and from the archipelago. Regular cruises from Arkhangelsk were popular and in demand in the Soviet times, when a regular connection was provided by the domestic cruise vessels *Bukovina* and *Tataria* (Maksimova 2016). Nowadays, the majority of passenger transportation is provided by private companies in the Republic of Karelia that offer daily marine voyages to and from the archipelago (Tsvetkov 2011). During the last decade, overseas cruise vessels have also started to visit Solovetsky (Table 17.1 and Fig. 17.3) but number of tourists that the cruise vessels bring is still small in comparison to the number of domestic tourists brought by the domestic passenger vessels.

17.3 Study Methods

Qualitative methods were chosen in order to facilitate an in-depth study of stakeholders' perspectives and to provide a more detailed analysis of the contextual attributes of sustainable development. The authors used in-depth interviews, a document review, and observations. Prior to the fieldwork, background information on cruise tourism development in the Arctic was collected, and the relevant stakeholders were mapped. With the help of this information, an interview guide was developed. The media (local and regional) and a review of documents (tourism strategies and development strategies) were used. To ensure the inclusion of relevant local stakeholders, the authors used a snowball technique (Blaikie 2010, p. 179) prior to and during the fieldwork, meaning that study participants were asked to link to and recommend other competent and knowledgeable people who could reflect on the main topics of this study.

A total of 20 stakeholders were interviewed for the study in the Arkhangelsk and Solovetsky archipelago in June 2017 (Table 17.2). The interview guide covered the topics of tourism and cruise trends, as well as the effects of such activities, stakeholders' perspectives and the role of stakeholders and the local population in overall regional tourism development. The interviews took place at the interviewees' work places and lasted for about 45–60 min. The selection process was based on these actors' involvement in tourism industry operations in Arkhangelsk and its archipelago. Preliminary data were collected with the help of semi-structured interviews. The interview guide was designed to cover questions on cruise tourism development, critical perspectives, constraints, opportunities, the role of stakeholders and cooperation between them, and the prospects for sustainability and future development. The interviews were conducted in Russian by two of the authors, then transcribed in Russian and thereafter translated into English. The study was approved by the Norwegian Centre for Research Data and, to ensure anonymity, the interviewees were classified only on the basis of their geographic location.

Since the interviews were conducted at the beginning of the tourism season, the presented empirical data cover the period 2008–2016. During these nine seasons, the number of port calls by passenger vessels and community visitors increased.

At the analysis stage, data were thematically analyzed using NVIVO software (Bazeley and Jackson 2013), based on pre-defined categories from the interview guide. The list of categories was extended during the analysis process to include emerging topics related to sustainability, such as the three main categories of sustainability (the environment, society, and the economy), practices, and contextual attributes.

Table 17.2 List of study participants (stakeholders) in Arkhangelsk and Solovetsky

Stakeholders	Description
Arkhangelsk: A1–A10	Four representatives from tourism companies
	Three representatives from regional authorities
	Three representatives from the shipping industry related to marine cruises
Solovetsky: S11–S20	Three representatives from local public bodies
	Three representatives from tourism companies
	One seasonal worker
	Three local residents who are partly employed in the tourism industry in the summer season

Despite the inclusion of a broad range of regional and local stakeholders in the study, one of the main limitations of the study involved mapping and reaching other potential key stakeholders. This was due to their geographical locations (i.e., some of them resided in towns other than Solovetsky or Arkhangelsk) and the seasonality of operations (i.e., some stakeholders were on summer vacation when the study took place). Another limitation for the study stems from the researchers' inability to be present in the community and observe the community dynamics and tourist behaviors when one of the cruise vessels approached the settlement. It was only during the fieldwork that port call information could be obtained for the first time.

17.4 Empirical Findings

The conceptual basis of the sustainability approach in cruise tourism and stakeholders' perspectives were applied to analyze the empirical data. For the purpose of this chapter, we present the regional and local stakeholders' observations on the sustainability of tourism activities for the Solovetsky archipelago and compare it to aspects of cruise tourism development. First, we assess the three categories of sustainability: economy, environment, and society. Then, we apply a stakeholder approach to gain a detailed understanding of how the various stakeholders perceived the sustainability of the cruise arrivals.

17.4.1 The Economy

At the regional level, regional authorities and tourist companies describe tourism, including cruise tourism, as a major industry for the Arkhangelsk region. Despite the fact that, historically, the region was well known as a "wood province," this region is now trying to find a new economic direction (A6). It is argued that the tourism industry "the future economy of the Arkhangelsk region" (A3). The Solovetsky archipelago is usually described as one of the region's main attractions (A3). However, from a local perspective, one of the main issues related to this sector is the way it fosters local value creation. The three main attributes of value creation were identified by the stakeholders to describe the local economic situation during the marine tourism season: employment, income generation, and income distribution.

17.4.1.1 Employment

When describing employment opportunities, both local and regional interviewees refer to the local population who, in addition to their main work, have tourism-related jobs and/or provide tourism services during the summer months (S16), but also seasonal summer workers who are usually hired from Arkhangelsk and/or the neighboring Karelian Republic (A5). A majority of the local population (even teenagers) are involved in the provision of tourism-related services, including everything from housing to private excursions on land and sea. Even though several services have been officially registered in the last few years (A8), both local and regional authorities are concerned that most services are organized by private individuals and not by officially registered entrepreneurs or companies, resulting in the loss of benefits for the municipality or region (A9, S13).

Compared to the regular tourist flow, only a few locals (such as employees at the Solovetsky Museum) are involved in helping the tourist groups from international cruises. This was in most cases due to language difficulties (A5). The hosting activities, such as organized excursions for overseas cruise tourists, require knowledge of at least the English language, especially among the main tour guides. The potential for adapting tourism services to international tourists was stressed by stakeholders in Arkhangelsk, with one of them explaining that "descriptions in the local Gulag Museum are available only in Russian" (A5). It was reported that seasonal workers who work as guides are required to attend training courses prior to their employment in the archipelago. However, these courses are organized outside the archipelago (Arkhangelsk and Petrozavodsk), which makes it difficult for locals to attend (some locals, especially among the younger population, do speak English). Even though this is a popular job opportunity, the stakeholders from Arkhangelsk said that because of the seasonality of the job, it might be difficult to plan for the availability of guides for overseas cruise vessels. The question of guides and their availability always came up when employment was discussed. For example, one interviewee said, "Will there be a suitable amount, will there be at least ten guides, as required for a large liner? … Sometimes guides do not have time to eat, because they are very busy. But also, there is some turnover of employees … someone leaves, someone gets married" (A5).

17.4.1.2 Income Generation

According to a local representative of the tourism industry, tourism services on the archipelago have expanded dramatically during the past decade: "Everything has been changing here during the last few years. The accommodation capacity has increased by 80% [referring to hotels and private rental services], we now have mobile services and vehicles, private businesses have been founded, and three private excursion bureaus have opened" (S20).

Both local and regional stakeholders identify the tourism industry as the main source of income in the summer months for the inhabitants and those involved in tourism organizations (S11, S18). By providing tourism-related services (housing, excursions, transportation services, and the selling of souvenirs and local products) in the summer months, the local population earns additional income to supplement their income from their permanent jobs (A5, S11). This income helps to subsidize the otherwise low wages and rather high prices of foodstuffs in the local shops in the archipelago during the winter (S11).

The kind of income generated from the overseas cruise tourists differs from that of regular tourism activities. Due to draft limitations, cruise vessels have to anchor in Prosperity Bay and use tender boats to bring groups of tourists to the community (A5, S16). Usually, these groups participate in some excursions provided by the state-owned Solovetsky Museum during the day. Moreover, the accommodation and food services are provided onboard the cruise vessels, and not much food is consumed locally. Hence, the only income that is generated by the cruise tourists is derived from their participation in the excursions organized by the tour guides working at the Solovetsky Museum or from the purchase of souvenirs at the souvenir shops (some of these are registered outside the region, however). As one of the stakeholders in Arkhangelsk noted, in reference to 2015, when Solovetsky was reached by six overseas vessels carrying a total of 3000 foreign tourists: "Each of them bought an excursion, and several bought two excursions offered by the Solovetsky Museum [the price included the bus services and guides]" (A5).

17.4.1.3 Income Distribution

During the Soviet era, the city of Arkhangelsk was a point of departure for the Solovetsky archipelago (A6, A16). Nowadays, the neighboring Karelia Republic accounts for most of the marine traffic to the archipelago. Another link to the archipelago is by air directly from Arkhangelsk. Some of the stakeholders are concerned that the majority of tourists travel to Solovetsky through the neighboring Karelia Republic because it means that a portion of the income generated by tourism goes to the neighboring region and not to Arkhangelsk. Others believe that "Karelia promotes Solovki and helps to develop it. It is much faster and cheaper to get there from Karelia. Furthermore, Arkhangelsk can provide a year-round connection with Solovetsky by plane." (A8).

Not all of the existing private businesses are registered in the municipality, but outside of the region, meaning that the dominant part of income and, eventually, the tax income leave the archipelago without any contribution to the local society and community (S15). Hence, according to representatives from the local administration, the income from the tourism industry (including taxes) does not stay locally. In addition, only some locally provided tourism services are properly taxed (S13). One of the local tourism entrepreneurs, said, "Not everyone understands that only a small part of income goes to the local budget…. Only the minimum part of the generated tourism income remains locally in the settlement. Hence, local entrepreneurs are willing to pay taxes as long as they remain in the settlement. People are willing to contribute if it leads to the development of the local community" (S18).

The extent of local value creation from the overseas cruise industry is still unclear. Overseas cruise tourists still represent a small percentage of the overall total tourist flow to the archipelago (S19), and much of the income generated from their visits goes to the Solovetsky Museum, which provides the excursions and transportation services for the cruise tourists (A5). Another regional stakeholder pointed out the local population sometimes has little or no knowledge or practical experience about how to benefit economically from the cruise industry. He illustrated the point with the following example to emphasize the economic potential that remains unused in the absence of proper management practices: “My colleague at X [a White Sea coastal community] said that a cruise vessel anchored near their settlement and the passengers arrived by tender boats to their community. They [the tourists] were looking at us, and we were looking at them” (A3).

17.4.2 The Environment

Increased visitation has resulted in visible changes in the archipelago’s natural environment. Several of those interviewed for this study expressed their concerns about the environment (A3, S20), but it should be noted that there is disagreement between the stakeholders on this matter. At the regional level, the public bodies work with the number of visitors provided by the Solovetsky Museum and believe that there are “currently about 30,000 tourists per season, which means that we could accommodate an increase in tourism of 70–80%” (A8). However, one of the community representatives noted that the statistics provided by the museum’s authorities do not take into account the number of individual tourists (those not in organized groups), which according to some estimates may be as high as 80,000 per season (S11). Thus, locally, the management practices are developed to accommodate a greater amount of community visitors.

Regardless of this difference in figures, one tourism sector representative underlines that the impact on nature is directly related to the state of development of the local infrastructure, such as access to fresh water and proper sewage and waste treatment systems. According to the local stakeholders, the current facilities are meant to facilitate the local population of 900, and certainly not the additional 30,000 tourists that come to the archipelago (according to the numbers provided by the museum). The necessary improvements should be made to the public infrastructure, in order to be able to accommodate the needs of up to 100,000 people per season (S20). Waste management is identified as one of the major concerns for the local stakeholders (S11, S12, S14, S20). Most of the waste collected on the island also remains there, because only a small portion of the waste generated is transported to the mainland (S13).

With reference to cruise tourism, one of the local stakeholders thinks it is an optimal business segment for the archipelago: “There is nothing wrong with it. It will not do any harm. They go out, participate in excursions, buy souvenirs” (S16). The same interviewee stated, with regard to the waste management concerns, “Cruise vessels have the infrastructure necessary for their voyage on board” (S16), meaning that they do not generate a considerable amount of waste on the island. Individual tourists are described as having the greatest impact on the vulnerable environment of the island due to their access and mobility on the island, either on foot or by rented bicycle.

17.4.3 The Society

Both regional and local stakeholders describe the archipelago as "more than a regular tourist destination." It is an isolated, holy island with a church community [monks living at the monastery]. It is a historical, spiritual, and natural heritage site that is highly valued, and there is a desire to have it protected (A3, S13), especially in the context of increasing tourism and tourism-related infrastructure development. At the same time, tourism is not a new development trend on the archipelago. Those who lived on the archipelago during the Soviet era recall that the number of tourists to the archipelago was much higher (S12), with tourism at that time being dominated by domestic cruise tourism (S14, A9, A10).

Given this historical experience with domestic cruises and the current trends in overseas cruises, the stakeholders are trying to integrate industry growth with the Solovetsky heritage and to take advantage of its remoteness (A3, S11). Some are aware that the growth of tourism will disturb the isolated church community (A8). Others point out that the current infrastructure does not support the projected tourism growth (S13) and that it is the local population that will be exposed to the overloading of infrastructural capacity.

At the same time, when describing the overseas cruises, some interviewees point out that "groups are more organized; they come to the island in tender boats, and not altogether. Otherwise, there would be queues everywhere" (S13). Moreover, recent developments show that the church is becoming more welcoming and even more flexible toward tourists: "The monks have agreed to open the church earlier on some occasions when organized groups wish to visit outside the scheduled hours. Previously, this was impossible" (A5). The regional stakeholders who deal with the organization of the cruise visits point out that cruise tourism follows the operator's recommendations and is not characterized as tourism with inappropriate behavior (A5). At the same time, cruise tourism is mostly aimed at the elderly, who require special safety conditions and infrastructure (A6). To facilitate the elderly, certain forms of infrastructure should be developed, such as rails, ramps, stairs, and minor piers.

17.4.4 Stakeholders' Perspectives

The tourism development on Solovetsky involves a broad range of stakeholders: "Solovki includes so many different aspects [referring to multiple stakeholders], and all of them should be taken into account" (A3). We have noticed a clear duality in the answers on tourism development (including cruises) on the Solovetsky archipelago. Some believe that more investment is needed, especially in infrastructure, to support tourism growth (A3, A13, A20). That, in turn, will be beneficial for the local population (A3, S20). Others are concerned that the current growth and expected growth are not balanced with the island's natural capacity. This point was described in the following way: "Another question is the pressure [referring to the environmental pressures] on Solovki. There may need to be restrictions. We can organize an opportunity to visit Solovki, but another question is how much or how many tourists will Solovki be able to accommodate?" (A3) The same interviewee asks, "Who will regulate the flow?" (A3).

Among the interviewees, there is still no clear vision of the goals for tourism development. The regional stakeholders support the growth of the cruise industry to Solovetsky: "We can accommodate 40 overseas vessels, but we are also interested in them visiting Arkhangelsk as well, not just Solovki" (A8). While local stakeholders still have little experience with overseas cruises and point to potential improvements in order to meet the growth, such as better language skills, local involvement in hosting activities, and infrastructure development (S11, S13). Moreover, regional-local tourism management is described as a system with an unclear vision with regard to tourism development: "We do not have an approved program for the development of tourism" (A10). A representative from the tourism industry pointed out that they were not invited to participate when the existing Solovetsky strategy was developed: "When this program was developed, we were not invited to any of the meetings" (A9).

Despite the wide range of stakeholders mentioned in the previous section, the locals describe the management of tourism in the following way: "We have this interesting management model: the management bodies are there [Arkhangelsk], while the island is here" (S20). "Decision making takes place in Arkhangelsk, some important people visit us, check things out here and then leave … We have a local council that discusses the relevant issues with the mayor, who is closely connected with Arkhangelsk" (S11). Another resident mentioned that there is a need for open public meetings where the local population could be informed about the local development plans (S18).

By contrast, when asked about the current communication methods in the tourism industry, our interviewees from Arkhangelsk described them as follows: "I would say that the museum is monopolistic. In addition, everything operates around the museum. It cannot be otherwise. It is the main local tourist attraction. Thus, we contact them not as a tourist center but as a service provider … The tour operator is an intermediary between the museum and the tourist. The only thing that the museum can develop is a program for the tourists" (A8). Another interviewee agreed, saying that if there are any tourism-related issues, they contact the museum administration (A5): "We contacted the Solovetsky Development Agency when there were a couple of emergencies and we needed help with the infrastructure for cruise vessels, and they agreed to help us. Otherwise, we contact the museum administration" (A5). It is important to note that the monastery and the museum have the same administrative director, making this a unique situation, according to one of our interviewees: "It is quite unusual when the church is involved in commercial activities" (S3).

Interviewees also mentioned other relevant stakeholders: researchers and search-and-rescue services. Several stakeholders pointed out that because overseas cruise vessels are a new development trend in the area, more research is needed to understand the threats and opportunities that tourists pose, and to assess the carrying capacity of several destinations. (A3, A8): "Research should be carried out by ecologists and historians to avoid turning Solovki into a trampled space" (A3). The efficiency of emergency preparedness services in the Arkhangelsk region still raises doubts (A2), mainly because of the large distances involved. Even in the summer, a fire station with an emergency team is operating in Solovetsky. On certain occasions, a search-and-rescue boat, usually located in Arkhangelsk, is sent to the archipelago (A7).

17.5 Discussion

In this paper, we have aimed to discuss and understand how cruise tourism on the Solovetsky archipelago is managed. In particular, we focused on sustainability concerns faced by the regional and local stakeholders. Our data indicate that the stakeholders and how they interact both play a crucial role in developing sustainable cruise tourism. Therefore, the stakeholders are an important aspect of tourism development. We first summarize our findings according to the three sustainability pillars and then discuss the need for and potential of the communication model.

17.5.1 The Economy

The potential economic value creation from cruise arrivals, when properly managed, can result in increased local employment and local economic ripple effects. Our study indicates that achieving this positive effect from cruise arrivals is challenging on the Solovetsky archipelago because of the low-level involvement of the residents and private businesses in hosting cruise tourists. This is mainly due to the lack of foreign language skills. This reduces the potential for local value creation. This finding is in line with one of the most essential problems mentioned by Pashkevich and Lamers (2015), who reported that the unavailability of skilled workers in the local tourism industry becomes a hindrance to the expansion of cruise tourism operations. The lack of language skills makes it difficult to engage international tourists with the cultural heritage of the archipelago and to co-create the tourism experience (Mossberg et al. 2014). Our case provides evidence that without interpretation, most of the traditions, habits, and cultural practices of this place cannot be understood or experienced. There is a need for the establishment of guides and staff who are able to speak foreign languages. However, if the tour guides are not local and come from outside the community, there will be an outflow of the generated income. In addition, residents are seen as carriers of local knowledge and traditions, and those who already have experience in hospitality work for both domestic and international tourists could be involved in hosting activities for the cruise industry if courses for (tour) guides are organized locally at Solovetsky. It is for these reasons that we suggest that the local residents are a valuable resource and key to hosting cruise activities.

Despite certain limitations in the taxation system, the value creation and distribution of income from the regular tourist flow is sufficient for the local residents and businesses. The museum (which is also managed by the monastery leader) provides the focal point for all touristic experiences (Tsvetkov 2011). It is also described as a monopolist of the cruise services offered on the Solovetsky. The income from the cruise industry is unequally distributed between the state-owned museum (which owns the leading attraction on the archipelago) and the local businesses (who mostly earn through domestic tourism). Several destinations around Europe (e.g., Longyearbyen on Svalbard) have implemented a visitor tax that generates income and secures investments in local development. Otherwise, unorganized tourism will challenge the sustainability of local tourism development.

Moreover, our findings illustrate regulatory challenges and limitations for continued cruise development. In particular, Solovetsky is a part of the Russian Arctic and thus is governed by Russia's national and regional institutions, which makes operations in the area more complex (see Pashkevich and Stjernström 2014; Pashkevich et al. 2015). At the regional level, an example of such complexity is a visa regulation that creates an institutional barrier for cruise development on the Solovetsky. Vessels are required to visit the border control officers in the port of Arkhangelsk prior their visit to Solovetsky. In 2016, the port of Arkhangelsk received permission to welcome foreign visitors, who could be granted a 72 h visa-free visit (Russian Federation Government 2016) which made it possible for them to visit other destinations as part of an organized cruise activity (of particular interest is the archipelago of Franz Jozef Land). However, given the two-year planning period for cruise routes, the consequences of this visa status will not be noticeable until the summer season of 2019.

In summary, our study present evidence that cruise tourism provides an opportunity for economic value creation, especially when properly managed and regulated. At the same time, our findings indicate that the growth in cruise activities brings along environmental and social challenges for sustainable tourism development.

17.5.2 The Environment

The Solovetsky archipelago is a location of unique natural and cultural heritage. An increase in anthropogenic impact on the archipelago may have negative consequences for its fragile nature and for its cultural and historical monuments. As was argued by Moore and Carter (1993 in McElroy and Potter 2006), there is no example of tourist use that is completely without impact. At the same time, when compared to individual tourism on the archipelago, overseas cruise tourism is described as having a more limited impact on the local natural environment. This is due to the way the cruise excursions are organized. Anchoring at a distance from the settlement, the organized excursions along specified routes limit their impact on sites and lead to almost no waste being generated locally.

Thus, the cruise tourism practices on the Solovetsky archipelago are seen as more environmentally conscientious than land tourism. Moreover, despite the reported negative environmental effects of cruise tourism (e.g., Lück 2010; McElroy and Potter 2006), it still accounts for a small portion of the tourism flow to the archipelago (Table 17.2). The planned increases in tourism activities, together with the potential need to minimize environmental risks and effects, may require the development of ecological and eco-cultural tourism practices and strategies for the preservation of the natural landscape (Grushenko and Vasiliev 2013). It is also necessary to improve the system for estimating the flow of visitors to the community. At present, this information is very fragmented and is held by different stakeholders without providing a clear overall picture of tourism flow in the area (Maksimova 2016).

17.5.3 The Society

As with other Arctic destinations, the case of Arkhangelsk indicates that one of the main concerns raised locally and regionally is the proper regulation of the tourism flow, which challenges infrastructure and the island's natural capacity, as well as creating a disturbance for the isolated religious sites (Olsen and Nenasheva 2018). The crowding became a concern for several small communities in the Arctic that experienced a growth in cruise tourism (see also Chap. 14). As identified earlier (e.g. Tsvetkov 2011), opinions differ about limiting of the number of tourists to the archipelago and about the flow of tourism there (see also Nevmerzhitskaya 2006). The more recently suggested measure of giving the archipelago the status of a natural reserve may limit the volume of human activities on the archipelago altogether (Olsen and Nenasheva 2018).

At the same time, having successful tourism practices requires and provides the opportunity to have a vibrant living community, which in turn contributes to the development of the archipelago (Tsvetkov 2011) by, for example, providing infrastructure, hosting services, and disseminating local knowledge and traditions. Thus, there is a need to involve the local population actively in decision making regarding the archipelago and in discussions about the plans and strategies for tourism development in the region.

17.5.4 Stakeholder Communication

On the basis of our findings, we conclude that sustainable cruise tourism development in the Arkhangelsk region, and especially in the Solovetsky archipelago, depends on both regional and local stakeholders, their cooperation, and a shared, clear vision on the growth of tourism. Our data show that cruise tourism management and decision making mostly take place in Arkhangelsk, where several stakeholders (e.g., the Agency for Tourism, private travel companies, and guide services) work on the development of destinations. At the same time, local stakeholders play a crucial role in the development of destinations (Baum 2006). Therefore, we argue that there is a need to strengthen collaboration across the regional-local nexus in order to secure the sustainability of operations and develop the conditions under which the income generated on the island will remain and be invested locally.

At the local level, the Solovetsky Museum is one of the main providers of tourism experiences for domestic and overseas cruise ship visitors. However, while the provision of services for cruise ships is characterized by closer cooperation with regional stakeholders, the role of the local population as a potential stakeholder in cruise tourism development is yet to be addressed under the current system. On the basis of our discussion, we argue that there is a need to develop an effective management model for the cruise industry that secures information sharing between regional and local stakeholders.

Among the obvious positive effects of the communicative model is the building of partnership relations between all interested parties. This partnership would provide opportunities for the local stakeholders' self-realization and involvement in the process of managerial decision making. At the same time, the case examined in this chapter illustrates the necessity of modernizing the economic management system, by forming an open public space, and creating an effectively functioning information and communication system that would provide a platform for interaction between all stakeholders (Krainova 2012).

We argue that an ideal model of communicative management for sustainable tourism development can be defined as *a system of information exchange and interaction among authorities, businesses, and local stakeholders (including residents) at all stages of development of tourism activities for the purpose of choosing strategic alternatives and ensuring sustainable practices for short- and long-term development.*

Thus, this study suggests the potential application of the communication model. However, further studies are necessary in order to find the best solution for the implementation of the suggested improvements to aid in better planning, better collaboration, and, ultimately, the more sustainable development of tourism in the archipelago.

17.6 Conclusion

The increase in cruise tourism to the Solovetsky archipelago challenges the unique environment and puts increasing pressure on the local society. Tourism development is a potential source of employment and income for the local community, but local stakeholders are beginning to realize that there is also a significant negative impact on the local community and its natural operating capacity (Olsen and Nenasheva 2018). This study stresses the importance of regional and local collaboration among the stakeholders involved in the provision of cruise tourism experiences and services to co-create value for both individual businesses and the destination.

Our findings suggest that sustainable tourism development of the Solovetsky archipelago is challenging without the establishment of improved communication channels and methods among the stakeholders whose interests are affected by tourism development. Private companies, public bodies, and the local people are the actors who should be involved in the process of planning, development, and management of cruise tourism in the region, as well as in the distribution of income derived from it. The current lack of communication among these stated stakeholders leads to the negative effects on the environment of the archipelago and the potential economic loss of the much-needed additional income to be gained from tourism development. That is why the development of tools to aide in cooperation can help to find a successful balance between economic benefits, societal development, and environmental protection.

Acknowledgements We would like to thank the participants involved in this study for sharing their views and experiences, as well as Maksim Iliin for sharing the picture of the cruise ship for the purpose of this study. We are particularly appreciative of the comments from an anonymous reviewer that helped to improve this chapter.

References

Baikina, O., & Valkova, T. M. (2011). Kruiznyj turizm: geograficheskie i metodicheskie rekomendacii po formirovaniyu individual'nyh kombinirovannyh programm (Cruise tourism: Geographical and methodological recommendations for the formation of individual combined programs). *Modern Problems of Service and Tourism, 1*, 88–95.

Baum, T. (2006). Human resource issues. In G. Baldacchino (Ed.), *Extreme tourism: Lessons from the world's cold water islands* (pp. 41–49). Amsterdam: Elsevier.

Bazeley, P., & Jackson, K. (2013). *Qualitative data analysis with NVIVO*. London: SAGE Publications.

Blaikie, N. (2010). *Designing social research*. Cambridge, UK: Polity Press.

Bulkeley, H. (2005). Reconfiguring environmental governance: Towards a politics of scales and networks. *Political Geography, 24*(8), 875–902.

Butler, R. W. (1999). Sustainable tourism: A state of the art review. *Tourism Geographies, 1*(1), 7–25.

Byrd, E. T. (2007). Stakeholders in sustainable tourism development and their roles: Applying stakeholder theory to sustainable tourism development. *Tourism Review, 62*(2), 6–13.

Bystrowska, M., Wigger, K., & Liggett, D. (2017). The use of information and communication technology (ICT) in managing high Arctic tourism sites: A collective action perspective. *Resources, 6*, 33.

Castro, A. P., & Nielsen, E. (2001). Indigenous people and co-management: Implications for conflict management. *Environmental Science & Policy, 4*(4–5), 229–239.

Dawson, J., Johnston, M. E., & Stewart, E. J. (2014). Governance of arctic expedition cruise ships in a time of rapid environmental and economic change. *Ocean and Coastal Management, 89*, 88–99.

Dawson, J., Pizzolato, L., Howell, S. E. L., et al. (2018). Temporal and spatial patterns of ship traffic in the Canadian Arctic from 1990 to 2015. *Arctic, 71*(1), 15–26.

Dumanskaya, I. O. (2014). *Ledovye usloviya morej evropejskoj chasti Rossii (Ice conditions of the seas in Russian European North)*. Moscow: Hydrometeorological Centre of Russia.

Gairabekov, U. T., Dashkova, E. V., Miroshnichenko, P. N. (2017, March 2). *Ustojchivoe razvitie turizma: sostoyanie i problem (Sustainable tourism development: State and problems)*. Paper presented at the 6th Annual Final Conference of Professors and Teachers of the Chechen State University, Chechen State University, Grozny.

Gibson, P. (2012). *Cruise operations management: Hospitality perspectives*. New York: Routledge.

Gomilevskaya, G. A., & Petrova, G. A. (2017). Morskoj turizm kak sostavlyayushchaya turistskogo brenda Vostochnoe kol'co Rossii (Marine tourism as an element of the tourist brand Eastern ring of Russia). *Bulletin of the Vladivostok State University of Economics and Service, 3*, 71–83.

Grushenko, E. (2009). Strategiya razvitiya morskogo turizma v Evroarkticheskoj zone (Strategy of marine tourism development in the Euro-Arctic zone). *Economic and Social Changes: Facts, Trends, Forecast, 1*(5), 74–79.

Grushenko, E. (2012, November 14–16). *Razvitie morskogo turizma v Arktike (Development of maritime tourism in the Arctic). Development of the North and the Arctic: Problems and prospects*. Paper presented at the Interregional Science and Practical Conference, Apatity.

Grushenko, E. (2014). Razvitie morskogo kruiznogo turizma v portah Zapadnoj Arktiki (Development of cruise tourism in the ports of the Western Arctic). *Arctic and North, 14*, 26–32.

Grushenko, E. B., & Vasiliev, A. M. (2013). *Turizm na Evropejskom Severe Rossii i v Zapadnoj Arktike (Tourism in the European North of Russia and in the Western Arctic)*. Apatity: Kola Science Center of the Russian Academy of Sciences.

Hall, C. M., James, M., & Wilson, S. (2010). Biodiversity, biosecurity, and cruising in the Arctic and sub-Arctic. *Journal of Heritage Tourism, 5*, 351–364.

Healey, P. (1996). The communicative turn in planning theory and its implications for spatial strategy formation. *Environment and Planning B: Planning and Design, 23*, 217–234.

Ho, J. (2010). The implications of Arctic Sea ice decline on shipping. *Marine Policy, 34*(3), 713–715.

Howard, R. (2009). *The Arctic gold rush: The new race for tomorrow's natural resources*. London/New York: Continuum.

Huse, M., Gustavsen, T., & Almedal, S. (1998). Tourism impact comparisons among Norwegian towns. *Annals of Tourism Research, 25*(3), 721–738.

Huxley, M., & Yiftachel, O. (2000). New paradigm or old myopia? Unsettling the communicative turn in planning theory. *Journal of Planning Education and Research, 19*(4), 333–342.

Johnston, M. E., Viken, A., & Dawson, J. (2012). Firsts and lasts in Arctic tourism: Last chance tourism and the dialectic of change. In R. H. Lemelin, E. Stewart, & J. Dawson (Eds.), *Last-chance tourism: Adapting tourism opportunities in a changing world* (Contemporary geographies of leisure, tourism and mobility) (pp. 10–24). London: Routledge.

Jovicic, D. Z. (2014). Key issues in the implementation of sustainable tourism. *Current Issues in Tourism, 17*(4), 297–302.

Karkkainen, B. C. (2002). Collaborative ecosystem governance: Scale, complexity, and dynamism. *Virginia Environmental Law Journal, 21*, 189–243.

Kiyakbaeva, E. G. (2014). Indikatory ustojchivogo razvitiya turizma i ih ispol'zovanie v federal'nyh programmah razvitiya turizma v Rossii (Indicators of sustainable tourism development and their use in Federal programs of tourism development in Russia). *News of the Sochi State University, 1*(29), 78–80.

Klein, R. A. (2011). Responsible cruise tourism: Issues of cruise tourism and sustainability. *Journal of Hospitality and Tourism Management, 18*(1), 107–116.

Krainova, K. A. (2012). Vzaimodejstvie vlasti I obchchestva v politico-kommunikativnom processe na urovne subjekta Rossijskoj Federacii (The interaction of authority and society in the political and communication process at the level of the subject of the Russian Federation). *Politbook, 4*, 78–86.

Krasilnikov, D. G., Sivinceva, O. V., & Troitskaya, E. A. (2014). Sovremennye zapadnye upravlencheskie modeli: sintez New Public Management i Good Governance (Modern Western management models: Synthesis of new public management and good governance). *Ars Administrandi, 2*, 45–62.

Lamers, M., & Pashkevich, A. (2015). Short-circuiting cruise tourism practices along the Russian Barents Sea coast? The case of Arkhangelsk. *Current Issues in Tourism, 21*, 440–454. https://doi.org/10.1080/13683500.2015.1092947.

Logunova, N. A. (2013). Razvitie kruiznogo sudohodstva kak sostavnoj chasti kruiznogo turizma (Development of cruise shipping as an integral part of cruise tourism). *Journal of Transport Economics and Industry, 43*, 89–94.

London, W. R., & Lohmann, G. (2014). Power in the context of cruise destination stakeholders' interrelationships. *Research in Transportation Business & Management, 13*, 24–35.

Lück, M. (2010). Environmental impact of polar cruises. In *Cruise tourism in polar regions. Promoting environmental and social sustainability?* London: Earthscan.

Maksimova, T. (2016, April 8–10). *Tourist flow in the Solovetsky islands: Dynamics and contemporary perspectives*. Paper presented at the Festival of the Russian Geographical Society, Saint Petersburg.

Marsh, E. A. (2012). The effects of cruise ship tourism in coastal heritage cities. *Journal of Cultural Heritage Management and Sustainable Development, 2*(2), 190–199.

Mason, P. (1997). Tourism codes of conduct in the Arctic and sub-Arctic region. *Journal of Sustainable Tourism, 5*(2), 151–165.

McElroy, J., & Potter, B. (2006). Sustainability issues. In G. Baldacchino (Ed.), *Extreme tourism: Lessons from the world's cold water islands* (pp. 31–40). Amsterdam: Elsevier.

Melnik, S. A. (2015). Local community as a subject of municipal management. *Bulletin of the Kostroma State Technological University, 1*(5), 20–22.

Moller, H., Berkes, F., Lyver, P. O. B., et al. (2004). Combining science and traditional ecological knowledge: Monitoring populations for co-management. *Ecology and Society, 9*(3), 2.

Moore, S., & Carter, B. (1993). Ecotourism in the 21st century. *Tourism Management, 14*(2), 123–130.

Mossberg, L., Hanefors, M., & Hansen, A. H. (2014). Guide performance: Co-created experiences for tourist immersion. In *Creating experience value in tourism* (pp. 234–247). Wallingford: CABI.

Nekrasova, E. A. (2016). *Mestnoe samoupravlenie v sovremennyh zarubezhnyh sociologicheskih teoriyah (Local self-government in modern foreign sociological theories)*. Novosibirsk: Center for the Development of Scientific Cooperation.

Nenasheva, M., & Olsen, J. (2018). Water transport in the European North of Russia: Social significance, challenges and perspectives of development. *Arctic and North, 32*, 49–62.

Nenasheva, M., Bickford, S. H., Lesser, P., et al. (2015). Legal tools of the public participation in the environmental impact assessment process and their application in the countries of the Barents Euro-Arctic. *The Region, 1*(3), 13–35.

Nevmerzhitskaya, J. (2006). The Solovetsky Archipelago, Russia. In G. Baldacchino (Ed.), *Extreme tourism: Lessons from the world's cold water islands* (pp. 159–166). Amsterdam: Elsevier.

Olsen, J., & Nenasheva, M. (2018). Adaptive capacity in the context of increasing shipping activities: A case from Solovetsky, Northern Russia. *Polar Geography, 41*, 241–261. https://doi.org/10.1080/1088937X.2018.1513960.

Pashkevich, A., & Lamers, M. (2015). Coastal tourism in the Northwest Russian Arctic: Current situation and future prospects. *Current Issues in Tourism, 21*, 440–454. https://doi.org/10.1080/13683500.2015.1092947.

Pashkevich, A., & Stjernström, O. (2014). Making Russian Arctic accessible for tourists: Analysis of the institutional barriers. *Polar Geography, 37*(2), 137–156.

Pashkevich, A., Dawson, J., & Stewart, E. (2015). Governance of expedition cruise ship tourism in the Arctic: A comparison of the Canadian and Russian Arctic. *Tourism in Marine Environments, 10*(3–4), 225–240.

Popov, G., & Davydov, R. (2003). *Morskoe sudohodstvo na Russkom Severe v XIX—nachale XXvv. (Marine ship traffic in the Russian North XIX-beginning XX century)*. Ekaterinburg/Arkhangelsk: Ural Branch of Russian Academy of Sciences.

Rassokhina, T. V., & Seselkin, A. I. (2015). Analiz ehffektivnosti realizacii sistemy upravleniya ustojchivym razvitiem turizma v Rossijskoj Federacii na principah gosudarstvenno-chastnogo partnerstva (Analysis of efficiency of implementation of the system of management of sustainable tourism development in the Russian Federation on a public-private partnership). *Bulletin of the Russian International Academy of Tourism, 2*, 62–68.

Russian Federation Government. (2014). *Russian tourism development strategy for the period up to 2020 (2014)*. Approved by the Decree of the Russian Federation Government of 31 May 2014 No 941-R.

Russian Federation Government. (2016). *O vnesenie izmenenij v prerechen portov, cherz kotorye dopuskaetsja vezd v Rossiiskuu Federaciu inostrannyx grazhdan I lic bez grazhdanstva, pribyvajushix v Rossiiskuu Federaciu v turisticheskix celyax na paromax, imeushix razreshenija na passazhirskie perevozki*. The Decree of the Russian Federation Governmnet of 22 July 2016 No 707.

Sautter, E. T., & Leisen, B. (1999). Managing stakeholders a tourism planning model. *Annals of Tourism Research, 26*(2), 312–328.

Scott, D. (2011). Why sustainable tourism must address climate change. *Journal of Sustainable Tourism, 19*(1), 17–34.

Seselkin, A. I. (2014). Ustojchivoe razvitie turizma kak prioritetnoe napravlenie deyatel'nosti vsemirnoj turistskoj organizacii: postanovka problemy issledovaniya (Sustainable tourism development as a priority direction of activities of the world tourism organization: The problem of the study). *Bulletin of the Russian International Academy of Tourism, 1*(10), 9–14.

Simpson, K. (2001). Strategic planning and community involvement as contributors to sustainable tourism development. *Current Issues in Tourism, 4*(1), 3–41.

Smirennikova, E. V. (2009). Faktory, vliyayushchie na ocenku turisticheskogo potenciala Arhangel'skoj oblasti (Factors influencing the evaluation of the tourist potential of the Arkhangelsk region). *Pomor University Bulletin, 4*, 23–27.

Stewart, E. J., & Draper, D. (2006). Sustainable cruise tourism in Arctic Canada: An integrated coastal management approach. *Tourism in Marine Environments, 3*(2), 77–88.

Stewart, E., Howell, S., Draper, D., et al. (2007). Sea ice in Canada's Arctic: Implications for cruise tourism. *Arctic, 60*(4), 370–380.

Stewart, E., Dawson, J., & Draper, D. (2011). Arctic tourism and residents in Arctic Canada: Development of a resident attitude typology. *Journal of Hospitality and Tourism Management, 17*(1), 95–106.

Stewart, E., Dawson, J., & Johnston, M. (2015). Risk and opportunities associated with change in the cruise tourism sector: Community perspectives from Arctic Canada. *The Polar Journal, 5*(2), 403–427.

Toskunina, V. E., & Smirennikova, E. V. (2011). Prirodnye turisticheskie resursy kak strategicheskij faktor razvitiya maloosvoennyh i slabovovlechennyh v hozyajstvennyj oborot territorij Rossijskogo Severa (Natural tourist resources as a strategic factor in the development of underdeveloped and poorly involved in the economic turnover of the territories of the Russian North). *Regional Development Strategy, 48*(231), 13–19.

Tsvetkov, A. (2011). Strategicheskoe upravlenie ustojchivym razvitiem territorii Soloveckogo arhipelaga (Strategic management of the Solovetsky island's sustainable development). *Arctic and North, 2*, 97–116.

UNEP-WTO. (2005). *Making tourism more sustainable. A guide for policy makers*. United Nations Environment Programme and World Tourism Organization. http://www.unep.fr/shared/publications/pdf/DTIx0592xPA-TourismPolicyEN.pdf

Van Bets, L. K. J., Lamers, M. A. J., & van Tatenhove, J. P. M. (2017). Collective self-governance in a marine community: Expedition cruise tourism at Svalbard. *Journal of Sustainable Tourism, 25*(11), 1583–1599.

Viken, A. (2004). *Turisme: Miljø og utvikling*. Oslo: Gyldendal akademisk.

Viken, A. (2011). Tourism, research, and governance on Svalbard: A symbiotic relationship. *Polar Record, 47*(4), 335–347.

Viken, A., & Aarsaether, N. (2013). Transforming an iconic attraction into a diversified destination: The case of North Cape tourism. *Scandinavian Journal of Hospitality and Tourism, 13*(1), 38–54.

Yakovenko, I. M., & Lazitskaya, N. F. (2014). Geograficheskie tendencii razvitiya morskogo kruiznogo turizma (Geographical trends in development of sea cruise tourism). *Scientific Notes of Taurida National University Named After V.I.Vernadsky, 27*(66), 238–248.

Chapter 18
Finnish Sámi: Is Tourism a Preservation of Indigenous Culture?

Eda Ayaydın and Samim Akgönül

Abstract The Sami "minority" of Finland is the smallest indigenous community of this specific Arctic group in Nordic countries. Finnish Sami constitute a cultural, linguistic and territorialized minority. Finland recognized Sami as a "people" in 1995, nevertheless without ratifying the ILO Convention 169 Concerning Indigenous and Tribal Peoples. Besides the fact that Finnish Sami Parliament (Saamelaiskäräjät) has been recognized since 1973, and the Sami linguistic rights have been established since 1982, Sami do not possess territorial rights, especially at economic level. One of the main economic sectors where Sami are active is tourism in Lapland. The debate among the defendants of indigenous rights but also among some Sami prominent leaders are today about the effectiveness of the tourism in the survival of Sami way of life and culture. While some observers denounce the folklorization process of "saminess" through exasperated touristic exploitation, others see in tourism the only way to prevent complete assimilation and fade-out. This chapter will explore the role of tourism in the preservation of Sami culture in Finnish context, by using a field research conducted in July–August 2018 in Inari, Ivalo and Rovaniemi.

Keywords Sami · Finland · Indigenous peoples · Minority · ILO · Territorial rights · Tourism

E. Ayaydın
University of Versailles, Science Po Bordeaux, Paris, France
e-mail: eda.ayaydin@uvsq.fr

S. Akgönül (✉)
University of Strasbourg - CNRS / DRES, Strasbourg, France
e-mail: akgonul@unistra.fr

E. Pongrácz et al. (eds.), *Arctic Marine Sustainability*,
Springer Polar Sciences, https://doi.org/10.1007/978-3-030-28404-6_18

18.1 Introduction

This chapter explores the current situation of Sámi[1] culture within two theoretical frameworks: minority rights on the one hand and cultural tourism on the other. The Sámi people are a group of specific culture -with some internal differences- living in the Sápmi (Sámi region), at the northern part of Norway, Sweden, Finland and the Northwestern part of the Russian Federation. The specificity of the group is at the same time linguistic and cultural, in the sense of living behaviours and everyday life habits, including economic activities, arts and artefacts, cuisine or music. This group of approximately 100,000 people lives sparsely over the Arctic Circle (in majority) and is in a strong process of consciousness of identity and search of legitimacy in the Simmelian sense of the word. If in four different countries Sámi have different status, in at least three of them (Norway, Sweden and Finland) they live at the heart of a region discovered recently by mass tourism. Tourists from all over the world are attracted in the region by the atmospheric phenomena, the taste of adventure, by ice and snow, by *aurora borealis*, by the wild life, fishing and hunting and, in the specific case of Finland, by the "invented tradition" of Santa Claus. The chapter aims to contribute to the debate on the effects of tourism on Finnish Sámi[2] culture. It is based on secondary sources on the one hand, especially for the conceptualisation of minority rights and tourism, but also on primary sources such as official documents, grey literature and two months of field research between Rovaniemi and Sevettijärvi, including observation and *touristic* stays in Sodankylä, Ivalo and Inari. The chapter will first examine the Sámi's specificity within the general minority and indigenous peoples' rights standards in Western Europe before being dedicated to the analyses of the role of touristic industry in the preservation (or not, through folklorization[3]) of the Sámi culture.

18.2 Minority Regime of Finland and Sámi

18.2.1 International Standards

In minority studies, the complex system of minority rights of Finland is seen as a model because of the diversity of minority characteristics present on the Finland's territory but also because of the nature of rights provided to minorities by Finland.

[1] Sámi name is spelled **Sami**, Saami, **Same,** or **Sabme**. **For this chapter, we have chosen the spelling of Sámi following the spelling of the** Sámi parliament of Finland.

[2] This chapter being on specifically Sámi living in Finland, hereafter the term Sámi will correspond only to this population.

[3] In this chapter, the folklorization concept is used to underline a process where cultural behaviours and objects are becoming touristic, superficial, falsely authentic wonders in the framework of *exoticification*. For this process see (Hafstein 2018).

Nevertheless, if the internal system is the consequence of a long process related to the nation building history of the country, the legal and political standards present some differences with international texts.

On the international scale, in the field of protection of minorities, one may distinguish three periods that the vision and aims vary substantially. During the first period, that can be considered as the system of the League of Nations, (1920–1945), 13 texts concern directly or indirectly minorities as bilateral and multilateral agreements, conventions and treaties. During this period, the main aim of the League of Nations' system is to protect the stability of the Nation State, by providing some collective and territorial rights to minorities that cannot be included in the given nation dominating the state. Thus, this system privileges the state's survival more than minority protection. Nevertheless, Finland being historically a constant periphery between two centres, Stockholm and Saint Petersburg (then Moscow), none of these texts concerns directly the minorities in Finland, the main issue then being the emigration and immigration waves between Sweden and Finland and between Russia and Finland (Similä 2003).

During the second period, between 1945 and 1992, "minorities" disappear from the international standards, seen as destabilizing the national coherence. During the UN system of the cold war, emphasis is given to individual human rights. In this understanding, "minority rights" are potentially dangerous because they create political and legal contexts that are ripe to separatist movements. The philosophy of the new system was to underline fundamental human rights without mentioning groups. The only exception of a text mentioning minorities during this period is the *International Covenant on Civil and Political Rights* (ICCPR) of 1966 (entry into force in 1976) that mentions minorities in a very shy and cautious way in its article 27:

> *In those states in which ethnic, religious or linguistic minorities exist, persons belonging to such minorities shall not be denied the right, in community with the other members of their group, to enjoy their own culture, to profess and practise their own religion, or to use their own language*[4]

The ICCPR is signed by Finland in 1967 and ratified in 1975. It is important to underline the fact that unlike many other signatory states, Finland did not formulate a declaration or reservation on the article 27 of this exceptional text.

At the end of this period, during a transitory junction with the third period of "minority protection", another text concerns directly Finland and the topic of this article. *The Convention concerning Indigenous and Tribal Peoples in Independent Countries* of International Labour Organization of 1989 (entry into force in 1991) is still not ratified by Finland (nor Sweden) unlike Denmark (in force since 1996) and Norway (in force since 1990) (Semb 2012). Finland, following the insistent recommendations of the UN Human Rights Council (Koivurova and Stepien 2011), declared that it would finally ratify the convention, nevertheless for internal reasons the parliament of Finland refused to ratify the document in 2015 (Schönfelt 2017).

[4] On the specificity of this text compared to the other Human rights texts at international level see (Keith 1999).

This situation must be seen as an anomaly, especially related to debates on the Sámi communities in Finland. The non-ratification of the ILO convention 169 is directly related to the land use claims (and resistances) that will be in the heart of our reflection on the role of tourism in Finland's Lapland.

The third period of minority protection at international law level begins in 1989–1990, with the collapse of the bipolar world and the rebirth of minority conflicts especially in Eastern Europe. This period is characterized on the one side by the UN "Declaration on the Rights of Persons Belonging to National or Ethnic, Religious and Linguistic Minorities" and on the other side, by two regional legal frameworks of the Council of Europe that are the "Framework Convention for the Protection of National Minorities » of 1995 (entry into force in 1998) and "The European Charter for Regional or Minority Languages" of 1992 (entry into force in 1998). Unlike the ILO convention 169, the Finland's attitude towards these two major documents has been very positive and reactive. Finland signed and ratified the FCNM in 1997[5] without any reservation or declaration,[6] and signed the ECRML in 1994 and ratified in 1994. The declarations formulated by Finland for this document recognizes specifically Sámi language as a "regional or minority language in Finland" and the Swedish language as "less widely used official language in Finland".[7]

As it is clear, except for the ILO convention 169, Finland followed a liberal policy in adopting international standards for the minority protection, at least concerning "old minorities". When it comes to "new minorities", *i.e.* descendants of immigrants, the situation is a bit different. For example, Helsinki did not sign nor ratified the UN *International Convention on the Protection of the Rights of All Migrant Workers and Members of Their Families* of 1990, (entry into force in 2003).[8] The number of immigrants in Finland has increased tremendously since 1990s. "The number of foreign citizens legally residing in Finland increased six-

[5] As for 2018, 39 member states of the COE are parties to the convention. Belgium, Greece Iceland and Luxemburg have signed the text without ratifying yet. Andorra, Monaco and especially France and Turkey have neither signed nor ratified the Convention.

[6] Sweden, on the other hand, specified explicitly "national minorities in Sweden are Sami, Swedish Finns, Tornedalers, Roma and Jews". Russia's declaration touches Finland too. Moscow formulated its declaration not towards its own minorities but in order to protect the situation of Russian minorities in other countries, especially in the Baltic countries and Finland, that are deprived from nationality: « The Russian Federation considers that none is entitled to include unilaterally in reservations or declarations, made while signing or ratifying the Framework Convention for the Protection of National Minorities, a definition of the term "national minority", which is not contained in the Framework Convention. In the opinion of the Russian Federation, attempts to exclude from the scope of the Framework Convention the persons who permanently reside in the territory of states parties to the Framework Convention and previously had a citizenship but have been arbitrarily deprived of it, contradict the purpose of the Framework Convention for the Protection of National Minorities. ».

[7] https://www.coe.int/en/web/conventions/full-list/-/conventions/treaty/148/declarations?p_auth=1iCIUXyM&_coeconventions_WAR_coeconventionsportlet_enVigueur=false&_coeconventions_WAR_coeconventionsportlet_searchBy=state&_coeconventions_WAR_coeconventionsportlet_codePays=FIN&_coeconventions_WAR_coeconventionsportlet_codeNature=10

[8] One must admit that no migrant-receiving state in Western Europe or North America has ratified the convention.

fold, from 26,300 to 155,700. Out of the total population of 5.3 million, approximately 300,000 people in Finland, or 5 %, claim a foreign background (having been foreign born, speaking a foreign language, or having foreign citizenship)" (Tanner 2011). Although Finland presents itself as the most welcoming Scandinavian country at immigration policies[9] tensions occur, especially in Lapland. Indeed, as a sparsely populated area, Finland's Lapland is new immigrant-receiving region (Nafisa 2016). Therefore, to debates on the rights of Sámi compared to non-Sámi population of Lapland (in majority Finnish but also Russian speakers) is added a new issue on integration of immigrants and their descendants, insofar as these immigrants and their descendants are step by step introduced in several economic and cultural sectors of Lapland, including tourism, a sector long time disputed between Sámi and Finns (Carson and Carson 2018).

18.2.2 Finland's Linguistic, Religious Minorities and Indigenous Peoples

These parallelism and specificities of the Finland's minority regime with international standards are due to the local political and social history, creating a *sui generis* context. Sociologically speaking, the largest minority of Finland is an ethnolinguistic group, Swedish speakers – self-presented as *finlandssvensk* – that is not considered as a minority legally speaking but as a constitutional linguistic group at section 17 of the Constitution of 1999. The same section defines Sámi as an "indigenous group' and Sámi language, Roma language and "other groups" (without specification), have the right to "maintain and develop" their own language:

Section 17 Right to one's language and culture

> *The national languages of Finland are Finnish and Swedish. The right of everyone to use his or her own language, either Finnish or Swedish, before courts of law and other authorities, and to receive official documents in that language, shall be guaranteed by an Act. The public authorities shall provide for the cultural and societal needs of the Finnish-speaking and Swedish-speaking populations of the country on an equal basis.*
>
> *The Sami, as an indigenous people, as well as the Roma and other groups, have the right to maintain and develop their own language and culture. Provisions on the right of the Sámi to use the Sámi language before the authorities are laid down by an Act.*

Swedish linguistic (and political) rights are at the same time territorial, in Åland, where around 25,000 Swedish speakers form the majority, and non-territorial in other parts of the country especially from Ostrobothnia to the southern coast for around 270,000 Swedish speakers residing in bilingual municipalities with Finnish as the majority language and bilingual municipalities with Swedish as the majority language.

[9] "Ministry of the Interior, International Migration 2016–2017 Report for Finland" Ministry of the Interior Publications, 2017, p. 31, https://reliefweb.int/sites/reliefweb.int/files/resources/SM_28_2017%20%281%29.pdf

The second largest minority of the country recognized as a "linguistic minority" is Roma, or more specifically Kale,[10] that are around 10,000 people in Finland (and around 3000 in Sweden). In 1968 an Advisory Commission on Gypsy Affairs was set up at governmental level.

A part from Swedish speakers and Roma, two ethno-religious minorities are also present at equal quantity. Finnish Jews (around 1200 people) and Tatar Muslims (around 1000 people) are fully integrated in Finnish society and protect their religious affiliation and ethno-religious culture. As for the Muslims, since 2000's there is a regular growth, thanks to the immigration from mostly Muslim countries. And finally, the Finnish orthodox church (liturgy in Finnish), recognized as a "national church alongside with the Finnish Lutheran Church, regrouping around 60,000 ethnic Fins, is attached to the Oecumenical patriarchate of Constantinople in order to cut ties with the Russian orthodox church after the collapse of the Soviet Union.

As for the 2012, the religious affiliations of the Finland's population is as follows.[11]

Total:	5,426,674
Lutheran National Church:	4,147,371
Other Lutheran:	1276
Greek Orthodox Church in Finland:	58,705
Other Orthodox:	2801
Jehovah's Witnesses:	18,826
Free Church in Finland:	14,932
Roman Catholic Church in Finland:	11,530
Islamic congregations:	10,596
Pentecostal Church in Finland:	7445
Adventist churches:	3474
Church of J.Chr. of Latter-day Saints:	3181
Baptist congregations:	2332
Methodist churches:	1352
Jewish congregations:	1188
Buddhist congregations:	538
Anglican Church in Finland:	91
Other:	1306
No religious affiliation:	1,139,730

[10]The Finnish Gypsies have adopted the self-designation of *kaale*, a word which is derived from the Romani adjective *kaló* 'black'. When speaking Finnish, they use either this word or the Finnish adjective *tumma* 'dark'. More rarely will they use *mustalainen* ('black person'), which is the most widespread Finnish word for Gypsy. Recently, however, the term Romani has gained currency in official Finnish publications. (Vuorela and Borin 1998).

[11]Official Statistics of Finland (OSF): Population structure [e-publication]. ISSN = 1797-5395. annual review 2012, Appendix table 7. Religious affiliation of the population by age 31.12.2012. Helsinki: Statistics Finland [referred: 31.10.2018]. Access method: http://www.stat.fi/til/vaerak/2012/01/vaerak_2012_01_2013-09-27_tau_007_en.html

After this very quick presentation of different minorities in Finland, we can now look at the specific situation of Sámi people, considered by Finland's constitution as "indigenous peoples".

According to José Martinez Cobo the rapporteur of the Sub-Commission, on the « Problem of Discrimination against indigenous Populations », (1986), indigenous peoples (that he qualifies as "communities, peoples and nations") are: *"those which, having a historical continuity with pre-invasion and pre-colonial societies that developed on their territories, consider themselves distinct from other sectors of the societies now prevailing in those territories, or parts of them. They form at present non-dominant sectors of society and are determined to preserve, develop and transmit to future generations their ancestral territories, and their ethnic identity, as the basis of their continued existence as peoples, in accordance with their own cultural patterns, social institutions and legal systems. There is historical continuity that may consist of the continuation, for an extended period reaching into the present of one or more of the following factors: a) Occupation of ancestral lands, or at least of part of them; b) Common ancestry with the original occupants of these lands; c) Culture in general, or in specific manifestations (such as religion, living under a tribal system, membership of an indigenous community, dress, means of livelihood, lifestyle, etc.); d) Language (whether used as the only language, as mother tongue, as the habitual means of communication at home or in the family, or as the main, preferred, habitual, general or normal language); e) Residence on certain parts of the country, or in certain regions of the world; f) Other relevant factors. On an individual basis, an indigenous person is one who belongs to these indigenous populations through self-identification as indigenous (group consciousness) and is recognised and accepted by these populations as one of its members (acceptance by the group). This preserves for these communities the sovereign right and power to decide who belongs to them, without external interference."* (Cobo 1981)

This working definition, inspired extensively by the definition of "minorities" by the UN Rapporteur Francesco Capotorti (Capotorti 1979) of the same period, suits in several domains Sámi group in Finland. Sámi form a specific group living mostly above the Arctic circle (66°34′N) in the territories four different states, Norway, Sweden, Finland and the Kola Peninsula of Russia, in a region called, sometimes pejoratively Lapland, especially in the Finnish context, called also Sápmi. This population of approximately 100,000 "members" (in Finland they are estimated around 8000[12]) are descending from semi-nomadic reindeer herders traditions but not only more than a half of Sámi lives nowadays outside of the Sápmi but also only 10% of Sámi still lives through reindeer herding (Tuulentie 2017). The economic specificity of Finland's Sámi is the fact that unlike Norway and Sweden, reindeer herding in Finland is not an exclusivity of Sámi people, creating an economic (and identity)

[12]The number of Sámi in general and in Finland in particular depends on the definition given to Sámi. If, according to the Finnish Statistical institute there are in 2017, around 2000 Sámi speakers in Finland, their total number in the country, including those who don't live in Lapland and those who don't declare speaking in Sámi, the number is estimated 7–8000 persons. http://www.stat.fi/tup/suoluk/suoluk_vaesto_en.html#populationbylanguage

context, more depended from other resources like tourism. Thus, Sámi of Finland have linguistic rights[13] -especially in the municipalities of Enontekiö, Inari, and Utsjoki and also reindeer herding region of Sodankylä- and possess a Sámi Parliament in Inari,[14] land rights concerning traditional fishing and reindeer herding activities are not provided. Therefore, Sámi cultural and traditional material and immaterial heritage is more threatened in Finland than in other Sápmi regions, and more fragile, creating a dependency from commercialization of cultural objects and activities.

18.2.3 Diversity in Lapland/Diversity in Finland/Diversity of Attitudes: Between Militantism, Resignation, Opportunism and Assimilation

Finland is home to 8000 Sámi people out of 10%[15] of indigenous peoples of the total Arctic population.[16] According to *Statistics Finland*, the number of persons with Finnish background who were born in Finland and whose mother tongue is Sámi is 1918 and the number of persons with Finnish background born out of Finland and whose mother tongue is Sámi is 32.[17] Therefore, according to 2017 census, there are 1950 Sámi people using Sámi language as a mother tongue within Finnish context. These numbers accentuate a serious loss in culture.

Language carries important indicators, characteristics of the culture. For example, this *sui generis* culture is not familiar with violence or revolt. In Sámi language, there is no word corresponding to "murder". Sámi people use the word *kotti* in order to say, "to kill". But, there is no word *murder!* According to Nuccio Mazzulo it is completely different understanding compared to the western culture. Killing is coming from hunting.[18] The Sámi have used to solve their identity and self-determination problems in reciprocal negotiations with Finnish decision-makers. Nevertheless there are also Sámi protest groups such as Suohpan Terror using irony to defend

[13] The linguistic debate in Finland has always been a delicate subject, which has poisoned the political debate until 1922, or even beyond. In these circumstances, granting special rights to the Sami in this area was a significant step forward.

[14] *Sajos*, meaning « the basis » in Inari Sámi, is situated in Inari, is not only a political centre regrouping 21 elected representatives since 1996 (The law establishing the Finnish Sámi Parliament was passed in 1973 and amended in 1995 and 1996), but also a strong touristic attraction, renting the building for festivals, conferences and other private activities. It is important to underline the will of Sámi Parliament to rent this wooden building achieved in 2012 for public and private activities, desacralizing the political and identity symbolism of the place.

[15] Arctic Council. https://arctic-council.org/index.php/en/our-work/arctic-peoples

[16] The total population of the Arctic region is 4 Million.

[17] Statistics Finland, 2017, https://www.stat.fi/index_en.html

[18] Interview with Nuccio Mazzulo, Anthropology Professor at the Arctic Centre, University of Lapland, September 7th 2018 in Rovaniemi.

Sámi interests or opposition groups for concrete causes such as Ellos Deatnu against Tana River Regulations.

It is possible to see the specificities of the Sámi culture in local Sámi History. During the Sámi History after the Scandinavian, Finnish or Russian domination, it is quite difficult to find violent episodes in the entire Sápmi. Even the event called "Kautekeino Revolt" in 1852, is nothing else then a group of Sámi rebels who killed two people. This passivity is all the more surprising when one studies many injurious experiences that Sámi faced in their past. Especially, the period between 1850 and World War II is the century of Sámi assimilation policies in Scandinavia however the Sámi did not raise any strong voice at structural level. (Herb and Kaplan 1999).

In the specific case of Finland, it might be a useful component to outline the milestones of Finnish Sámi history and development of their self-determination demands. World War II was a crucial watershed for Sámi mobilization because of the Finnish-USSR agreement in 1944 and its result as Rovaniemi War. While Nazis were retreating from Finland through Lapland, they burnt whole Rovaniemi and the region (that has a direct effect on the lack of architecturally attractive "old towns" in the region). As a result, there was not left any single construction and infrastructure, since, the Sámi people lost their houses. Reconstruction took a long time, therefore economic recovery as well. This started to move of Sámi mobilization in Finland.

There are three "types" of Sámi in Finland; North Sámi, Inari Sámi and Skolt Sámi. By the cession of Petsamo to the USSR in the end of the WWII, the Skolt Sámi who were used to live in Petsamo were resettled by Finnish government around Sevettijärvi. *Sámi Litto* (Sámi Union) was founded in 1945 however it was not efficient as in Sweden and Norway.

Moreover, a "Sámi Delegation" was organized by Finnish initiative in 1973 as a judicial state authority. However, as it was a Finnish state initiative, this delegation authority became a simple representative body of Sámi instead of being decision-making power. By changing its name in 1996, this "delegation" became an elected body as "Sámi Parliament" or *Sámediggi*. It has a Parliament composed by 21 representatives elected for four years. This parliament is likely an advisory body, as it has no coercive power such as decision-making related with Sámi issues (Müller-Wille 1979). In other words the *Sámediggi* advises and orients the Finnish government on Sámi related issues, may object to central decisions, but has not a right of veto.

The question that Sámi Parliament also deals with is the identification of Sáminess *i.e.* to define "who is Sámi?". This question has a direct effect on the identification of Ethnic Sámi and of who can vote in Sámi elections. According to the new legislation: *That he himself or at least one of his parents or grandparents has learnt Sámi as his first language; that he is a descendent of a person who has been entered in a land, taxation or population register as a mountain, forest or fishing Lapp; or that at least one of his parents has or could have been registered as an elector for an election to the Sámi Delegation or the Sámi Parliament.*[19]

[19] Sámi Parliament web-site https://www.finlex.fi/fi/laki/kaannokset/1995/en19950974.pdf

Rain or shine, the existence of Sámi Parliament is socially, politically and culturally highly significant vested right that contributes to the promotion of self-determination and Sámi nation building process. (Wessendorf 2001). Thanks to *Sámediggi*, Finnish Sámi asked for land rights and recognition which are important steps in terms of territorialisation of the Sámi identity, besides cultural autonomy. (Herb and Kaplan 1999).

Today, Finnish Sámi do not have land determination. They have "linguistic self-determination" since 1994. As for the economy, Finnish Sámi community has four economic activities: reindeer-herding, forestry, fishing and gathering. Apart from being an economic activity, reindeer husbandry has always played an important role in Sámi culture. However in Finland, Sámi do not have an exclusive right on reindeer-herding, the Finnish are as well reindeer-herders. Moreover, industrialization of Lapland and climate change play negative role on reindeer-herding in the Northern Finland. Forestry as well, equally affected by climate change and industrialization, is very important for the economy. These activities are all interconnected to each other. The entire region of the Sámi culture is currently under another threat of a new railway project, which will be constructed through Lapland. The $3.4 billion, 526 km project would run north from Rovaniemi, Finland, to the Barents Sea port city of Kirkenes in Norway's far eastern Arctic.[20]

This foreseen but not accepted yet project will for sure destroy the nature of the region with bringing the danger of dying reindeers on the railways and harming the reindeer herding areas. According to Finnish authorities, it is easy to find investors for the project; however, there is a strict opposition from Sámi and international environmentalists.[21] Even if the existing railway project concerns only cargo transportation (and not passengers), one must underline the fact that there is still room for negotiation to improve safety and herds. It would also be useful to note that the train is likely to help open up the region and is a relatively clean and relatively safe means of transport. On the other hand, the debate on the railway construction, more than rational and concrete results, is mainly due to the emotional and symbolic reasons: who will decide on the future of region? Finnish state, Lapland in its entirety or Sami?

Fishing salmon is another very important economic activity for Sámi. Not only there are difficulties in fishing because of climate change and industrialization, but also new policies of Finland's government threatens the way of life in Lapland. New policy of Finnish government in 2017 was made in order to revive the weaken stock of salmon in Tana (Deatnu) River.[22] The river in question is the border between in the northern part of Finland and Norway. According to the governance of the river,

[20] « Proposed Arctic Railway Would Cut Through Lapland Reindeer Habitat » https://www.newsdeeply.com/arctic/articles/2017/08/03/proposed-arctic-railway-would-cut-through-lapland-reindeer-habitat

[21] « Industrial railway line and logging threaten the Sámi homeland » https://www.greenpeace.org/international/press-release/18253/railway-logging-threaten-sami/

[22] Centre for Economic Development, Transport and the Environment, https://www.ely-keskus.fi/documents/10191/23117928/Teno+info+English.pdf/1a185302-fa5d-4ba6-bd90-17183fefdf21

fishing rights are given to private ownership of land and people who live permanently in the valleys of Tana. New agreement of Tana, brings restrictions on fishing methods, tools and licences while forbidding fishing for non-permanent residents of Tana river valley[23] (Holmberg 2018).[24] Paradoxically, according to this new regulation, cabin owners who buy land and build cabins in the Tana Valley would have right for fishing. The cabin owners would also get licence for their guests (tourists) very easily and with cheap prices, which would allow everybody to fish in this fragile river. This would end up even in change in biodiversity, change in distribution of salmon and change in Sámi economy.

After all, coercive policies in the past, new assimilative regulations today and climate change frustrated the Sámi. As the use of traditional knowledge (*árbediehtu* in North Sámi), especially traditional ecologic knowledge and traditional skills (*árbemáhttu*) (Holmberg 2018), are in their way of becoming obsolete, the economic option that Sámi take refuge in tourism with the remaining culture and, therefore, are obliged to transform the knowledge as marketable products and activities. The dilemma is the fact that some Sámi cultural products are not Sámi monopolies (Thuen 2004).

18.3 Does Tourism Protect Identity?

18.3.1 What Do We Mean by "Tourism"?

In 1900, the future father of Inuitology Knud Rasmussen, was writing the following lines, after visiting Swedish Lapland:

> *"There is war now in Lapland, and the combatants are two cultures. And the new must prevail, because it carries with the future. But any victory causes death. The Lapps will be conquered in their own land... and they will die just as quietly and unobtrusively as they have always lived up there"* (Brown 2015).

As many people who visited Finnish Lapland, the authors of this chapter have, on their wall, a small Noaidi (Sámi 'shaman") drum, bought in a touristic shop between Saariselkä and Inari, on the way to Sámi Music Festival of August 2018, held in the gardens of Sajos, the Finnish Sámi Parliament. During this trip in the heart of Lapland, where they spent approximately more than 500 € for 48 h, they met 1,5 Sámi: one man, in his 40s, a rich souvenir and necklace merchant from Norway who came to Inari for the festival (mainly to sell his products with his wife), and the guardian and ticket seller of the Skolt Sámi Heritage House, a young woman in her

[23]"General fishing rules for visitors anglers in Tana River System » http://tanafisk.no/wp-content/uploads/2012/06/Fiskeregler-for-tilreisende-fiskere-2012-PDF-engelsk.pdf

[24]Aslak Holmberg (or Aslat Niilas Aslat) is a young Sámi fisherman who is affected by the new policy of Finnish government. He explains in detail Salmon fishing of Sámi and the impact of the new regulation on Sámi culture.

30s, married to a Sámi man, who were struggling to have her word at the municipal council because she was only a wife of Sámi and not Sámi herself.

Was the very stereotypical "touristic trip" of the authors of these lines profitable for the protection of Sámi culture or not? We should discuss.

According to the United Nations World Tourism Organization, "Tourism is a social, cultural and economic phenomenon, which entails the movement of people to countries or places outside their usual environment for personal or business/professional purposes. These people are called visitors (which may be either tourists or excursionists; residents or non-residents) and tourism has to do with their activities, some of which imply tourism expenditure".[25]

Thus, the touristic attractiveness can be related to the climate, geography, history, entertainment activities, sportive activities, and architecture and living cultural "exotic" activities. Many of these pull factors coexist in one locality, or at least, one main factor provokes several others. A region attracting tourists, for example, for its geographic specificity (beach, mountain…) develops other pull factors such as sports, or entertainment. In other words, one or several pull factors interact and/or create other pull factors. The question is, of course, if the search (and attraction) of cultural tangible and intangible richness and values are threatening the very same richness and values by exhausting objects and architectures on the one hand, and by degenerating behaviours and customs on the other hand, mainly due to the unhealthy social, financial and symbolic interaction. The symbolic interaction is the interrelationships within the social interactions in which actors of the relation behave as a response of the behaviours of others (Blumer 1994). Especially in minority groups such as Sámi, the social identity is constructed by interpreting, how others perceive and respond to the group and individuals actions. This response being communicated through interpretative symbols (adopting the receiving group's behaviours or not, language, food, etc.). The immediate use of the symbolic interaction in tourism is its contribution (or not) to an understanding (or the opposite, folklorization) within the tourist-host unequal relationship due to the variety of expectations (Sharpley 2014). Therefore, the authenticity and inauthenticity issues in tourism may affect actors' behaviours (Sharpley 1999) (in both sides, and in the Sámi case in three sides: Sámi, non-Sámi actors in Finnish Lapland and tourists). In other words, Finns and foreigners visiting for specific purposes Lapland (Northern lights, snow activities, Santa, fishing…) have a direct and indirect effect on Sámi culture.

On the one hand, there is no doubt that there is a positive effect of tourism on Sámi cultural specificities. Many of the cultural behaviours (reindeer herding, husky farms), or traditional objects (clothes, music instruments) could have been disappeared (only "could" because history cannot be written with "if") without being touristic and therefore financial attractions. On the other hand, one may con-

[25] http://www2.unwto.org/

sider that the survival of these tangible and intangible heritages[26] is far from the authenticity and the objects and traditions that "remain" are "something else", a weak imitation or worst, the results of an invented tradition (Hobsbawm and Ranger 1983).

18.3.2 The Role of Tourism in Regional Development: Folklorization or Preservation?

Tourism is basically founded on *differences*, at least on the perception and expectation of differences, real or imaginary but especially at certain degree. If during the first half of the twentieth century, the "tourism" in the Arctic searched radical differences between civilization/primitiveness (Hinch and Butler 2007) urbanization/emptiness, usual conditions/extreme conditions, the mass tourism as an industry of the second half of the twentieth century and especially after 1990's must limit in time and in quality the sharpness of these dichotomies. Indeed, mass tourism as an industry requires equipment, accommodation, transport, food, trade and guides. Tourists are not explorers. They need quick and authentic-kind features in acceptable life conditions. For example, in their study on perceptions, Triväinen *et alii*, demonstrated how foreign and Finnish tourists in Lapland require emptiness (expectation), but do not sacrifice comfort (condition) (Tyrväinen et al. 2014). Thus, the participation of Sámi in the Lapland touristic sector is not only a cultural protection issue but also related implicitly to the question of power, self-determination and autonomy, inclusion in the national and international market, and especially legitimacy at economic activities (Viken and Müller 2006): who can fish in Lapland? Sámi? Finns? Americans? Everybody? If, for the identity protection we accept the postulate that Sámi need financial strength, then the answer is the opening of the traditional economic sectors to tourism, taking the risk that these traditional economic sectors become not traditional anymore. This is the touristic paradox.

In Finnish Lapland, tourism is mainly attracted by allegedly "extreme" climatic conditions (long sunny days in summer, long nights in winter, northern lights, cold…), sportive entertainment activities (hiking, skiing, fishing, hunting…),

[26] Here one must underline the fact that none of the Sámi cultural behaviours or traditions is listed by UNESCO's *List of Intangible Cultural Heritage* and only Swedish Lapland (called "Laponian Area") is awarded by «The world heritage » label of UNESCO with the following description: *The Arctic Circle region of northern Sweden is the home of the Sámi, or Lapp people. It is the largest area in the world (and one of the last) with an ancestral way of life based on the seasonal movement of livestock. Every summer, the Sámi lead their huge herds of reindeer towards the mountains through a natural landscape hitherto preserved, but now threatened by the advent of motor vehicles. Historical and on-going geological processes can be seen in the glacial moraines and changing water courses.* https://whc.unesco.org/en/list/774

invented traditions (mainly Santa village in Rovaniemi) and, finally, animal linked attractions (reindeers that the herding is not Sámi exclusivity in Finnish Lapland,[27] Husky Farms). It will not be wrong to assess that "cultural tourism" related to Sámi in the region has become secondary, a supporting sector more than the main attraction.

First, one must underline the philosophical difference between "ethnic tourism" that is related to the curiosity towards primitiveness and "cultural tourism" that is more related to the attraction of cultural activities (Smith 1978). In both cases, as one of the main definitions of contemporary tourism is the "quest of the other" (Scott and Selwyn 2010), "tourism transforms difference into the global discourse of consumerism, a process by which the 'otherness' becomes a commodity to be consumed" (Cole 2006). Specifically for Sámi in Finnish Lapland, if the ethnic tourism occurred in the beginning of twentieth century, as Sámi are not so "primitive" to be attractive anymore, the shift is towards "cultural tourism" where, cultural behaviours and products are becoming emboldened. There is no doubt that the number of Sámi knives on the walls of tourists' houses is much superior to Sámi using these knives in their everyday lives. Nevertheless, the search of the otherness operated by tourists in Sápmi, could have the opposite impact also, provoking consciousness. The affirmation of a specific identity, and as a matter of fact, an "indigenous identity" and the (re)building process of the 'difference" can be a reaction to/a consequence/a response of otherness imposed as consumable tourist commodities. This, while reinforcing the group identity and claims, may also provoke changes in social roles. For example, according to Cole, in Finnish Lapland, women were empowered through their participation in a craft cooperative (Cole 2006). In the same vein, Miettinnen defends strongly the idea that in Inari, Sámi found themselves empowered in front if non-Sámi of the region, thanks to tourism, and thus could enlarge their new power to other segments of the social life such as local politics or environmental issues. According to the authors –who are actors of the "cultural tourism" more than observers– projects such as EUROTEX between 1997 and 1999, or the tourism development project of 2002–2003 called *"there is more than snow"* created a fruitful atmosphere for the general empowerment of Sámi, in this very sparsely populated area of Inari[28] where communities are very small and fragmented (Miettinnen 2006). Thus, if tourism *gives* identity, it also gives power to protect, develop and enrich this identity even if the risk of folklorization is always present.

[27] It is very interesting to underline the touristic success of reindeer related attractions, (herding, products) that provoked recently the need of more reindeer products and their importation from Russia to Finland, especially meat and hides: « Finland imports Reindeer Hides from Russian Arctic »

http://www.rusexporter.com/news/detail/5474/?sphrase_id=97018 or « Demand for reindeer meat up, export markets look tempting »

https://yle.fi/uutiset/osasto/news/demand_for_reindeer_meat_up_export_markets_look_tempting/6962600

[28] 7500 inhabitants in an area of 17,400 km^2

18.3.3 Tourism for Sámi vs *Sámi for Tourists*

18.3.3.1 Tourists for Sámi

It is inevitable to talk about indigeneity and ethnic issues, while analysing tourism in Lapland as the region is home to Finnish Sámi and has crucial cultural touristic attractions as reindeer farms, husky farms, festivals and cultural heritage museums. Thus, this section analysis the impacts of ethno-political factors on tourism in Lapland, specifically the cities Rovaniemi, Sodankylä, Ivalo, Inari and Sevettijärvi.

Even not all the Finnish Sámi live in Lapland, even not all Finnish Sámi speak Sámi, the language, symbols, traditions play an important (but not primary) role in the region's tourism, therefore – in regional development. Reindeer-herding is one of the Sámi cultural habit even if the number of herders has been decreasing dramatically. However, it is not a barrier for Sámi to be associated with reindeers including by tourists. This might be called *selfdom* that for centuries the Sámi are known as masters of reindeer herding, therefore it is not easy to separate this traditional way from them.

The assimilation process of Sámi in Finland was much stronger than in Scandinavia. Thus, the similarity of Sámi with Finnish people is much more visible than the similarity with Sámi in Scandinavian countries (Viken and Müller 2006).

During the post-cold war era, the reconstruction of the roads in Lapland, which were destroyed by the German army during the war, paved the way for tourism in the region. Tourism is a year-round industry in the region as northern lights, dog-sledging, reindeer sledging, ski in winter season and fishing, hunting, lake cabins, festivals in summer time increased the nature based tourism in Lapland. Sámi traditional culture has become a complementary point to Santa Claus Village in Rovaniemi, ice and snow hotels in Lapland and all the seasonal touristic nature attractions above. Thus, the Sámi symbols became largely visible in Lapland. It is ordinary to see Sámi tents or Sámi handicraft shops very often. On the one hand, while the cultural items are sold to tourists by locals as authentic objects; on the other hand, cultural attractions are also present in the region: *Ijahis idja* (Sámi music festival organised in Inari since 2004[29]), Siida museum in Inari presenting Sámi history and ethnology,[30] or Skolt Sámi heritage house in Sevettijärvi.[31] For example, *Ijahis idja* festival is a cross-regional Sámi music festival that all Sámi from Finland, Scandinavia and Russia attend, but also this organization brings the domestic and international tourists to the city. Cultural authenticity reached to such commercial dimension that the Sámi parliament building *Sajos* in Inari might be rented for congress, conferences, meetings, and business events. Therefore, the parliament, sensitively *holly* place, which was gained in the end of a very difficult process might be rented for non-Sámi events for money and the people who rent might buy the most expensive touristic handicrafts.

[29] http://www.ijahisidja.fi/en/updates.php

[30] http://www.siida.fi/contents/sami-museum

[31] http://www.nationalparks.fi/skoltsamiheritagehouse

18.3.3.2 Sámi for Tourists

As making handicrafts requires time, patience and artisanship, the prices of the products rise. In Lapland, concerning touristic attractions, expensiveness comes to mind before Sámi colours, symbols and activities. Finland is one of the most expensive countries in the world but how come Lapland, least populated northern part of Finland might be that expensive?

The culture of Sámi is perceived as otherness by foreigners-tourists, yet it is actually selfdom. The Sámi use this misperception as a tool and gain economic profits, recognition by touristification of their own culture. And, perhaps, gaining the *habitus* of selling the culture is becoming one feature of the Sámi selfdom.

Taking the Fennoscandinavian cultures as the centre and identifying the other cultures as the deviances of them is very normative as in the case of seeing Sámi culture as authentic. (Viken and Müller 2006) This normative approach or othering increases the tourist flow in Lapland and more tourists do not mean interiorization or homogenization. On the contrary Sámi remains the "eternal" other, instrumentalizing this otherness by highlighting specificities. Otherness is constructed on dissimilarity (Clock and Little 1997). However, this otherness takes another form when there is a minority involved. The majority sees minority subconsciously as a threat. Its intolerance is expressed by suppression, assimilation and folklorization, even sometimes by force. Thus, this marginalization increases the authentification of the minority culture and the entire region as in Sámi and Lapland.

The tourists want to see Sámi colours, figures, and tents, when they go to Lapland and they encourage the folklorization of the culture. The people measure the Sáminess and Sámi places with culture-related facts (symbols, colours, animals, costumes). Therefore, Sámi people use their traditional symbols very well in tourism to increase the attraction of their location, as it is the case in Rovaniemi, Sodankylä, Ivalo, Inari and Sevettijärvi. For sure, using symbols is not the evidence of Sáminess, however, it is the demand from the foreigners. Actually, this might be useful for Sámi to make handicrafts, tents, costumes that prevent from forgetting traditional facts and objects.

The Sámi sell cultural products, however, they do not approve non-Sámi people to use their cultural materials such as costumes, hats or singing yoiks (Thuen 2004). Apart from authentification of tradition and economic development, tourism increased the questioning of identity by/of Sámi people.

18.4 Conclusion

Following Knud Rasmussen, Jean Malaurie, while visiting Greenland all his life, remained in a hesitation. On the one hand, he always regretted the contamination of the Inuit culture by contact with *Qallunaat*,[32] thinking that they were loosing their

[32] Literally "men with thick eyebrows": white men.

authenticity. On the other hand, he noticed that these contacts were facilitating Inuits' hard life conditions and, therefore, there was no reason to deprive them in a colonialist approach (Malaurie 1989). What non-Sámi people visiting Lapland are excepting from Sámi? To *be* authentic? To *look* authentic? And the opposite side, what Sámi expect from tourism? *Money*? *Recognition* or *respect*? Of course, "tourists" and even "Sámi" are ideal-typical categories and there is no doubt that expectations vary according other factors such as socio-professional category, educational level or individual political orientation. But all in all, this relationship between tourists and Sámi, is a two-way relationship. We should first underline the very positive effect of tourism in Sámi culture. Such a house abandoned by Hestia looses its fire and falls into ruin, a culture without life in it, becomes a museum at best and disappears at worst. It is impossible to know if without any touristic value, Kosa or Guksi, these wooden Sámi cups used all life long, would be still produced, at least in such quantity or they would become cute heritage of "ancient times", exhibited in a museum. The authors of this chapter have witnessed, as many of their colleagues do frequently, a Sámi civil society representative in an academic meeting, flaunting proudly her Guksi, in her traditional costume, while reading her communication on an I-Pad. This is the paradox of tourism: tourists come to see untouched cultural behaviours and objects in search of authenticity and, at the same time, it gives life but it mummifies. That is why the Sámi folk singer with strong yoik influences, Mari Boine, feels the need to sing:

> *Let language and culture take their place in the museum, as research object and tourist attraction.*
> *Give lively speeches on each festive occasion.*
> *Let it disintegrate and die that which was a nation.*[33]

Nevertheless, as mentioned before, the Sámi culture is not the main touristic attraction of the Finnish Lapland. Therefore, the benefits or damages of tourism on Sámi culture must be seen as collateral. What would happen, if one day Santa Claus dies? In other words, what could be the effects of the absence or drastic decrease of tourism in the region, on Sámi culture? Climate change, uncontrolled industrialisation, construction of the railway may change in the future the values of the attraction of Lapland. Or, if according to economist Arthur Laffer, too much tax kills tax, too much tourism may kill tourism. Tourism preserves the past in some cases, but does not protect the future for sure. In any case, land use rights claimed by Finnish Sámi seem indispensable to insure a future rich cultural life, where behaviours and objects are not only genuine but also examples for other cultures on how to deal with nature, in a context where these experiences become vital.

---

[33] *Oppskrift for herrefolk* (Recipe for a master race) https://ourochreway.wordpress.com/2010/02/12/oppskrift-for-herrefolk-recipe-for-a-master-race/

References

Blumer, H. (1994). Society as a symbolic interaction. In H. Nancy & R. Larry (Eds.), *Symbolic interaction and introduction to social psychology* (pp. 263–267). New York: General Hall Publishers.

Brown, S. (2015). *White Eskimo: Knud Rasmussen's fearless journey into the heart of the Arctic.* Boston: De Capo Press.

Capotorti, F. (1979). *Study on the rights of persons belonging to ethnic, religious and linguistic minorities.* New York: United Nations.

Carson, D., & Carson, D. (2018). International lifestyle immigrants and their contributions to rural tourism innovation: Experiences from Sweden's far north. *Journal of Rural Studies, 64*, 230–240. https://doi.org/10.1016/j.jrurstud.2017.08.004.

Cloke, P., & Little, J. (1997). *Contested countryside cultures: Otherness, marginalisation, and rurality.* London: Routledge.

Cobo, J. M. (1981). *Study of the problem of discrimination against indigenous populations: Final report* submitted by the Special Rapporteur «Introduction» 30 July 1981E/CN.4/Sub.2/476.

Cole, S. (2006). Cultural tourism, community participation and empowerment. In S. Melanie & R. Mike (Eds.), *Cultural tourism in a changing world: Politics, participation and (re)presentation* (pp. 89–103). Clevendon: Channel View Publications.

Hafstein, V. (2018). Intangible heritage as a festival; or, folklorization Revisited. *The Journal of American Folklore, 131*(520, Spring), 127–149.

Herb, H. G., & Kaplan, H. D. (1999). *Nested identities: Nationalism, territory, and scale.* Lanham: Rowman & Littlefielld Publishers.

Hinch, T., & Butler, R. (2007). *Tourism and indigenous peoples: Issues and implications.* Oxford: Butterworth-Heinemann.

Hobsbawm, E., & Ranger, T. (Eds.). (1983). *The invention of tradition.* London: Cambridge University Press.

Holmberg A. (2018). *To ask for salmon. Saami traditional knowledge on salmon and the river Deatnu: In research and decision making*, Master Thesis, The Arctic University of Norway.

Keith, C. L. (1999). The United Nations international covenant on civil and political rights: Does it make a difference in human rights behavior? *Journal of Peace Research, 1*(1), 95–118.

Koivurova, T., & Stepien, A. (2011). How international law has influenced the national policy and law related to indigenous peoples in the Arctic. *Waikato Law Review, 19*, 123–143.

Malaurie, J. (1989). *Les derniers rois de Thulé* (5th ed.). Paris: Plon. 1st edition in 1954.

Miettinnen, S. (2006). Raising the status of Lappish communities through tourism development. In M. Smith & M. Robinson (Eds.), *Cultural tourism in a changing world: Politics, participation and (re)presentation* (pp. 159–174). Clevendon: Channel View Publications.

Nafisa, Y. (2016). The determinants of sustainable entrepreneurship of immigrants in Lapland: An analysis of theoretical factors. *Entrepreneurial Business and Economics Review, 4*(1), 125–159.

Schönfelt, K. (Ed.). (2017). *The Arctic in international law and policy.* Oxford: Hart Publishing.

Scott, J., & Selwyn, T. (2010). Thinking through tourism – Framing the volume. In J. Scott & T. Selwyn (Eds.), *Thinking through tourism* (pp. 1–26). New York: Berg.

Semb, A. J. (2012). Why (not) commit: Norway, Sweden and Finland and the ILO convention 169. *Nordic Journal of Human Rights, 30*(2), 122–147.

Sharpley, R. (1999). *Tourism, tourists and society.* London: Elm Publications.

Sharpley, R. (2014). Host perceptions of tourism: A review of the research. *Tourism Management, 42*, 37–49.

Similä, M. (2003). Immigrants and minorities in Finland: Problems and challenges. In T. David & G. Julia (Eds.), *Immigration in Europe: Issues, policies and case studies* (pp. 97–112). Bilbao: University of Deusto.

Smith, V. (Ed.). (1978). *Hosts and guest: Anthropology of tourism.* Philadelphia: University of Pennsylvania Press.

Tanner, A. (2011). *Finland's balancing act: The labor market, humanitarian relief, and immigrant integration.* https://www.migrationpolicy.org/article/finlands-balancing-act-labor-market-humanitarian-relief-and-immigrant-integration

Thuen, T. (2004). Culture as property? Some Sámi dilemmas. In K. Erich (Ed.), *Properties of culture – Culture as property. Pathways to reform in post-Soviet Siberia* (pp. 87–108). Berlin: Dietrich Reimer Verlag.

Tuulentie, S. (2017). Destination development in the Middle of Sápmi: Who's voice is heard and how ? In V. Arvin & M. Dieter (Eds.), *Tourism and indigeneity in the Arctic* (Tourism and cultural change, n° 51). Bristol: Chanel View Publications.

Tyrväinen, L., Uusitalo, M., Silvennoienen, H., & Hasu, E. (2014). Towards sustainable growth in nature-based tourism destinations: Clients' views of land use options in Finnish Lapland. *Landscape and Urban Planning, 122*, 1–15.

Viken, A., & Müller, D. (2006). Introduction: Tourism and the Sámi. *Scandinavian Journal of Hospitality and Tourism, 6*(1), 1–6.

Vuorela, K., & Borin, L. (1998). Finnish Romani. In A. Ó Corráin & S. Mac Mathúna (Eds.), *Minority languages in Scandinavia, Britain and Ireland* (Studia Celtica Upsaliensia n°3) (pp. 51–76). Uppsala: Acta Universitatis Upsaliensis.

Wessendorf, K. (2001). *Challenging politics: Indigenous peoples' experiences with political parties and elections*. Copenhagen: International Work Group for Indigenous Affairs (IWGIA) Publishing.

Wille, M. L. (1979). The Sami Parliament in Finland: A model for ethnic minority management? *Études/Inuit/Studies, 3*(2), 63–72.

Part V
Sustainable Governance

Chapter 19
Regulation of Cargo Shipping on the Northern Sea Route: A Strategic Compliance in Pursuing Arctic Safety and Commercial Considerations

Antonina Tsvetkova

Abstract This study aims to explore how the regulatory process is shaped under the influence of interaction between the regulator and the most powerful actors involved. It presents an overview of the historical development of cargo shipping regulation in the Russian Arctic during 2001–2018 to illustrate key events that have influenced the existing legislation. Data from 22 semi-structured interviews and archival materials are interpreted through the institutional logics approach.

The study reveals how strategic actions, including lobbying and information manipulation, by most powerful actors affect regulation, political initiatives, and commercial outcomes. The findings further reveal how contextual and institutional circumstances make business companies reconsider their core competencies and supply chain practices by seeking to prevent the regulatory burden that results in non-compliance. Distortions between the interest groups' expected benefits and political initiatives caused a change in the existing legislation and shipping traffic in the Russian Arctic. The study provides an understanding of how regulation is shaped as a co-produced process due to interactions between the regulator and all the players involved and how it changes during implementation in practice.

Future research should include the influence of external agents like international law on regulation of cargo shipping in the Russian Arctic.

Keywords Regulatory process · Supply chain management · Institutional logics · Arctic shipping

---

A. Tsvetkova (✉)
Faculty of Logistics, Molde University College, Molde, Norway
e-mail: antonina.tsvetkova@himolde.no

E. Pongrácz et al. (eds.), *Arctic Marine Sustainability*,
Springer Polar Sciences, https://doi.org/10.1007/978-3-030-28404-6_19

19.1 Introduction

The maritime transport sector has been always facing commercial pressure and an ever-increasing demand to optimize logistics management systems and improve connectivity of all maritime operations. The need for enhanced cooperation in maritime surveillance and enforcement between all the actors in the maritime domain is addressed by such an order-preserving mechanism as regulation. Regulatory principles and legislation constrain or promoting the behaviour of all the actors involved in cargo shipping by determining the rules of inter-organizational relationships, contractual terms of maritime transport and even specifying payment options at sea. At the same time, maritime transport services have always been a sensitive area for the application of regulatory rules because many different stakeholders are involved.

In literature, regulation is commonly viewed as politically formalized mechanisms derived from governmental policies and legislation. A number of previous research has primarily focused on how these regulatory mechanisms coercively pressure an organization to behave in a certain way, engage in specified supply chain practices and adapt existing practices to new technologies in order to retain legitimacy (Yaibuathet et al. 2008; Shook et al. 2009; Williams et al. 2009; Bhakoo and Choi 2013; Sodero et al. 2013; Moxham and Kauppi 2014). However, the idea that we are living in an era of the "regulatory state" where the government has an exclusive right to exercise power and control has been challenged. Several research has identified that there are some interest groups of actors, which are able to affect state regulation to gain benefit and even take some regulatory functions into their own hands in order to act, to some extent, as rule makers (Stigler 1971; Peltzman 1976; Becker 1983; Veljanovski 2010). The interactions between those who establish regulation and those who are subject to it have acquired special significance for the regulatory process. Thus, the regulatory process contains mostly two closely related but distinct issues of the actual behaviour of the actors under and in so-called the regulatory game and the economic purpose of the process itself. It refers to how the game works and why it is played (Owen and Braeutigam 1978). However, this theoretical dispute has been weakly supported by empirical research of how various regulatory participators interact and affect each other in real practice; especially the scarce attention paid to the transportation area (Baldwin et al. 2010).

Motivated by the above-mentioned shortcomings in the literature, *the present study aims to provide deeper insights into how the regulatory process is shaped under the influence of interaction between the regulator and the most powerful actors involved.*

In doing so, the study presents an empirical case of regulatory processes, which have affected and guided the development of cargo shipping along the Northern Sea Route (NSR). The NSR is a crucial artery in the transport system of the Arctic for the social, economic and cultural development of the remote Arctic regions and global trade (Høifødt et al. 1995; Hong 2012). The recent potential for maritime activities on the NSR includes offshore petroleum resources extraction, intra-regional transportation of extracted minerals and onshore energy resources, and international transit, though it remains limited in volume for the last few years. Thus, the range of actors with interests in the Arctic maritime domain has diversified considerably to include mining industry, oil/gas companies, transnational logistics providers, the Russian government, national authorities and other stakeholder organizations. Some of them act as powerful players able to resist and affect the Russian governmental enforcement. The main challenge relating to industrial activities in the High North is that the Arctic nature is vulnerable to external influences and recovers slowly from encroachments and accidents. There is widespread concern that increased interest in oil and gas exploration in both Arctic seas and regions will unleash unsafe and high-risk projects that could harm the Arctic environment and indigenous people.

The study employs the institutional logics approach in order to address the changes in the regulation of cargo shipping in the Russian Arctic and the role of the most powerful actors involved. Through the institutional lenses, this study provides an overview of the historical development of cargo shipping regulation in the Russian Arctic over the past 17 years (2001–2018) to illustrate the main events that have influenced the existing legislation.

The study is organized as follows. The next section presents academic knowledge about the regulation concept. Then the theoretical framework is described. This is followed by the research method. The fifth section presents the research context and empirical case. The findings are discussed in the following section. The study concludes with theoretical and practical implications, as well as suggestions for future research.

19.2 Understanding of Regulation as a Strategic Game in Literature

Regulation has been often viewed in literature as something that comes from the coercive actions of the government and authorities where

> the government is the rule maker, monitor and enforcer, usually operating through a public agency. (Black 2001)

Broadly accepted is Selznick's (1985, p. 363) definition of regulation as

> sustained and focused control exercised by a public authority over activities valued by the community.

Regulation includes an intentional activity of attempting to control, order and alter the behaviour of others to comply with the law, defined standards, norms and ultimately to achieve the desired outcomes by involving mechanisms of standard-setting, information-gathering and behaviour modification (Black and Kingsford 2002; Baldwin et al. 2010). However, the dominant view that government has a monopoly on the exercise of power and control makes the understanding of the regulatory process fragmented and limited. It neglects fact that organizations and regulators not only operate within the regulatory framework or the rules of the game but also are able to change those rules (Veljanovski 2010).

Several earlier research has recognised that regulation inherently derived from government and/or public agencies can be affected by those organisations that are subject to it (Stigler 1971; Peltzman 1976; Becker 1983). Regulators stay out of day-to-day operational activities of practitioners and then the existing legislation is often imperfect for addressing their needs and desired outcomes. Interest groups like big manufacturers, businesses and consumers can influence regulation through different strategic actions like the lobbying and the manipulation of the information in trying to gain more favourable and beneficial regulation, e.g. lower tariffs, subsidies, occupational licensing and fees (Stigler 1971).

Further, regulation is intended to get private organizations to modify their behaviour to comply with the law and ultimately to achieve the desired outcomes (Veljanovski 2010). Thus, regulation should be a process of coordinating, influencing and balancing interactions between the actors and the regulatory framework to develop the existing patterns and/or create new patterns of interaction, which enable organisations to organize themselves. The new understanding of regulation as a co-produced process embraces that interdependencies and interactions or so-called "strategic game" exist between its various players like regulators, organizations and interest groups (Black 2001). However, there is still a lack of understanding of how interactions and interdependencies between both regulator and other players can influence regulatory and even commercial outcomes by acting strategically (Black and Kingsford 2002; Veljanovski 2010).

Further, the lobbying and seeking to prevent regulatory burden by big powerful companies can be costly and cause distortions between the companies' expected benefits and political initiatives. Resolving the contradiction between public and private goals is a fundamental task for government intervention (Trebing 1987). If the regulatory development is in stagnation and the government coercive enforcement increases, companies implement strategies that can affect the existing practice and institutional factors, thereby making resistance to state coercive pressures and constraints a viable option (Tsvetkova and Gammelgaard 2018). The results of this process are well illustrated by Stigler's words:

> The innumerable regulatory actions are conclusive proof, not of effective regulation, but of the desire to regulate. And if your wishes were horses, one would buy stock in a harness factory. (Leube and Moore 1986, p. 224)

Thus, powerful companies and regulator can be uncooperative and even antagonistic, resulting in frequent non-compliance and judicial processes. However, the literature's understanding of under what circumstances interactions between regulator and companies turn into non-compliance and opposition is limited. Further, the companies' strategic actions have different pay-offs and can unintentionally produce the change of regulation over time. This may lead again to an unforeseen discrepancy between the companies' desired outcomes and legal purposes within the existing rules.

To address the change in the regulatory process and the role of all the actors involved, this study applies the lenses of institutional logics perspective as outlined in the section below.

19.3 Theoretical Framework: The Regulatory Process Through Institutional Logics

SCM literature applying institutional theory has mostly focused on what effects different institutional pressures – coercive, normative, and mimetic – have on supply chain operations (Huo et al. 2013), strategic decision making (Kinra and Kotzab 2008; Doha et al. 2013), as well as the adoption of new SCM practices and technologies in existing and new contexts (Bello et al.2004; Zhang and Dhaliwal 2009; Lee et al. 2013; Doha et al. 2013; Hoejmose et al. 2014). This literature, however, neglects the processes of how regulatory principles change when companies face a choice of new strategic actions. The institutional logics approach, as applied to this study, discusses, in particular, the meaning and behavioural motive that social actors invest their strategic actions with not only to gain compliance with the existing regulation but also to achieve desired outcomes. It demonstrates how regulatory and institutional principles, via logics, affect organizational strategies, structures and practices (Thornton and Ocasio 1999; Thornton et al. 2012). Further, it highlights that actors can translate logics at the local level and then new logics become manifest and shape a new meaning (Battilana et al. 2009) in regulation. In this way, the application of this institutional approach provides unique insight into regulatory process change and dynamics of interactions between regulator and organizations within a single legislative framework.

Institutional logics are generally understood as macro-belief systems that shape historical patterns of practices, values, and rules by which social actors produce and reproduce their operational activity, organize time, and space, and provide meaning to their social reality (Thornton and Ocasio 1999). Further, institutional logics act as "taken-for-granted social prescriptions" for social actors (Battilana et al. 2009); influence decision-making processes in organizational fields; define goals, expectations and legitimate activity (Thornton and Ocasio 1999); and often become embodied in organizational structures and practices (Thornton 2004).

Regulation constitutes a number of institutionalized norms, laws, rules and logics that guide a legitimate order of the institutional environment. Legitimacy means that organizational actions are "desirable, proper, or appropriate within some socially constructed systems of norms, values, beliefs, and definitions" (Suchman 1995, p. 574). It affects not only how organizations act but also how they understand themselves to be legitimate (Suchman 1995). Organizations need legitimacy as well as technical efficiency to survive and thrive in their environments. The institutionalized norms, practices and logics, which structure organizational fields, exert isomorphic pressures, forming an "iron cage", which constrains, in turn, organizational actions. Organizations are seen as legitimate when they conform to field structures and operate within the iron cage (DiMaggio and Powell 1983).

At the same time, logics established by the regulatory principles and norms to organize the behaviour of organizations in a certain way can be translated into action that either reinforces or reconstitutes the logics themselves (Thornton et al. 2012); e.g. lobbying business interests in trying to change the existing legislation, or social pressures to increase normative beliefs and conscientious societal values in managerial ideologies. During interaction between social actors and institutions, one institutional logic may be undermined and replaced by another. This process encompasses institutional change and deinstitutionalization when regulatory rules weaken and disappear (Scott 2014). Following this view, this study proposes that logics serve as tools, which can be used by social actors in a contested environment to affect decisions, justify their strategic actions and advocate for change of the regulatory framework. Further, the same logic can be used in different situations to achieve various goals, and the same actor can choose to employ different logics at different times depending on the perceived needs of the immediate situation (McPherson and Sauder 2013).

19.4 Method

19.4.1 Data Collection

The qualitative historical approach was employed to investigate the changes in the regulation of cargo shipping in the Russian Arctic over the past 18 years (2001–2018), as well as examine the behaviour and motives of the key actors that could affect the NSR regulation. Data collection was based on multiple data sources, including 22 semi-structured face-to-face interviews with the representatives of 11 organizations (see Appendix 1), the Russian legislation on cargo shipping along the NSR, as well as archival materials. The interviewees were selected on the basis of their involvement in the regulatory processes and cargo shipping in the Russian Arctic. The interviews focused on tracing the main historical events of the regulatory development and strategic actions and role of most powerful organizations that could affect the regulatory processes and behaviour of other actors involved. The interviews were conducted in Murmansk and St. Petersburg during three periods: May 2014, November 2014, May 2015. Three rounds of conducting the interviews allowed increasing the data reliability. The interviews were conducted in Russian and then translated into English. All the interviews were hand-written and recorded with the consent of each interviewee to be transcribed later. To avoid idealization or, in contrast, ambiguity of subjective criticism at some events in the regulatory process, the empirical data was crosschecked through asking similar questions to different interviewees repeatedly and over several periods of time.

Various secondary data were collected, mainly from the Russian legislative rules and laws on cargo shipping in the Russian Arctic (see Appendix 2) and in the waters area of the NSR, official annual reports of industrial organizations involved in shipping in the Russian Arctic, press releases and official websites.

This study contains the official names of agencies, business companies and other organizations involved in the regulatory process because they are explicitly outlined in Russian legislation.

19.4.2 Data Analysis

The content analysis was employed to evaluate the meaning of a great number of data collected from multiple sources. It made possible to generate reliability and traceability of my arguments. The data analysis was guided by identifying the keywords and interpretations provided by the Russian legislation and used by the interviewees during their day-to-day routines. These keywords were considered influential words in different texts of the collected data for further determining what effect they have on the strategic actions of the actors concerning the regulation of cargo shipping. It allowed creating communication between various interpretations of how business companies perceive and apply the governmental regulation in practice. When analyzing the empirical data, the meaning of the interview contents, the laws, and different texts of the archival materials played an important role in the interpretation of organizational actions.

The overview of the key historical events in the regulatory process over the past 18 years (2001–2018) starts with the description of the historical development of the NSR to proceed then the current state of affairs.

19.5 Definition and Historical Development of the NSR

According to the Russian legal regulations, the NSR stretches from Novaya Zemlya in the west (meridian 168 degrees 58 minutes 37 seconds West) to the Bering Strait in the east (parallel 66 degrees North), including internal sea waters, territorial seas, contiguous zone and exclusive economic zone of the Russian Federation (NSR Law, 2012). Thus, the area of the NSR does not cover all the waters of the Russian Arctic. When examining the regulation processes, this study encompasses cargo shipping both in the water area of both the NSR and the Barents Sea between the Kara Strait and the Murmansk port.

Unlike the most other sea routes, the NSR is a series of different sailing lanes, mostly trafficked along the Russian Arctic coast. It consists of the Kara Sea, the Laptev Sea, the East Siberian Sea and the Chukchi Sea (Østreng et al. 1999). Together with the numerous rivers flowing into the Arctic Ocean, the NSR constitutes an integral water-transport system. There are two different ways of using the NSR as: (1) international use for transit commercial operations through the whole length of the NSR from/to Europe and to/from Asian regions; and (2) domestic use for cabotage operations by local industries primarily located in the western part of the NSR.

Vessels encounter various natural challenges to their movements and navigation along the NSR. This route is characterized by harsh climate conditions (with polar nights, a long, cold winter, snowstorms, spray icing of ships), long distances between ports, limited transportation choice, the necessity to use icebreaker assistance and a short period of Arctic navigation per year. The route covers some 2200–2900 nautical miles of ice-infested waters (Østreng et al. 1999). These factors make cargo shipping along the NSR vulnerable to possible disruptions and increase transportation costs.

Cargo shipping along the NSR has been pushed ahead by the development of the basic industrial complexes: the mining-metallurgical, mining-chemical, and oil and gas industries (Høifødt et al. 1995; Østreng et al. 1999; Hong 2012). Bringing out the output from the economic base industries and bringing in input to the industries and consumer goods to the inhabitants, maritime industrial flows within the NSR provide not only the maintenance of industrial needs but also the ability of Arctic remote regions to survive. They include fish and fish products; timber; iron-ore and iron-ore semi-products; nickel-ore and metals from mineral industries; oil and gas products; industrial equipment, food products and other inputs (Høifødt et al. 1995; Granberg 1997).

In the history of commercial use of the NSR, three distinct phases can be identified:

– *1932 to the early 1950s: organization of regular navigation and construction of special fleet and port*

The Soviet government placed significant emphasis on the Arctic and the NSR, making large-scale investment in infrastructure and allocating considerable funds for scientific research from the 1930s onwards. In 1932, the Chief Administration of the NSR was established with wide-ranging authority "to develop a sea route along the sparsely populated coast of Northern Siberia" (Bulatov 1997). The first shipping with industrial cargo was implemented in the summer of 1939. However, for a long period, navigation was carried out only during summer navigation.

– *1950s–1970s: completion of the NSR development*

Ensuring reliable communications and infrastructure allowed transforming the NSR into a normally operating main traffic line during the summer-autumn seasons of navigation.

– *From the 1970s: transition to year-round navigation along the NSR*

Extending the summer-autumn periods of navigation to year-round use of the NSR was achieved after powerful atomic and diesel-electric icebreakers, ice-strengthened vessels for Arctic navigation were put into operation. Since 1978, year-round voyages have been regular within the western section of the NSR:

Murmansk-Dudinka due to an increased demand for establishing a dependable transport scheme of interaction between the Kola and Norilsk industrial complexes (Høifødt et al. 1995; Østreng et al. 1999). Director General of FSUE "Rosatomflot" emphasized:

> In the 1970s the development of the Norilsk industrial region became a starting point for the creation of a powerful nuclear icebreaker fleet. We call MMC "Norilsk Nickel" our "daddy" for fun.

Until the dissolution of the USSR in the 1990s, the volume of shipments increased steadily (Granberg 1995). The maximum volume 6,578,000 tones was reached in 1987. In comparison with 1945, the volume of shipments by 1987 increased 14.8 times, and in comparison with 1960 – 6.8 times (Østreng et al. 1999). However, the reforms of the 1990s were marked by a huge economic decline in production (Shleifer and Treisman 2005; Granberg 1997) and change in the Russian institutional environment (Polterovich 1999). The economic crisis particularly affected the regions along the NSR (Granberg 1997), which were characterised by weak administrative management and a tendency for rapid degradation (Utkin and Denisov 2001; Kuznetsova and Nikiforov 2013). Industries reduced their operations due to the general economic downturn in the domestic market. This caused a decrease in cargo shipping activities and the interruption of regularity in cargo shipping in the Russian Arctic because freight traffic has been one of the most vital components behind the existence of the NSR (Granberg 1997). In 1996 the total volume of shipments totalled just 1,642,000 tonnes (Østreng et al. 1999).

The Russian government was late in understanding that the NSR could not adapt to new market conditions and overcome crisis without the change in regulations and adjustments in the financial-economic mechanisms (Østreng et al. 1999).

19.6 Regulation Processes Between 2001 and 2018

19.6.1 Period 1 (2001–2005): Trapped Inside the "Iron Cage" of the Old Legislation

In the early 2000s, the starting point in updating the existing regulation was the adoption of the Marine Doctrine in 2001 that became a key document for maritime legislation to govern the national marine policy until 2020. Marine Doctrine (2001) emphasised that

> maritime transport is crucial in ensuring the domestic traffic, especially in regions where sea transportation is the only mode of transport, as well as in foreign economic activity.

The role of maritime transport was determined as decisive for the maintenance and development of the Arctic and sub-Arctic regions.

Navigation and safety of cargo shipping along the NSR were regulated primarily by the Rules of 1990 and other legislative procedures adopted during the 1990s and outdated to new market realities. Those norms aimed at ensuring the safety of navigation, as well as at preventing, reducing, and controlling marine environment pollution from vessels.

The management of the NSR was effectuated by the NSR Administration, together with other structures within the Service of Marine Transport of the Ministry of Transport of the Russian Federation. Transportation was regulated by the following joint-stock companies that owned the shipping fleet: Murmansk Shipping Company (Murmansk), the Northern Shipping Company (Arkhangelsk), the Arctic Shipping Company (Tiksi), the Far Eastern Shipping Company (Vladivostok), and the Primorsk Shipping Company (Nakhodka). Transportation in the winter navigation was carried out only with icebreaker assistance, according to the old legislation and outdated technical parameters of vessels. The icebreakers used on the NSR have been owned by the state and entrusted to the shipping companies and ports. Two icebreaker operators were empowered to carry out traffic management of the icebreaker fleets: the fleet of JSC "Murmansk Shipping Company" operated under extreme winter conditions – with heavy ice and temperatures down to −50 °C, the fleet of JSC "Northern Shipping Company" was designed for operations under less extreme ice conditions (Høifødt et al. 1995). JSC "Murmansk Shipping Company" was also a major logistics provider in cargo shipping in the Russian Arctic waters and provided the bulk of all Arctic shipments (Granberg 1995). Mining and Metallurgical Company "Norilsk Nickel" (further MMC "Norilsk Nickel") was the largest cargo owner in the region, and its freight traffic on the western section of the NSR comprised about 45 per cent of the profits of JSC "Murmansk Shipping Company".

To compensate for the financial losses after the economic crisis and the significant decrease in shipping activities, the government changed the regulatory policy by constantly increasing tariffs on icebreaker services from 2003 (Order #69, 2000). Tariffs increased every year due to inflation and rapid growth of prices for atomic fuel. The government subsidies for the maintenance, construction and operation of the icebreaker fleet were eliminated since 2003. So financing of the operating costs of the icebreaker fleet carried out exclusively through payments of cargo owners for the icebreaker services. Consequently, the icebreaker per ton fee increased from $5 to $70 per ton of cargo between 2003 and 2008 and then amounted approximately $80–$100 per ton of cargo in 2011. The NSR policy was particularly influenced by the actions of the operator of the icebreaker fleet.

Further, the situation was complicated by the fact that the state fleet of ice-strengthened vessels and atomic icebreakers became obsolete and required renovation and technical innovation. Any accident on the atomic icebreaker, e.g. some technical malfunction or a fire, could become crucial for the major owners of cargoes. The disruption of cargo delivery could violate the manufacturing process, cause significant economic losses for the northern enterprises, and increased transportation costs due to demurrage. The Russian government planned the building of new icebreakers but had to stop it due to financial issues. The operational activity of industrial complexes was challenged by a shortage of icebreakers and ice-class vessels, constantly increasing tariffs for icebreaker services and the limited choice of suppliers and sea carriers.

Additionally, the NSR policy, in particular, the Rules of 1990, could not provide appropriate norms of behaviour and interactions between all the actors involved. In 2004, the NSR Administration was eliminated. Then, JSC "Murmansk Shipping Company" was privatised. The management and maintenance of the NSR required the state support that was acknowledged by not only federal authorities and agencies but also major business companies.

19.6.2 Period 2 (2006–2008): Attempts to Escape from the Clutches of the "Iron Cage"

MMC "Norilsk Nickel" searched for several alternatives including negations with the government to get tariffs on icebreaking services lower as being the largest cargo owner, but lobbying of the company's interests was not successful. As Transport Minister S. Frank announced at a press conference in Krasnoyarsk on February 10, 2003:

> To ensure shipping of its own cargoes, MMC "Norilsk Nickel" was going to invest $15 million in the Federal program for the development of the NSR, especially for the building of a new icebreaker.

However, collaboration with the state was weak.

Between 2006 and 2008, MMC "Norilsk Nickel" launched its own Arctic fleet, consisted of five container vessels with ice-class Arc7 according to the Russian Maritime Register of Shipping (DNV Ice-15 + DAT −30 °C) and one universal tanker able to overcome up to 1.7 m thick Arctic ice without icebreaker assistance. Having the own fleet ensured regular cargo transportation along the NSR without icebreaker assistance all year round and transport independence from the government policies in order not to pay the obligatory icebreaker fees when sailing along the NSR. It allowed reducing transportation costs significantly. The new Arctic fleet of MMC "Norilsk Nickel" with the application of new technologies changed the historically established cargo transportation in the Arctic waters. New supply chain strategic practices affected multiple aspects of existing cargo transportation and marked a new development of Russian Arctic shipping in general (Tsvetkova and Gammelgaard 2018).

However, the new independent position of MMC "Norilsk Nickel" caused discontent among its former suppliers – the state sea carrier and the state operator of icebreaker services. As they lost their major customer and cargo owner, their profits suffered considerably. The complete non-acceptance and discontent manifested themselves through claims, complaints and several legal proceedings for not paying icebreaker services by MMC "Norilsk Nickel" due to conflicting economic interests between this company and other most influential actors, presented mainly by the state. As noted by one of the interviewees:

> The new ice-class vessels became a new phenomenon in Arctic maritime transportation. The previous regulatory framework did not determine procedures for them because they were incompatible with the new technologies.

The claims to recover fines from MMC "Norilsk Nickel" were called unjustified enrichment. The latest claim initiated by the state operator of the icebreaker fleet was in 2008 and amounted to 10,641,265.45 roubles (about $212,825) but the court recognised it as unsubstantiated.

As one of the interviewees highlighted:

> The issue of not paying tariffs for icebreaker services by MMC "Norilsk Nickel" that they did not use remained actually hot-arguing for a long time like a bleeding wound.

According to the data presented on the official website of FSUE "Rosatomflot", the refusal of MMC "Norilsk Nickel" to pay icebreaker services resulted in a shortage of income for maintaining the state atomic icebreaker fleet in the amount of 238,500,000 roubles (about $4,770,000) in 2006; 329,800,000 roubles (about $6,596,000) in 2007; 757,800,000 roubles (about $15,156,000) in 2008 and a predictable shortfall of payments for 2009 was estimated in the amount of 1,608,400,000 roubles (about $32,168,000). Thus, the dynamics of these numbers highlighted that the regulation of cargo shipping continued to be based on the steady growth of tariffs on icebreaker services (from 2003) rather than to strengthen the regulatory policy. As added by one of the interviewees:

> The icebreaker operator also condemned the lobbying actions of MMC "Norilsk Nickel" in that they might become one of the main reasons for delaying the adoption of necessary laws and regulations on the NSR.

MMC "Norilsk Nickel" was not alone in promoting transport independence. Companies "Sovcomflot" (in 2010) and JSC "LUKOIL Oil Company" (1999–2002) purchased tankers of high ice-class for sailing in Arctic waters without icebreaker assistance. Thus, business companies had to take strategic initiatives for ensuring the reliability of cargo transportation by own efforts due to the inappropriate NSR policy and government enforcement. The role of the state icebreaker operator as a dominant player reduced.

19.6.3 Period 3 (2008–2011): Prerequisites for Weakening the Clutches of the "Iron Cage"

This period could be identified as the implementation of a new state ideology for the further development of Arctic cargo shipping. In 2008, the Russian government initiated a number of federal programmes and strategies for the NSR revitalisation. A new Russian Arctic state policy until 2020 outlined new strategic priorities and emphasized the dependence of the Russian northern regions on delivering suppliers from the other regions for their surviving and further developing. The NSR was identified as a national integrated transport-communications system of the Russian Federation in the Arctic," specifically, an "active coastguard system" in the Russian Arctic. (Arctic state policy 2008)

One of the main challenges was how to increase freight traffic along the NSR and create a monitoring system for maintaining navigational safety and managing cargo flows.

As emphasized by D. Medvedev, President of the Russian Federation, during the Security Council of the Russian Federation "On the protection of the national interests of the Russian Federation in the Arctic" on September 17, 2008:

> Modernization of the NSR infrastructure needs an advanced system of navigation, search and rescue. It is necessary to expand the port network and develop the NSR as a strategic national highway of Russia. This is our national priority.

Further, the Russian government planned to provide support for the construction of new icebreakers, day-to-day rescue/auxiliary vessels and the coastal infrastructure.

In 2008, the government appointed the new icebreaker operator, FSUE "Rosatomflot" through transferring back responsibilities from JSC "Murmansk Shipping Company". As officially announced on the website of FSUE "Rosatomflot", its mission was "to support the intensification of Arctic shipping as the key factor of growth and development of the Russian North".

In 2009–2010, several transit commercial voyages were successfully carried out along the NSR with the icebreaker assistance from the Asian regions to Europe in order to attract foreign partners for international transit use of the NSR. FSUE "Rosatomflot" became a new powerful actor in Arctic cargo shipping that started playing a dominant role among the state actors involved.

In 2011, the Russian government changed the framework of tariff policy for the icebreaker services (Order #122-T/1, 2011). The rates of tariffs for icebreaker assistance remained intact but were newly defined as "maximum" in order to offer the possibility of bargaining between ship owners and the icebreaker operator within certain limits. While, since the 1990s until 2011, tariff regulation was characterised by constantly growing rates, this legislative innovative initiative allowed for adjusting tariffs to match the current parameters of the shipping market and take into account the ship owners' interests. According to the Executive Director of the Non-commercial Partnership on Coordination of the NSR Use, the new tariff policy made the NSR more attractive with record figures on the East-West cargo transit totalling 835,000 tons in the summer and autumn of 2011.

This regulatory change indicated that the role of the Russian government enforcement remained strictly coercive, albeit somewhat more loyal to the participants involved in the day-to-day practice of cargo shipping. The efforts of the Russian government on the NSR development for the last years was apparently insufficient to create a high-quality legislative framework, to make actions of the actors involved coordinated, as well as attract investments.

19.6.4 Period 4 (2012–2013): The New Legislation – Well, Has the "Iron Cage" Been Exposed?

Several improvements in legislation took place in the regulation of cargo shipping in the Russian Arctic waters. Since January 27, 2013, the NSR Federal Law presented a number of important amendments to the existing laws and legislative acts. The major amendments warranted the reestablishment of the NSR Administration and the adoption of new navigation rules in the water area of the NSR.

The NSR Administration aimed at organising navigation in the water area of the NSR, and ensuring its safe navigation and protection of the marine environment from pollution but not including plans for further development. Its main functions included the organisation of icebreaking, navigational, hydrographic and legal security of navigation, conducting work on the prevention and elimination of oil spills on the slopes of the NSR, and interaction with emergency services during work on the prevention and mitigation of emergencies that involved natural and man-made disasters on the NSR. In practice, the NSR Administration performed only general coordination and document circulation, primarily granting permissions for the navigation of a vessel in the water area of the NSR and receiving daily reports by shipmasters via the online system. Traffic of vessels, sailing nearby and within the water area of the NSR, were monitored electronically.

The Rules of 2013 determined a new order to the organisation of navigation of vessels along the NSR, the icebreaker assistance and ice pilotage of vessels, provision towards the navigational-hydrographic and hydro-meteorological support of the navigation of vessels, radio communication, as well as special requirements to vessels. Unlike the Rules of 1990, the new rules did not provide for the establishment of Marine Operations Headquarters as a special navigational service to carry out actual operational management of navigation. According to the new NSR Law, the master of a vessel had to choose a particular route of sailing on his own, based just on ice information from relevant agencies.

However, the new Rules of 2013 caused doubts among some experts in ensuring the safety of navigation and proper control of navigation. As highlighted by Head of the NSR Administration who previously worked as the master of a nuclear icebreaker at the II international conference "The Northern Sea Route: State, Problems and Prospects" in St. Petersburg, 2013:

> The role of ship owners and shipmasters significantly increased in making decisions on sailing a vessel. However, the forecast for ice conditions covers up to half the Arctic waters. This may increase risks for ensuring safe navigation. It is reasonable to re-establish Marine Operations Headquarters. Further, legal responsibility for the violation of the Rules was not ordered. Non-compliant vessels should be refused next permissions. There is no procedure developed for cooperation between the NSR Administration and the icebreaker operator.

Thus, the Rules of 2013 contained some legal shortcomings to make day-to-day routines appropriate to the challenges of cargo shipping in Arctic waters that took place in practice.

Further, the NSR Law raised the debate about what organisation - either FSUE "Rosatomflot" or the NSR Administration - would be responsible for the collection of payments for navigation along the NSR according to the rate of tariffs. Initially, the NSR Administration was intended to accumulate different charges on the NSR. But FSUE "Rosatomflot" managed to protect its interests to be responsible for taking fees for icebreaker services. Additionally, FSUE "Rosatomflot" lobbied but unsuccessfully for introducing into the NSR Law an order that icebreaking services were mandatory for all vessels, including high ice-class vessels able to sail without icebreaker assistance (like vessels owned by MMC "Norilsk Nickel"). This debate emphasised again that receiving payments for ice-breaking services according to the fixed rates of tariffs continued to play an important role in the strategic game of regulation.

The tariff policy changed and provided the abolition of the obligatory icebreaker fees. The payments were understood as charges only for actual services of icebreaker assistance and ice piloting to be provided, in fact, with the maximum tariff determined by the Federal Tariff Service. These tariff adjustments, however, caused complaints of FSUE "Rosatomflot". According to Director General of FSUE "Rosatomflot", the new system of tariff collection resulted in considerable losses in profits to the company and became a heavy burden for the maintenance of the icebreaker fleet.

Further, the activities like "icebreaker support, ice pilotage of vessels in the area of the Northern Sea Route" were identified as the services of natural monopolies, which "cannot be replaced in consumption for other goods...and the demand ...for such goods [services] depends to a lesser degree on changes in the price for these goods" (Law #147-FZ, 1995; NSR Law). This meant that the demand for these activities was identified as "more efficient in the absence of competition due to specific technical features of production" (Law #147-FZ, 1995). The following organisations were registered as being able to provide the services of icebreaker assistance: FSUE "Rosatomflot", FSUE "Rosmorport", JSC "Far Eastern Shipping Company", JSC "Murmansk Shipping Company", Murmansk Transport Branch "MMC "Norilsk Nickel", JSC "LUKOIL Oil Company." Thus, some ice-class vessels, owned by business companies, were also recognised as performing icebreaker functions in the Russian Arctic. Thereby, the role of these business companies took on special significance in the regulatory process. At the same time, the governmental intervention substantially increased with the legislative change.

The regional authorities provided no implication in the regulatory process because the NSR is a Federal highway of Russian national interests. They focused on the creation of a certain information field for mutual discussion and dialogue between all the stakeholders through organising conferences, forums, and seminars. Thus, the regional authorities played a role of a mediator between the federal government policy and business companies.

19.6.5 Period 5 (2014–2017): Still in the "Iron Cage"? – The Rivalry for Opening a New Regulatory Game

Despite the amendments of 2013 in the regulatory framework of navigation along the NSR, that made it more favourable for business companies, and several successful transit commercial voyages in 2009–2010, the total volumes of cargo remained quite small in the following years. It was estimated 2.8 mil tons in 2013, 3.7 mil tons in 2014 and 5.15 mil tons in 2015. Then, Yamal LNG project started contributing to internal Russian traffic delivering natural gas via the NSR. It allowed increasing the total volume of cargo from 7.5 mil tons in 2016 to 10.2 million tons in 2017 (by nearly 40%). While internal traffic has been escalating, transit volume, however, fell sharply in 2014 and it has still been at a trickle by 2017: 1.3 mil tons (71 vessels) in 2013; 0.24 mil tons (31 vessels) in 2014; 0.04 mil tons (18 vessels) in 2015; 0.21 mil tons (19 vessels) in 2016; 0.19 mil tons in 2017. This decline in transit coincided with a sharp drop in the price of bunker fuel on the world market in 2014, unfavourable freight rates, difficult ice conditions during several years, a scarcity of commercial ice-strengthened ships and geopolitical tensions due to USA-EU sanctions against Russia. These factors and the uncertainty markedly reduced the economic value of the time saved by using the NSR in comparison to the Suez Canal, through which approximately 18,000 vessels make their way every year.

To improve this situation, the Russian government began to discuss the reorganization of competencies for the further development of the NSR. New governmental efforts aimed at improving the effectiveness of the NSR management and increasing the economic value of the NSR, including the volumes of internal and international freight traffics. This process started in 2016 with a proposal from the Arctic Commission led by Deputy Prime Minister Rogozin to create a unified logistics operator for the NSR, primarily to exploit infrastructure, including the icebreaker fleet, more efficiently. That proposal did not provide details how this body would be organized but assumed the creation of a new independent organization.

The proposed reorganization signified the change in the administrative structure. By 2016, the administrative structure was fragmented when different government-owned agencies were responsible for different competencies on the NSR. FSUE "Rosmorport", a government-owned corporation created in 2002 and part of the Ministry of Transportation, tasked with running the NSR infrastructure, including ports and conventional icebreakers. The NSR Administration, also part of the Ministry of Transport, was created in 2013 to ensure the safe operations in the water area of the NSR, including issuing of permits, information about ice conditions, and coordination of icebreaker usage and search and rescue operations. FSUE "Rosatomflot", also a government-owned corporation, was responsible for the running and maintenance of the atomic icebreaker fleet since 2008. According to the current Russian legislation, the NSR Administration was not authorized up to 2018 to engage in promoting commercial activities; predicting future demand for NSR traffic, cargo volume, and subsequent demand for icebreaker assistance and other support services (in contrast to the functions of the previous NSR Administration formed in 1932).

In 2017, the Russian government received two rivalling proposals for the reorganization of the NSR management. One proposal suggested upgrading the NSR Administration and making the Ministry of Transport responsible for all operations, including the atomic icebreaker fleet. Another proposal considered that FSUE "Rosatomflot" would consolidate all competencies of the NSR, including the NSR infrastructure, communications, navigation, and scientific issues in order to become a key and single agency of future NSR development policy. This ongoing rivalry for influence led to a conflict between FSUE "Rosatomflot" and the NSR Administration.

19.6.6 Period 6 (2018): Arctic Safety or Commercial Considerations? – Resuming the Battle Between Powerful Agencies

The violation of Arctic safety rules by the Cyprus-flagged LNG tanker "Boris Vilkitsky" in April 2018 escalated a growing conflict between FSUE "Rosatomflot" and the NSR Administration. The vessel, owned by Greek company "Dynagas", suffered the malfunction in March on its way from Rotterdam to Sabetta port. After getting damage to stern thrusters and port steering column, the vessel's ice capabilities reduced from Arc7 to Arc4 (from DNV Ice-15 to DNV Ice-05) that made it illegal for the vessel to enter the NSR waters either alone or with icebreaker escort. However, the "Boris Vilkitsky" entered the route, being escorted by the Rosatomflot icebreaker "Taimyr" via Cape Zhelaniya into the Kara Sea and, thereby, violated the NSR legislation. Further, the vessel's master did not notify the NSR Administration officials and the port authorities in Sabetta about the vessel's condition and faulty mechanisms, violating, thereby, also rule #19 of navigation. Officials called the incident a gross violation of NSR rules and *"a threat to the safety of navigation, as well as the protection of the marine environment"*.

The NSR Administration officials became aware of the damage when the vessel experienced difficulties navigating in heavy ice on way to Sabetta while being escorted by the Rosatomflot icebreaker "Taimyr". On the vessel's arrival, officials revealed a host of additional violations, including the absence of accurate ice charts and the lack of required ice navigation experience by the master and crew. The "Boris Vilkitsky" remained in port for more than a week before it was permitted to leave after direct intervention by Russian President Vladimir Putin, not to constrain the development of Russia's largest Arctic natural resource project Yamal LNG, strategically important for the state. However, even being escorted by Russia's most powerful atomic icebreaker "50 Years of Victory" it took almost a week to get the "Boris Vilkitsky" out of the ice-covered waters of the NSR.

Actually, every year a number of vessels violated the safety rules of the NSR, e.g. in 2017 nearly 100 vessels. It highlighted the inability of the NSR Administration to properly enforce its own rules and regulations. But the incident in April 2018 was considerably intensified because FSUE "Rosatomflot", the state-owned corporate, knew the malfunction of "Boris Vilkitsky" and concealed it from the NSR Administration. The officials blamed FSUE "Rosatomflot" for prioritizing commercial considerations from escorting "Boris Vilkitsky" through the NSR waters rather than upholding the safety rules. Thus, the incident exposed the struggle for control over the NSR, which simmered below the surface for at least the past 2 years since the Russian government initiated plans to reorganize the NSR's competencies. As highlighted by a senior researcher at the Fridtjof Nansen Institute:

> This looks like a very hard battle between different agencies and personalities. Large sums of money in the form of state investments and subsidies are involved.

In June 2018, the struggle for control over the NSR between "Rosatomflot" and the Ministry of Transport was preliminarily resolved by the government for further sharing competencies.

From December 27, 2018, new Federal Law #525-FZ entered in force and defined powers of FSUE "Rosatom" in development and functioning of the NSR and adjacent territories. The Law says that "Rosatom" has acquired rights to develop proposals on forming the state policy for development and sustainable functioning of NSR. Besides, FSUE "Rosatom" together with authorized bodies makes navigation viable, safe, and accurate within the NSR. This work may include hydrographic and topographic surveys. Further, the Law delegates FSUE "Rosatom" powers to control the budget money for development and sustainable functioning of the NSR infrastructure and seaports, ensuring of navigation and all-year-round towing of ships along the NSR. FSUE "Rosatom" will define a subordinate enterprise, which is authorized to grant permissions for navigation within the NSR.

As highlighted by one of the interviewees:

> FSUE "Rosatom" takes now the lead role in the NR management and the development of Russian Arctic logistics in general.

The recent legislative change aimed primarily at increasing shipping traffic along the NSR and attracting foreign companies. In 2018, more than 18 million tons of cargo was shipped along the NSR, a 68% growth from 2017, despite complicated ice conditions. Both in the Kara Sea and in the East Siberian Sea there was thick ice through major parts of early summer of 2018. But transit shipments on the route remain sparse.

The current reorganization of the NSR competencies is challenged by how to, on the one hand, arrange both commercial considerations to increase the freight traffic and maximize economic development, and, on the other hand, meet all the safety requirements and remain transparent.

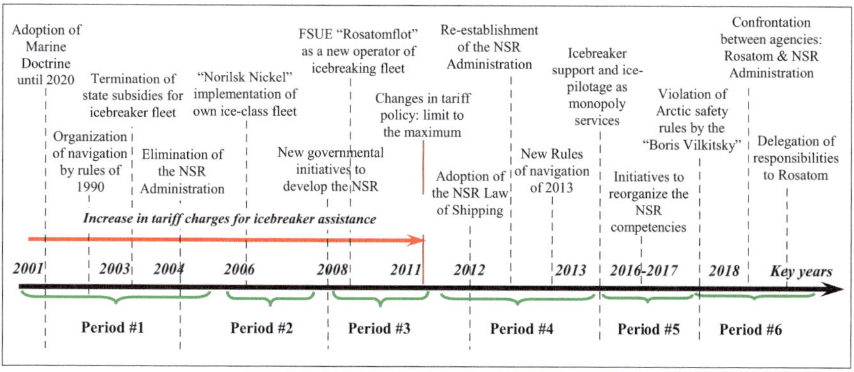

Fig. 19.1 Key historical events of the regulatory process of cargo shipping in the Russian Arctic between 2001 and 2018

19.6.7 Summary

The regulatory process between 2001 and 2018 encompassed six periods, each characterized by specific government initiatives, actions of business companies and the adoption of new legislative norms of shipping along the NSR (see Fig. 19.1).

The study illustrates how interactions between the regulator and key players affect the legislation and regulatory implications. The regulatory process illustrates a movement from the government coercive enforcement to the adoption of more favourable norms and rules for increasing shipping traffic along the NSR (see Table 19.1).

19.7 Discussion

The regulatory process of cargo shipping along the NSR between 2001 and 2018 constitutes a co-produced activity that encompasses interaction between the Russian government as the main regulator and all the players involved.

The institutional logics approach has been suggested by focusing on interaction between the regulator and actions of interest players involved that seems to be invisible in the regulatory process, but, under the certain circumstances and political purposes, can affect the existing legislation and SCM practices. The findings reveal at least three main institutional logics, which have participated in the regulatory process between 2001 and 2008:

- *Logic of state policy*: refers to the government and focuses on providing the rules and norms to order the behaviour and actions of all the players involved; ensuring the safety norms of navigation and increasing shipping traffic along the NSR;
- *Logic of stability*: refers to business companies and focuses on ensuring regular cargo deliveries and profitability (avoiding obligatory icebreaker fees);

Table 19.1 Interaction between the state policies and actions of the players involved in the regulatory process of cargo shipping along the NSR between 2001 and 2018

Period	State policies	Actions of the players involved	Regulatory implications	Shipping traffic along the NSR
2001–2005	Regulatory constraints: coercive unstable state policy; Constantly increasing tariffs to cover all costs of the NSR infrastructure and icebreaker maintenance; No financial support for building new icebreakers; Elimination of the NSR Administration	Decision to develop own transport infrastructure	Uncertainty over regular deliveries for cargo-owners; Lack of icebreakers; Obsolescence of ice-strengthened vessels; High risks of disruption; Obligatory payments for the icebreaker services	Decline in volume of both internal and transit traffics
2006–2008	Continuously increasing tariffs; Debates on the necessity to reform the NSR legislation	Commissioning own Arctic fleet with new technologies to overcome ice by the key cargo-owner (MMC "Norilsk Nickel") and not to pay icebreaker fees; Issuing by the icebreaker operator legal proceedings against the key cargo-owner	New SCM practices: navigation without obligatory icebreaker assistance; Improving shipping efficiency; Losing main profits by the state; Causing the discontent between the main cargo-owner, its former suppliers and the icebreaker operator	Very slight increase in both internal (2.0 mil tons in 2006) and transit traffic
2008–2011	New Russian Arctic state policy until 2020 highlighting the NSR importance; Appointment of the new operator of the atomic icebreaker fleet; Change in tariff policy: fixing maximum limits and adjusting tariff rates by agreement with the new operator of atomic icebreakers based on the vessel displacement	Organizing several commercial transit voyages by the icebreaker operator to attract foreign partners	More favourable state policy for business companies	Increase in internal traffic: 3.1 mil tons in 2011; Slight increase in transit traffic: 0.1 mil tons in 2010

(continued)

Table 19.1 (continued)

Period	State policies	Actions of the players involved	Regulatory implications	Shipping traffic along the NSR
2012–2013	Adoption of the new NSR Law: re-establishing the NSR Administration, new rules of navigation; Navigation without obligatory icebreaker assistance (depending on vessel type); Change in tariff policy: abolition of obligatory icebreaker fees and adoption of tariffs only for actual services; Recognizing MMC "Norilsk Nickel" ice-class vessels as legally capable of providing icebreaker services	Lobbying by the icebreaker operator to be responsible for collecting payments for navigation and icebreaker services instead of the NSR Administration (successfully); Lobbying by the icebreaker operator to make icebreaker services obligatory for all types of vessels (unsuccessfully)	More flexible and less coercive legislation for all players involved; Increasing role of ship owners and shipmasters in making decisions on sailing along the NSR (with icebreaker assistance or not)	Increase in internal traffic (2.8 mil tons in 2013); Increase in transit traffic from 0.1 million tons in 2010 to 1.35 million tons in 2013
2014–2017	Initiatives on the reorganization of the NSR competencies between two agencies: the icebreaker operator and the NSR Administration; External pressures on state policy: sharp drop in the price of bunker fuel, geopolitical tensions and USA-EU sanctions	Lobbying by the icebreaker operator for consolidation of all the NSR competencies and responsibilities to become a single logistics operator of the NSR	Regulation becomes a more co-produced activity	Growth of internal freight traffic mainly due to the construction of Sabetta port and oil/gas projects: 7.5 mil tons in 2016, 10.5 mil tons in 2017; Sharp decline in transit traffic: 0.24 mil tons in 2014; 0.21 mil tons in 2015; 0.21 mil tons in 2016; 0.19 mil tons in 2017

(continued)

Table 19.1 (continued)

Period	State policies	Actions of the players involved	Regulatory implications	Shipping traffic along the NSR
2018	Adoption of the new Law on transferring all the NSR competencies and responsibilities to the operator of the atomic icebreaker fleet	Violation by the "Boris Vilkitsky" of Arctic safety rules and illegal entering the NSR with icebreaker assistance; Concealment of this violation by the icebreaker operator for own commercial interest; Struggle for control over the NSR between the icebreaker operator and the NSR Administration	Strengthening state intervention; Single logistics operator of the NSR	Growth of internal freight traffic app. 18 mil tons, mainly due to the export of Yamal LNG; Record growth of transit traffic app. 0.5 mil tons, mainly due to shipping Cosco (five tranits) and Maersk (first-ever container ship)

- *Logic of priority*: refers to the operator of atomic icebreaker fleet and focuses on gaining control over all the NSR competencies and increasing profitability of the icebreaker use along the NSR.

The links between the use of these logics and regulatory outcomes become clearer when examining the interaction between the state policies, business companies' expected benefits and the interests of the icebreaker operator. This study argues that institutional logics are tools used purposefully to push the players' decisions in desired directions. Thus, the actions of the key cargo-owner to launch its own Arctic fleet in 2006–2008 manifested themselves as a strategic response by this organization to state coercive unstable policy and constraints. Contextual and institutional circumstances made the key cargo-owner reconsider its core competencies and the role of SCM practices. In this sense, logics resemble instruments that can be implemented by whoever picks them up and used in ways that suit the purpose at hand. At the same time, the key cargo-owner's new SCM practice of sailing without icebreaker assistance broke the historically established cargo transportation in the Arctic waters. During its implementation and adoption, the new SCM practice met resistance from other players in the Russian Arctic, particularly from the icebreaker operator, who problematized the legitimacy of this practice.

During 2001–2012, the icebreaker operator manipulated the tariff rates on icebreaker services and the tariff policy in general as a tool of the logic of priority through the lobbying and pressure on the government. However, these actions considerably decreased shipping traffic along the NSR and, thereby, the profitability of the NSR use. Due to key cargo-owner's new SCM practice and unsuccessful manipulation of the icebreaker operator, the government had to make significant changes in the Russian legislation on sailing in Arctic waters and tariff regulation of icebreaker assistance in 2012–2013. The obligatory use of icebreaker assistance was abolished: ship-owners could now decide for themselves whether they needed icebreaker services or not according to ice conditions. Further, the key cargo-owner's ice-class vessels were recognized as legally capable of providing icebreaker services. In this way, the key cargo-owner's new SCM practice of sailing without icebreaker assistance became legitimate to new legislation and re-embedded in the Russian Arctic institutional environment (Tsvetkova and Gammelgaard 2018). Thus, the logic of stability and the logic of priority confronted each other before the government had to change the legislation of navigation along the NSR. This confrontation between both logics caused a significant change of the logic of state policy referring to the tariff policy and the rules of navigation along the NSR. The theoretical and practical focus of regulation is often the role of government and regulatory agencies in setting and implementing policies. This study shows that there are more aspects in the regulatory implementation. Strategic actions of most powerful actors, e.g. big manufacturers and the icebreaker operator, as well as possible outcomes of interaction between these actors should be taken into consideration when setting political goals and implementing regulatory procedures.

After the legislation change in 2012–2013, the logic of state policy transformed from coercive unstable enforcement into state policy based on market conditions, contractual arrangements, as well as coordinating and balancing patterns of interaction, which could enable the players involved to organize their behaviour in a legitimate way. The legislation became more flexible and less coercive for the business companies in the Russian Arctic waters. At the same time, commercial outcomes were quite unsatisfactory - internal shipping traffics along the NSR remained small in the following years; transit traffic, however, sharply fell due to mainly additional external factors. According to a common view in the literature on regulation, actors that are regulated suppress their private interests to be aligned with the public interest (Veljanovski 2010). This study, however, shows that private benefits and collective gains can diverge in the regulatory process.

Further, the adoption of the new NSR legislation in 2012–2013 caused a battle between two administrative agencies – the icebreaker operator and the NSR Administration – for control over the NSR. The icebreaker operator was not satisfied with changes in the legislation and continued pursuing self-interest referring to the tariff policy. The lobbying actions demanded the development of more effective regulation for increasing cargo shipping - internal traffic and particularly transit volumes along the NSR. It should be stressed that strategic actions by the key cargo-owner in 2006–2008 were the legitimate pursuit of self-interest just to ensure regular cargo deliveries and mitigate negative coercive pressures of unstable state policy to survive rather than were intended to violate the law. In contrast, the lobbying and strategic gaming by the icebreaker operator in 2012–2018 were quite manipulative to influence politicians and even illegal in the situation when the agency concealed the violation of Arctic safety rules by the "Boris Vilkitsky" in pursuing commercial interests. Government intervention through direct participation of the Russian president in resolving this incident revealed another side of the regulatory process. The study illustrates that not only the players that are regulated can manipulate information and influence the regulator, but the regulator can also make strategic use of the formal and informal procedures available to it, particularly when it concerns the national values. The law cannot be comprehensive, e.g. the previous legislation did not determine procedures for new technologies the key cargo-owner's ice-class vessels that caused claims of other actors who argued them illegal. But in the incident with the "Boris Vilkitsky" the government revealed wide discretion to apply the law. This sometimes leads to extra-legal or illegal actions. This may be exercised in order to deal with the uncertainty due to weak laws and regulations, which are imperfect for desired political and public purposes.

The competing struggle for control over the NSR between two agencies - the icebreaker operator and the NSR Administration - contributed to a new significant change in legislation in 2018. At the same time, preliminary agreements with the government to make the icebreaker operator a single logistic operator of the NSR facilitated the attraction of major foreign partners like Cosco and Maersk companies for transit traffic along the NSR. This allowed reaching record growth in transit volumes in 2018 over the past decade. This finding is consistent with theoretical assumptions by Becker (1983) that as a result of pressure by several competing groups, the government actually pursues more efficient redistribution of policies and competencies in order to minimize potential losses.

Although the government put forward various initiatives to further development of the NSR, the demand for the legislation change came twice from interest groups, typically industry and the administrative agencies, involved in day-to-day routines. The study emphasizes that regulation can bring commercial outcomes when it is shaped strategically. The time of regulatory innovation in strategic use comes when the government highlights the importance of interest players. Thus, knowledge of how the regulatory process is shaped as a strategic game under the influence of interaction between most powerful actors involved contributes to a deeper understanding of the regulation of cargo shipping in the Russian Arctic.

19.8 Conclusion and Implications for Theory and Practice

Through the lenses of the institutional logics approach, this study explores how the regulatory process of cargo shipping in the Russian Arctic has been shaped during 2001–2018. Regulation is viewed as a co-produced activity under the influence of interaction between the regulator and most powerful actors involved. The study shows that distortions between the interest groups' expected benefits and political initiatives cause the change in the existing legislation and shipping traffic in the Russian Arctic.

Through the analysis of institutional logics in the implementation of regulation, the findings reveal how strategic actions, including lobbying and information manipulation, by most powerful actors affect regulation, political initiatives and commercial outcomes. Further, the study shows that not only government policies influence and constrain the behaviour of other actors but actions of these actors may, in turn, affect them as a strategic response to the regulatory uncertainty and imperfection in order to gain expected benefits or mitigate unfavourable burden. This may happen more often in real-life settings than usually anticipated in the literature.

Further, the findings provide in-depth understanding about under what circumstances strategic actions of most powerful actors involved can influence political initiatives by making resistance to state coercive pressures legitimate or by turning it into non-compliance with the existing legislation. The literature is often limited by calls only for more government regulation to reassert control through more sustained oversight (Baldwin et al. 2010). This study emphasizes that regulation is also part of doing business and has a major impact on commercial outcomes. It is obvious that where possible industries will seek to influence regulation and exploit the existing policies in order to gain more favourable outcomes.

Investigation of the institutional challenges and interactions between the regulator and other most powerful actors may be crucial in choosing a set of subsequent strategic actions. By considering these interactions and interdependencies, managers will gain a better understanding of how to use strategic responses to regulation and gain more favourable outcomes for their companies.

Finally, previous research on the NSR development has focused on the use of international law for the NSR only as a transit shipping route (Moe 2014) and disregarded the Russian internal legislation, which constitutes the main rules of navigation in the Russian Arctic. The findings may be valuable for managers responsible for developing supply chains and cargo shipping along the NSR.

19.9 Limitations and Further Research

The findings provide deep insights into the real-life interactions between the regulator and other players involved in the implementation of internal regulation during 2001–2018. However, the international law, including Polar Code, also participates in shipping regulation along the NSR. Further research should focus on how the regulatory process of polar shipping is developed in accordance with global, regional and national policies and can be influenced by a wider scope of the actors involved like military forces, external agencies, as well as foreign partners and customers of the NSR.

Further, regulation of cargo shipping along the NSR is a specific phenomenon, not at least because of Arctic harsh natural conditions and the need to use atomic icebreaker escort. More in-depth empirical studies on regulation as a co-produced process in interaction with commercial interests, contextual factors and political initiatives in other contexts are suggested. To understand the regulatory process even better, it should be studied how industry seeks to influence regulation and exploits the latitude that the existing legislation allows it due to its imperfection in order to gain more favourable outcomes (Veljanovski 2010). Therefore, future empirical research may also extend knowledge about how the players' manipulations affect the rules and under what circumstances interactions between the regulator and other players can turn into non-compliance and/or opposition.

Appendices

Appendix 1: List of Organizations Participated in Interviewing

#	Organizations	Periods and number of interviews		
		May 2014	November 2014	May 2015
1	Transport branch of MMC "Norilsk Nickel" in Murmansk	4	2	
2	Murmansk shipping company (MSCO)	1		
3	Government of the Murmansk region		3	
4	Federal Assembly of the Russian Federation		1	
5	Northern Sea Route Information Office (Murmansk) created by Center for High North Logistics (Kirkenes)	2	1	1
6	NPO Association "Murmanshelf" – Association of suppliers for oil-and-gas industry	1		
7	Federal State Institution Administration of Murmansk seaport	1		
8	Lusin Institute for Economic Studies of Kola Science Centre of the Russian Academy of Sciences (Apatity town)		1	
9	Central Scientific Research and Design Institute of the Marine Fleet – CNIIMF in St. Petersburg; Laboratory of Icebreaking technologies and ice performance of ships		1	1
10	Akvaplan Niva AS (Tromsø)	1		
11	FSUE "Rosatomflot"			1
Total amount of interviews per month:		**10**	**9**	**3**
Total amount of interviews:		**22**		

Appendix 2: The Russian Legislation of Shipping on the Water Area of the NSR Used for Data Collection

The title of the Law	Reference in the text
Federal Law No. 132-FZ dated July 28, 2012 "On Amendments to Specific Legislative Acts of the Russian Federation related to Governmental Regulation of Merchant Shipping in the Water Area of the Northern Sea Route"	*NSR Law, 2012*
Federal Law No.147-FZ dated August 17, 1995 "On Natural Monopolies"	*Law #147-FZ, 1995*

(continued)

The title of the Law	Reference in the text
Order #45-T/1 dated March 04, 2014 "On approval of the tariff rates for provision of icebreaking pilotage services provided by the FSUE «Atomflot» on the Northern Sea Route water area" Published: 09.04.2014 at http://www.rg.ru/2014/04/09/suda-dok.html	
Federal Law No. 81-FZ dated April 30, 1999 "The Merchant Shipping Code of the Russian Federation"	
Federal Law No. 155-FZ dated July 31, 1998 "On Inland Sea Waters, Territorial Sea and Contiguous Zone of the Russian Federation"	*Law #155-FZ, 1998*
Permissions for navigation on the water area of the Northern Sea Route, issued by Federal State Institution "The Northern Sea Route Administration"	
Maritime Doctrine of the Russian Federation for the period up to 2020, approved by President of the Russian Federation Vladimir Putin on July 27, 2001, Pr-1387	*Marine Doctrine, 2001*
"The rules of navigation: Regulations for navigation on the seaways of the Northern Sea Route", dated September 14, 1990	*Rules of 1990*
The Requirements for the Design, Equipment and Supply of Vessels Navigating the NSR, adopted in 1990	
The rules of navigation in the Northern Sea Route water area, dated January 17, 2013	*Rules of 2013*
"Basics of the state policy of the Russian Federation in the Arctic for the period till 2020 and for a further perspective", adopted by the President of the Russian Federation, D. Medvedev, on September 18, 2008	*Arctic state policy, 2008*
"The Strategy for the Development of the Arctic Zone of the Russian Federation and National Security up to 2020" (Strategy-2013) approved by President of Russian Federation Vladimir Putin on February 20, 2013	
The Order of the Government of the Russian Federation #358-p dated March 15, 2013	
Order of the RF #1528-R dated October 31, 2002	
Order of the Ministry of Economic Development #69 dated October 31, 2000 "On the change of tariffs for icebreaking fleet on the Northern Sea Route"	*Order #69, 2000*
Order of Federal Tariff Service dated June 07, 2011 #122-T/1 "On setting of rates for services of the icebreaker fleet on the Northern Sea Route"	*Order #122-T/1, 2011*
Communications instructions for Arctic navigations for Arctic navigation 2012–2013 on the Northern Sea Route, issued by Federal Agency of Maritime and river Transport	
Federal Law No. 525-FZ dated December 27, 2018 "On Amendments to Specific Legislative Acts of the Russian Federation"	*Law #525-FZ*

References

Baldwin, R., Cave, M., & Lodge, M. (2010). *The Oxford handbook of regulation*. New York: Oxford University Press.
Battilana, J., Leca, B., & Boxenbaum, E. (2009). How actors change institutions: Towards a theory of institutional entrepreneurship. *The Academy of Management Annals, 3*, 65–107.
Becker, G. S. (1983). A theory of competition among pressure groups for political influence. *Quarterly Journal of Economics, 98*(3), 371–400.
Bello, D. C., Lohtia, R., & Sangtani, V. (2004). An institutional analysis of supply chain innovations in global market channels. *Industrial Marketing Management, 33*, 57–64.
Bhakoo, V., & Choi, T. (2013). The iron cage exposed: Institutional pressures and heterogeneity across the healthcare supply chain. *Journal of Operations Management, 31*, 432–449.
Black, J. (2001). Decentring regulation: Understanding the role of regulation and self regulation in a 'post-regulatory' world. *Current Legal Problems, 54*, 103–147.
Black, J., & Kingsford, S. (2002). Critical reflections on regulation. *Australian Journal of Legal Philosophy, 27*(2002), 1–46.
Bulatov, V. (1997). Historical and current uses of the Northern Sea route. In *Part IV: The Administration of the Northern Sea Route (1917–1991)* (INSROP Working Papers 84–1997, IV.1.1).
DiMaggio, P. J., & Powell, W. (1983). The iron cage revisited: Institutional isomorphism and collective rationality in organizational fields. *American Sociological Review, 48*(2), 147–160.
Doha, A., Das, A., & Pagell, M. (2013). The influence of product life cycle on the efficacy of purchasing practices. *International Journal of Operations and Production Management, 3*(4), 470–498.
Granberg, A. G. (1995). *The significance of the Northern Sea route for regional development on Arctic areas of Russia* (INSROP Working Papers 19–1995, III.01.1).
Granberg, A. G. (1997). *Selected studies in regional economic development along the Northern Sea route* (INSROP Working Papers 74–1997, III.02.3).
Hoejmose, S. U., Grosvold, J., & Millington, A. (2014). The effect of institutional pressure on cooperative and coercive 'green' supply chain practices. *Journal of Purchasing and Supply Management, 20*, 215–224.
Høifødt, S., Nygaard, V., & Aanesen, M. (1995). *The Northern Sea route and possible regional consequences* (INSROP Working Papers 16–1995, III.02.1).
Hong, N. (2012). The melting Arctic and its impact on China's maritime transport. *Research in Transportation Economics, 35*, 50–57.
Huo, B., Han, Z., Zhao, X., Zhou, H., Wood, C. H., & Zhai, X. (2013). The impact of institutional pressures on supplier integration and financial performance: Evidence from China. *International Journal of Production Economics, 146*, 82–94.
Kinra, A., & Kotzab, H. (2008). A macro-institutional perspective on supply chain environmental complexity. *International Journal of Production Economics, 115*, 283–295.
Kuznetsova, T. E., & Nikiforov, L. V. (2013). About the strategy for the use of the spatial potential of Russia. *Journal of Public Administration, 2*, 51–64.
Lee, S. M., Rha, J. S., Choi, D., & Noh, Y. (2013). Pressures affecting green supply chain performance. *Management Decision, 51*(8), 1753–1768.
Leube, K. R., & Moore, T. G. (1986). *The essence of Stigler*. Stanford: Hoover Institution Press, Stanford University.
McPherson, C. M., & Sauder, M. (2013). Logics in action: Managing institutional complexity in a drug court. *Administrative Science Quarterly, 58*(2), 165–196.
Moe, A. (2014). The Northern Sea route: Smooth sailing ahead? *Strategic Analysis, 38*(6), 784–802.
Moxham, C., & Kauppi, K. (2014). Using organizational theories to further our understanding of socially sustainable supply chains: The case of fair trade. *Supply Chain Management: An International Journal, 19*(4), 413–420.

Østreng, W., Brigham, L., Brubaker, D., Gold, E., Granberg, A., Grishchenko, V., Jernsletten, J., Kamesaki, K., Kolodkin, A., Moe, K., Semanov, G., & Tamvakis, M. (1999). *The challenges of the Northern Sea Route. Interplay between Natural and Societal Factors*. INSROP Working Paper 167–1999.

Owen, B. M., & Braeutigam, R. (1978). *The regulation game: Strategic use of the administrative process*. Cambridge, MA: Ballinger Publishing Company.

Peltzman, J. (1976). Toward a more general theory of regulation. *Journal of Law and Economics, 19*, 211–240.

Polterovich, V. (1999). Institutional traps and economic reforms. *Economics and Mathematical Methods, 35*(2), 3–19.

Scott, W. R. (2014). *Institutions and organizations: Ideas, interests and identities* (4th ed.). Thousands Oaks: Stanford University.

Selznick, P. (1985). Focusing organizational research on regulation. In R. Noll (Ed.), *Regulatory policy and the social sciences* (pp. 363–367). Berkeley: University of California Press.

Shleifer, A., & Treisman, D. (2005). A normal country: Russia after communism. *The Journal of Economic Perspectives, 19*(1), 151–174.

Shook, C. L., Adams, G. L., & Ketchen, D. J. (2009). Towards a "theoretical toolbox" for strategic sourcing. *Supply Chain Management International Journal, 14*(1), 3–10.

Sodero, A. C., Rabinovich, E., & Sinha, R. K. (2013). Drivers and outcomes of open-standard interorganizational information systems assimilation in high-technology supply chains. *Journal of Operations Management, 31*, 330–344.

Stigler, G. J. (1971). The theory of economic regulation. *The Bell Journal of Economics and Management Science, 2*(1), 3–21.

Suchman, M. C. (1995). Managing legitimacy: Strategic and institutional approaches. *The Academy of Management Review, 20*, 571–610.

Thornton, P. H. (2004). *Markets from culture: Institutional logics and organizational decisions in higher education purchasing*. Stanford: Stanford University Press.

Thornton, P. H., & Ocasio, W. (1999). Institutional logics and the historical contingency of power in organizations: Executive succession in the higher education publishing industry, 1958–1990. *The American Journal of Sociology, 105*(3), 801–843.

Thornton, P. H., Ocasio, W., & Lounsbury, M. (2012). *The institutional logics perspective: A new approach to culture, structure, and process*. Oxford: Oxford University Press.

Trebing, H. M. (1987). Regulation of industry: An institutional approach. *Journal of Economic Issues, 21*(4), 1707–1737.

Tsvetkova, A., & Gammelgaard, B. (2018). The idea of transport independence in the Russian Arctic: A Scandinavian institutional approach to understanding supply chain strategy. *International Journal of Physical Distribution & Logistics Management, 48*(9), 913–930.

Utkin, E., & Denisov, A. (2001). *State and municipal management*. Moscow: Association of Authors and Publishers "Tandem".

Veljanovski, C. (2010). Chapter 5: Strategic use of regulation. In R. Baldwin, M. Cave, & M. Lodge (Eds.), *The Oxford handbook of regulation*. New York: Oxford University Press.

Williams, Z., Lueg, J. E., Taylor, R. D., & Cook, R. L. (2009). Why all the changes? An institutional approach to exploring the drivers of supply chain security (SCS). *International Journal of Physics Distribution & Logistics Management, 39*(7), 595–618.

Yaibuathet, K., Enkawa, T., & Suzuki, S. (2008). Influences of institutional environment toward the development of supply chain management. *International Journal of Production Economics, 115*, 262–271.

Zhang, C., & Dhaliwal, J. (2009). An investigation of resource-based and institutional theoretical factors in technology adoption for operations and supply chain management. *International Journal of Production Economics, 120*, 252–269.

Chapter 20
Resource Use Conflicts in Arctic Waters: A Legal Perspective

Amber Rose Maggio

Abstract This chapter seeks to place the resource use conflicts discussed in this book in the wider legal framework of Arctic and oceans governance. The chapter briefly sketches out the regimes for the exploration and exploitation of natural resources in the various maritime zones as laid out in the UNCLOS; it then goes on to examine the legal framework for environmental protection in the Arctic region, particularly focussing on shipping, oil and gas, marine scientific research and the regime of fisheries. It then gives an overview of some elements of Arctic governance from an institutional perspective.

Keywords Oceans governance · Environmental protection · Regional seas · Arctic governance

20.1 Introduction

The use of resources in Arctic waters is regulated internationally by the United Nations Convention on the Law of the Sea (UNCLOS 1982), supplementary hard and soft law agreements and customary international law. The legal framework recognises the rights of states to explore and exploit natural resources, while at the same time providing an overarching obligation to protect and preserve the marine environment. The international law of the sea separates ocean space into different maritime zones, which provide for differing rights and obligations of coastal and flag states depending on the zone in question. This paper will give an overview of the legal regime regulating the exploration and exploitation of natural resources and marine environmental protection with specific reference to issues that are relevant

A. R. Maggio (✉)
Centre for International Law, National University of Singapore, Singapore, Singapore
e-mail: cilarm@nus.edu.sg

E. Pongrácz et al. (eds.), *Arctic Marine Sustainability*,
Springer Polar Sciences, https://doi.org/10.1007/978-3-030-28404-6_20

for Arctic waters. It will then examine some issue-based regimes and highlight some of the Arctic governance mechanisms currently in place. It aims to provide the legal background to frame the other papers in this volume for an interdisciplinary audience.

The Arctic is an area of abundant resources, but it is also an area where the threat of environmental damage to ecosystems and biodiversity is particularly severe. The rights of Arctic states to exploit their natural resources and the freedoms of all states to fish on the high seas are explicitly protected in the UNCLOS, as is the general obligation on all states to protect and preserve the marine environment. Thus, the legal framework itself reflects the conflict in resource development and protection of the marine environment. The more specific provisions of Part XII UNCLOS lay out an environmental protection regime that governs uses of the oceans including resource exploitation and is supplemented by further more specific regional and international agreements. Natural resources on land are governed largely by domestic law, with reference to the international environmental obligations of the state in question, and are beyond the scope of this work.

The Arctic is not subject to a specific legal regime under international law in the same way, for example, as the Antarctic (see Antarctic Treaty 1959). Nevertheless, the so-called "Arctic resources race" due to newly available resources in Arctic waters makes it important to understand the legal regime governing the exploitation of such resources. The UNCLOS –the general international framework treaty on the law of the sea- and a range of more specific issue-based and regionally centred hard and soft law instruments provide a fairly comprehensive legal regime governing most aspects of ocean resource uses in the Arctic area. This paper will give an overview of the general provisions regarding resource use in the various maritime zones of the Arctic and the general obligations contained in the UNCLOS on the protection and preservation of the marine environment (Part XII UNCLOS), followed by a more specific examination of the rules concerning some of the most relevant issues such as shipping, offshore oil and gas development, marine scientific research (MSR), and fisheries.

First, however, a small introduction to the international legal system may be helpful. In international law, on the basis of the doctrine of sovereignty, in most cases states are only bound by what they have explicitly agreed to. In practice, this means that for the rules contained in an international agreement to be binding on a state, it must become party to the agreement by signing and ratifying it. Soft law agreements do not bind states, but rather provide for a non-binding agreement that states will act in a certain way. Breach of an international binding agreement produces consequences in the law of state responsibility (see ILC 2001), whereas breach of soft law agreements may produce political and reputational effects but no legal consequences. Customary international law is law that is binding on all states, provided that both the requisite state practice and the belief that the action/practice was carried out in fulfilment of a legal obligation (*opinion juris*) are evidenced. With regard to Arctic governance, customary international law is of significant importance, as one of the "Arctic Five" states (Russia, United States, Norway, Canada and Denmark) -the United States- is not party to the UNCLOS. This means that the so-called "constitution for the oceans" (Koh 2015) is not directly binding on the US. However, the US, the international community, and various international courts and tribunals have attested to the customary status of many of the rules contained within the UNCLOS, meaning that they are also binding on non-state parties.

This has great importance when considering Arctic governance, particularly with regard to the general environmental protection obligations contained in the UNCLOS, which now can be seen to represent customary international law, at least in part (McConnell and Gold 1991, p. 84). However, the specific rules contained in the UNCLOS on the (partially) compulsory settlement of disputes do not have customary status. This means that it may not be possible to bring a claim against a non-state party regarding a dispute that would ordinarily be justiciable within the dispute settlement system provided for in the UNCLOS. In this case, there would need to be an additional consent of the parties to resolve the dispute before an international court or tribunal.

In addition, it is important to note that there is no legal definition of 'the Arctic', and thus which coastal states are relevant when discussing Arctic issues can be a little challenging. The 'Arctic Five' are the coastal states bordering the Central Arctic Ocean. However, the Arctic Council, discussed below, has eight permanent members and includes Iceland, Sweden and Finland. Neither Sweden nor Finland have coastlines bordering the Arctic, although they both have land territory that falls in the Arctic Circle. Iceland's mainland is not within the Arctic Circle, but it does have maritime zones that extend into it (see further Roach 2018, p. 5 *et seq.*).

20.2 Maritime Zones

The UNCLOS is the main international tool for oceans governance; it lays out the specific rules governing ocean uses in a sectoral manner. States are categorised as either coastal or flag states (the rights of landlocked states are also addressed in the Convention but regarding their uses of the ocean they would then also be categorised as flag states). Coastal states have coastlines along which they draw the base-points from which the various maritime zones are measured. Coastal states have varying rights, obligations and jurisdictional nexuses depending on the maritime zone in question. Flag states are states that have registered ships in their flag registries, which allows the ship to fly their flag and means the flag state must exercise jurisdiction over the ship, regardless of the maritime zone in question. Flag states also have varying rights in each maritime zone. This section will briefly sketch out the rights of coastal and flag states in each maritime zone. The following sections will explore the Part XII duties on the protection and preservation of the marine environment and then the selected resource use conflict issues in more detail.

20.2.1 The Territorial Sea

The territorial sea is the furthest landward maritime zone and can extend up to 12 NM from the baselines established by the coastal states along their coasts (Art. 3 UNCLOS). Waters landward of the baseline constitute internal waters (Art. 8 UNCLOS). As the name suggests, the territorial sea is the maritime zone where the coastal state exercises territorial sovereignty (Art. 2 UNCLOS), meaning that all resource use and exploitation is subject to the exclusive jurisdiction of the coastal state. The rights of other states are limited to innocent passage of ships (Art. 17 UNCLOS) and do not extend to resources, marine scientific research (Art. 245 UNCLOS) or fisheries (Arts. 19 (2)(i)).

20.2.2 The Exclusive Economic Zone

The exclusive economic zone, or EEZ, extends up to 200 NM (Art. 57 UNCLOS) from the baselines from which the territorial sea is drawn. It is a zone in which the coastal state has 'sovereign rights for the purpose of exploring and exploiting, conserving and managing the natural resources, whether living or non-living, of the waters superjacent to the seabed and of the seabed and its subsoil, and with regard to other activities for the economic exploitation and exploration of the zone, such as the production of energy from the water, currents and winds' (Art. 56 (1)(a) UNCLOS) and jurisdiction with regard to '(i) the establishment and use of artificial islands, installations and structures; (ii) marine scientific research; (iii) the protection and preservation of the marine environment; (c) other rights and duties provided for in this Convention' (Art. 56 (1)(b) UNCLOS). 'Sovereign rights' is not defined in the convention but should be understood to be exclusive rights held only by the coastal state, whose authorisation is required for other states to exercise them. Art. 56 (1)(a) makes it clear that the exploration and exploitation of natural resources of the waters, seabed and subsoil of the EEZ are reserved exclusively for the coastal state. EEZ fishing and other exploitation of resources is possible for other states only with the permission of the coastal state. There is a detailed regime on EEZ fisheries laid out in Arts. 61–69 UNCLOS, which will be addressed below.

The EEZ legal regime is *sui generis* (Nelson 2008, para. 14) and the high seas freedoms (see below section on high seas) continue to apply in this area in as far as they do not conflict with the specific rules laid out in Part V UNCLOS. This means that the rights of 'navigation and overflight and of the laying of submarine cables and pipelines, and other internationally lawful uses of the sea' related to these rights are permissible for all states (Art. 58 UNCLOS). Flag states may continue to exercise these rights in the EEZ, so long as they 'have due regard to the rights and duties of the coastal State and shall comply with the laws and regulations adopted by the coastal State in accordance with the provisions of this Convention and other rules of international law' (Art. 58 (3) UNCLOS).

Jurisdiction regarding the establishment and use of artificial islands, installations and structures is regulated in detail in Art. 60, which lays out that in the EEZ the coastal state has the exclusive right to construct and to authorize and regulate the construction, operation and use of: artificial islands for all purposes; installations and structures for exploring and exploiting, conserving and managing the natural resources, whether living or non-living, of the waters superjacent to the seabed and of the seabed and its subsoil, and with regard to other activities for the economic exploitation and exploration of the zone and other economic purposes; and installations and structures which may interfere with the exercise of the rights of the coastal State in the zone.

With regard to marine scientific research, in the EEZ coastal states have the right to 'regulate, authorize and conduct' MSR in their EEZs (Art. 246 (1)) and any MSR conducted by another state 'shall be conducted with the consent of the coastal State' (Art. 246 (2)), whose consent should generally be granted (Art. 246 (3)). Consent may be withheld *inter alia* in cases where the MSR '(a) is of direct significance for the exploration and exploitation of natural resources, whether living or non-living; (b) involves drilling into the continental shelf, the use of explosives or the introduction of harmful substances into the marine environment; (c) involves the construction, operation or use of artificial islands, installations and structures referred to in articles 60 and 80.'

20.2.3 The Continental Shelf

The continental shelf regime applies to the 'legal continental shelf', rather than any scientific definition. The UN Commission on the Outer Limits of the Continental Shelf (CLCS), the body tasked with making recommendations on extended outer continental shelf claims made by states under the UNCLOS, states that 'the continental shelf of a coastal State comprises the submerged prolongation of the land territory of the coastal State - the seabed and subsoil of the submarine areas that extend beyond its territorial sea to the outer edge of the continental margin, or to a distance of 200 nautical miles where the outer edge of the continental margin does not extend up to that distance. The continental margin consists of the seabed and subsoil of the shelf, the slope and the rise. It does not include the deep ocean floor with its oceanic ridges or the subsoil thereof' (CLCS 2012). Coastal states may claim an extended continental shelf on the basis of one of the two complex formulae set out in Art. 76 UNCLOS (see Parson 2017; Golitsyn 2009). If no extended shelf is claimed, it will extend to 200 NM from the baselines from which the breadth of the territorial sea is measured, but if an extended shelf is claimed it may reach out further than the maximum limit of the EEZ (meaning the superjacent waters would be high seas, see below Sect. 20.2.4).

The rights of the coastal state over the continental shelf are set out in Art. 77 UNCLOS. It exercises 'sovereign rights' for the purpose of exploring and exploiting the natural resources of the continental shelf (Art. 77 (1) UNCLOS). These sovereign rights are exclusive and inherent; they do not depend on any proclamation by the coastal state (see Maggio 2017, pp. 612–613). As laid out above, the EEZ regime regulates natural resources in the water column and on the seabed and its subsoil, whereas the continental shelf regime regulates 'the mineral and other non-living resources of the seabed and subsoil together with living organisms belonging to sedentary species, that is to say, organisms which, at the harvestable stage, either are immobile on or under the seabed or are unable to move except in constant physical contact with the seabed or the subsoil' (Art. 77 (4) UNCLOS). The mineral resources mainly include oil and gas (de La Fayette 2008, p. 538). The coastal state has the exclusive right to conduct drilling on the continental shelf for all purposes (Art. 81 UNCLOS).

Both artificial islands, installations and structures on the continental shelf (Art. 80 UNCLOS) and marine scientific research on the continental shelf (Art. 246 UNCLOS) are regulated according to the same rules as those in the EEZ. This means that the coastal state has the same rights regarding these activities in this maritime zone and consent is required for other states to engage in related activities.

20.2.4 The High Seas

The high seas are regulated by Part VII of UNCLOS. They are an area beyond national jurisdiction (ABNJ) and comprise 'all parts of the sea that are not included in the exclusive economic zone, in the territorial sea or in the internal waters of a State, or in the archipelagic waters of an archipelagic State' (Art. 86 UNCLOS). As the high seas are ABNJ, they are open to all states, whether coastal or landlocked, and the basic principle governing them is that of the freedom of the high seas (Art. 87 UNCLOS).

Art. 87 UNCLOS gives a non-exhaustive list of high seas freedoms states enjoy, which include navigation, overflight, fishing, construction of artificial islands and structures and scientific research. The freedoms enjoyed by all states on the high seas are limited by Art. 87 (2), which states that '[t]hese freedoms shall be exercised by all States with due regard for the interests of other States in their exercise of the freedom of the high seas, and also with due regard for the rights under this Convention with respect to activities in the Area'. The meaning of 'due regard' is indeterminate (Sohn et al. 2014, p. 79) but has been described as signifying legal equality of states and lack of presumption of priority in ocean use conflicts (Gaunce 2018, p. 55). This means that the exercise of high seas freedoms by one state does not automatically enjoy priority over the exercise of high seas freedom by another state, regardless of the conduct involved -if it is otherwise lawful and in line with the Art. 88 clause that reserves the high seas for peaceful purposes. The exercise of high seas freedoms is not, however, completely uncurtailed, as the later examination of the environmental provisions of UNCLOS and other more specific issue regimes will demonstrate.

With regard to high seas fisheries, the UNCLOS lays out more specific obligations on states in Arts. 116–118. This is elaborated on in the 1995 United Nations

Agreement for the Implementation of the Provisions of the United Nations Convention on the Law of the Sea of 10 December 1982 relating to the Conservation and Management of Straddling Fish Stocks and Highly Migratory Fish Stocks (UNFSA), which provides a more specialised regime for straddling fish stocks (i.e. fish stocks that straddle maritime zones, the EEZ or one or more states and the high seas) and highly migratory species. The regime for biodiversity in areas beyond national jurisdiction (BBNJ) is currently being negotiated.

20.2.5 The Area

UNCLOS defines 'the Area' as 'the seabed and ocean floor and the subsoil thereof, beyond the limits of national jurisdiction' (Art. 1 (1)(1)), meaning beyond the outer limits of the continental shelves of coastal states. It is an area of ocean space where activities of exploration for and exploitation of natural resources, meaning 'all solid liquid or gaseous mineral resources *in situ* in the Area at or beneath the seabed, including polymetallic nodules' (Art. 133 (a)), are carried out for the benefit of mankind as a whole (Art. 140 UNCLOS), and where no state may purport to exercise sovereignty over any part or resources (Art. 137 UNCLOS). There is a complex regime regarding the exploration and exploitation of these resources and licensing and mining regulated by Part XI UNCLOS and the Agreement relating to the implementation of Part XI of the United Nations Convention on the Law of the Sea of 10 December 1982 (1994 Implementation Agreement). The International Seabed Authority (ISA or the Authority), a treaty body set up by the UNCLOS, has the responsibility to 'organize and control activities in the Area, particularly with a view to administering the resources of the Area' (Art. 157 (1) UNCLOS). It has developed a Mining Code, which comprises a comprehensive set of rules, regulations and procedures issued by the International Seabed Authority to regulate prospecting, exploration and exploitation of marine minerals in the international seabed Area. As mentioned above, the regime for BBNJ is currently being negotiated.

Although the designation of the resources of the Area as the common heritage of mankind, incapable of claims of sovereignty by any state, arguably has the status of customary international law, the regulatory and benefit sharing regime of the Authority is only applicable to UNCLOS states parties. This leaves some difficult questions regarding how non-states parties can participate in deep seabed mining. The US, for example, is of the position that it is within its high seas freedoms to mine the deep seabed and no agreement between other states may impinge upon this freedom (see Groves 2012).

Marine scientific research in the Area is regulated by Art. 256 of the Convention, which states that all states and competent international organizations have the right, in conformity with Part XI, to conduct MSR in the Area. In Part XI, Art. 143 states that MSR in the Area shall be carried out exclusively for peaceful purposes and for the benefit of mankind as a whole.

20.2.6 *Resource and Maritime Boundary Disputes*

These descriptions of the natural resources rights regimes in the various maritime zones clearly demonstrate that there should be no conflict between states concerning the right to explore and exploit natural resources in the Arctic. What is potentially problematic, in practice, however, is the determination of the precise boundary between states with opposite or adjacent coasts. States sharing a maritime boundary must agree as to its position, either through negotiation or though third party dispute settlement such as though a court or tribunal. There are still overlapping claims and unsettled boundaries in the Arctic, for example between the United States and Canada in the Beaufort Sea or the disputed sovereignty of Hans Island in the Davis Strait between Canada and Denmark (see de La Fayette 2008, p. 541 *et seq.*; Roach 2018, p. 65 *et seq.*) and the exploration or exploitation of resources in areas with overlapping claims by either state could give rise to a dispute. The UNCLOS provides from the delimitation of maritime zones in Arts. 15, 74, and 83. Difficulties may also arise for States that have made claims for an extended continental shelf under Art. 76 but have not received the recommendations from the CLCS upon which they can make their delineation like 'final and binding'.

20.3 The Protection and Preservation of the Marine Environment

The above section sketched out the rights of states in the various maritime zones relating to the exploration and exploitation of natural resources, the construction, installation and use of artificial islands, installations and structures, marine scientific resources and fisheries. Such resource related uses of the oceans may be seen to conflict with environmental protection measures. This issue is balanced in the UNCLOS in Part XII through a general obligation on all states, in Art. 192, to protect and preserve the marine environment; Art. 193 states 'states have the sovereign right to exploit their natural resources pursuant to their environmental policies and in accordance with their duty to protect and preserve the marine environment.' This provision is central to the international legal regulation of resource use conflicts from an environmental perspective. It reinforces the sovereign rights of states laid out in the Convention to explore and exploit their natural resources in the various maritime zones, while at the same time clearly stating how these rights are to be exercised: in conformity with the general obligation to protect and preserve the marine environment.

Part XII contains both more specific obligations on environmental protection, some of which will be laid out below, and a framework for states to create more specific standards and rules regionally or through the competent international organization. This dual system -which both binds states and ensures that more specific regimes can develop within the framework architecture- ensures that the UNCLOS is a living treaty (see eg Barrett and Barnes 2016) that can adapt (and indeed has adapted) to developments in our understanding of the marine environment and its degradation through human activities. This has developed significantly since the UNCLOS was concluded in 1982 and this framework, underpinned by general obligations, has ensured that the UNCLOS system remains relevant in issues of marine environmental protection today. This has been bolstered by a tendency towards dynamic interpretation by *inter alia* the International Tribunal for the Law of the Sea (ITLOS) in environmental matters (see Proelss 2016) that has ensured the continued relevance of the UNCLOS in current issues regarding the protection and preservation of the marine environment.

The provisions of Part XII are split into sections on: general provisions, providing the general obligations detailed below; global and regional cooperation obligations; technical assistance for developing states; monitoring and assessment; international rules and national legislation for the prevention, reduction and control of pollution of the marine environment, obliging states to adopt laws and regulations with regard to pollution from land-based sources, seabed activities subject to national jurisdiction, activities in the Area, by dumping, from vessels, and from or through the atmosphere; enforcement; safeguards; ice covered areas; responsibility and liability; sovereign immunity; and obligations under other conventions on the protection and preservation of the marine environment. Art. 194, in the general provisions section, is very detailed and warrants reproduction in its entirety:

Article 194
Measures to prevent, reduce and control pollution of the marine environment

1. States shall take, individually or jointly as appropriate, all measures consistent with this Convention that are necessary to prevent, reduce and control pollution of the marine environment from any source, using for this purpose the best practicable means at their disposal and in accordance with their capabilities, and they shall endeavour to harmonize their policies in this connection.
2. States shall take all measures necessary to ensure that activities under their jurisdiction or control are so conducted as not to cause damage by pollution to other States and their environment, and that pollution arising from incidents or activities under their jurisdiction or control does not spread beyond the areas where they exercise sovereign rights in accordance with this Convention.
3. The measures taken pursuant to this Part shall deal with all sources of pollution of the marine environment. These measures shall include, inter alia, those designed to minimize to the fullest possible extent:

(a) the release of toxic, harmful or noxious substances, especially those which are persistent, from land-based sources, from or through the atmosphere or by dumping;
(b) pollution from vessels, in particular measures for preventing accidents and dealing with emergencies, ensuring the safety of operations at sea, preventing intentional and unintentional discharges, and regulating the design, construction, equipment, operation and manning of vessels;
(c) pollution from installations and devices used in exploration or exploitation of the natural resources of the seabed and subsoil, in particular measures for preventing accidents and dealing with emergencies, ensuring the safety of operations at sea, and regulating the design, construction, equipment, operation and manning of such installations or devices;
(d) pollution from other installations and devices operating in the marine environment, in particular measures for preventing accidents and dealing with emergencies, ensuring the safety of

operations at sea, and regulating the design, construction, equipment, operation and manning of such installations or devices.

4. In taking measures to prevent, reduce or control pollution of the marine environment, States shall refrain from unjustifiable interference with activities carried out by other States in the exercise of their rights and in pursuance of their duties in conformity with this Convention.
5. The measures taken in accordance with this Part shall include those necessary to protect and preserve rare or fragile ecosystems as well as the habitat of depleted, threatened or endangered species and other forms of marine life.

This provision creates obligations on states to take all necessary measures to prevent, reduce and control pollution from any source -a high standard- and to ensure that activities under their jurisdiction do not cause transboundary harm to the environment. These measures shall deal with all sources of pollution, including the non-exhaustive list of measures regarding dumping, pollution from vessels, pollution from installations and devices used in exploration or exploitation of the natural resources of the seabed and subsoil and other installations and structures. These obligations are further conditioned in that when states carry them out they shall refrain from unjustifiable interference with lawful activities of other states. The measures taken shall include those necessary to protect and preserve rare or fragile ecosystems as well as the habitat of depleted, threatened or endangered species and other forms of marine life, arguably paving way for an ‘ecosystem approach’ (De Lucia 2015, p. 108).

Part XII provides a comprehensive suite of obligations that function to ensure that the protection of the marine environment is an overarching obligation that applies in all maritime zones and should be taken into account in all uses of ocean space, including resource extraction. The obligation applies to states, who must create rules and regulations to ensure that activities by ships flying their flag, land-based activities on their territory or seabed activities under their jurisdiction are carried out in accordance with this general obligation. Many of the provisions in Part XII contain a clause that obliges states, acting through the ‘competent international organization/s’ (generally the International Maritime Organization) or ‘diplomatic conference’, to ‘endeavour to establish’ (or in the case of Art. 208 simply ‘to establish’) global and regional rules relating to the particular issue (ie pollution from land-based sources (no current binding rules), pollution from seabed activities under national jurisdiction (currently no specific global instrument), pollution by dumping (regulated by the London Dumping Convention and its 1996 Protocol), pollution from vessels (regulated by the International Convention for the Prevention of Pollution from Ships, 1973 as modified by the Protocol of 1978 (MARPOL 73/78) and its 6 Annexes, with the International Maritime Organization as the competent international organization), and pollution through the atmosphere). This demonstrates the framework nature of Part XII; while it does contain general substantive obligations, it also provides for more specific agreements and even obliges states to establish or endeavour to establish them. Furthermore, the UNCLOS also obliges states to ensure that their national laws and regulations relating to environmental protection in these areas are ‘no less effective in preventing, reducing and controlling pollution than the global rules and standards’. This means that even where states parties to the UNCLOS do not adopt eg one of the Annexes to MARPOL 73/78 or an international environmental treaty within the framework of Part XII, then they would nevertheless be obligated to ensure their national legislation is no less effective. Such a provision is contained in Arts. 208 (3), 209 (2), and 210 (6), while Art. 211 stipules that such national rules shall have 'the same effect as internationally accepted rules and standards'.

Section 6 of Part XII provides for issue specific enforcement obligations on states and for the differing enforcement regimes for flag, port and coastal states. States are obliged to enforce their laws adopted in accordance with Part XII with regard to pollution from land-based sources, for example agricultural run off or marine plastic pollution (Art. 213), pollution from seabed activities, for example oil and gas drilling (Art. 214), pollution from activities in the Area (Art. 215), and dumping (Art. 216). Flag states are obliged to ensure compliance by vessels flying their flag with applicable international rules or standards and take other measures necessary for their implementation (Art. 217 (1)). Enforcement of environmental obligations is permitted in certain circumstances; when a vessel is voluntarily within a port, the port state can undertake investigations and where warranted institute proceedings against vessels found to have committed discharge violations of applicable international rules and standards in certain circumstances (Art. 218). Coastal state enforcement regarding protection and preservation of the marine environment is possible in certain constellations: if a vessel is voluntarily within a port or at an offshore terminal, proceedings may be brought by the coastal state for violations of its laws and regulations where the violation has occurred in its territorial sea or the exclusive economic zone (Art. 220 (1)). In the territorial sea, where there are clear grounds for believing that a vessel has, during its passage, violated coastal state laws implementing its Part XII obligations, the coastal state may undertake a physical inspection of the vessel and may when so warranted detain the vessel (Art. 220 (2)). If such a violation occurs in the EEZ, the coastal state may 'require the vessel to give information regarding its identity and port of registry, its last and next port of call and other relevant information required to establish whether a violation has occurred' (Art. 220 (3)). There is no right to detain the vessel unless there is 'clear objective evidence' that the vessel has committed a violation resulting in discharge causing 'major damage or threat of major damage' to the coastline or other interests of the coastal state, including any resources in the territorial sea or EEZ. In this case, the Convention provides for the institution of proceedings and possible detention of the vessel (Art. 220 (6)). There is no corresponding provision for environmental violations relating to the continental shelf –which may be an oversight by the drafters of the convention, especially as there are enforcement provisions for violations of the laws and regulations of the coastal state on the continental shelf contained in Art. 73 of the Convention. This brief overview of enforcement provisions demonstrates the primacy of flag state jurisdiction and enforcement, with port and coastal states enforcement being permissible in limited circumstances.

Part XII has a further provision relevant for the Arctic: Art. 234 on ice-covered areas. It states that coastal states have the right to adopt and enforce non-discriminatory laws and regulations for the prevention, reduction and control of marine pollution from vessels in ice-covered areas in the EEZ, 'where particularly severe climactic conditions and the presence of ice-covering of such areas for most of the year create obstructions or exceptional hazards to navigation, and pollution of the marine environment could cause major harm to or irreversible disturbance of the ecological balance. Such laws and regulations shall have due regard to navigation and the protection and preservation of the marine environment based on the best available scientific evidence.' This article is essentially an authorisation for coastal states in ice-covered areas to create a special, enforceable rules regime regarding pollution in such areas. The due regard obligation contained in Art. 234 is expanded on through the rules contained in the Polar Code (Franckx and Boone 2017, p. 1585), discussed below. To date, only Canada and Russia have adopted regulations under Art. 234 *inter alia* with regard to navigational issues (Russia: Rules of navigation in the water area of the Northern Sea Route), and discharge rules stricter than that required by MARPOL 73/78 (Canada: Arctic Waters Pollution Prevention Act). The exact scope of Art. 234 is unclear, and it has been questioned whether the Canadian and Russian regulations are fully in conformity with UNCLOS (see Thorén 2014, p. 30 *et seq.*, 41).

20.4 Issue Based Regimes

As elaborated above, the environmental provisions of Part XII include both substantive and framework provisions. The treaties adopted under the Convention as part of the framework system also have relevance for resource use conflicts. They provide international rules and standards, enforceable through the dispute settlement system of the Convention, which govern ocean uses in various areas. This overview of the issue based regimes makes no claim of completeness, but rather attempts to highlight some of the most important issue-based rules applicable to resource use and environmental protection in the Arctic. Always to be held in mind when thinking about both restrictions on resource uses and environmental protection measures and mechanisms are the sovereign right of states to explore and exploit natural resources in their maritime zones and high seas freedoms, particularly related to navigation. The exact scope of these rights vary depending on maritime zones, as do the restrictions, but the inherent conflict in resource based ocean uses is centred on these rights –based on the sovereignty of states- that are inherent in the very notion of statehood.

20.4.1 Shipping

With regard to shipping in the Arctic, there are rules and regulations contained in the International Code for Ships Operating in Polar Waters (Polar Code 2017), a mandatory code adopted under the auspices of the International Maritime Organization and effective since 2017. The Polar Code contains mandatory measures on safety and pollution (Parts I-A and II-A) that are supplemented by non-binding recommendatory guidance (Parts I-B and II-B). The Polar Code is designed to 'cover the full range of shipping-related matters relevant to navigation in waters surrounding the two poles – ship design, construction and equipment; operational and training concerns; search and rescue; and, equally important, the protection of the unique environment and eco-systems of the polar regions' (IMO Polar Code website). This description again reflects the conflict between oceans uses, in terms of navigation and of environmental protection.

The very detailed rules on these issues contained in the Code include those on 'ship structure; stability and subdivision; watertight and weathertight integrity; machinery installations; operational safety; fire safety/protection; life-saving appliances and arrangements; safety of navigation; communications; voyage planning; manning and training; prevention of oil pollution; prevention of pollution form from noxious liquid substances from ships; prevention of pollution by sewage from ships; and prevention of pollution by discharge of garbage from ships.' They supplement the more general rules of the IMO on maritime safety (The International Convention for the Safety of Life at Sea 1974) and pollution from shipping (MARPOL 73/78, including Annex I: Regulations for the prevention of pollution by oil; Annex II: Regulations for the control of pollution by noxious liquid substances in bulk; Annex III: Regulations for the prevention of pollution by harmful substances carried by sea in packaged form; Annex IV: Regulations for the prevention of pollution by sewage from ships; Annex V: Regulations for the prevention of pollution by garbage from ships; Annex VI: Regulations for the prevention of air pollution from ships, and other IMO conventions including International Convention on the Control of Harmful Anti-fouling Systems on Ships (AFS), 2001, and the International Convention for the Control and Management of Ships' Ballast Water and Sediments, 2004), which are applicable everywhere -not only in polar regions.

The focus of the Code on the particular maritime safety issues faced in polar regions and the environmental sustainability of their unique ecosystems demonstrates how the particularities of the polar regions and the Arctic have necessitated a more detailed regulatory regime to govern ocean uses. Freedom of navigation, required for shipping and resource exploration and exploitation in the Arctic, is further restricted by detailed rules on maritime safety and environmental protection in order to ensure protection of life, property and the environment at sea.

20.4.2 Oil and Gas

Oil and gas development in the Arctic on the continental shelf is governed by Part VI of UNCLOS. As outlined above, the natural resources of the continental shelf consist of *inter alia* 'the mineral and other non-living resources of the seabed and subsoil' and the coastal state maintains the sovereign right to explore and exploit them (Art. 77), and exclusive rights to construct authorize and regulate artificial islands installations and structures (Art. 80) and to drill (Art. 81). These rights must be exercised in line with the obligations laid out in Part XII UNCLOS and subsequent related treaties. Thus, oil and gas development is also governed by the IMO conventions mentioned above and the Polar Code, although the International and Convention on Oil Pollution Preparedness, Response and Co-operation (OPRC), 1990 with its Protocol on Preparedness, Response and Co-operation to pollution Incidents by Hazardous and Noxious Substances, 2000 (OPRC-HNS Protocol) are also of particular relevance.

Arctic states regulate oil and gas development on their continental shelves in their national law in a way that must reflect their international obligations. In addition, they have adopted regional solutions to deal with certain issues. The Arctic Council, discussed below, has adopted an agreement on Cooperation on Marine Oil Pollution Preparedness and Response in the Arctic between the Arctic Council members and has in place an Emergency Prevention, Preparedness and Response (EPPR) programme.

Although there was previously a lot of optimism that oil and gas development from the continental shelf of the Arctic would be significant for the Arctic states (a 'bonanza', see eg Morgunova and Westphal 2016, p. 15) the current long term slump in oil prices has led to greatly reduced profitability and hence interest in the growth of this development sector, which some have seen as an opportunity to improve governance and environmental pollution control (Gulas et al. 2017) However, there have been some licenses granted by the United States for exploratory drilling off Alaska (see America-First Offshore Energy Strategy Executive Order 2017).

20.4.3 Fisheries

The exploitation of living resources in the Arctic is governed internationally by the provisions of the UNCLOS, especially Art. 61–69 and 116–118. In the EEZ, these provide for a regime whereby coastal states must determine the total allowable catch, based on the maximum sustainable yield (Art. 61). If the total allowable catch exceeds the maximum fishing capacity of the state or a provides a surplus, coastal states shall give other states access to the surplus (Art. 62). On the high seas, states are obliged to take measures or cooperate to take measures for their nationals as necessary for the conservation of the living resources of the high seas (Art. 117), including an obligation to cooperate in the conservation and management of living resources and where appropriate establish regional fisheries management organizations (RFMOs) (Art. 118). Further, much more detailed, provisions are laid out in the UNFSA.

With regard to the Central Arctic Ocean (CAO), there are several relevant RFMOs: the North-East Atlantic Fisheries Commission (NEAFC); Joint Norwegian-Russian Fisheries Commission; North Atlantic Salmon Conservation Organization

(NASCO); International Commission for the Conservation of Atlantic Tunas (ICCAT). In addition, the Arctic Five-plus-Five (Iceland, EU plus China, Japan and South Korea as high seas fishing nations) have entered into a treaty that seeks to 'prevent unregulated fishing in the high seas portion of the central Arctic Ocean through the application of precautionary conservation and management measures as part of a long-term strategy to safeguard healthy marine ecosystems and to ensure the conservation and sustainable use of fish stocks' (Art. 2 Agreement to Prevent Unregulated High Seas Fisheries in the Central Arctic Ocean (CAOF Agreement)). This is a significant restriction on one of the high seas freedoms contained in Art. 87 UNCLOS. The application of precautionary conservation and management measures in the Arctic waters demonstrates a strong commitment by these high seas fishing nations to protecting the resources in the Arctic waters. However, the CAOF Agreement makes an exception for fisheries already regulated by an RFMO (Art. 3 (1)(a)) and interim conservation and management measures that may be established by the parties (Art. 3 (1)(b)). There is also an exception for exploratory fishing (Art. 3 (3)) under very limited circumstances. Marine scientific research is encouraged under the Joint Scientific Research and Monitoring (Arts. 3 (2), 4), although parties shall 'ensure that their scientific research activities involving the catching of fish in the Agreement Area do not undermine the prevention of unregulated commercial and exploratory fishing and the protection of healthy marine ecosystems' (Art. 3 (4)). It should also be pointed out that this agreement only applies to parties, and not all states.

In the wider region, there are other relevant RFMOs engaging in fisheries management. RFMOs are organisations or arrangements through which states manage fish stocks in a particular area. They can be advisory or have management powers, such as setting catch limits, and represent an important –if perhaps ineffective (Cullis-Suzuki and Pauly 2010, p. 1042) – means of conservation. In the Arctic, the relevant RFMOs are the Central Bering Sea (CBS) Convention; International Pacific Halibut Commission (IPHC); North Pacific Anadromous Fish Commission (NPAFC); Yukon River Panel to Pacific Salmon Treaty; Intergovernmental Consultative Committee (ICC); Western and Central Pacific Fisheries Commission (WCPFC); and Northwest Atlantic Fisheries Organization (NAFO) (for further detail see Molenaar 2013, p. 250 *et seq.*). All eight 'Arctic states' are parties to the UNFSA, which provides that states with a 'real interest' have a right to participate in relevant RFMOs (Art. 8 (3) UNFSA). This stresses the 'open and non-discriminatory' character of RFMOs, indicating in particular that 'the terms of participation shall not be used to prevent interested States from becoming members in such organizations and arrangements' (DOALOS/UNITAR 2002, p. 10). China, a non-Arctic distant waters fishing state (Molenaar 2013, p. 246), is not a party to the UNFSA and would not be bound by any RFMO restrictions on fisheries in the Arctic region, with the exception of the CAO which is now covered by the CAOF Agreement. Other significant distant water fishing states are parties.

20.4.4 MSR

MSR includes physical oceanography, marine chemistry, marine biology, scientific ocean drilling and coring, geological/geophysical research, as well as other activities with a scientific purpose (Roach 2018, p. 157). MSR in the Arctic is governed internationally by the provisions of the UNCLOS, as outlined above. For areas under national jurisdiction, coastal states are responsible for creating the appropriate rules and regulations in line with their international obligations. This appears to have been done, though in varying ways (Takei 2013, p. 350). With regard to fisheries and MSR in the CAO, the CAOF Agreement provides for a Joint Scientific Research and Monitoring obligation for states parties. The general high seas freedom to conduct MSR otherwise applies. There have also been some efforts towards regional or bilateral cooperation or coordination that includes MSR (eg the Joint Norwegian-Russian Fisheries Commission), including the possibility that the Arctic Council could develop a set of international rules and principles on MSR in the Arctic (see Takei 2013, p. 359).

20.5 Arctic Governance Regimes

The above -necessarily brief- overview of the issue-based regimes involved in resource uses in the Arctic should provide some idea about the complexity of governance in the Arctic. In addition to issue based agreements and arrangements, there are also governance regimes (at times independent or with some interaction) that should be addressed in order to give a broader –though by no means complete-introduction to institutional arrangements.

20.5.1 The Arctic Council

The Arctic Council had its genesis in the 1991 Arctic Environmental Protection Strategy (AEPS), a non-binding declaration on cooperation for the protection of the Arctic environment. On the initiative of Finland, the governments of Canada, Denmark, Finland, Iceland, Norway, Sweden, the Union of Soviet Socialist Republics, and the United States of America signed a joint Action Plan on environmental protection in the Arctic including: cooperation in scientific research; assessment of potential environmental impacts of development activities; and full implementation and consideration of further measures to control pollutants and reduce their adverse effects to the Arctic environment. Then, in 1996 these states (now with the Russian Federation) established the Arctic Council by way of the Ottawa Declaration. The preamble to the declaration contains the following two paragraphs:

> AFFIRMING our commitment to sustainable development in the Arctic region, including economic and social development, improved health conditions and cultural wellbeing;
>
> AFFIRMING concurrently our commitment to the protection of the Arctic environment, including the health of Arctic ecosystems, maintenance or biodiversity in the Arctic region and conservation and sustainable use of natural resources;

These clearly show the Arctic Council's governance model continues to embrace the right of states to explore and exploit their natural resources (albeit here using the term 'sustainable development'), while 'concurrently' having a responsibility to protect and preserve (or protect and maintain, with conservation and sustainable use of natural resources) the marine environment. This clearly reflects the international legal regime set out in Part XII of UNCLOS that recognizes the rights of states to explore and exploit resources while also obliging them to protect and preserve the marine environment.

The Arctic Council is an intergovernmental forum that works to promote cooperation, coordination and interaction among its members, 'in particular on issues of sustainable development and environmental protection in the Arctic' (Arctic Council website). It is made up of working groups and although it is not a fully-fledged international organization, it has a permanent secretariat. Alongside the permanent members, the Council also provides for permanent participant status for indigenous peoples in the Arctic and observer status for non-Arctic states. It is not a law-making forum, but rather can be seen as a way in which members are fulfilling their obligation of cooperation obligation under Art. 197 UNCLOS. It has also served as a forum for the negotiation of legally binding instruments, such as the Agreement on Cooperation on Aeronautical and Maritime Search and Rescue in the Arctic (2011), the Agreement on Cooperation on Marine Oil Pollution Preparedness and Response in the Arctic (2013), and the Agreement on Enhancing International Arctic Scientific Cooperation (2017).

20.5.2 Ilulissat Declaration 2008

The Illulissat Declaration is a soft law declaration (ie non-binding in a strictly legal sense) issued by the Arctic Five states at the 2008 Arctic Ocean Conference in Ilulissat, Greenland. It is a statement of intent regarding their conduct in the future governance of the Arctic Ocean. Although the Arctic Five do not constitute any kind of formalised governance arrangement, the important of this declaration for Arctic governance should not be understated.

In the Declaration, these states recognise that '[t]he Arctic Ocean stands at a threshold of significant changes'. They make reference to climate change and melting sea ice, and especially their potential impact on 'vulnerable ecosystems, the livelihoods of local inhabitants and indigenous communities, and the potential exploitation of natural resources', and their 'unique position' to address possibilities and challenges. The Declaration is an expression of their commitment to the current legal framework governing the Arctic and recognition that it is 'extensive' and 'provides a solid foundation for responsible management.' The Declaration clearly rejects the formation of a new comprehensive legal order for the Arctic and reaffirms the Arctic Five states' position that the current legal framework is an appropriate basis for oceans governance in the field of resource development and environmental protection. It further lays out the commitment to strengthen existing measures to ensure the protection and preservation of the marine environment, including in conjunction with the IMO. There is an additional commitment to continued close cooperation, including regarding 'the collection of scientific data concerning the continental shelf, the protection of the marine environment and other scientific research.'

A further meeting of the Arctic Five in 2010 did not lead to a further declaration but rather raised issues of exclusivity, the undermining of the Arctic Council forum and insensitivity to the rights of indigenous peoples.

20.5.3 Other and Possible Governance Regimes

There are other relevant governance regimes in the Arctic that, for reasons of space, cannot be included in any detail here. They also contribute to ocean use governance in the Arctic including resource uses. They include *inter alia* the Barents Euro-Arctic Council, the Arctic Economic Council, the Conference of Parliamentarians of the Arctic Region and the various RFMOs mentioned above. There are suggestions that Arctic governance should develop further than the current regime, perhaps with the establishment of an Arctic Ocean regional seas programme, the formalisation of the Arctic Council into an international organization or an Arctic Treaty system akin to that of the Antarctic (see Exner-Pirot 2012, p. 229 *et seq.*). Although the Ilulissat Declaration appears to make such endeavours less or unlikely, it should be noted that any possible future efforts would almost certainly maintain the balance of recognition of the right of states to explore and exploit resources and concurrent commitments to protect and preserve the marine environment.

20.6 Conclusion

This chapter aimed to give an overview of the legal regime of the Arctic as it applied to resource use conflicts. To this end, it began by giving some essential background on the functioning of international law, followed by an explanation of the maritime zones and the rights and obligations of states in each zone, especially with regard to natural resources and their exploration and exploitation, and MSR. This was then expanded upon with an overview of the most pertinent issue-based regimes and concluded with a select review of elements of the Arctic governance regime(s).

Each element brought forth and examined in this paper has demonstrated the internalisation of resource use conflicts into the Arctic legal governance regime. The right of states to explore and exploit their natural resources is repeatedly recognised –in regard to maritime zones, issue based regimes and governance models- along with the concurrent and conditioning obligation to protect and preserve the marine environment. The exact balance to be struck is left to states, but the overarching obligation contained in Art. 192 and expanded upon in the following sections of Part XII demonstrates that whatever the ocean use envisaged, licensed or permitted by states, it must be in accordance with their duty to protect and preserve the marine environment.

Bibliography

Documents

Agreement to Prevent Unregulated High Seas Fisheries in the Central Arctic Ocean. (2018).
America-First Offshore Energy Strategy, Executive Order 13795 of April 28, 2017. Available via https://www.federalregister.gov/documents/2017/05/03/2017-09087/implementing-an-america-first-offshore-energy-strategy
Arctic Environmental Protection Strategy (AEPS). (2001).
Arctic Waters Pollution Prevention Act (AWPPA). (R.S.C., 1985, C.A-12) (Canada). Available via https://laws-lois.justice.gc.ca/eng/acts/a-12/
CLCS. (2012). The continental shelf. Available via http://www.un.org/depts/los/clcs_new/continental_shelf_description.htm
DOALOS/UNITAR. (2002). *Briefing on developments in ocean affairs and the law of the sea 20 years after the conclusion of the United Nations convention on the law of the sea*. Available via http://www.un.org/depts/los/convention_agreements/convention_20years/Information%20Note.pdf
ILC. (2001). Responsibility of states for internationally wrongful acts, GA Res. 56/83, Annex.
Ilulissat Declaration. (2008). Available via: https://cil.nus.edu.sg/wp-content/uploads/formidable/18/2008-Ilulissat-Declaration.pdf
International and Convention on Oil Pollution Preparedness, Response and Co-operation (OPRC). (1990). Protocol on Preparedness, Response and Co-operation to pollution Incidents by Hazardous and Noxious Substances (OPRC-HNS Protocol) (2000).
International Convention for the Control and Management of Ships' Ballast Water and Sediments. (2004).
International Convention on the Control of Harmful Anti-fouling Systems on Ships (AFS). (2001).
International Convention for the Safety of Life at Sea. (1974). UNTS 1184, 1185, 2.
Mining Code. Available via https://www.isa.org.jm/mining-code
Ottawa Declaration. (1996). Declaration on the establishment of the Arctic council. Available via https://oaarchive.arctic-council.org/bitstream/handle/11374/85/EDOCS-1752-v2-ACMMCA00_Ottawa_1996_Founding_Declaration.PDF?sequence=5&isAllowed=y
Polar Code. Available via http://www.imo.org/en/MediaCentre/HotTopics/polar/Pages/default.aspx
Rules of Navigation in the Water Area of the Northern Sea Route. (2013, January 17). Approved by the Order of the Ministry of Transport of Russia No 7. Available via http://www.nsra.ru/en/pravila_plavaniya/
The Antarctic Treaty. (1959). UNTS 402, 71.

The International Convention for the Prevention of Pollution from Ships, 1973 as modified by the Protocol of 1978 (MARPOL 73/78).
The London Dumping Convention: Convention on the Prevention of Marine Pollution by Dumping of Wastes and Other Matter. (1972). UNTS 1046, 120; 1996 Protocol to the London Convention (1996), 36 ILM 1 (1997).
The United Nations Agreement for the Implementation of the Provisions of the United Nations Convention on the Law of the Sea of 10 December 1982 relating to the Conservation and Management of Straddling Fish Stocks and Highly Migratory Fish Stocks. (1995). UNTS 1833, 397.
The United Nations Convention on the Law of the Sea. (1992). UNTS 1833–1835, 3.

Books and Articles

Barrett, J., & Barnes, R. (Eds.). (2016). *Law of the sea: UNCLOS as a living treaty*. Cambridge: Cambridge University Press.
Cullis-Suzuki, S., & Pauly, D. (2010). Failing the high seas: A global evaluation of regional fisheries management organizations. *Marine Policy, 34*, 1036–1042.
de La Fayette, L. A. (2008). Oceans governance in the Arctic. *IJMCL, 23*, 531–566.
De Lucia, V. (2015). Competing narratives and complex genealogies: The ecosystem approach in international environmental law. *Journal of Environmental Law, 27*, 31–117.
Exner-Pirot, H. (2012). New directions for governance in the Arctic region. *Arctic Yearbook, 1*, 224–246.
Franckx, E., & Boone, L. (2017). Article 234. In A. Proelss (Ed.), *The United Nations convention on the law of the sea: A commentary* (pp. 1566–1585). Oxford: Beck/Hart.
Gaunce, J. (2018). On the interpretation of the general duty of "due regard". *Ocean Yearbook, 32*, 27–58.
Golitsyn, V. (2009). Continental shelf claims in the Arctic Ocean: A commentary. *IJMCL, 24*, 401–408.
Groves, S. (2012). The U.S. can mine the deep seabed without joining the U.N. convention on the law of the sea. *Background, 2476*.
Gulas, S., Downton, M., D'Souza, K., Hayden, K., & Walker, T. R. (2017). Declining Arctic Ocean oil and gas developments: Opportunities to improve governance and environmental pollution control. *Marine Policy, 75*, 53–61.
Koh, T. T. B. (2015). 'A constitution for the oceans': Remarks by Tommy T.B. Koh, President of the Third United Nations Conference on the Law of the Sea. Available at: https://cil.nus.edu.sg/wp-content/uploads/2015/12/Ses1-6.-Tommy-T.B.-Koh-of-Singapore-President-of-the-Third-United-Nations-Conference-on-the-Law-of-the-Sea-_A-Constitution-for-the-Oceans_.pdf
Maggio, A. R. (2017). Article 77. In A. Proelss (Ed.), *The United Nations convention on the law of the sea: A commentary* (pp. 605–614). Oxford: Beck/Hart.
McConnell, M., & Gold, E. (1991). The modern law of the sea: Framework for the protection and preservation of the marine environment. *Case Western Reserve Journal of International Law, 23*, 83–105.
Molenaar, E. J. (2013). Arctic fisheries management. In E. J. Molenaar, A. G. Oude Elferink, & D. R. Rothwell (Eds.), *The law of the sea and the polar regions* (pp. 243–266). Leiden: Brill.
Morgunova, M., & Westphal, K. (2016). Offshore hydrocarbon resources in the Arctic: From cooperation to confrontation in an era of geopolitical and economic turbulence? *SWP Research Paper*. Available at: https://www.diva-portal.org/smash/get/diva2:914598/FULLTEXT01.pdf
Nelson, D. (2008). Exclusive economic zone, MPEPIL. Available at: www.opil.ouplaw.com
Parson, L. (2017). Article 76. In A. Proelss (Ed.), *The United Nations convention on the law of the sea: A commentary* (pp. 587–600). Oxford: Beck/Hart.
Proelss, A. (2016). Naturschutz im Meeresvölkerrecht. *Archiv des Völkerrechts, 54*, 468–492.

Roach, J. A. (2018). A guide to Arctic issues for Arctic Council observers. Available at: https://cil.nus.edu.sg/publication/a-guide-to-arctic-issues-for-arctic-council-observers/

Sohn, L. B., Noyes, J. E., Franckx, E., & Juras, K. G. (Eds.). (2014). *Cases and materials on the law of the sea* (2nd ed.). Leiden: Martinus Nijhoff.

Takei, Y. (2013). Marine scientific research in the Arctic. In E. J. Molenaar, A. G. Oude Elferink, & D. R. Rothwell (Eds.), *The law of the sea and the polar regions* (pp. 343–365). Leiden: Brill.

Thorén, A. M. (2014). Article 234 and the polar code: The interaction between regulations on different levels in the Arctic Region. Available at: https://www.duo.uio.no/bitstream/handle/10852/42108/5071.pdf

Chapter 21
The Red Dragon in Global Waters: The Making of the Polar Silk Road

Liisa Kauppila and Tuomas Kiiski

Abstract China's rise to a global economic superpower is one of the most significant megatrends of current times. As a result of the country's latest global strategy, the Belt and Road Initiative (一带一路), China's geo-economic outreach is particularly visible in maritime transport. This chapter discusses the making of the Polar Silk Road (冰上丝绸之路)—a shipping corridor consisting of three major sea lanes running through Arctic waters—as an empirical case study to explore China's involvement in global shipping. In particular, the chapter identifies four categories of corridor-making practices, ways through which Chinese maritime actors are advancing the emerging of a geo-economic space that connects China with Arctic localities. Ranging from physical facilitation to enabling a smooth flow of traffic, these sets of practices are shaped by both the external politico-economic environment and the domestic interplay between the Chinese central government, local governments, and Chinese companies and scholars.

Keywords China · Belt and Road Initiative · Polar Silk Road · Arctic Sea routes · Arctic shipping · Geo-economics

21.1 Introduction

China's rise to the status of a global economic superpower is one of the most significant megatrends of current times. The country's expanding outreach is altering regional landscape worldwide and enhancing China's connectivity to new frontiers,

Liisa Kauppila wishes to thank Jenny and Antti Wihuri Foundation for the generous grant that enabled her to write this chapter.

L. Kauppila (✉)
Centre for East Asian Studies, University of Turku, Turku, Finland
e-mail: liisa.kauppila@utu.fi

T. Kiiski
Department of Marketing and International Business, University of Turku, Turku, Finland

E. Pongrácz et al. (eds.), *Arctic Marine Sustainability*,
Springer Polar Sciences, https://doi.org/10.1007/978-3-030-28404-6_21

from Africa to the Arctic. In many ways, it is fair to argue the country is gradually re-establishing itself as the 'primary node' (Womack 2014) of global flows and thereby altering the existing Western-centric readings of geo-economic space and globalisation. One domain in which China's economic presence is particularly strong is maritime transport. The country is increasingly an influential node of sea-borne trade and energy flows, and its fast-developing shipbuilding industry is now the largest in the world (e.g., BRS Group 2018). At the same time, its sizable middle class is altering the dynamics of global tourism and reinvigorating the popularity of cruise shipping (e.g., Jiang 2017). For these reasons, there is an increasing need to understand how China is reshaping the future of global shipping. This chapter answers this call by presenting an empirical case study of China's entry into the northernmost waters of the world and analysing the making of the *Polar Silk Road,* a shipping corridor that consists of three major sea lanes that run through Arctic waters.

The Arctic—the land and waters above the Arctic Circle (approximately 66° North)—is becoming an increasingly attractive region for the geo-economic calculations of states. As a result of climate change, the development of shipping and ice management technology, this region is gradually being unlocked for maritime activities, especially the shipping of the natural resources found in that area. Although the results have been contested, the report published by the United States Geological Survey points out that the Arctic region may hold up to 30% of the world's undiscovered natural gas and 13% of its oil resources (U.S. Geological Survey 2008). Given the importance of having effective logistics for maintaining national competitiveness in global business (e.g., Arvis et al. 2018), and with the value of sea routes for securing China's energy and mineral resources (e.g., Hong 2012), it makes sense that the Chinese interest in potentially opening Arctic sea routes has grown substantially during the 2010s. The most attractive option currently, the Northeast Passage (NEP) links China with European markets through a waterway that is around 30% shorter than the alternative route using the Suez Canal. In addition to potential savings in time and especially in fuel costs, this route offers a relatively secure operating environment, as piracy and terrorism have, so far, remained totally absent from Arctic waters. Further still, a large section (around 5600 km) of the Northeast Passage—the Northern Sea Route (NSR)—runs through Russian waters, granting Chinese actors a relatively favourable position to use to develop the sea lane—given that Ukrainian crisis caused Russia to pivot toward Asia (cf. e.g., Gabuev 2016, p. 3, 20; Sørensen and Klimenko 2017, p. 23, 26).

The Northwest Passage (NWP), in turn, connects China with North America through a route that runs mainly through Canadian and American waters. Although the legal status of large sections of the NWP remains disputed due to unresolved territorial claims, the Chinese are less likely to utilise that route in the near and mid-term future. Not only are there potential political barriers resulting from the power balancing between the United States and China (e.g., economic nationalism) but also the operating conditions along this route are typically challenging in that it runs through a number of narrow straits that are often packed with ice (e.g., Chircop 2007). These issues, however, have not prevented China's largest shipping company, the China Ocean Shipping Company (COSCO), from estimating the feasibility of the route, as evidenced by a seminar held in China in March of 2018 (COSCO 2018). Finally, the Transpolar Passage (TPP) remains, at least for the time being, the least likely option; although located mainly in international waters—with fewer disputed areas—the TPP may not be seasonally navigational until the middle of this century (Aksenov et al. 2017). For this reason, its value to Chinese actors lies in its long-term potential. Together, these three shipping lanes constitute what in the Chinese political jargon has become known as the 'Polar (or Ice) Silk Road' (冰上丝绸之路) (State Council of the People's Republic of China 2018)—a status further cemented in the country's very first Arctic policy paper published in January 2018.[1]

This chapter traces the practices through which the Chinese maritime actors—the central government, local governments in the Northeastern provinces, companies, and scholars—all together are advancing the making of the Polar Silk Road. Although the project is coordinated and guided by the Chinese central government, recognising the role and contributions of all the above-mentioned actor groups is important since China is not a monolith. In other words, what makes up *China's practices for maritime corridor building* can only be understood by taking into account the actions, priorities, and agendas of not only the central government but also companies and local governments, etc.[2] By focusing on the level of practice—the everyday logic of building the Polar Silk Road—this current analysis seeks to shed light on what the Chinese actors *actually do* to make the corridor emerge, instead of testing or drawing insights from any existing theory that tries to explain how functional regions like transport corridors come into being. Having said this, however, it should also be noted that the focus on the local context does not rule out the opportunity to produce 'analytically general insights' (Pouliot 2015, p. 237) and thereby create categories of analysis that are useful in enhancing the understanding of the general mechanisms through which Chinese actors are building maritime corridors that link China with the rest of the world. It is this latter aim that gives the analysis its bite as a practice-tracing *case study* (see, Pouliot 2015).

The structure of this chapter is thus the following. The second section offers some general background information on the role of maritime corridors in China's geo-economic calculation and introduces the country's latest global strategy, the Belt and Road Initiative (一带一路). The third section introduces the case study setting by discussing the current state of China's engagement in Arctic shipping. Section Four presents a practice-tracing analysis, our main argument being that it is possible to distinguish four different categories of analytical mechanisms based on their performative roles in the corridor-making process, namely, physical facilitation, enhancing maritime preparedness, influencing the regulatory framework, and

[1] Originally, in June 2017, only the passage 'leading up to Europe via the Arctic Ocean' was explicitly stated to be part of China's twenty-first Century Maritime Silk Road (State Council of the People's Republic of China 2017a).

[2] The focus on China's practices is a research strategy; it does not imply that the making of the Polar Silk Road—or any other maritime corridor—is a process undertaken and dictated by Chinese maritime actors only.

enabling the smooth flow of traffic. Finally, in our conclusions we summarise the findings and discuss the future prospects for the emerging Polar Silk Road.

21.2 Maritime Corridors in China's Geo-economic Calculation

Maritime corridors can be viewed as geo-economic spaces that link different localities—or nodes—through the flows of energy and goods. They are socially constructed manifestations of connectivity, pace, networks, and trans-border linkages (Sparke 2007, p. 340) that emerge as certain actors seek to accumulate wealth from the global markets (Cowen and Smith 2009, p. 42). In contrast with geopolitical *places*, geo-economic spaces are not 'territorial expressions of power' (Cowen and Smith 2009, p. 42); they are spheres of economic integration, (asymmetric) interdependence and advancing globalisation. This point is not to argue that their construction is a cosmopolitan act: geo-economic space-making may—and indeed often *does*—serve the geostrategic interests of national governments. However, as Cowen and Smith (2009, p. 42) argue, in terms of geo-economic logic, 'control of territory' is not a 'strategic necessity' but an option that a government may—or may not choose to take. In other words, geo-economic space-making is not automatically a (new) form of interstate rivalry through economic competition, as the classical 'Luttwakian' (see, Luttwak 1990, 1993) reading maintains; rather it opens a door for relative power gains in addition to economic benefits.

According to the geo-economic calculations of the Chinese government, the making of maritime corridors—first and foremost—serves the purpose of maintaining domestic economic growth, an essential precondition for social stability in the country and the core pillar of the Chinese Communist Party's legitimacy. Waterways are vital for China's foreign trade, energy security, and the overall security of supply (cf. Hong 2012). In particular, the Chinese economy is highly dependent on shipments of raw materials and semi-finished goods, given that approximately 30% of the country's GDP derives from manufacturing (World Bank 2017). Indeed, despite the ongoing structural transformation of the Chinese economy, the country continues to be dependent on large amounts of intermediate components, raw materials and energy, most of which are transported in bulk by sea. The Chinese seaborne trade in 2017 amounted to around 3 billion tonnes; imports comprised 4/5 of that amount and exports around 1/5 (Hellenic Shipping News 2018). China's energy consumption, the world's highest, is also facilitated by imports—accounting for 420 million tonnes (MT) in 2017 (approximately 19% of the world's total) (BP 2018), thus making maritime corridors essential for the country's security of supply. This reliance also explains the Chinese government's urgent desire to diversify the country's energy-transportation routes and avoid having to rely on chokepoints, such as the Strait of Malacca.

President Xi Jinping's Belt and Road Initiative (BRI), launched in 2013, underscores the importance of waterways for the Chinese economy. As a megaproject that aims to redraw the global geo-economic landscape, BRI purports to connect China with European, Central Asian and African markets by re-establishing the country's historic trade routes. In Chinese political jargon, the 'Belt' (一带) consists of land routes, whereas the 'Road' (一路) refers to waterways. Perhaps best defined as a strategy aimed at placing China in the centre of global flows and advancing the country's status as a primary node, the BRI is a project that is in the process of being crafted. Thus, it is almost impossible to anticipate which parts of the world will be included and which will be excluded. According to a recent estimation, however, the territories and waters of 72 sovereign states currently (as of October 2018) appear to constitute the Belt and the Road (Fu 2018). In terms of the Twenty-first Century Maritime Silk Road, China's presence currently extends all the way from its east coast to the Strait of Malacca, the Horn of Africa, the Suez Canal and Southern Europe. In addition to these corridors, which mainly cut through the developing world, the Polar Silk Road connects China with European and North American markets and runs mainly through Russian, Canadian, and American waters.

Given the vast geographical coverage of the BRI, it is not surprising that the question of whether it is, in essence, an economic or a political project has been widely debated (see, e.g., Cheng 2016; Yu 2017). As often is the case in real life, it may be impossible to formulate an exhaustive 'either-or' answer since the target is not only moving, but also versatile, and the relative weight of different economic/political drivers may vary across the included countries and regions. In any case, an economic rationale *does* direct certain strategy-making in that all the chosen localities possess natural resources that are essential for China's economic development. However, it is also a fact that the construction of geo-economic spaces will always alter the political landscape as well (Cowen and Smith 2009, p. 25). Moreover, often the line between economic and political presence—and even influence—is a fine one, as foreign investments in critical infrastructure may create unhealthy power relations that give the investor country political leverage over smaller states (cf. Conrad and Kostka 2017). As for China's BRI, fears of and speculations about 'economic imperialism'—no matter how ungrounded—may, in fact, become the most severe challenge threatening China's global advancement. Indeed, small countries' new-found economic dependency on China and China's potential military strategic use of the maritime corridors have received a considerable amount of media attention during the past few years. For example, China's very first overseas military base in Eastern Africa's Djibouti and the 99-year-long port lease in Sri Lanka are often viewed as the most alarming examples of the potential risks of Chinese engagement (see, e.g., Cheng 2018; Kuo and Kommenda 2018).

Finally, as noted here in the introduction, it should be pointed out that geo-economic space-making cannot be understood as a process in which China acts as a monolith. National interests naturally guide the actions of the Chinese government in its facilitating and overseeing role; yet Chinese companies and local governments are also participating in the making of these maritime corridors. In this sense, the idea that the logic of the markets underpins geo-economic projects (cf. Luttwak 1990, 1993; Cowen and Smith 2009) makes sense—although with Chinese characteristics: it is clear that no maritime corridor will become fully operational—a scene of frequent traffic—if there is no potential to achieve commercial gains at least in the mid- to long-term future. In addition, being located along a major maritime route is highly beneficial for the Chinese coastal cities in that the status of a seaborne trade hub could invigorate the provincial economy. For local governments and opportunistic companies operating in these localities, the national government's known interest in developing a certain maritime corridor makes participation in the project more lucrative. This is simply because such national-level in-principle interest guarantees both financial and governmental support.

21.3 The Case Study Setting: China and Arctic Maritime Engagement

Analysing the making of the Polar Silk Road brings under the microscope an empirical case that is both highly futures-oriented and characterised by uncertainties. Indeed, for the time being, Arctic maritime activities in general and China's engagement in that industry in particular, are widely discussed, but remain still only *emerging* trends. As indicated in Table 21.1, which presents recent traffic statistics for the most utilised Arctic shipping lane NSR, the route has so far been utilised mainly for Russian-oriented destinational traffic (activity to and from the route's ports)—

Table 21.1 Total cargo volumes and Chinese engagement in Northern Sea Route traffic, 2012–2018

	NSR cargo volume by traffic type (million tonnes)		Number of NSR voyages** by ship type		
Year	Total	Transit	Owned/operated by COSCO	Other Chinese-flagged	Total
2018*	13.7	n/a	7	1	*8*
2017	10.7	0.2	9	3	*12*
2016	7.5	0.2	6	2	*8*
2015	5.4	0.04	3	0	*3*
2014	4.0	0.3	0	0	*0*
2013	4.2	1.4	1	0	*1*
2012	3.8	1.3	0	2	*2*
		Total	*26*	*8*	*34*

The authors' compiled calculations based on data from NSRA (2018), NSRIO (2018), and Portnews.ru (2018)
* = Details as of 10/2018
** = a voyage refers to one consecutive sailing through the NSR or one prolonged stay within the NSR water area

instead of transit (activity that crosses the area without making calls to NSR ports)—that comprises the transportation of oil and gas or the supply for settlements and industrial sites. In terms of volume, shipping gradually increased between 2012 and 2018, from around 3 MT to over 13 MT (as of 1 October 2018) (Portnews.ru 2018). As for reported transit traffic, a peak was reached in 2013, when around 1.4 MT were transported; however, since then, the volumes have remained minuscule.

Economic realities largely explain the modest performance of transit traffic, as a number of hurdles must be overcome before large-scale Asia-Europe shipping can commence. The cost of shipping is increased by harsh and unpredictable conditions that call for expensively built and operated ice-classed ships that are limited in their number; in most cases, icebreaker escorts are also needed (Solakivi et al. 2017, 2018). Also the revenue-making side, which is especially relevant for scheduled container shipping, is adversely affected by a thin cargo base, seasonality, reduced sailing speeds and potential delays (e.g., Kiiski et al. 2018). For these reasons, the business potential remains contested, and shipping companies have signaled the difficulty of fitting Arctic ventures into their corporate strategies (Lasserre et al. 2016).

Still, despite these challenges, a number of companies have shown interest in *testing* the viability of the NSR. Although Asian ship-owners have generally expressed only modest interest in Arctic ventures, and even then almost solely in destinational shipping related to the transportation of natural resources (Beveridge et al. 2016), Chinese companies have excelled in conducting Arctic test sailings. The country's shipping activity on the NSR officially commenced in 2012 when the country's first icebreaker, *Xuelong*, made a round trip to explore the conditions on the route. A year later, China's largest shipping company, COSCO, became engaged in Arctic shipping, as its general cargo ship, *Yongsheng,* traversed the NSR en route from Busan to Rotterdam. At the time of this writing, the tally for COSCO's sailings on the NSR had reached 26, and respectively, 8 for other ships flying the Chinese flag. There are no exact cargo details available for the most recent voyages, but the ship types used—mostly general cargo and heavy lift vessels—does suggest that the loads are probably related to break bulk, project shipments, and general cargo.

Overall, the total number of Chinese voyages at 34 may not be much compared to annual activity on the NSR at around 1500 sailings (Balmasov 2018), nor does the number correspond with the bold predictions, which do suggest that by 2020 around 5–15% of China's international trade will use the NSR, as predicted by the head of the Polar Research Institute of China, Yang Huigen in 2014 (The Economist 2014). This figure does, however, indicate that a growing degree of Chinese interest will be expressed in Arctic shipping—and that it is possible to treat the emerging Polar Silk Road as an empirical setting for exploring the various analytical mechanisms through which China becomes connected with the rest of the world through the use of waterways.

21.4 A Polar Silk Road in the Making

This section reports on a *practice-tracing* exercise carried out to meet the twofold purpose of this study: (1) to identify the *practices* through which the Polar Silk Road is created, and (based on this step) (2) to produce *analytical mechanisms* that are useful when seeking to understand the building of other maritime corridors that connect China with the world. A brief discussion of the core concepts—the practice and analytical mechanisms—as well as the 'methodological hybrid' of practice tracing (Pouliot 2015, 235), thus, seems necessary before proceeding into the actual analysis.

Practices can be described as 'ways of doing things' (Pouliot 2015, p. 241) or 'engines of social action' (Adler and Pouliot 2011, p. 7) that make 'other things happen' (Pouliot 2015, p. 241)—for example, letting maritime corridors emerge. According to Adler and Pouliot (2011, pp. 4–7),[3] practices differ from mere behaviour in that they are always *laden with meaning*, and mere action—meaningful behaviour—in that practices are *patterned* and *repeated* in the 'particular organized settings' in which they have evolved through cycles of iteration. From this viewpoint, it follows that (good) practices are also considered *competent* by an audience that has the capacity to judge whether an action makes sense—given its goal. What is also key here is that *background knowledge* is a prerequisite for practices to take actual form; they always act out a degree of practical understanding of the role of the performed action in the bigger picture. Given these features, practices clearly contain elements of *both material and discursive realms*, and discursive acts of communication play a crucial role in turning mere (material) behaviour into practice. To meet Purpose 1, this analysis delves into the empirical setting of China's Arctic shipping engagement and seeks to pinpoint the social processes that possess the above-mentioned characteristics: China's practices of maritime corridor building.

Then, in order to utilise the insights drawn from the single case of the Polar Silk Road, i.e. to meet the purpose two, the practice-tracing analysis groups the identified (concrete) practices into four abstract categories (Fig. 21.1). These categories can be viewed as mechanisms, analytical constructs that, in our view, are (to some extent) valid for explaining other occurrences of the same phenomenon, namely, the making of other maritime corridors to connect China with the rest of the world. Taken together, these mechanisms can be viewed as a kind of 'theory' that is comparable to a typology. However, such mechanisms cannot be demonstrated to be true or false through empirical testing, but they can aim to be *useful* in making sense of the chaos of the world in future prospective studies (Pouliot 2015, p. 239; italics added by the authors).

[3] Use of italics here differs from the original text.

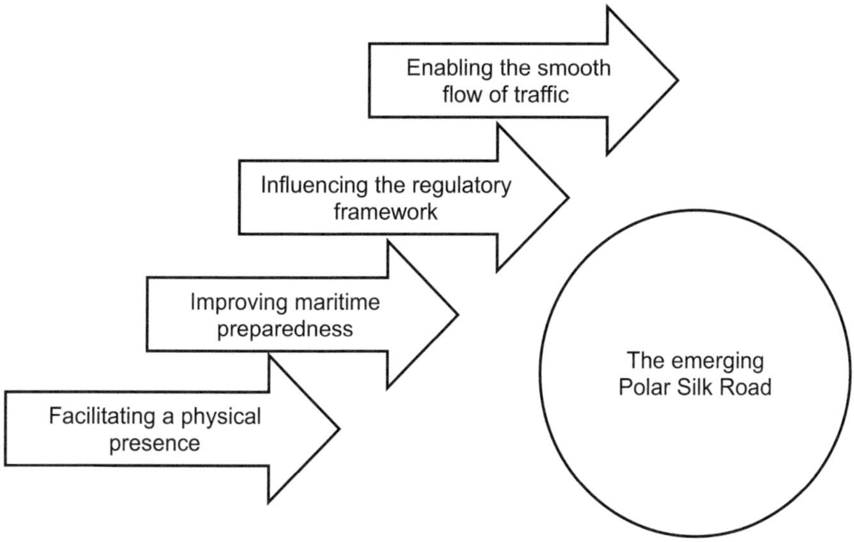

Fig. 21.1 The Making of the Polar Silk Road

21.4.1 Facilitating a Physical Presence

It is possible to identify at least three practices that Chinese maritime actors—the central government, Northeastern local governments, and shipping and shipbuilding companies—are advancing together to facilitate China's physical presence along the Polar Silk Road.

Setting up China's Overseas Maritime Hubs Over the past two decades, Chinese companies have acquired port infrastructure along several major maritime corridors worldwide, including 10 harbours[4] within the European Union. Most of these acquisitions are container terminals although (parts of) some rail ports have also been acquired (European Parliament 2018). These investments grant Chinese companies long-term access to critical chokepoints, such as the Suez Canal and the Strait of Malacca, and, thus, they serve the needs of China's export-oriented economy. As for the High North, Chinese companies have expressed their interest in investing in the Port of Kirkenes in Northern Norway and the Port of Akureyri in Northern Iceland (e.g., Liang and Zhang 2018). If realised, these acquisitions will serve as significant hubs for cargo and possibly polar cruise shipping on the NSR. Whereas the Port of Kirkenes is a rather undeveloped harbour, the Port of Akureyri has raised its profile as a Northern cruise-shipping hub over the past 6 years (2013–2018), and the number of cruise-ship visits there rose from only 63 to 179 (Port of Akureyri 2019). In both cases, Chinese capital can perhaps improve the locality's socio-economic-development

[4]Rotterdam, Antwerp, Zeebrugge, Bilbao, Valencia, Madrid, Zaragoza, Marseilles, Vado Ligure, and Piraeus.

prospects. Moreover, investments are generally welcomed, as evidenced by the various public presentations offered by local decision-makers and port officials.

In addition, port leasing deals at Russia's Zarubino and North Korea's Rason have been connected to Arctic shipping (e.g., Zhong and Liu 2015). In particular, actors from the Chinese Province of Jilin have shown an interest in developing a cross-border hub that alters the landlocked county's position vis-à-vis the Sea of Japan and the NSR (Chosun Media 2012; RIA Novosti 2014). The local economy of Jilin Province could benefit greatly from this new access to maritime trade routes via some of Northeast Asia's northernmost ice-free ports.

Developing the Infrastructure of Non-Chinese-Controlled Overseas Ports In addition to gaining (partial) control over the harbour infrastructure in certain ports, taking part in the development of port infrastructure is a common practice in China's maritime corridor-making processes across the world. In the Arctic, Chinese companies and the central government have played an indirect role in the development of the Russian port of Sabetta, a harbour that was built only to facilitate the transportation of liquefied natural gas (LNG) produced in the Peninsula's gas plants. One of China's largest state-owned energy companies, the China National Petroleum Corporation (CNPC), joined the pioneering Arctic LNG project *Yamal LNG* in 2013 with an ownership share of 20%, whereas the central government's investment vehicle, the New Silk Road Fund (丝路基金), obtained a share of 9.9% in 2016. The other owners of the project are the Russian privately-owned Novatek (50.1%) and the French Total (20%) (Yamal LNG 2018). Shipments of LNG started in late 2017 with the first load to China arriving in July 2018 (Novatek 2018). In the future, the port will also serve the needs of another LNG scheme with Chinese involvement, *Yamal LNG 2*, which is planned to be launched in 2023. Given the fact that large areas of the Arctic region lack any major basic (harbour) infrastructure, it is likely that Chinese capital will play an increasing role in the development of ports across the High North—especially when full acquisition of the infrastructure is not possible for political reasons.

Domestic Production of a Specialised Fleet In addition to safeguarding the access of Chinese companies to functional ports with a developed infrastructure, it has become increasingly important for the Chinese government to advance domestic production of the fleet utilised in Arctic ventures and, thus, decrease the country's dependence on foreign production. This goal is in line with the main targets of two national-level strategies. The country's *13th Five-Year Plan* (2016–2020) emphasises the role of technological innovation in driving China's economic growth (State Council of the People's Republic of China 2016), whereas the *Made in China 2025* (中国制造2025) initiative seeks to transform China into a world-class manufacturing power whose production will be directed toward high-tech and knowledge-intensive fields (State Council of the People's Republic of China 2015).

In terms of constructing the Polar Silk Road, the term 'specialised fleet' refers mainly to icebreakers and ice-classed ships. China started to build its first domestically produced icebreaker, *Xuelong 2*, in 2016, adding to a fleet that currently consists of just one vessel, *Xuelong*, which was bought from Ukraine in the early 1990s. Constructed by the Jiangnan Shipyard Group in collaboration with Finnish Aker Arctic, the new icebreaker is scheduled to be operational in 2019 (Zhou 2018). In addition, in June of 2018 the China National Nuclear Corporation, a state-owned corporation responsible for overseeing China's nuclear projects, opened a bidding process for shipbuilding companies to construct the country's very first nuclear-powered icebreaker, with the explicit aim to provide nuclear technological support for the Chinese government (Chen 2018). It is, thus, clear that the domestic production of nuclear-powered icebreakers makes the Chinese government much better positioned to shape the construction of the geo-economic space of the Polar Silk Road and meet the needs of the country's economy. Indeed, given the thickness of Arctic multi-year ice, the season for shipping with only non-nuclear-powered escort vessels is likely to be short for decades to come. In addition, Russia has thus far been the sole producer of nuclear icebreakers: national legislation bans domestic production of these types of vessels in countries such as Finland.

21.4.2 Improving Maritime Preparedness

At least three types of Chinese practices appear to be improving the context-specific maritime preparedness, and, thus, advancing the creation and eventual development of the Polar Silk Road.

Test Sailings Undertaken by a State-Owned Fleet In order to estimate the feasibility of an emerging maritime corridor, it is vital to conduct test sailings to explore that route and obtain knowledge of the environment-specific critical skills. Such test sailings allow actors to discover the conditions under which it is possible to conduct safe, sustainable and profitable maritime activities. As discussed in Sect. 21.3, Chinese test sailings on the NSR started in 2012 with a round trip made by a non-commercial vessel, *Xuelong*, and then, they have continued with a series of COSCO's sailings. Generally, it can be argued that the *Xuelong* experience prepared the ground for voyages that can focus more on exploring the business aspects of Arctic shipping. Despite the fact that the company's suggested plans have not yet been realised (as of 13 November 2018), a significant watershed in this testing process was reached in October of 2015 when COSCO announced that it would indeed open a frequent route on the NSR (SCMP 2015).

It should also be noted that in addition to improving China's maritime preparedness, test sailings may serve two additional purposes in the corridor-making process. First, the ship types used in COSCO's sailings provide the grounds to suggest that 'test' voyages are utilised for shipping materials to the Yamal Peninsula—an observation that is backed up by media reports (e.g., Humpert 2018a). Second, test sailings—and their active advertising—are a way to attract attention, reach potential customers, and construct an image of a waterway that has a genuine Chinese presence.

Overseas Training of Chinese Crews Each maritime environment calls for different kinds of skills from the ships' crews. Crews operating in the area of the Suez Canal, for example, and in particular in African coastal areas, must be capable of piracy-deterrent measures. As for the High North, the challenging operating conditions of the Arctic waters call for exceptionally specific knowledge for anyone venturing into that area. In addition to the extreme cold, with record temperatures reaching −60 °C, and thick ice coverage, floating icebergs and polar winds also make it extremely challenging to navigate these waters. Indeed, the closer the North Pole gets, the harder it is for an inexperienced crew to operate effectively. In addition to facing these basic challenges, crews must also know how to behave in the event of an accident. This skill is particularly important in Arctic waters, where an accident can always have severe consequences both for the crew and the fragile marine environment.

Given the fact that Arctic shipping has never been activity with widespread global involvement, very few countries and companies possess the necessary knowledge to be able to cope successfully with Arctic-specific navigational and ice-management conditions. Currently, Finland and Russia are the only Arctic countries that possess ice simulators in which related training can be given. Complementing the use of simulators, on-site training also helps crews to develop these skills. The Chinese interest in collaborating with Arctic countries in these fields has reportedly been expressed in several Sino-Arctic workshops.

Setting Up Joint Ventures to Gain Environment-Specific Skills and Technologies In addition to training crews, Chinese shipping companies also need to obtain environment-specific technologies and acquire the skills to use them. Setting up joint ventures with foreign shipping companies is a typical way of transferring technology and exchanging ideas. For their Arctic skills, China Shipping, Ltd. made a deal with Japanese Mitsui O.S.K Lines in 2014 that involved collaboration on the transportation of LNG from the Yamal Peninsula to Chinese markets. This joint venture is vital for the Chinese company, which lacked the experience and technical knowledge necessary for transporting LNG in the challenging operating conditions of the Arctic. The Japanese side, in turn, gained access to the sizable Chinese market (Interview with a representative of a Japanese shipping company, Tokyo, 26 Oct 2016).

From the point of view of certain stakeholders and observers of this process, the Sino-Japanese arrangement could merely be viewed as a form of pragmatic business-to-business collaboration (Interview with a representative of a Japanese shipping company, Tokyo, 26 Oct 2016; Interview with a representative of a Chinese energy company (personal views), Beijing, 26 Feb 2016). However, it is hard to ignore the exceptional political context of these activities. Sino-Japanese business collaboration is generally known to suffer from the troubled political relationship between the governments of these two countries. Since political frictions were not allowed to hinder this Sino-Japanese Arctic shipping deal, it seems clear that maritime technology and skill transfer—and the country's overall participation in the economic development of the Arctic—rank high in the Chinese government's geo-economic calculations and priorities. Consequently, these aims can be secured through exceptional and unconventional arrangements and partnerships (see also Sect. 21.4.3).

21.4.3 Influencing the Regulatory Framework

The Chinese government is also participating in the making of the Polar Silk Road by seeking to shape the regulatory framework of Arctic shipping. To meet this goal, three types of practices are being performed. Here the role of Chinese scholars should not be ignored, as the background knowledge and policy recommendations produced by them are important in shaping the form that the practices will take. Naturally, the Chinese government's scope of action is limited by the fact that none of the three major Arctic sea lanes—NEP, NWP or TPP—actually run through Chinese waters. In addition, large stretches of these waterways belong to the national jurisdiction of Arctic countries, granting them the exclusive right to enact any specific regulations. For example, Russia is planning by 2019 to introduce a new domestic law banning the transportation of hydrocarbons along the NSR for vessels that are not Russian-flagged or domestically built (Staalesen 2017). On the other hand, the country is reportedly also planning to soften the NSR admittance criteria with regard to ice-class demands (Humpert 2018b).

Participation in the Existing Institutions of Regional Governance Regional governance forums across the world are platforms on which local maritime-related regulations are debated and negotiated. To be able to participate in these discussions and shape the regulations, China seeks to become actively engaged in the efforts of these regional governance bodies, as has become evident in Africa (the African Union) and Latin America (Union of South American Nations/UNASUR). As for the Arctic, the Arctic Council is currently the most influential intergovernmental discussion forum. China began to attend the meetings of the Council in 2008 and joined the organization as a permanent observer in 2013. As a non-Arctic state, China does not have voting rights, but it may now participate in the tasks of the six working groups. Their mandates vary from improving the protection of the Arctic marine environment (e.g., from sea-based activities) to promoting sustainable development (e.g., of the infrastructure). In addition, and despite the fact that the Arctic Council is mainly a discussion forum rather than a decision-making body, two significant agreements have been successfully negotiated under its auspices: The *Agreement on Cooperation on Marine Oil Pollution Preparedness and Response in the Arctic* (signed in 2013) and the *Agreement on Cooperation on Aeronautical and Maritime Search and Rescue in the Arctic* (signed in 2011) (Arctic Council 2018). Both of these agreements are likely to shape the regulatory framework of the Polar Silk Road in the future.

Setting up Parallel Institutions of Governance Despite China's active participation in the existing institutions of regional governance across the world, the country's scope of action still remains limited in these forums. For this reason, it has become customary for the Chinese government to set up parallel institutions to facilitate the discussion on regional issues, including seafaring-related topics. Illustrative examples of this tendency include the China-Africa and China-Arab Cooperation Forums (see, FOCAC 2018a; FMPRC 2018), where such issues as piracy, training of shipping professionals, and cooperation between shipping companies are discussed and included in action plans (FOCAC 2018b; Li 2014). As a result, the Chinese government is better positioned to shape the building of maritime corridors that link the country with the rest of the world. In terms of Arctic affairs, China has participated in establishing a Sino-Korean-Japanese trilateral Arctic forum, *Trilateral High-Level Dialogue on the Arctic,* and it holds annual meetings in each country's capital on a rolling basis. Following each meeting, the three countries issue joint statements on themes that relate to their collaboration and participation in the High North. In their joint statement published in Tokyo in 2017 the governments explicitly emphasised the importance of developing a 'rule-based maritime order' (Governments of Japan, the People's Republic of China, and the Republic of Korea 2017). The trilateral Arctic forum is exceptional when compared to other parallel institutions of regional governance wherein China collaborates with the governments of the region in question. This forum was set up by the three Arctic Council observer states, and participation by Arctic countries is limited to observation only.

Emphasising the Role of International Maritime Governance In the age of globalisation, it is not always obvious which issues should be debated and decided locally within just the scope of regional governance forums, and which issues call for global participation under the auspices of international organisations, such as the IMO. In short, most shipping-related regulations are international, whereas some national rules have global implications. The Chinese government, for example, is planning to ban the import of international ship scrap to its yards (Zeng 2018), and that decision is likely to have substantial effects on the global ship-scrapping market.

Generally, the Chinese government emphasises the role of international organisations in shaping maritime governance because that makes the process more democratic. In other words, it gives all countries the chance to express their concerns (cf. Liu 2014). Naturally, bringing a wide array of issues up for global—instead of regional—discussion also makes the Chinese government better positioned to shape the making of maritime corridors. The Chinese government has actively stressed the role of international law and international organisations in Arctic maritime affairs. In its first official Arctic policy paper, it strongly emphasises its adherence to international treaties, such as the *United Nations Convention on the Law of the Seas* (UNCLOS). However, it also underscores the need to respect the rights of non-Arctic countries to navigate in Arctic waters (State Council of the People's Republic of China 2018). This rhetoric advances the geo-economic reading of space in that it actively seeks to blur the boundaries between countries.

Chinese Arctic experts have also praised the fact that the Polar Code (*International Code for Ships Operating in Polar Waters*) was negotiated in the IMO instead of in regional forums (Interview with a Chinese expert on Arctic policy, Shanghai, 2 March 2016). According to these experts, as an international regulation that allows stricter environmental and safety standards to be imposed on shipping in polar waters, the Polar Code is a global, not a regional, issue. In a similar vein, some Chinese scholars have questioned the role of the Arctic Council as a major intergovernmental discussion forum on Arctic maritime issues on the grounds that this forum does not include all major actors sufficiently or comprehensively enough (see, e.g., Guo and Yao 2015).

21.4.4 Enabling the Smooth Flow of Traffic

In addition to performing practices that either create the basic preconditions for maritime corridors to emerge or shape the regulatory framework along these routes, the Chinese government is also taking certain actions that enable the *smooth* flow of cargo and passenger traffic. These practices could potentially make the Arctic shipping more attractive—or at least less troublesome—to Chinese companies.

Issuing Favourable Investment Regulations Given China's status as a socialist market economy, capital does not flow freely in and out of the country. Outbound foreign direct investments are regulated through guidelines issued by the central government, the Ministry for Trade and Commerce and the National Development Council in particular. Similarly, the government regularly issues a list of targets that it supports and favours. Through such mechanisms, the creation of geo-economic spaces is harmonised and guided in a direction that serves the country's broader national development aims. As for the Arctic, the official attaching of the NEP, NWP, and TPP to the BRI as the Polar Silk Road has been a significant decision in that investments related to the BRI are generally favoured and allowed over other targets, as duly announced in August of 2018 (State Council of the People's Republic of China 2017b). Naturally, this known support for Arctic projects is likely to make many opportunistic actors in China reconsider their business strategies.

Building Trust Trust is an essential precondition for the smooth transnational flow of traffic to emerge and continue and international business to flourish. From a geo-economic perspective, diplomatic activities can be assumed to increase mutual trust and, thus, pave the way for making spaces that cut across national borders. Trust creation is a particularly important goal for Chinese actors, whose global economic outreach is followed closely due to the country's rising power status. Moreover, the country's reputation in localities such as Africa adds to the general speculation over any geo-economic project that China advances and increases the pressure worldwide to focus actively on tackling the issue of mistrust.

In the Arctic context, various actions have been taken to build greater trust. The most significant act so far was the publishing of the country's Arctic policy paper in January 2018 (State Council of the People's Republic of China 2018). This much-anticipated document was drafted under clear pressure from the Arctic community in order to increase the level of transparency of China's Arctic engagement (Interview with a Chinese expert on Arctic policy, Qingdao, 8 March 2016), with researchers and non-governmental organisations having expressed the most vocal opinions (see, e.g., Brady 2014; Braw 2017). In addition, Chinese state leaders have made exceptionally many official visits to the small Arctic countries over just a short period of time. In April of 2017, for example, President Xi Jinping visited Finland, which chairs the Arctic Council during the period 2017–2019. Arctic issues were an important item on this meeting's agenda (Niinistö 2017). Moreover, China normalised its relations with Norway—a major Arctic shipping country—in late 2016 after a diplomatic breakdown due to the Nobel Prize controversy (Governments of the People's Republic of China and the Kingdom of Norway 2016). Finally, along with representatives of other East Asian governments, Chinese delegations have played an active role at informal forums of Arctic governance, namely the Arctic Circle Assembly and the Arctic Frontiers. In particular, Iceland's Arctic Circle Assembly has become a major forum for China to manage its Arctic relations and build trust.

21.5 Conclusion and Final Thoughts

China's increasing maritime presence—and its geo-economic space-making in general—has aroused curiosity worldwide. Attempts to make sense of China's gradual rise to become a primary node of the global economy are often challenged by the Western-centrism of existing theories. In this chapter, we proposed that understanding this transformation and the forces by which it unfolds call for adopting a case-study approach and delving deeper into the level of concrete practices to find and understand what the Chinese maritime actors actually do.

The Arctic case demonstrates that China's practices of maritime corridor making and building of geo-economic spaces stem equally from the dynamics of the country's political economy and the external operating environment. The role of the Chinese central government is to advance practices that facilitate, enable, and encourage Chinese companies to become engaged in regular seafaring, wherein these companies must gain sufficient context-specific skills, obtain the right technologies, and assess whether or not they have the economic rationale to launch frequent traffic. Local governments may connect the development goals of their provincial economies to an overall geo-economic space-making project, and thus, they play a viable role in advancing practices such as setting up Chinese overseas maritime hubs. Scholars provide background information, which is always important for any practice to take form.

Some of these practices, such as trust creation, develop directly in response to external pressures and, thus, manifest the importance of the context, when choosing the best way to advance certain goals. It could be argued, for example, that building trust is particularly important in the Arctic context—not the least of which is because Chinese actors operate in the High North with developed countries, most of which are democracies with active civil societies fully capable of influencing national policies. In fact, it is not an impossible scenario to suggest that general opinion in Arctic countries could turn against practices like setting up China's overseas maritime hubs—despite the Chinese government's attempts to manage Sino-Arctic relations by very distinct acts like paying a presidential visit to Finland and publishing one of the very few official Chinese policy documents on extra-regional affairs.

The Arctic case also provides grounds to propose that four types of analytical mechanisms may explain the development of China's maritime corridors in more general terms. It may be useful to treat these mechanisms—i.e., facilitating a physical presence, improving maritime preparedness, influencing the regulatory framework, and enabling the smooth flow of traffic—as particular categories where more context-specific and concrete practices could be grouped in prospective empirical studies. Naturally, these mechanisms/categories are only heuristic tools that might—and indeed *should*—be developed so as to better understand the complex set of practices that allow China's maritime corridors to emerge across the world.

Practices do not represent forces of destiny. Despite the increasingly active creation of maritime corridors that link China with the world, it is not at all clear to what extent—and indeed precisely how—the Chinese presence will continue to alter the dynamics of global seafaring. As for the Polar Silk Road, a high number of economic and political uncertainties make it particularly difficult to anticipate what kind of role, if any, the NEP and the NSR—let alone the two other Arctic sea routes NWP and TPP—would play in China's Belt and Road Initiative and geo-economic calculus in general. Although destinational traffic related to the transportation of LNG from Yamal is highly likely to continue for decades, the level of activity—at least in the near to mid-term—may be rather different for cruise and Asia-Europe transit shipping.

Global shipping patterns are subject to various changes, such as improvements in existing transport infrastructure like the Panama Canal expansion, the potential opening of alternate corridors like the Canals of Kra and Nicaragua (cf. Zeng et al. 2018; Yip and Wong 2015), and other developments that clearly influence transport demand and the price of shipping. With regard to costs, tightening the environmental standards for shipping may turn out to be highly influential. Indeed, such regulations as the global sulphur cap, which will enter into force by 2020, will add extra costs to maritime transport. As for the Polar Silk Road, some of these new regulations may add to the relative attractiveness of the NSR compared to principal canal routes, while others may diminish it (e.g. the proposed ban on heavy fuel oil).

Finally, forces inside China may equally both decrease and increase the country's maritime engagement globally and in the Arctic. The slowing down of Chinese economic growth, as well as the country's ongoing structural and energy transformation, may also decrease the need for cargo shipping. As for cruise shipping, increasing environmental consciousness in Chinese middle class may kill the emerging interest, at least in the long term.

References

Adler, E., & Pouliot, V. (2011). International practices. *International Theory, 3*(1), 1–36.

Aksenov, Y., Popova, E. E., Yool, A., et al. (2017). On the future navigability of Arctic Sea routes: High-resolution projections of the Arctic Ocean and sea ice. *Marine Policy, 75*, 300–317.

Arctic Council. (2018). *Agreements*. https://www.arctic-council.org/en/our-work/agreements. Accessed 29 Oct 2018.

Arvis, J. F., Ojala, L., Wiederer, C., et al. (2018). *Connecting to compete: Trade logistics in the global economy. The logistics performance index and its indicators*. World Bank. Available at: https://openknowledge.worldbank.org/bitstream/handle/10986/29971/LPI2018.pdf. Accessed 6 Nov 2018.

Balmasov, S. (2018). Detailed analysis of ship traffic on the NSR in 2017 based on AIS data. In Paper presented at the Arctic Shipping Forum 2018, Helsinki, Finland, 17 Apr 2018. Available at: http://arctic-lio.com/?p=1215. Accessed 8 Jan 2019.

Beveridge, L., Fournier, M., Lasserre, F., et al. (2016). Interest of Asian shipping companies in navigating the Arctic. *Polar Science, 10*, 404–414.

BP. (2018). *BP statistical review of world energy. June 2018*. 67th edn. Available at: https://www.bp.com/content/dam/bp/en/corporate/pdf/energy-economics/statistical-review/bp-stats-review-2018-full-report.pdf. Accessed 6 Nov 2018.

Brady, A. M. (2014). *China's undeclared Arctic foreign policy. Polar initiative policy brief series. Arctic 2014: Who gets a voice and why it matters*. Wilson Center. Available at: https://www.wilsoncenter.org/sites/default/files/chinas_undeclared_arctic_foreign_policy.pdf. Accessed 10 Jan 2019.

Braw, E. (2017). China needs to spell out its Arctic ambitions, to ease suspicions. *South China Morning Post (21 Nov)*. Available at: https://www.scmp.com/comment/insight-opinion/article/2120850/china-needs-spell-out-its-arctic-ambitions-ease-suspicions. Accessed 10 Jan 2019.

BRS Group. (2018). *Annual review 2018*. Available at: https://it4v7.interactiv-doc.fr/html/brsgroup2018annualreview_pdf_668. Accessed 6 Nov 2018.

Chen, Y. (2018). 我国首艘核动力破冰船揭开面纱——将为海上浮动核电站动力支持铺平道路 (*China's first nuclear-powered icebreaker unveiled – Paving the way for powering offshore floating nuclear power plants*) 科技日报 (Science and technology daily)/Xinhuanet (27 June). Available at: http://www.xinhuanet.com/politics/2018-06/27/c_1123041028.htm. Accessed 8 Nov 2018.

Cheng, L. K. (2016). Three questions on China's 'Belt and Road Initiative'. *China Economic Review, 40*, 309–313.

Cheng, A. (2018). Will Djibouti become latest country to fall into China's debt trap? *Foreign Policy (31 July)*. Available at: https://foreignpolicy.com/2018/07/31/will-djibouti-become-latest-country-to-fall-into-chinas-debt-trap/. Accessed 12 Nov 2018.

Chircop, A. (2007). Climate change and the prospects of increased navigation in the Canadian Arctic. *WMU Journal of Maritime Affairs, 6*(2), 193–205.

ChosunMedia. (2012). *China gains use of another N. Korean port*. 11 Sept. Available at: http://english.chosun.com/site/data/html_dir/2012/09/11/2012091101347.html. Accessed 12 Sept 2014.

Conrad, B., & Kostka, G. (2017). Chinese investments in Europe's energy sector: Risks and opportunities? *Energy Policy, 101*, 644–648.

COSCO. (2018). *COSCO SHIPPING SPE. Organized seminar for navigation along Arctic Northwest Passage*. 14 Mar. Available at: http://www.coscol.com.cn/En/News/detail.aspx?id=11720. Accessed 6 Sept 2018.

Cowen, D., & Smith, N. (2009). After geopolitics? From the geopolitical social to geoeconomics. *Antipode, 41*(1), 22–48.

Economist. (2014). *China and the Arctic: Polar bearings. China pursues its interest in the frozen north*. Jul 12. Available at: https://www.economist.com/china/2014/07/12/polar-bearings. Accessed 24 Sept 2018.

European Parliament. (2018). *China's maritime silk road initiative increasingly touches the EU. Briefing of the European Parliament*. Mar 2018. Available at: http://www.europarl.europa.eu/RegData/etudes/BRIE/2018/614767/EPRS_BRI(2018)614767_EN.pdf. Accessed 29 Oct 2018.

FMPRC. (2018). *The 8th ministerial meeting of the China-Arab States Cooperation Forum (CASCF) held in Beijing*. Available at: https://www.fmprc.gov.cn/mfa_eng/zxxx_662805/t1576621.shtml. Accessed 29 Oct 2018.

FOCAC. (2018a). *About FOCAC*. https://www.focac.org/eng/ltjj_3/ltjz/. Accessed 29 Oct 2018.

FOCAC. (2018b). *Forum on China-Africa Cooperation Beijing Action Plan 2019–2021*. https://www.focac.org/eng/zfgx_4/zzjw/t1594399.htm. Accessed 10 Jan 2019.

Fu, C. (2018). *Opening session speech given at the 6th Arctic Circle Assembly, Reykjavik, Iceland*, 19 Oct 2018.

Gabuev, A. (2016). *Friends with benefits? Russian-Chinese relations after the Ukraine crisis*. Carnegie Moscow Center. Available at: https://carnegieendowment.org/files/CEIP_CP278_Gabuev_revised_FINAL.pdf. Accessed 9 Jan 2019.

Governments of Japan, the People's Republic of China and the Republic of Korea. (2017). *Joint statement. The Second trilateral high-level dialogue on the Arctic. Tokyo, Japan*. 8 June. Available at: http://www.mofa.go.jp/mofaj/files/000263104.pdf. Accessed 3 Nov 2018.

Governments of the People's Republic of China and the Kingdom of Norway. (2016). *Statement of the government of the People's Republic of China and the government of the Kingdom of Norway on normalization of bilateral relations*. Beijing, China. 19 Dec 2016. Available at: https://www.regjeringen.no/globalassets/departementene/ud/vedlegg/statement_kina.pdf. Accessed 9 Jan 2019.

Guo, P., & Yao, L. (2015). 北极治理模式的国际探过及北极治理实践的发展 (International discussion on Arctic governance models and new advances of Arctic governance practice). 国际观察 *(International Observations), 5*, 56–70.

Hellenic Shipping News. (2018). *China's seaborne trade: A spectacular upwards trend*. 26 Apr. Available at: https://www.hellenicshippingnews.com/chinas-seaborne-trade-a-spectacular-upwards-trend/. Accessed 3 Nov 2018.

Hong, N. (2012). The melting Arctic and its impact on China's maritime transport. *Research in Transportation Economics, 35*(1), 50–57.

Humpert, M. (2018a). *Record traffic on Northern Sea Route as COSCO completes five transits*. High North News. 3 Sept. Available at: https://www.highnorthnews.com/en/record-traffic-northern-sea-route-cosco-completes-five-transits. Accessed 2 Nov 2018.

Humpert, M. (2018b). *Economic interests may trump shipping safety as Russia seeks to reduce ice-class requirements*. High North News (12 Nov). Available at: https://www.highnorthnews.com/en/economic-interests-may-trump-shipping-safety-russia-seeks-reduce-ice-class-requirements. Accessed 13 Nov 2018.

Interview with a Chinese expert on Arctic policy, Qingdao, 8 Mar 2016.

Interview with a Chinese expert on Arctic policy, Shanghai, 2 Mar 2016.

Interview with a representative of a Chinese energy company (personal views), Beijing, 26 Feb 2016.

Interview with a representative of a Japanese shipping company, Tokyo, 26 Oct 2016.

Jiang, J. (2017). *Polar attraction*. People's Daily Online (25 Sept). Available at: http://en.people.cn/n3/2017/0925/c90000-9273275.html. Accessed 12 Nov 2018.

Kiiski, T., Solakivi, T., Töyli, J., et al. (2018). Long-term dynamics of shipping and icebreaker capacity along the Northern Sea Route. *Maritime Economics & Logistics, 20*(3), 375–399.

Kuo, L., & Kommenda, N. (2018). What is China's *Belt and Road* Initiative? *The Guardian* (30 Jul). Available at: https://www.theguardian.com/cities/ng-interactive/2018/jul/30/what-china-belt-road-initiative-silk-road-explainer. Accessed 10 Nov 2018.

Lasserre, F., Beveridge, L., Fournier, M., et al. (2016). Polar seaways? Maritime transport in the Arctic: An analysis of shipowners' intentions II. *Journal of Transport Geography, 57*, 105–114.

Li, C. (2014). *One Belt and One Road. Asharq Al-Awsat (13 Apr)*. English translation of the article (full text) written by China's ambassor to Saudi-Arabia. Available at: https://www.fmprc.gov.cn/mfa_eng/wjb_663304/zwjg_665342/zwbd_665378/t1115855.sht. Accessed 10 Jan 2019.

Liang, Y., & Zhang, S. (2018). *Feature: Norway's Arctic town envisions gateway on Polar Silk Road with link to China*. New China/Xinhuanet (10 Mar). Available at: http://www.xinhuanet.com/english/2018-03/10/c_137029993.htm. Accessed 29 Oct 2018.

Liu, W. (2014). *China in the United Nations*. Singapore: World Century Publishing Corporation.

Luttwak, E. (1990). From geopolitics to geo-economics: Logic of conflict, grammar of commerce. *The National Interest, 20*, 17–24.

Luttwak, E. (1993). The coming global war for economic power: There are no nice guys on the battlefield of geoeconomics. *The International Economy, 7*(5), 18–67.

Niinistö, S. (2017). *Press statement by the President of the Republic of Finland, Sauli Niinistö, on the state visit by the President of the People's Republic of China on 5 April*. 6 Apr. Available at: https://www.presidentti.fi/en/news/press-statement-by-the-president-of-the-republic-of-finland-sauli-niinisto-on-the-state-visit-by-the-president-of-the-peoples-republic-of-china-on-5-april/. Accessed 10 Jan 2019.

Novatek. (2018). *NOVATEK shipped first LNG cargos to China*. Press release (19 Jul). Available at: http://www.novatek.ru/en/press/releases/index.php?id_4=2528. Accessed 5 Sept 2018.

NSRA. (2018). *Northern Sea Route Administration. Traffic statistics 2011–2018*. Available at: http://www.nsra.ru/en/operativnaya_informatsiya/grafik_dvijeniya_po_smp. Accessed 10 Nov 2018.

NSRIO. (2018). *Northern Sea Route Information Office. Transit statistics 2011–2018*. Available at: http://arctic-lio.com/?cat=27. Accessed 10 Nov 2018.

Port of Akureyri. (2019). *Cruise ships 2018 – Statistics for visits*. https://www.port.is/index.php?pid=65&w=st. Accessed 24 Jan 2019.

Portnews.ru. (2018). *Cargo transportation via Northern Sea route can reach 17 million tonnes in 2018*. 18 Oct. Available at: http://en.portnews.ru/news/266264/. Accessed 3 Nov 2018.

Pouliot, V. (2015). Practice tracing. In A. Bennett & J. T. Checkel (Eds.), *Process tracing: From metaphor to analytic tool* (pp. 237–259). Cambridge: Cambridge University Press.

RIA Novosti. (2014). *'Сумма' привлечет СМНI к строительству порта Зарубино (Summa to work with CMHI in constructing the port of Zarubino)*. 11 Nov 2014. Available at: https://ria.ru/east/20141111/1032705539.html. Accessed 12 Sept 2016.

SCMP. (2015). *Chinese shipping firm COSCO plans to launch services to Europe through Arctic Northeast Passage, saving days in travel time*. 27 Oct. Available at: https://www.scmp.com/news/china/economy/article/1872806/chinese-shipping-firm-plans-launch-services-through-arctic. Accessed 30 Oct 2018.

Solakivi, T., Kiiski, T., & Ojala, L. (2018). The impact of ice class on the economics of wet and dry bulk shipping in the Arctic waters. *Maritime Policy & Management, 45*(4), 530–542.

Sørensen, C., & Klimenko, E. (2017). *Emerging Chinese-Russian cooperation in the Arctic. Possibilities and constraints*. SIPRI Policy paper 46. Available at: https://www.sipri.org/sites/default/files/2017-06/emerging-chinese-russian-cooperation-arctic.pdf. Accessed 8 Jan 2019.

Sparke, M. (2007). Geopolitical fears, geoeconomic hopes, and the responsibilities of geography. *Annals of the American Association of Geographers, 97*(2), 338–349.

Staalesen, A. (2017). *Russian legislators ban foreign shipments of oil, natural gas and coal along Northern Sea Route*. The Barents Observer. 26 Dec. Available at: https://thebarentsobserver.com/en/arctic/2017/12/russian-legislators-ban-foreign-shipments-oil-natural-gas-and-coal-along-northern-sea. Accessed 13 Nov 2018.

State Council of the People's Republic of China. (2015). 国务院关于印发《中国制造*2025*》的通知 *(Announcement on the publication of 'Made in China 2025' report)*. 8 May 2015. Available at: http://www.gov.cn/zhengce/content/2015-05/19/content_9784.htm. Accessed 2 Nov 2018.

State Council of the People's Republic of China. (2016). *The 13th five-year plan for economic and social development of the People's Republic of China 2016–2020*. Available at: http://en.ndrc.gov.cn/newsrelease/201612/P020161207645765233498.pdf. Accessed 2 Nov 2018.

State Council of the People's Republic of China. (2017a). *Vision for maritime cooperation under the Belt and Road Initiative*. 20 June 2017. Available at: http://english.gov.cn/archive/publications/2017/06/20/content_281475691873460.htm. Accessed 2 Nov 2018.

State Council of the People's Republic of China. (2017b). 国务院办公厅转发国家发展改革委商务部人民银行外交部关于进一步引导和规范境外投资方向指导意见的通知 *(Announcement on regulations regarding overseas investments by Chinese companies)*. 4 Aug 2017. Available at: http://www.gov.cn/zhengce/content/2017-08/18/content_5218665.htm. Accessed 2 Nov 2018.

State Council of the People's Republic of China. (2018). *China's Arctic Policy*. 26 Jan 2018. Available at: http://www.xinhuanet.com/english/2018-01/26/c_136926498.htm. Accessed 7 Mar 2018.

U.S. Geological Survey. (2008). *Circum-Arctic resource appraisal: Estimates of undiscovered oil and gas north of the Arctic Circle*. Available at: http://pubs.usgs.gov/fs/2008/3049/fs2008-3049.pdf. Accessed 8 Jan 2019.

Womack, B. (2014). China's future in a multinodal world order. *Pacific Affairs, 87*(2), 265–284.

World Bank. (2017). *Data catalog. Table 4.2: World development indicators: structure of output*. Available at: http://wdi.worldbank.org/table/4.2. Accessed 3 Nov 2018.

Yamal LNG. (2018). *About the project*. Available at: http://yamallng.ru/en/project/about/. Accessed 28 Oct 2018.

Yip, T. L., & Wong, M. C. (2015). The Nicaragua Canal: Scenarios of its future roles. *Journal of Transport Geography, 43*, 1–13.

Yu, H. (2017). Motivation behind China's 'One Belt, One Road' initiatives and establishment of the Asian Infrastructure Investment Bank. *Journal of Contemporary China, 26*(105), 353–368.

Zeng, X. (2018). *Beijing curbs ship scrapping, leaving market to Indian subcontinent*. IHS Fairplay (3 May). Available at: https://fairplay.ihs.com/markets/article/4300856/beijing-curbs-ship-scrapping-leaving-market-to-indian-subcontinent. Accessed 13 Nov 2018.

Zeng, Q., Wang, G. W. Y., Qu, C., et al. (2018). Impact of the Carat Canal on the evolution of hub ports under China's Belt and Road initiative. *Transportation Research Part E, 117*, 96–107.

Zhong, N., & Liu, M. (2015). *Jilin province proposes more transport links across region*. 3 Sept. China Daily/Information Office of the People's Government of Jilin Province. http://www.chinadaily.com.cn/m/jilin/2015-09/03/content_21782513.htm. Accessed 4 Jan 2019.

Zhou, X. (Ed.) (2018). *First Chinese-built polar icebreaker to come into use in 2019. Xinhuanet*. Available at: http://www.xinhuanet.com/english/2018-04/04/c_137088340.htm. Accessed 2 Nov 2018.

Index

E. Pongrácz et al. (eds.), *Arctic Marine Sustainability*,
Springer Polar Sciences, https://doi.org/10.1007/978-3-030-28404-6

MIX
Papier aus verantwortungsvollen Quellen
Paper from responsible sources
FSC® C105338

FSC
www.fsc.org

If you have any concerns about our products,
you can contact us on
ProductSafety@springernature.com

In case Publisher is established outside the EU,
the EU authorized representative is:
Springer Nature Customer Service Center GmbH
Europaplatz 3, 69115 Heidelberg, Germany

Printed by Libri Plureos GmbH
in Hamburg, Germany